Transformer Ageing

Transformer Ageing

Monitoring and Estimation Techniques

Edited by

Tapan Kumar Saha
The University of Queensland
St. Lucia, Brisbane, Australia

Prithwiraj Purkait
Haldia Institute of Technology
West Bengal, India

WILEY

IEEE PRESS

This edition first published 2017

Registered Offices
John Wiley & Sons, Inc., 111 River Street, Hoboken, NJ 07030, USA
John Wiley & Sons Singapore Pte. Ltd, 1 Fusionopolis Walk, #07-01 Solaris South Tower, Singapore 138628

Editorial Office
1 Fusionopolis Walk, #07-01 Solaris South Tower, Singapore 138628

For details of our global editorial offices, customer services, and more information about Wiley products visit us at www.wiley.com.

Wiley also publishes its books in a variety of electronic formats and by print-on-demand. Some content that appears in standard print versions of this book may not be available in other formats.

Library of Congress Cataloging-in-Publication Data

Names: Saha, Tapan Kumar, 1959– editor. | Purkait, Prithwiraj, 1973– editor.
Title: Transformer ageing : monitoring and estimation techniques / edited by Tapan Kumar Saha, Prithwiraj Purkait.
Other titles: Transformer aging
Description: Chichester, West Sussex : John Wiley & Sons, Inc., 2018. | Includes bibliographical references and index.
Identifiers: LCCN 2017003580 (print) | LCCN 2017004685 (ebook) | ISBN 9781119239963 (cloth) | ISBN 9781119239994 (Adobe PDF) | ISBN 9781119239987 (ePub)
Subjects: LCSH: Electric transformers–Maintenance and repair–Handbooks, manuals, etc. | Electric lines–Maintenance and repair–Handbooks, manuals, etc. | Electric insulators and insulation–Testing. | Electric power distribution.
Classification: LCC TK2551 .T7635 2018 (print) | LCC TK2551 (ebook) | DDC 621.31/4–dc23
LC record available at https://lccn.loc.gov/2017003580

Cover image: © gjp311/Gettyimages
Cover design: Wiley

Set in 10/12pt Warnock by SPi Global, Pondicherry, India

10 9 8 7 6 5 4 3 2 1

Contents

Preface

The transformer is one of the most important pieces of equipment in a power grid. The condition of large power transformers has a significant impact on the reliability of the power grid. Large power transformers are expensive and complex in design and operation. Transformer condition monitoring and assessment of their remaining life is an important task for transformer owners/operators. Transformer condition monitoring covers many areas closely related to transformer structure and operation. The condition of the insulation system plays a major role in determining the life of a transformer. Similarly, winding/core integrity, bushing, and tap changer health are also important in maintaining the overall reliable operation of a transformer.

Throughout the life of transformer operation, the insulation system degrades and the degradation mechanism depends on the operating conditions inside the tank. Thermal, hydrolytic, and oxidation processes are the main causes of ageing of a transformer. Many diagnosis techniques have been in use for several decades, and their interpretation tools have always been the focus of improvements over the years. Many new diagnosis tools are being investigated continuously by researchers and engineers in the field. New insulation systems for solid and liquid insulations are being proposed and investigated for power transformers. This book will provide fundamental knowledge of transformer insulation materials, their ageing mechanisms, traditional as well as advanced condition monitoring techniques, and interpretation techniques. Basic knowledge of the transformer will be a prerequisite for readers.

The research work presented in this book was conducted with funding support from the Australian Research Council and the Australian electricity supply industry, with collaboration from several transmission and distribution utilities in Australia over a period of 25 years. Our expectation is that this book will provide state-of-the-art knowledge about transformer ageing, condition monitoring, and fault diagnosis. No single book is currently available that provides such an important knowledge base for transformer condition monitoring and life assessment. The authors hope that this book will be a "one-stop" information provider for engineering students, practicing engineers, and researchers. We believe anyone working in transformer condition monitoring – particularly engineers working in electricity utilities, graduate or senior undergraduate students and researchers, postdoctoral fellows, and academics – will benefit from this publication.

The authors of this book have published scores of journal and conference articles via the IEEE, IET, and CIGRE. Many of these will provide an additional knowledge base resource for the reader. Many diagnostic algorithms have been developed throughout this journey, and they are currently available from the University of Queensland research team.

This book is organized into ten chapters. **Chapter 1** discusses the sources, properties, and applications of insulating materials used in transformers, along with an overview of ageing of oil–paper insulation systems. This chapter also provides a foundation for understanding insulation diagnostic tools, which can assist the reader in relating the diagnosis with the cause of insulation ageing. **Chapter 2** explains comprehensively the dissolved gas analysis (DGA), furan analysis, and degree of polymerization (DP), and their relevant international standards. In addition, this chapter introduces electrical-based traditional diagnoses, which include insulation resistance (IR), polarization index (PI), dielectric dissipation factor (DDF), capacitance and power factor, dispersion factor, and partial discharge (PD).

Chapter 3 provides theoretical explanations of polarization–depolarization current (PDC), recovery voltage measurements (RVM), and frequency domain dielectric spectroscopy (FDS), along with their interpretation schemes. The effects of moisture, ageing, temperature, and insulation geometry on the interpretation of PDC, RVM, and FDS measurements are also described in this chapter.

Chapter 4 outlines commonly used interpretation techniques of dissolved gas analysis (DGA), with a comprehensive review and illustration of machine learning-based DGA interpretation techniques, with specific focus on artificial neural networks (ANNs), fuzzy logic systems, expert systems, decision-making algorithms, and support vector machine (SVM) and population-based algorithms. This chapter also provides some insights into training dataset construction and data quality improvement, and discusses approaches to classification accuracy and generalization capability validation.

Chapter 5 provides a detailed analysis of partial discharge (PD) measurement and interpretation tools for transformer condition monitoring. This chapter primarily highlights advanced signal processing techniques, with focus on wavelet transform (WT), empirical mode decomposition (EMD), ensemble EMD (EEMD), and mathematical morphology (MM) methods. Special techniques developed for multiple-PD source separation and their PD feature extraction and recognition are also discussed in this chapter.

Chapter 6 concentrates on frequency response analysis (FRA) for transformer winding mechanical deformation/displacement analysis. A number of international standards are discussed in this chapter, along with a novel statistical approach.

Chapter 7 primarily explains moisture measurements by online sensors, with a comprehensive guideline for practicing engineers to estimate the remaining life of insulation as a function of the water content of paper.

Chapter 8 presents biodegradable oil fundamentals and their impact on paper insulation ageing. Then, oil chemical/physical measurements and PDC/FDS interpretation schemes for biodegradable oil-filled transformers are presented, with a comparison of currently used condition monitoring interpretation techniques for mineral oil-based transformers.

Chapter 9 provides an intelligent framework for transformer condition monitoring using online sensors, along with the importance of numerical modeling to assist fault detection in transformers, statistical learning for dealing with measurement uncertainties, and data and information fusion for transformer condition assessment. A hardware and software platform for implementing a smart transformer condition monitoring system and a concept of health index and their interpretation are also discussed in this chapter.

Chapter 10 highlights the limitations of current condition monitoring techniques and the need for future research.

Acknowledgments

Many people have supported this work, directly or indirectly, throughout our involvement with transformer research. We would like to acknowledge some of the key personnel without whose contributions this publication would never have reached this point.

1) Emeritus Professor Mat Darveniza for introducing the topic of transformer insulation ageing and life assessment during Tapan Saha's Ph.D. research.
2) Honorary Reader David Hill and Dr. Tri Li from the School of Chemistry and Molecular Biosciences, University of Queensland for helping to understand chemistry of insulation materials and some chemical-based diagnosis concepts.
3) Mr. Richard Marco, Mr. Brian Williams, and Dr. Zheng Tong Yao during the initial hardware/software design of PDC-RVM equipment at the University of Queensland.
4) A number of research fellows who worked at the University of Queensland with Tapan Saha during the last 25 years need to be mentioned specifically: Dr. Abbas Zargari, Dr. Prithwiraj Purkait, Dr. Manoj Pradhan, Dr. Chandima Ekanayake, Dr. Hui Ma, and Dr. Dan Martin.
5) Tapan Saha has been fortunate to advise numerous Ph.D. students in this area. Their contributions are worthy of note: Dr. Zheng Tong Yao, Dr. Karl Mardira, Dr. Jing Haur Yew (Kelvin), Dr. Raj Jadav, Dr. Mohd Fairouz, Dr. Jeffery Chan, Dr. Yi Cui, and Dr. Kapila Bandara. Thanks to a number of Masters by research students for their contributions to this publication.
6) Dr. David Allan from Powerlink Queensland, who provided extensive industry collaboration throughout Tapan Saha's research.
7) Mr. Bryce Corderoy, Mr. Vic Galea, and Dr. Frances Mitchell from TransGrid New South Wales for providing industry-oriented transformer research opportunities.
8) Numerous undergraduate and Masters students for their contributions through their thesis projects.
9) The Australian Research Council for providing several funding supports through the ARC Linkage Project Scheme, without which this volume of work would never have been possible to conduct.
10) Industry support from Powerlink Queensland, Energex, Ergon Energy, TransGrid, Ausgrid, and Aurecon through extensive industry collaborations.

11) CIGRE Australian Panel Members A2 and D1 for providing extensive knowledge in the area of transformer and insulation diagnostics.
12) The authors of many papers and books, from which we have continuously benefitted in our journey. If we have inadvertently missed any referencing or acknowledgment of these authors, we sincerely apologize.
13) Special thanks to Mr. Steven Wright for his help during many experiments in the intelligent equipment condition monitoring laboratory and for proof-reading the book.
14) Special thanks to Dr. Hui Ma for reading many chapters of this book throughout the last 12 months of manuscript preparation.
15) Sincere thanks to the University of Queensland for providing the facilities and opportunities to carry out research in this area.
16) Thanks to our families for understanding and support throughout our research career.

Contributing Authors

A number of our current colleagues and Ph.D. students at the University of Queensland have contributed directly in preparing the manuscript of this book. Their contributions are greatly appreciated.

Chapter 1: Prof. Tapan Saha & Prof. Prithwiraj Purkait

Chapter 2: Prof. Tapan Saha & Prof. Prithwiraj Purkait

Chapter 3 Part A: Prof. Tapan Saha & Prof. Prithwiraj Purkait

Chapter 3 Part B: Prof. Tapan Saha, Prof. Prithwiraj Purkait, & Dr. Chandima Ekanayake

Chapter 4: Dr. Yi Cui, Prof. Tapan Saha, & Dr. Hui Ma

Chapter 5: Dr. Jeffery Chan, Prof. Tapan Saha, & Dr. Hui Ma

Chapter 6: Dr. Mohd Fairouz, Prof. Tapan Saha, & Dr. Chandima Ekanayake

Chapter 7: Dr. Dan Martin & Prof. Tapan Saha

Chapter 8: Dr. Kapila Bandara, Prof. Tapan Saha, & Dr. Chandima Ekanayake

Chapter 9: Dr. Hui Ma & Prof. Tapan Saha

Chapter 10: Prof. Tapan Saha & Prof. Prithwiraj Purkait

Research colleagues: Dr. Chandima Ekanayake, Dr. Hui Ma, & Dr. Dan Martin.

Former Ph.D. students: Dr. Jeffery Chan, Dr. Mohd Fairouz, Dr. Yi Cui, & Dr. Kapila Bandara.

1

Transformer Insulation Materials and Ageing

1.1 Introduction

The primary and secondary coils of a transformer are the key components in performing its basic function of transforming voltage and current. Materials are used to insulate the primary and secondary coils. In transformers, in addition to the primary and secondary coils, there are several other important components and accessories. The insulating material is one of the most critical components of a transformer. Sufficient insulation between different active parts of the transformer is necessary for its safe operation. Adequate insulation is not only necessary to isolate coils from one another, or from the core and tank, but also ensures the safety of the transformer against accidental overvoltages.

The insulation system in a transformer can be categorized as follows.

- Major insulation:
  - between core and low-voltage (LV) winding;
  - between LV and high-voltage (HV) winding;
  - between top and bottom of winding and yoke;
  - between HV and tank;
  - bushings.
- Minor insulation:
  - between conductors;
  - between turns;
  - between layers;
  - between laminations;
  - between joints and connections.

The insulation material commonly used between the grounded core and the LV coil to ground, and also between HV and LV coils, is oil-impregnated solid pressboard. Solid insulation, including pressboard or paper, can have small internal air voids. This reduces the insulating strength of the solid insulation as well as reducing its heat dissipation capacity. When transformer oil is used to impregnate solid insulation, the voids are filled with oil, resulting in an improvement of both the insulation strength and the heat dissipation capacity of the solid insulation. In

Transformer Ageing: Monitoring and Estimation Techniques, First Edition.
Edited by Tapan Kumar Saha and Prithwiraj Purkait.

larger transformers, cellulose-based paper tape is usually wrapped over individual conductors. Layer-to-layer or disc-to-disc insulation is mostly provided by oil-impregnated Kraft paper or even thick pressboard or transformer board in case of higher-rating transformers.

1.2 Solid Insulation – Paper, Pressboard

The solid insulation materials widely used in the transformer are paper, pressboard, and transformer board, which are formed from the cellulose found in plants. Cellulose insulation with mineral oil has played a major role as the main insulation system for transformers for a very long time. Cellulose paper, tapes, and cloths have also been widely used. They provide excellent dielectric strength and low dielectric loss, and hence impregnated paper is now widely accepted as the insulation pillar of the electricity industry. Paper and pressboard insulation derived from pure cellulose have an excellent capacity for being impregnated with oil, thereby improving their insulation properties many times over. In addition, such solid insulating materials are easy to mold and wrap around coils, and can be made of various dimensions as per requirements.

The main concern with using dry paper as an insulating material is that it is very hygroscopic (i.e., it readily absorbs moisture). In order to overcome this deficiency, it must be dried and treated (impregnated) in some liquid (oil, varnish, resins) to reduce moisture ingress and maintain its dielectric strength. Such treatments fill the spaces between fibers and increase the dielectric strength. Nowadays, other synthetic insulating materials are used to insulate areas where the operating temperature is designed to be high (hybrid insulation), or for entire transformers specially designed to operate at high temperatures (artificial polymer – e.g., Aramid paper). Max Schaible summarized several lists of desirable qualities of high-voltage insulation, as shown in Table 1.1.

1.2.1 Cellulose Structure

Natural cellulose comes from plants. Paper and pressboard insulation is generally made from a "Kraft" process. The typical composition of unbleached softwood Kraft is as follows:

75–85%	cellulose
10–20%	hemicelluloses
2–6%	lignin
<0.5%	inorganic

The structures of glucose and cellulose are shown in Figures 1.1 and 1.2.

Cellulose is a linear polymer of anhydro-glucose elements connected together by glycosidic bonds as shown in Figure 1.1.

Table 1.1 Desirable qualities for electrical grade fibers

Natural pulp (wood, cotton, hemp, etc.)	Synthetic pulp (polymers)
Chemical properties • High cellulose content • Low lignin content for reduced specific conductivity • Low hemicellulose content for low p.f. • Low ash for low inorganic ion content [low ionic impurities; replace mono- with bivalent ions (Ca++, Ba++ for Na+), wash with deionized water to reduce dielectric losses] • High molecular weight for good mechanical properties • Low shieve or knot content • Blocked hydroxyl groups to lower the dielectric constant of paper w.r.t. that of oil • Low carboxyl number for lower high-temperature losses	*Chemical properties* • Select crosslinking to depolymerizing polymers • Select materials with high T_g for thermal stability • Optimize mechanical strength through adjustment of molecular weight and molecular distribution • Reduce polar group content for better electrical properties • Choose chemically inert aromatic polymers with little swelling in oil at service temperatures • Avoid adding hydrogen or halogen groups to the carbons next to the carbon with characteristic functional moiety (6)
Physical properties • High density for good electrical strength • Low moisture content after oil impregnation • High oil penetration (low density) for low overall dielectric constant • High air resistance (beat pulp to optimize conduction through inter-fiber bonding and mechanical strength)	*Physical properties* • Minimize fiber-to-film void content • Optimize cross sectional shape of fiber for good mechanical strength • Maximize bonding between polymer fibers by heat, mechanical entanglement, or chemical means • Raise air resistance

Source: Ref. [1].

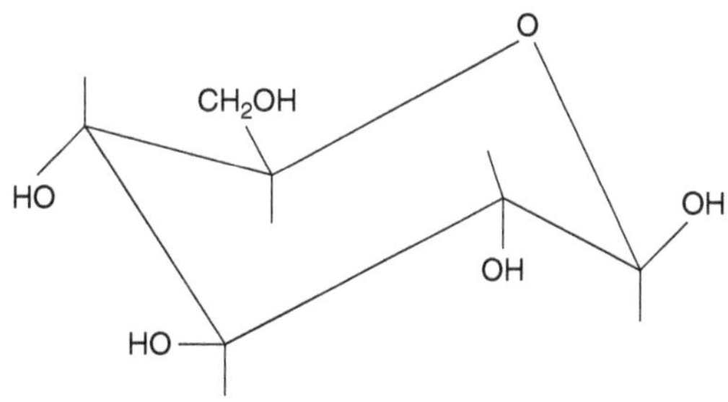

Figure 1.1 Chemical structure of glucose.

The most commonly used terminology to indicate a polymer structure is the degree of polymerization (DP), defined as the average number of glucose rings joined together to form a cellulose macromolecule. The DP value may be as high as 14,000 in naturally occurring cellulose, but after purification it usually reduces to the order of 1000 to 1500 [2].

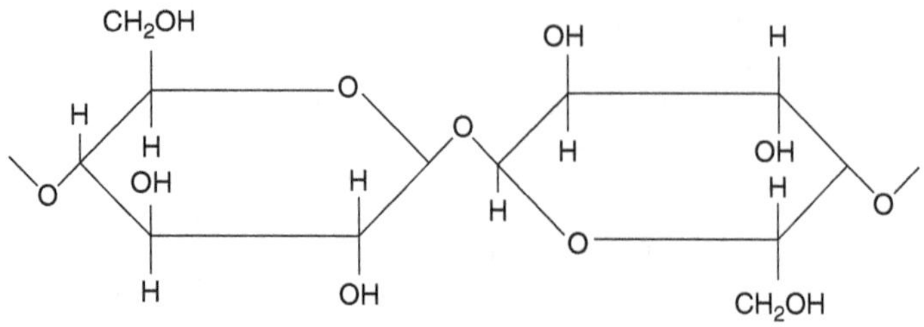

Figure 1.2 Chemical structure of cellulose (polymer chains of glucose).

1.2.2 Commercial Cellulose Insulations

Cellulose-based electrical insulation has evolved over a long period and widely used cellulose-based paper is derived from chemical wood pulp. Papers made from wood pulp are available in a wide range of thicknesses. The electrical and mechanical properties of such wood pulp papers have been found acceptable for meeting the requirements of electrical insulation. Such insulation contributes much more than a mere mechanical separation between conductors; it contributes to the total dielectric properties of any composite insulation of which it may be part. The presence of cellulose insulation affects the voltage distribution in heterogeneous dielectric systems: it contributes to the dielectric loss; determines the short-term dielectric strength; and, to a major extent, the life of the total insulated system.

Cellulose insulation, however, has defects which are concerned with the following:

- its hygroscopic properties;
- its reactivity with oxygen;
- its thermal instability at high temperature.

1.2.3 Moisture Absorption by Paper and Pressboard

Even though paper is treated or impregnated, it can still absorb moisture if the air or oil surrounding it contains moisture. In oil-filled transformers, dry paper will slowly absorb moisture from the oil. Moisture will distribute between paper and oil in a definite ratio in the final state of equilibrium in which the moisture content of the paper is much more than that of the oil. Ingress of moisture in insulation increases the dielectric loss and decreases the effective dielectric strength of paper/pressboard. The quantity of water that paper can absorb in a large power transformer can be as high as several hundred liters. Koestinger *et al.* [3] gave an example: "a 300 MVA Transformer, with 10000 kg insulation with a moisture level of 3% represents 300 litres water. The amount in the oil is normally negligible (300 MVA Transformer with 60000 kg oil and 10 ppm H_2O at 30 °C represents 0.6 litre water)." Reduction of the moisture level in an oil–paper insulation system is not a trivial task, and significant research has been in progress over the years to determine the best method. The importance of a drying process and some key techniques are described in the following sections.

1.2.4 Drying Paper and Pressboard

The previous discussion has indicated the importance of removing the moisture from the pressboard and paper of oil-filled transformers. Cellulose insulations are traditionally dried by oven heating at atmospheric pressure, oven heating at reduced pressure, or a combination of both. The problem is to remove moisture from the insulation to an extent consistent with the engineering requirements without affecting the chemical properties. The removal of moisture from cellulosic insulation requires the application of high temperatures, in the range of 100–120 °C. Under excessive heating, however, it is extremely difficult to avoid degradation of the cellulose and yet secure complete removal of absorbed and adsorbed moisture from the insulation.

In the drying of cellulose structures, the moisture moves by diffusion from the innermost layers of the assembly to the outside surface. It is removed from this surface by diffusion into the surrounding medium. The practical problem involved in the factory drying of cellulose insulated apparatus therefore involves two distinct steps:

- the problem of accelerating the movement of the moisture from the inner layers to the solid–air (or vacuum) interface;
- the actual diffusion of the moisture from that surface into the surrounding space.

Because of the presence of metallic conductors and insulating varnish or resin films, the migration of moisture from the innermost parts to the diffusing surface is largely impeded. In addition to that, the dense and compact assembly of the cellulose insulation layers presents an increasingly difficult arrangement for easy and efficient drying.

When a cellulose sheet is heated in air at a constant temperature, the loss of its water content is rapid at first. Eventually, however, the rate of water removal slows and finally an equilibrium condition is reached with the surrounding atmosphere. During the period when this moisture equilibrium exists, there is no further moisture change in the insulation. Increasing the temperature may allow further moisture to be removed, but that may lead to degradation of the cellulose.

Depending on the transformer capacity, several such drying cycles of heating followed by vacuum are carried out. When the moisture content value drops to less than 0.5% in the solid insulation, it is conventionally accepted to be fully dried. Several commonly used drying mechanisms are briefly described.

1.2.4.1 Drying by Conductor Heating

When insulation is dried by the passage of current through the copper conductor, it supplies a source of heat to the innermost layers of the insulation. Heat is then transferred from the conductor to the insulation with which it is in direct contact. This heat is then transferred through the layer of the insulation to the outer surface layers, thereby heating the total thickness of insulation, with a heat gradient. In this process of drying, the advantage is that the heat transfer from the hot conductor to the moist, deep-seated layers of the insulation is efficiently accomplished. However, when the insulation adjacent to the conductor is substantially free from moisture, a thermal gradient may be established under some conditions, leading to a rapid degradation of the cellulose contained in that area. The heat source must be so controlled that even with thick layers of insulation, the thermal drop across the dried mass of the insulation does not cause any degradation.

1.2.4.2 Drying by Vacuum Treatment

The equilibrium between moisture in paper and moisture in surrounding air depends on the vapor pressure of the moisture and the temperature. To increase the rate of drying, some means must be provided to reduce the partial vapor pressure of the moisture. This can be done by applying vacuum or low pressure. When the moisture present within the insulation turns to vapor, the vapor pressure inside the tank will increase. This vapor is to be removed by passing hot, dry air at low pressure through the tank in a closed circulation cycle.

1.2.4.3 Heating by Circulation of Air Through the Tank

Suitable ducts and blowers should be assembled to blow air into the transformer, either at the bottom or at the top. A controlled amount of heat needs to be supplied in large quantities by passing dry, hot air at 90–120 °C. Electric heaters with precise temperature control are usually the most convenient.

1.2.4.4 Heating by Circulation of Hot Oil

The method of heating with a circulating current has the practical disadvantage that all the oil must be drained off before the transformer can be further dried. This can be avoided by heating the oil in a separate compartment and passing the hot oil over the top of the core and coils, pumping it out of the bottom and back through the heater.

1.2.4.5 Hot Oil Spray Drying

Vacuum is maintained on the transformer under ambient temperature conditions for a period of 48 hours. Moisture is removed via the cold-trap condensation technique. Then hot (80 °C) transformer oil is circulated as a spray internally on the tank cover. The hot oil falls by gravity on the winding insulation. Moisture is collected via the cold-trap method. At completion of the drying operation, the transformer is filled with transformer oil under vacuum.

The cold-trap condensation technique consists of circulating liquid nitrogen within a cold-trap condenser, where water vapor drawn by the vacuum pump condenses in the form of ice in the cold trap.

1.2.4.6 Kerosene Vapor Phase Drying (KVPD)

A much faster and more efficient process that also ensures improved insulation characteristics of paper is the vapor phase drying (VPD) method using kerosene. In this method, the heating medium used is a high-temperature kerosene vapor. A special grade of kerosene is converted into vapor form by heating it to temperatures around 130 °C. The high-temperature vapor is then injected into the transformer tank containing insulated coils. When the hot vapor flows through the coil insulation, it will condense and release latent heat of condensation, which will heat the insulation. Condensed kerosene will be collected and retransmitted to the evaporator to be heated again. In this way, after several cycles, the temperature of the coil insulation increases continuously and moisture inside the insulation evaporates to water vapor. The water vapor is taken out by vacuum, thereby achieving uniform and effective drying of the insulation.

Compared with conventional hot-air drying, where the vapor temperature cannot be raised beyond 110 °C due to the presence of oxygen, kerosene vapor can be raised to temperatures around 130 °C without the risk of ageing the insulation.

Condensed kerosene that may accumulate on winding and insulation surfaces is also a very good cleaning solvent. It can wash out dust and dirt from the insulation surfaces. The KVPD process also takes approximately half the time for drying compared with the conventional process.

1.2.4.7 Low-Frequency Heating [3, 4]

Low-frequency heating (LFH) is one of the most modern techniques of drying transformer insulation by supplying a controlled, low-frequency current to the transformer windings instead of using hot air or solvent vapor as the main heat transfer medium. The LFH system is much faster and more cost-effective than conventional hot oil and vacuum drying processes. The National Industry of Norway built the first LFH drying plants for distribution transformers between 1984 and 1987. ABB Switzerland Ltd. Micafil further developed the LFH drying method for small power transformers in production, and large power transformers for onsite drying (Figure 1.3).

The process heats the transformer windings uniformly from inside by passing a low-frequency (0.4–2 Hz) current at low voltage through the HV windings, while the LV windings are kept short-circuited. In this process, HV and LV windings can be made to heat up to drying temperatures of 110–120 °C. Special controls must be used to closely monitor the drying process, in particular the winding temperature, so that hot spots do not form and damage to the insulation is avoided.

As in any conventional drying system, an LFH heating system is part of a drying installation consisting of a vacuum autoclave or a transformer tank under vacuum, a vacuum pumping system, etc.

Given a moisture level in the insulation of between 3% and 1.5%, LFH offers a drying speed that is eight times faster than conventional methods, resulting in huge savings in energy, personnel, and equipment.

To reduce the moisture extraction process and also to enhance the quality of insulation, often the LFH drying method is used in combination with conventional methods, especially with "hot oil spray." The combination of the hot oil and LFH heating ensures uniform heating of all parts of the transformer. Reference [3] provides a good explanation of this methodology. Once the temperature inside the transformer tank is stable, the oil is drained out to a separate tank followed by application of vacuum. For the next heating cycle, only electrical heating (LFH) is used to heat the windings. During this period, the vacuum level is removed. As the preset temperature is reached, the LFH heating is stopped and only a partial vacuum of <1 mbar is applied. This process makes most of the water to evaporate. The duration of these heating cycles is dependent on the size of the transformer, the humidity level, and the total weight of insulation inside the transformer. These periodic cycles are repeated until the rate of moisture extraction

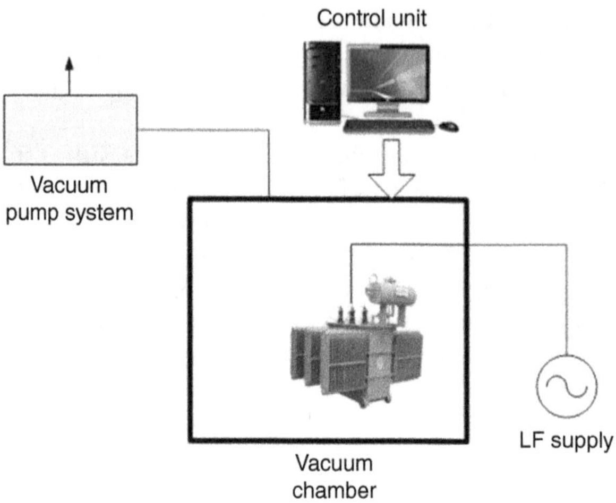

Figure 1.3 Schematic diagram of LFH [3].

becomes less than a set point [3]. Once the desired level of drying is reached, the transformer is filled with oil again.

1.2.5 Special Treatments of Insulation Paper/Pressboard – Thermal Upgrading

"Thermal upgrading" or "thermal uprating" is the process by which the rate of thermal decomposition of paper insulation is reduced over the lifespan of a transformer. Tom Prevost [5] defined thermally upgraded paper as:

> Cellulose based paper which has been chemically modified to reduce the rate at which the paper decomposes. Ageing effects are reduced either by partial elimination of water forming agents (as in cyanoethylation) or by inhibiting the formation of water through the use of stabilizing agents (as in amine addition, dicyandiamide). A paper is considered as thermally upgraded if it meets the life criteria as defined in ANSI/IEEE C57.100; 50% retention in tensile strength after 65,000 hours in a sealed tube at 110 °C or any other time/temperature combination given by the equation
>
> $$\text{Time}\,(\text{hrs}) = e^{[15{,}000/(T+273)-28.082]}$$
>
> Because the thermal upgrading chemicals used today contain nitrogen, which is not present in Kraft pulp, the degree of chemical modification is determined by testing for the amount of nitrogen present in the treated paper. Typical values for nitrogen content of thermally upgraded papers are between 1 and 4 percent, when tested per ASTM D-982.

A significant advantage of thermally upgraded papers is their better resistance to loss of physical strength in operation. Often, the expression "55 °C rise paper" is used for standard (plain, non-upgraded) paper, whereas "65 °C rise paper" stands for thermally upgraded paper. The numbers refer to the average oil rise temperature, indicating that the designed hotspot temperature with upgraded paper is higher than with untreated paper. This allows an increased continuous load rating of the transformer.

1.3 Liquid Insulation - Oil

Oil is an equally important part of a transformer's overall insulation. Oil, like paper/pressboard, is a product of nature containing a variety of impurities of speculative nature and amount.

1.3.1 Functions of Oil

1.3.1.1 Electrical Insulation

The main function of insulating oil in a transformer is to provide electrical insulation between the various energized parts; it also acts as a protective coating layer to prevent oxidation of the metal surfaces.

1.3.1.2 Heat Dissipation

Another important function of the oil is to enhance heat dissipation. Transformer cores and windings get heated up during operation due to various power losses. Oil takes heat away from the core and windings by the process of conduction and carries heat to the surrounding tank, which is then radiated out to the atmosphere. In order that the mineral oil can dissipate the heat away effectively, certain specifications – including viscosity, pour point, and flash point – need to be maintained.

1.3.1.3 Diagnostic Purposes

The third (very useful) function of insulating oil in a transformer is that it acts as a health indicator for the device. Both the chemical and electrical conditions of the transformer can be monitored by examining the oil periodically. Oil samples are collected from designated sampling points of the tank and taken to laboratories for several tests to be performed.

When a fault develops within the transformer, the energy is dissipated through the oil, which causes chemical degradation of the liquid. Testing oil samples for degradation products can provide useful information about the nature and severity of possible faults inside a transformer.

1.3.2 Types of Oil

The insulating liquids that can be used for transformers are

mineral petroleum-based oils
askarels
vegetable oils
organic esters
polyhydrocarbon liquids
fluorinated liquids
silicone liquids.

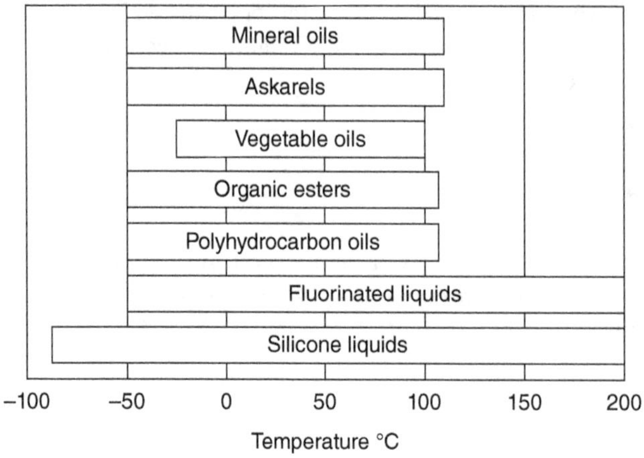

Figure 1.4 Temperature range of normal usability of various classes of insulation oil [6].

Table 1.2 Relative costs of insulating oils [6]

Type of liquid insulation	Relative cost
Mineral petroleum-based oils (also commonly known as mineral insulating oil)	1
Askarels	8
Vegetable oils	2
Organic esters	4–8
Polyhydrocarbon liquids	3–5
Fluorinated liquids	10–20
Silicone liquids	10

The temperature ranges over which these liquid insulants can operate are shown in Figure 1.4.

Petroleum-based mineral oil is by far the most common and the cheapest liquid used as an insulating fluid, and the use of alternative types – which can be several times costlier – can only be justified economically in circumstances where technical advantages are to be gained. Table 1.2 gives the relative costs of various types of liquid insulant.

The ideal liquid for fulfilling the purposes of being used in a transformer can be listed as having

- high dielectric strength
- high impulse strength
- high volume resistivity
- low dielectric dissipation factor
- high specific heat and thermal conductivity

- good chemical stability
- low viscosity
- low volatility and high flash point
- good arc quenching properties

and being

- non-flammable and non-toxic
- cheap and easily available.

It is however, not possible that a single liquid will satisfy all of the above properties. It is also observed that some of the properties may – to some extent – contradict each other. It is thus necessary to have some compromise while selecting the liquid to be used for transformers, and suitable modifications need to be done in the design stage to cater for such deficiencies.

1.3.2.1 Mineral Insulating Oils

Until now, mineral oil has been the most widely used fluid in electrical equipment that serves the dual purpose of electrical insulation and heat dissipation. There are some synthetic liquids available as alternatives, but the enhanced cost of such liquids masks their performance in widespread uses. As a consequence, mineral oil still serves as the most popular liquid insulation for electrical equipment.

1.3.2.1.1 Composition of Mineral Oil

Mineral oils to be used for electrical apparatus are basically derived from crude petroleum through a distillation and refining process. The composition of such crude petroleum can vary depending on its geographical source. Basically, all mineral oils are mixtures of various hydrocarbon compounds with different chemical structures. Some crude oils can have, in addition to hydrocarbons of different molecular structure, some paraffinic as well as naphthenic compounds. In addition, some crude oils also contain some amounts of aromatic and poly-aromatic compounds which, besides carbon and hydrogen, contain other atoms such as nitrogen, sulfur, and oxygen [7]. While some poly-aromatic compounds that are oxidation inhibitors are beneficial, most are detrimental, since they often act as oxidation initiators and electrical charge carriers.

1.3.2.1.2 Types of Mineral Oil

As already mentioned, mineral insulating oils are derived from crude petroleum. In their preparation, the crude oil is distilled and separated into distinct commercial fractions from which a variety of industrial products can be obtained. The crude oil is passed through a series of physical and chemical treatments before producing the refined version of mineral oil to be used in transformers. The main constituents of transformer oil are as follows.

- *The saturated hydrocarbons (i.e., paraffins).* These are straight-chain hydrocarbons possessing the general formula C_nH_{2n+2}.
- *Iso-paraffins.* These are branched hydrocarbons possessing the same general formula as normal paraffins.

- *Naphthenes.* These are saturated cyclic hydrocarbons possessing the general formula C_nH_{2n} and having a closed ring structure.
- *Aromatics.* These are unsaturated hydrocarbons having a ring structure similar to that of benzene (e.g., naphthalene). The aromatic hydrocarbons have a high affinity for hydrogen.

1.3.2.1.3 Properties of Mineral Oil

Most of the transformer oils consist of a mixture of these compounds. The properties of the refined mineral oil to be considered in the development and application of mineral oil for use in transformers are listed in Table 1.3.

Details and acceptable limits of these parameters can be obtained from the standards, as described in Table 1.4.

1.3.2.1.4 Significance of Transformer Oil Properties

Physical Properties *Color and Appearance* Newly installed transformer oil is of light and clear appearance. With time and operation, the color of the mineral oil turns darker and in many cases this is the first indication of deterioration or contamination of the oil.

Table 1.3 Properties of mineral oil

Physical properties	Chemical properties	Electrical properties
Thermal conductivity	Oxidation stability	Breakdown strength
Specific heat	Gassing characteristics under electric stress	Impulse strength
Coefficient of volume expansion	Ionization conditions (silent discharge)	Dielectric dissipation factor
Density	Gassing characteristics under high-temperature pyrolysis conditions (thermal and disruptive discharge)	Volume resistivity
Viscosity	Neutralization value	Permittivity
Pour point	Saponification value	Contaminants
Density	Sulfur staining and corrosion	
Refractive index	Nitrogen content	
Molecular weight	Ionic contamination	
Solvent power	Water content	
Vapor pressure		
Flammability		
Interfacial tension		

Table 1.4 Oil properties in various standards

Properties	ASTM	ISO/IEC
Physical properties		
Color and appearance	D 1500 D 1524	ISO 2049 IEC 60296
Density/specific gravity	D 1298	IEC 60296
Viscosity	D 88 D 445 D 2161	ISO 3675 IEC 61868
Pour point	D 97	IEC 60296
Flash and fire point	D 92	ISO 3104 IEC 60296
Aniline point	D 611	
Interfacial tension	D 2285 D 971	ISO 3016 IEC 60296
Electrical properties		
AC breakdown strength	D 877 D 1816	IEC 60156
Impulse strength	D 3300	
Dissipation factor	D 924	IEC 60247
Chemical properties		
Acidity	D 974 D 664	IEC 62021
Oxidation stability	D 2112 D 2440	IEC 1125
Oxidation inhibitor content	D 2668 D 4768	IEC 60666
Water content	D 1533	IEC 60296 IEC 60814
Furan content	D 5837	IEC 61198
Corrosive sulfur content	D 1275	IEC 60733
Total gas content	D 2945 D 3612	IEC 60567

Source: IEEE C57.106-2006 – IEEE Guide for Acceptance and Maintenance of Insulating Oil in Equipment. Partly extracted from Table 4, p. 11.

Relative Density (Specific Gravity) The specific gravity of oil is defined as the ratio of weights of oil and water of equal volumes. It is important to determine the specific gravity of oil in certain applications, such as in cold climates, where ice may form inside the transformer tank if it is subjected to temperatures below freezing. In such cases, ice floating on the oil surface may cause flashover between conductors above the oil level.

Viscosity The viscosity of oil determines its tendency to resist flow of itself under specified conditions. Viscosity is calculated by measuring the time taken by a given volume of oil to flow through a calibrated tube. The heat transfer capacity of oil is strongly influenced by its viscosity. Highly viscous oil has much reduced cooling efficiency. Oil having low viscosity enhances cooling by easier circulation. Especially in cold climates, highly viscous oil will inhibit heat transfer from hot-spots due to its low circulation capacity, and also affects the speed of moving parts (like those in circuit breakers and tap changers). The viscosity of oil also affects the oil processing and cellulose impregnation time.

Pour Point The pour point of oil is defined as the temperature at which oil refuses to flow under the given test conditions. Knowledge of the pour point of oil is important to judge its suitability for use in a particular climate.

Flash and Fire Points The flash point of oil is the temperature to which, when the oil is heated, it produces enough vapor that it may form a flammable mixture with air.

In contrast, the fire point is the temperature at which oil vapors will ignite and sustain a fire for at least 5 seconds. The flash and fire points of oil are thus useful information in determining the volatility and fire resistance properties of oil. A low value of flash point will generally indicate that the oil has volatile combustible contaminants in it.

Interfacial Tension Interfacial tension is the property of oil that is a measure of the surface tension maintained against water under non-equilibrium conditions. This property indirectly indicates the amount of oil-soluble polar contaminants and oxidation products present in insulating oils. The film strength of the oil is further weakened when certain contaminants – such as paints, varnishes, soaps, and oxidation products – are present in the oil.

Electrical Properties *AC Breakdown Strength* The AC breakdown voltage of insulating oil is the upper limit of voltage stress that the oil can withstand without failure. A low value of breakdown voltage primarily indicates the presence of conductive contaminants in the oil, such as water, dirt, or moist cellulosic fibers.

Impulse Strength The impulse strength of insulating oil indicates its ability to withstand high-voltage transients of very short duration, such as those it may be subjected to during lightning strikes. The standard lightning impulse test described in IEEE Standard C57.12.90-1999 specifies a 1.2/50 μs negative polarity wave. The impulse strength of oil gets lowered with an increased concentration of aromatic and hydrocarbon molecules, and due to contact with materials of construction, ageing, and other impurities.

Dissipation Factor The dissipation factor is the measure of dielectric loss taking place in the insulating oil when it is subjected to an AC field. The dissipation factor generally increases with an increasing presence of contamination or ageing by-products such as moisture, carbon or other conducting matters, and oxidation by-products.

Chemical Properties *Neutralization Number (Acidity)* The neutralization number is a measure of acidic components present in the insulating oil. New and unused mineral oil is expected to be neutral and free from any acidic compound. During service, the neutralization number is the measure of acidic by-products of oxidation of the oil. The acidity and therefore the neutralization number increases as oils age during service. A high acid number indicates the presence of oxidation or contamination with foreign matter. The neutralization number may be used as a useful indicator of whether the oil should be reprocessed or replaced.

Oxidation Stability The oxidation stability of insulating oil is a property that indicates its resistance to oxidation during service. A high value of oxidation stability is a desirable property that ensures long service life of the oil. Otherwise, oxidation of oil gives rise to acid and sludge formation, resulting in increased electrical losses, metal corrosion, and electrical faults, thereby reducing insulation service life. Antioxidant additives such as 2-6 ditertiary-butyl phenol (DBP) and 2-6 ditertiary-butyl para-cresol (DBPC) are sometimes used as inhibitors that slow down oxidation in insulating oil, thereby increasing its service life.

Water Content The water content of the insulating oil to be used in electrical equipment must be low to ensure high electrical strength and low dielectric loss. Water may be present in insulating oils either as free water or in dissolved form. Measurement of moisture in oil is done in terms of milligrams/kilogram or ppm. New insulating oil will generally have a moisture content of less than 10 ppm.

Furans Content Furanic compounds are degradation by-products of cellulosic materials such as insulating paper and pressboard. These are often soluble in oil, and thus a measure of furanic compounds in oil indicates the extent of cellulose degradation. The furans content in oil is tested by high-performance liquid chromatography (HPLC). New oils should not have any significant quantity of furanic compounds dissolved in them.

Corrosive Sulfur Content The presence of sulfur or sulfur-containing compounds in oil can lead to corrosion of polished copper conductors. Sulfur may either come from the crude oil from which it is refined or from the rubber hoses that are used during oil processing and even from gasket materials.

Total Gas Content Under thermal and electrical stresses, oil and oil-immersed insulating materials may decompose and generate gaseous decomposition products that dissolve into oil. The amount, composition, and rate of generation of these dissolved gases serve as a very good indicator of the type and severity of faults or abnormalities taking place inside the transformer.

See Table 1.5 for an overview of the property limits for new mineral oil.

1.3.2.2 Askarels

Askarel is the general name given to a class of synthetic chlorinated hydrocarbon insulating liquids that are sometimes used in place of oil because of their non-

Table 1.5 Limits for new mineral oil

	ASTM/IEEE C57.106	
Property	**Specification value**	**ASTM no.**
Color	0.5 ASTM units max.	D 1500
Visual examination	Bright and clear	D 1524
Density/specific gravity at 15 °C	0.91 max.	D 1298
Viscosity (mm^2/s or cSt)		D 88
at 100 °C	3	D 341
at 40 °C	12	
at 0 °C	76	
Pour point	−40 °C	D 97
Flash point	145 °C min.	D 92
Aniline point	63–83 °C	D 611
Interfacial tension	40 mN/min.	D 971
Dielectric strength		
1 mm gap	20 kV	D 1816
2 mm gap	35 kV	D 1816
Dielectric breakdown, impulse, needle-negative	145 kV	D 3300
Dissipation factor		D 924
25 °C	0.05% max.	
100 °C	0.30% max.	
Neutralization number (acidity)	0.015 mg KOH/g max.	D 974
Oxidation stability		D 2440
Sludge after 164 hrs	0.30%	
Total acid after 72 hrs	0.5 mg KOH	
Total acid after 164 hrs	0.6 mg KOH	
Oxidation inhibitor content		D 2668
Type I oil % max.	0.08	
Type I oil % min.	0.0	
Type II oil % max.	0.3	
Type II oil % min.	>0.08	
Moisture content	35 ppm max.	D 1533
Free or corrosive sulfur	Noncorrosive	D 1275
Inorganic chlorides or sulfates	None	D 878

Source: IEEE C57.106-2006, partly extracted from Table 1.

flammable nature. The gases produced under arcing conditions in askarels consist predominantly of non-combustible hydrogen chloride and a much lesser amount of combustible gases. Besides its very high cost compared with oil, askarel has some disadvantages.

- Pressboard or paper impregnated with this liquid has a lower impulse strength compared with mineral oil, even though the dielectric strength of the liquid itself is comparable with that of mineral oil.
- The products of decomposition are generally more corrosive and damaging to insulation than those of oil.

Therefore, although this liquid has application for low-voltage transformers to be installed where oil is an intolerable fire hazard, it has not become generally popular for use.

1.3.2.3 Vegetable Oil

In line with the search for "green products," vegetable oils are being explored for their potential use as insulating oil as an alternative to mineral oils. Vegetable oil basically consists of tri-glycerides. Vegetable oils, being of non-fossil origin, pose fewer environmental, safety, and health problems. The relative superiority of vegetable oils compared with mineral oil and silicon oil in terms of biodegradability is shown in Figure 1.5. Some vegetable oils have shown promising dielectric characteristics for use as an insulating liquid in electrical equipment. In addition to AC breakdown strength, the viscosity and pour point of vegetable oils can be comparable to conventional mineral oil. The flash and fire point values of certain vegetable oils are found to be comparable to those of high-temperature hydrocarbons or silicones. Soya and rapeseed oil and ester-based fluids have shown superior environmental, health, and safety characteristics compared with conventional mineral oils for potential applications in transformers.

1.3.2.4 Synthetic Esters

The major concerns related to the use of mineral oil can be compensated for by the use of organic ester liquids due to their environmentally friendly biodegradable nature, high fire point, and non-flammability. Synthetic liquids like Midel 7131 are currently being used as insulating liquid for whole transformers, or only for the impregnation of the paper insulating material or sometimes as a mixture with mineral oil.

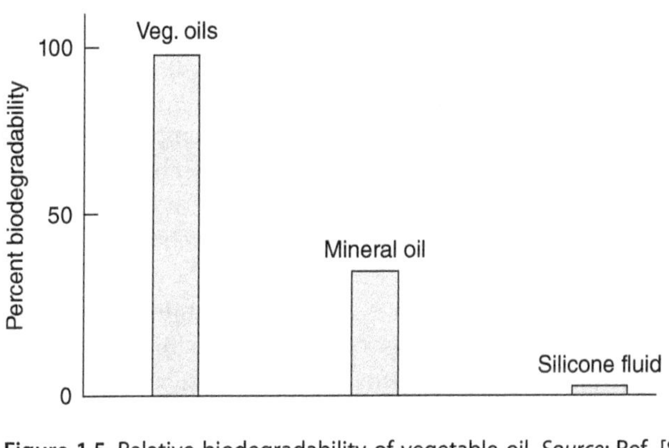

Figure 1.5 Relative biodegradability of vegetable oil. *Source:* Ref. [8], Figure 4.

Synthetic esters contain carbon, hydrogen, and oxygen. They are prepared from mono- or polyhydric alcohols and mono- or dibasic aliphatic or aromatic acids. Organic esters have high boiling points in relation to their viscosity and, therefore, have high fire points. The permittivity of selected esters is higher than that of mineral oils. However, dielectric losses are higher in organic esters compared with mineral oil.

1.3.2.5 Silicone Liquids

Silicone liquids were developed in the early part of this century as specialist liquids, which exhibit low toxicity and a high degree of fire resistance. Silicone transformer fluid is one of a family of fluids known chemically as polydimethylsiloxanes.

Silicone fluids are manufactured by the hydrolysis of alkyl or aryl-chlorosilanes. The use of silicone oil in transformers improves their performance. The DC volume resistivity and dielectric dissipation factor of silicone oils exhibit normal behavior between 20 °C and 100 °C, and are better than mineral oils. The permittivity of the silicone oil is low, ~2.7 at 90 °C, and offers little advantage over mineral oil. Stability is the most important chemical property of silicone oil, and is excellent for a higher-viscosity material at temperatures <150 °C under normal conditions. Because of their high cost, their use in large transformers is rather limited.

1.4 Insulation Ageing

1.4.1 Introduction

The insulation system of a transformer undergoes ageing due to transformer operating conditions. Thermal, electrical, and chemical stresses act on the insulation system during operation of the equipment to cause an irreversible change in the properties of the insulation system. Such ageing stresses may cause either intrinsic or extrinsic degradation of insulation materials. In most insulation systems, extrinsic ageing predominates because of the presence of imperfections and contaminants.

The solid insulation and mineral oil form an inseparable insulation system in a transformer. Cellulose-based insulation and mineral oil used in transformers degrade with time. The degree of insulation ageing depends on thermal, electrical, chemical, and mechanical stresses within the transformer. The temperature of the oil–cellulose insulation plays a major role in influencing the mechanical, chemical, and electrical properties of the material. The temperature of the insulation depends on the loading pattern of the transformer and on ambient operating conditions.

Other than normal ageing, operational stresses are usually dominated by lightning strikes and short-circuits. The windings of a new transformer are well clamped, so have sufficient mechanical strength. But as the insulation ages, the paper insulation shrinks, resulting in a reduction in clamping pressure; thus, the mechanical stability of the winding is hampered. Such mechanically weak windings suffer from enhanced mechanical stresses due to forces arising from short-circuit faults.

Under high operating temperatures, oxygen dissolved in oil also accelerates the rate of degradation of the solid insulation and the oil, particularly in the presence of water. The oxidation products of the oil are alcohols, organic acids, aldehydes, ketones, saturated esters, lactones, peroxides, unstable products, polymerized hydrocarbons, sludge, carbon monoxide, carbon dioxide, and water. As a consequence of oxidation of oil in the presence of copper, iron, and water, the amount of dissolved oxygen (coming from the dissolved peroxides and other products) increases during the service life of a transformer.

The third factor affecting the degradation of transformer insulation is the water content of the solid insulation and oil. The rate of thermal degradation of paper is usually directly proportional to its water content. At the time of manufacture of a transformer, the solid insulation is first dried out to a moisture level in the range of 0.3–1.0% and then impregnated and finally immersed in degassed and dried oil. The usually recommended upper limit for moisture level in the insulation of new units is about 0.5%. However, the water content of the solid insulation and of the oil increases during the life of the transformer. This increase in water comes partly from the degradation of cellulose caused by high temperature and oxidation, and partly from the ingress of water from the atmosphere.

As the transformer breathes under different loading cycles, there is a chance that atmospheric moisture may ingress. Power transformers are fitted with a conservator to accommodate the increased oil volume. The conservator then exhales, either directly to the atmosphere or through dehydrating breathers. As the transformer cools down during low load cycles, the oil volume in the conservator reduces and the transformer breathes in through the breather. Silica gel present in the breather chamber is used to absorb moisture from the ambient air as it flows in toward the conservator. The efficiency of silica gel-type breathers depends on the temperature and humidity of both air and desiccant, together with the rate of air flow. Another type of breather – known as a freezer dehydrator or drycol – dries the insulation automatically. This also allows continuous drying of the air within the conservator. The water content of the transformer may build up to 2–5% over its lifetime, depending on its breather system, loading cycle, and design characteristics.

1.4.2 Factors Contributing to Ageing

Ageing of transformer insulation can take place due to one or a combination of several factors that include electrical, thermal, chemical, mechanical, and environmental ageing mechanisms. These ageing factors may either act independently, or there may be direct interactions between the stresses. The ageing of a practical insulation system can be complex and failure is usually caused by a combination of ageing mechanisms, even though there may be only one dominant ageing factor. For consideration of transformer insulation, the following factors are important.

1.4.2.1 Thermal Ageing

The most dominant mechanism responsible for ageing of transformer insulation is thermal ageing, which involves chemical and physical changes in the insulation. Such ageing is brought about as a consequence of chemical degradation reactions, polymerization, depolymerization, diffusions, etc. Thermo-mechanical effects caused by forces due to thermal expansion and/or contraction are also a major factor responsible for insulation degradation. Both of these effects of chemical change and thermo-mechanical stress are highly influenced by the operating temperature.

1.4.2.2 Electrical Ageing

Electrical ageing, either under operating AC stress or impulse, also leads to insulation degradation over years of operation. These factors involve the influence of partial discharges in the liquid or gaseous dielectric adjacent to the discharge source, or the effects of tracking and treeing in the solid and liquid insulation under high electrical stresses. In addition, high dielectric losses and effects of space charges also contribute to insulation degradation. A combination of electrical and chemical degradation effects also takes place in the form of electrolysis of the liquid insulation, especially when it is contaminated with polar impurities.

1.4.2.3 Mechanical Ageing

Mechanical ageing of the insulation structure can originate from low levels of electro-mechanical or thermo-mechanical stress cycles, but can gradually escalate to even rupture the solid insulation under external or internal forces. Components involving moving parts, such as tap changers, may have abrasive wear take place in the insulation parts.

1.4.2.4 Environmental Ageing

In addition to the chemical and thermal degradation processes highlighted above, other external environmental factors include the influences of dust and other contamination on electrical behavior of the equipment.

1.4.3 Preventing the Sources of Contamination

External sources of contamination such as moisture, oxygen, and solid debris can be minimized by providing a sealed system as far as practicable during normal operation and also during maintenance or during oil processing.

Minimizing the internal sources of contamination is, however, not as straightforward as it is with external sources. Internal sources of contamination are caused by chemical reactions that take place inside the transformer tank and are greatly influenced by operating temperature, presence of moisture, oxygen, and other contaminants. Thus, if these internal sources cannot fully be stopped, their rate can definitely be reduced. Some other internal sources of contamination include metal particles originating from mechanical or electrical wear, nonmetallic particles such as cellulose particles from the paper and pressboard, and some chemical degradation products of the oil such as acids, aldehydes, and ketones.

1.4.4 Degradation of Cellulose

1.4.4.1 Factors Affecting Ageing of Solid Insulation

The overall life expectancy of a transformer is related – to a large extent – to the ability of the cellulose-based insulation to withstand mechanical forces generated by fault conditions. This withstanding capacity of the solid insulation degrades over years of operation. Such degradation mainly involves oxidation, hydrolysis, or breakage of polymer chains under the influence of heat.

Ageing of paper/pressboard/transformerboard used as electrical insulation is generally ascribed to thermal deterioration, where the average length of the cellulose polymer chain is shortened by breakages, which occur more frequently as the temperature is increased. The result of shortening the polymer chain is the diminishing of physical properties of the paper. These properties may be determined as tensile strength, elongation prior to tensile failure, tear strength, number of folds to cause a fracture or burst (or punch) strength. Alternatively, there are chemical analytical means that determine the average polymer chain length. The physical and chemical effects, unsurprisingly, may be modeled using the chemical rate laws (the first-order law seems to work best) and the Arrhenius relationship to model the rate of reaction at any temperature.

1.4.4.2 Thermal Degradation

Thermal degradation of cellulose in oil filled electrical equipment is primarily attributed to chemical reactions that take place due to oxidation and hydrolysis. The rate of these reactions is greatly influenced by the presence of oxygen, water, and acids in contact with the cellulose. The rate of degradation reaction depends on the availability of these items and the temperature. An increase in temperature of 6–8 °C will reduce the life of paper and oil by a factor of 50%.

While in normal operation, a transformer generally heats up to the range of 40–80 °C depending on the loading of the transformer. In that range, the cellulose insulation is found to degrade slowly by a variety of mechanisms, including oxidation and hydrolysis. However, operating temperatures above 110 °C are extremely harmful and can lead to premature failure of the transformer. The degree to which thermal degradation occurs is dependent on the nature of the cellulose, the presence of air and/or water, and whether or not the samples are allowed to remain in contact with the products of decomposition. As thermal degradation proceeds, both aldehydes and carboxyl groups develop and carbon dioxide evolution occurs. When cellulose is heated at high temperature, the heat energy tends to break the glycosidic bonds and give rise to degradation products such as carbon dioxide, carbon monoxide, water, hydrogen, and traces of methane.

Interaction of the remaining groups of cellulose chain with oil components can give rise to other substances such as sludge, acids, etc. Thermal degradation of cellulose can also result in polar groups getting formed on the cellulose chains.

1.4.4.3 Oxidative Degradation

Cellulose is highly susceptible to oxidation. The hydroxyl groups present in the chain lead to formation of carbonyl and carboxyl groups during degradation, which can eventually cause secondary reactions giving chain scission. However, no free oxygen is formed inside the transformer during the ageing process of either the oil or the cellulose. Thus, the only source of oxygen in a transformer is what enters the transformer from outside through the breather system. If the oxygen level in the oil is maintained below 2000 ppm, then the rate of degradation of the insulation system is 80% lower than that of a free-breathing transformer. Oxygen in transformer oil can reach about 30,000 ppm when the oil is fully saturated, but in reality most free-breathing transformers in service (warm) only contain 20,000 ppm as a result of dissolution of gas from the air. Oxidation accelerates the cellulose degradation process, leading to weakening of the glucose bonds, resulting in depolymerization. Oxidation promotes accumulation of additional ageing accelerators such as acids.

Oxidative depolymerization is catalyzed by hydroxyl-radicals (HO•), which are produced by decomposition of hydrogen peroxide, H_2O_2, and organic hydroperoxides (ROOH). Hydrogen peroxide can, for example, be formed from oxygen and water by reactions catalyzed by transition metal positive ions (such as Cu^+/Cu^{2+} or Fe^{2+}/Fe^{3+}).

In general, the hydroxyl groups are oxidized to form carbonyl and carboxyl groups. This leads to scission of chains, leading to weakening of the glycosidic bond and production of water. The oxidation reaction is subject to catalysis by transition metals such as cobalt, iron, and manganese, and to retardation by silver and organic antioxidants. Positive catalysis is accompanied by an increase in depolymerization per unit of oxygen absorption.

1.4.4.4 Hydrolytic Degradation

Water has a major influence on the degradation of cellulose-based insulation material. The presence of water will generally increase the rate of degradation. Cellulose is hygroscopic in nature. As already mentioned, at the beginning of a transformer's life, the Kraft insulation material contains less than 0.5% water and the oil is also dried. The water content levels within the transformer may increase up to 5% during its lifetime, depending on operating and other conditions. Not only that, water molecules also accumulate between the cellulose chains and thermal degradation of cellulose also produces water. The water causes further decomposition of the cellulose chain, thereby producing free glucose. Thus, the addition of a water molecule to the glycosidic bond results in a chain breakdown. Measuring the chain scissions can provide an indication of the rate of hydrolytic degradation. Otherwise, measuring the increase in the number of end groups produced by the chain scission can also provide information about the degradation rate. The rate of degradation of the paper at an initial value of 4% water content can be as high as 20 times greater than that at 0.5% water content. So, in principle, as the transformer ages, the rate at which the insulation deteriorates is expected to increase significantly.

Temperature is the main influence on the rate of ageing of wet cellulose material. Hence, a wet transformer must be operated at a lower temperature than a very dry transformer if accelerated paper ageing is to be avoided. This could mean derating older transformers in order to extend their service life.

As oil–paper insulation ages, there is a significant increase in moisture content of the system. Because paper has a high affinity for water, this moisture is mainly held in the paper rather than in the oil. When new, a transformer would normally contain less than 0.5% moisture by weight in its paper, but this could increase to as much as 4–5% after some years in service.

1.4.4.5 Acid

Chemical degradation of cellulose can also take place due to a hydrolysis process catalyzed by acids present in the oil. Acid content (carboxylic acids) of the oil increases as the oil oxidizes.

1.4.4.6 Measures to Minimize Degradation of Cellulose

The degradation of cellulose breaks the glucosidic bonds and produces glucose. Glucose can degrade to form water, carbon dioxide, carbon monoxide, small amounts of hydrocarbon, hydrogen, and other polar molecules. To reduce the rate of cellulose degradation, it is necessary that the ingress of moisture and oxygen from the atmosphere is minimized. However, even if external ingress is minimized, one cannot avoid the water produced internally by natural degradation of cellulose over the course of time. Hence, it is essential to have regular checking of oil and paper for moisture content; subsequently performing proper diagnostic tests to investigate the degree of degradation.

1.4.4.7 Influence from Oil

Liquids for transformers are mainly mineral oils with different composition and refining, containing different oil molecules (aromatics, naphthenic, and paraphinic), sulfur compounds, inhibitors, passivators, etc. Also, synthetic liquids like vegetable and mineral oil-based esters, silicone oils, etc. are used. Ageing by-products from these liquids may also influence the paper ageing to some extent.

Oxidation of oil produces both low molecular weight acids and higher molecular weight acids. Hydrolysis of cellulose is accelerated by low molecular weight acids coming from the oil. Oil testing according to IEC 61125 includes measurement of both high and low molecular weight acids in the aged oil and in a water trap, respectively. However, it seems that a significant fraction (or even most) of the low molecular weight acids found are paper degradation products.

1.4.4.8 Thermally Upgraded Paper

Thermal upgrading of paper may reduce the ageing rate by a factor of 1.5–3.0. The ageing acceleration from the presence of water is seen to be largely reduced for the upgraded paper, while ageing with access to oxygen seems to be less influenced. For upgraded paper there is still some uncertainty about the activation energies and which process dominates ageing. Humidity and oxidation greatly influence the ageing of both Kraft and thermally stabilized paper. Ingress of moisture and air not only reduces the mechanical strength of the paper, but also leads to degradation products that accelerate further degradation. Thermally stabilized paper is, however, less sensitive to moisture and oxidation than normal Kraft paper. It is thus common to use a combination of normal Kraft paper and thermally stabilized Kraft paper in a transformer design. For instance, while the winding insulation is often done with thermally stabilized Kraft product, normal Kraft pulp is used to manufacture spacers.

1.4.5 Transformer Oil Degradation

1.4.5.1 Factors Affecting Degradation of Liquid Insulation

Major factors influencing the degradation of transformer oil include excessive heating, oxidation, contamination, partial discharges, and related by-products.

1.4.5.2 Thermal Degradation

Excessive temperature rise is one of the most severe factors influencing the degradation of transformer oil, and it is estimated that an increase of 8–10 °C in temperature approximately doubles the rate of oil degradation. Oil decomposes, producing several by-products at elevated temperature, which further accelerates the ageing process of both liquid and solid insulation inside the transformer.

1.4.5.3 Oxidation

Oxidation of oil causes a breakdown of bonds, followed by chemical reaction. The main oxidation products of oil include water, carbon dioxide, carbon monoxide, acids, and sludge. In addition, the presence of dissolved metals such as iron, copper, etc. acts as a catalyst to accelerate the oil degradation process.

As oxygen has two free electrons, any ingress of air in oil makes it more susceptible to an auto-oxidation process. These free radicals can easily break the hydrocarbon chains in oil and produce peroxides. Under heat, these peroxides decay to produce more free radicals, and thus the oxidation process compounds as a chain reaction process.

1.4.5.4 Contamination

The presence of water in oil, either in dissolved form or suspended form, affects the dielectric properties of oil adversely. Thermal degradation of cellulose and oil will produce water internally. Water may also be absorbed from the atmosphere in free-breathing units.

1.4.5.5 Partial Discharges

Owing to excessive thermal stress or spurious electric discharges due to high electrical stress, gas may evolve in oil. These gas pockets contain hydrogen and hydrocarbon gases produced by decomposition of the oil. Because of the difference in permittivity of the gas and the liquid insulation surrounding it, the electrical stress across the gas voids will be very high. Such high electrical stress in the gas pockets will promote further partial discharges and can cause progressive deterioration of the oil.

1.4.5.6 Acid

Acids are produced as degradation products of cellulose that get dissolved in oil and also by oil oxidation. Acid levels higher than 0.6 mg KOH/g of oil are considered high and are detrimental to the operation of the transformer. Acids cause the formation of sludge in oil. Sludge is a solid product of complex chemical composition and gets deposited throughout the transformer. Deposition of sludge can adversely affect the heat dissipation process by blocking oil circulation through radiator pipes, leading to overheating and subsequent failure of the transformer.

1.4.5.7 Undesirable Effects of Moisture in Oil

All oil–paper transformers contain some water in their solid insulation. The paper in a transformer acts as the reservoir for major moisture content throughout the whole insulating system. Water remains in a continuous dynamic state within the complex insulation system of a transformer. Depending on the operating temperature, water exchange takes place between the liquid and solid insulation. High levels of moisture in transformers can accelerate the ageing process of solid insulation, reducing dielectric strength of the oil and forming bubbles.

At increased temperature, water has a tendency to move out of paper and into oil. However, since at higher temperature the solubility of water in aged oil increases, even with an increased amount of water, the percentage saturation of the oil remains low and the dielectric breakdown voltage of the insulation system remains high. Problems arise in the oil when the transformer cools down again. It is possible for excessive moisture to remain in the oil because it migrates much more slowly back into the paper. This causes the oil to get supersaturated with water, leading to the formation of free water in the transformer. This free water can settle where it will contribute to surface flashover or could collect in the bottom of the cooler bank. Sudden operation of the oil pumps can direct this water onto windings, causing insulation failure and resulting in major electrical breakdown or short-circuit failure of the windings.

In free-breathing transformers, a small quantity of water can always enter the transformer oil throughout its operational life, especially in humid atmospheric conditions. However, since paper has more affinity for water than oil, such moisture will primarily be stored in the paper rather than in the oil.

During the manufacturing process, the core and coil assembly is heated for many days under vacuum to drive out any possible moisture that may be present. Hot oil is then admitted to the tank under vacuum to ensure the removal of any gas or moisture remaining in the oil. The whole unit is then processed further by circulating hot oil through a filter plant. At this stage the new transformer has a moisture content of <0.5% by weight in the paper and <5 ppm in the oil. During operation of the transformer, thermal and chemical degradation of paper insulation also produces water as a by-product of chemical reactions and in a severely deteriorated system the moisture content of the paper may reach more than 4%.

This process is accelerated by the heat produced when the transformer is on load. If the transformer spends much of its life at low load, its useful life will be greater than a generator transformer, which runs mainly near its nameplate rating.

1.4.6 Measures to Minimize Oil Degradation

As discussed in the previous sections, the major cause of transformer oil degradation in free-breathing transformers is oxidation. To prevent the ingress of air inside the transformer during the breathing process, nitrogen gas cushion transformers can be used, where dry nitrogen is used to fill up the tank space above the oil so as to minimize the exposure of oil to oxygen. Also, adding oxidation inhibitors such as DBPC to new oil effectively prevents the ageing of the oil.

The ageing process of oil is accelerated by the presence of metals. Adding amino group inhibitors can eliminate this influence.

Temperature is one major dominant factor of oil degradation. The oil temperature can be limited to safe values by the use of cooling tubes that aid in the heat convection and dissipation process. Cooling efficiency can further be improved by artificial cooling techniques such as forced fan cooling, forced oil cooling, or water cooling. Large transformers are very often equipped with pumps in the oil line and also in the water line to enhance cooling.

The entire mechanism of degradation of an oil–paper insulation system can be summarized as in Figure 1.6 [9].

1.4.7 Degradation Products in Impregnated Insulation Systems

Water, oxygen, acids, and various other elements are responsible for insulation degradation that significantly shortens transformer life under different electrical, thermal, chemical, and mechanical stresses.

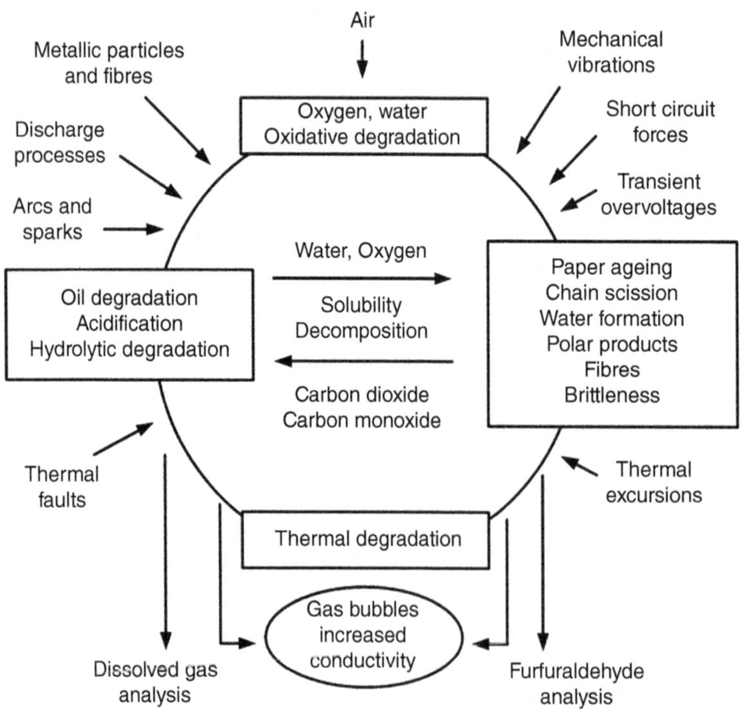

Figure 1.6 Factors influencing performance and degradation of transformer oil–paper insulation and resulting breakdown mechanisms. *Source:* Ref. [10], Figure 1.

The process of insulation deterioration basically involves slow diffusion of water, gases, and other ageing by-products during in-service operation of a transformer. The affected parts are thus primarily the thin structure of paper insulation of turns and coils, pressboard sheets, etc., which comprise typically 40–60% of the total mass. Oil is an integral part of the transformer, playing a dynamic role in the operation of the entire system. The ageing of paper and board leads to the formation of many different products. All the main constituents of paper (i.e., cellulose, hemicellulose, and lignin) will degrade. Many different degradation products will be formed. Some are fairly stable; others may degrade further. Some of the most important degradation products will be discussed briefly in this section, from different points of view.

In theory, the ultimate degradation products of cellulose are water and carbon oxides. Obviously, in real life no transformer can be operated up to the point where all paper components are totally degraded. To be truthful, in a real transformer, at the end of its technical life, the main degradation product is in fact slightly modified cellulose. However, this fills the same function as the unaged material, and what concerns us are mainly what could be described as by-products. A degradation product is discussed in this section if it meets one or more of the following criteria.

- Its presence may constitute a threat to the safe operation of the transformer.
- It affects the ageing rate of the insulating system.
- It can be used as an indicator of ageing, or has some other diagnostic value.

Another aspect that deserves to be stressed is that the oil and the solid insulation will age together. Sometimes a certain product can be formed from either of them. Furthermore, any degradation product, regardless of its origin – paper or oil – will be found in both (although the distribution may vary greatly from case to case). Also, the degradation products of one may affect the degradation of the other. For these reasons it is not possible, or meaningful, to keep at all times a strict distinction between paper/pressboard/transformerboard degradation products and oil degradation products.

1.4.7.1 Degradation Products from Cellulosic Insulation Materials

Water is the main degradation product of paper and pressboard. It is formed by dehydration reactions following hydrolysis, but is also an end product of the oxidation of oil as well as that of paper.

Most of the water that is formed due to cellulose degradation is retained in the solid insulation. Substantial amounts of water may also be the result of moisture ingress from outside.

Preliminary hydrolysis reactions cause scission of the cellulose chain present in paper. Hydrolysis of paper produces acids. These acids undergo further reactions, producing other acidic polymeric compounds, carbon monoxide, and water.

Free acids such as small carboxylic acids, as well as acidic groups attached to the cellulose molecular chain, are also produced due to oxidation of cellulose. Many of these acids are hydrophilic in nature and thus get dissolved well in paper.

The third significant by-product of paper degradation is the "furanic compounds," referring to a whole family of compounds, all of which could be described as furan derivatives. The most commonly found furan is 2-furfural (2-FAL), whereas 2-acetylfuran (2-ACF), 5-methyl-2-furfural (5-MEF), 5-hydroxy-methyl-2-furfural (5-HMF), and 2-furfurylalcohol (2-FOL) have also been traced in the oil–paper degradation process.

5-HMF and 2-FAL are formed by dehydration reactions following hydrolysis of the cellulose and hemicellulose. All the furanic compounds mentioned above are also formed by oxidative pyrolysis.

The measurement of furanic compound content, especially 2-FAL, has found some use in transformer diagnostics. There is believed to be some correlation between the degree of polymerization of paper and 2-FAL (or total furanic compound) content of the oil. This is discussed in more detail in Chapter 2.

Finally, CO and CO_2 are ultimate degradation products of all the constituents of paper and pressboard. In lesser amounts, CO and CO_2 are also oxidation products of the oil. The content of these gases is always measured by Dissolved Gas Analysis (DGA), and the production rates and relative amounts of CO_2 and CO are used in the interpretation of the degradation results. Very high CO_2 contents from paper degradation may influence the acidity of the oil, and may thus complicate the assessment of oil condition.

1.4.7.2 Major Degradation Products of Oil

Acids originate from the oxidation of oils, and the acids formed from paper and from oil are partially the same compounds. Acidic oxidation products from the oil may influence the degradation rate of paper. Part of the acidity of aged paper is due to the absorption of acids that are the product of oil degradation. Other polar oxidation products, such as alcohols, aldehydes, and ketones, may also be partially absorbed.

Various acids can be formed when oil is oxidized. Mineral insulating oil contains many thousands of different molecules that are susceptible to oxidation and (at least in theory) the number of different acids that can be formed is even larger. They will include both carboxylic acids and phenols, and range from the simplest possible representatives from both groups to very large and complex molecules. Likewise, they will represent a very wide range of acid strengths. However, it has been demonstrated that the strongest and most polar acids, such as low molecular weight carboxylic acids, are those that will have the greatest influence on paper degradation.

While acids are formed in a relatively late stage in the oxidation process and significant amounts of sludge even later, peroxides are among the first oxidation products to appear. They are very reactive and a key intermediary in the oxidation process of paper as well as oil.

In inhibited oils [i.e., oil that contains a synthetic antioxidant such as DBPC, sometimes also called butylated hydroxytoluene (BHT)], the production of acids and peroxides is very low as long as there is sufficient inhibitor left in the oil.

Another by-product of oil degradation is sludge; this consists of degradation products that have low solubility in the oil. Sludge may be present in the form of suspended particles or as deposits. Most of the sludge is due to oxidation of the oil, but degradation products from other materials (including paper) most likely contribute to the total amount. The low solubility of sludge is due to the high polarity and possibly to some degree to a high molecular weight, since some degradation products may polymerize.

Analytical methods exist for the determination of sludge in transformer oil (e.g., as described in IEC 60422). The amount of "sludge" determined by such a method may include not only sludge already present as a solid, but also sludge that is precipitated when the oil is diluted with a non-polar solvent (such as heptane). This could be seen as a kind of "potential" sludge. A rise in oil acidity can also be used as an early warning of sludge formation. For evaluation of the hazardous influence of sediment and sludge, as well as for estimating the effectiveness of the maintenance measures (e.g., reclaiming), it is favorable to determine the type of sludge: chloroform soluble or chloroform insoluble. The presence of chloroform-insoluble sludge (which is formed by further oxidation of chloroform-soluble sludge and points to progressive oxidation processes) can be estimated by washing the glass filter containing sediment with chloroform (see the procedure described in IEC 60422). If there are still particles present on the glass filter after washing with chloroform, chloroform-insoluble sludge is present.

Large sludge deposits may be an obstacle to oil flow and thus prevent proper cooling of all parts of the windings. Sludge is usually strongly acidic and contains peroxides, and has been shown to promote paper degradation. Since sludge is mainly oxidation products of oil, preventing the oil from becoming acidic also prevents the formation of sludge. Thus, any measures taken to prolong the life of the oil will also be measured against sludge formation.

Sludge can be at least partly removed by on-site treatment. One example is the use of special regenerative oils that have higher solubility for sludge than normal transformer oil. If a transformer is temporarily filled with such oil, some of the sludge will be dissolved. Oil reclaiming has the same effect. However, especially in the latter case, the efficiency of sludge removal will depend very much on oil temperature during the treatment. The higher the temperature, the more sludge will be dissolved. Sludge present as suspended particles and "potential" sludge (i.e., not yet precipitated sludge) will of course be removed whatever the temperature.

1.4.8 Reducing Ageing of Insulation

Various methods – such as completely sealing the units, using nitrogen blankets, and using flexible membranes – have been suggested as means of improving the lifetime of the oil. In addition, thermally upgraded papers have been devised and have been available for some years.

Various processes have been developed to prevent the rate of degradation of paper, or "upgrade" it. In some processes, the pulp is chemically modified by converting some of the OH radicals to more stable groups, but other simpler processes involve adding "stabilizers," such as a nitrogen-containing chemical called dicyandiamide.

If the ingress of oxygen and water is eliminated using a static system, the only further step that appears possible would be to remove the water generated by the paper degradation. If the regular measurements of water content show the need, this can be done by oil treatments on site.

1.5 Scope of the Book

This chapter has discussed the sources, properties, and applications of insulating materials used in transformers. The insulation system is primarily based on mineral oil and cellulose in the form of paper, pressboard, and sometimes selected natural wood. This combination of oil and fibrous cellulose materials has dominated the technology of power transformer insulation since electrification began about a hundred years ago.

Many new insulation technologies have been proposed and tried. Some examples are modern plastic foils and tapes instead of paper, and biodegradable and synthetic insulating liquids instead of mineral oil. Biodegradable oils are now increasingly used in distribution class transformers, and their ageing behavior and performance in transformers will be presented in detail in Chapter 8.

A comprehensive overview of the ageing of oil–paper insulation systems is also discussed in this chapter. Common ageing effects and their by-products for oil and paper are highlighted separately. Since oil–paper forms an inseparable insulation system, their mutual ageing effects are also described in this chapter. The introductory part of the chapter provides a foundation for understanding a number of chemical-based diagnostics and will also help the reader in relating the diagnosis to the cause of insulation ageing.

Once the ageing of insulation materials is understood, it is important to know what diagnostic techniques are available for assessing the ageing. These diagnostic techniques are generally classified as chemical and electrical based. Chapter 2 comprehensively explains the DGA, furan analysis, degree of polymerization (DP), and their relevant international standards. The chapter introduces electrical-based traditional diagnostics that have been in use for decades, including the insulation resistance (IR), polarization index (PI), dielectric dissipation factor (DDF), capacitance and power factor, dispersion factor, and partial discharge (PD).

During the last 25 years, significant research has been conducted on polarization-based diagnosis, which includes both time domain and frequency domain measurements. Chapter 3 first introduces the need and opportunities for polarization-based diagnosis techniques, followed by theoretical explanations of polarization–depolarization current (PDC) and recovery voltage measurements (RVM). Comprehensive interpretation schemes of these techniques are then described, along with influential effects of moisture, ageing, temperature, and insulation geometry on the interpretation of PDC and RVM. Practical case studies are presented, along with international practice recommended by CIGRE and other major research groups. Frequency domain dielectric spectroscopy (FDS) has been widely used for transformer moisture assessment, as explained from theory to practical demonstration, along with several case studies and their interpretation techniques.

DGA has been in practice for more than five decades, and several standards/working group recommendation documents are available for practicing engineers and researchers. Commonly used interpretation techniques include the key gas concentration method, Doernenburg ratio, Rogers ratio, IEC ratio, Duval triangle, and Duval pentagon schemes. Unfortunately, there is still significant uncertainty in DGA. In order to address these uncertainties, statistical learning-based intelligent diagnostic techniques have gained wide usage in recent years. Chapter 4 provides a comprehensive analysis of machine learning-based DGA interpretation techniques, with specific focus on artificial neural networks, fuzzy logic systems, expert systems, decision-making algorithms, support vector machines, and population-based algorithms. A number of other hybrid algorithms are being proposed by the authors of this book, including the synthetic minority oversampling technique (SMOTE), SMOTEBoost Technique, integrated boosting of SMOTE, and hybrid SMOTEBoost/bootstrap methods. A number of case studies and results are presented in this chapter, along with some insight into training dataset construction and a discussion of classification accuracy and generalization capability validation.

PD monitoring has been widely used for detecting incipient faults, in particular in recent times, with better hardware and software analytical tools. Chapter 5 provides a detailed analysis of PD measurements and interpretation tools. First, an overview of PD analysis and measurement methods is given. Then, advanced signal processing techniques are presented, with the focus on time-frequency (TF) analysis, wavelet transform (WT), empirical mode decomposition (EMD), ensemble EMD (EEMD), and mathematical morphology (MM) methods. With several case studies on these techniques, blind processing methodology, TF map, PD pattern representation techniques in phase-resolved pulse sequence (PRPS), and kurtogram construction using spectral kurtosis are also presented. Three other signal processing techniques – fractal dimension and entropy, pre-whitening, and MM for thresholding – are introduced for PD analysis. Finally, application of advanced signal processing techniques for PD analysis with special emphasis on PD signal de-noising is presented. The chapter concludes with special techniques developed for multiple PD source separation and their PD feature extraction for laboratory as well as field-based case studies.

Frequency response analysis (FRA) plays an important role in transformer winding mechanical deformation/displacement analysis. Chapter 6 concentrates on the FRA technique and interpretation for winding deformation. Fundamentals of FRA for measuring the response, how to perform the measurement, test connections, and mechanical faults commonly detectable with FRA are introduced first. A number of international standards are available for FRA use, and interpretation schemes based on these standards are presented, along with newer statistical approaches. Finally, FRA sensitivity to non-mechanical factors in power transformers – such as temperature, moisture content, presence of oil, residual magnetism, etc. – is described.

Chapter 7 systematically presents the impact of moisture and remaining life estimation of a transformer. This chapter is aimed at explaining how water exists in insulation around a transformer, including how the normal cyclic operation of a transformer causes migration of water around the oil–paper insulation. The chapter explains how water is measured in the oil and paper, in particular with Karl Fischer titration, vacuum drying method, moisture equilibrium curves, and using online sensors (e.g., Vaisala probes). Finally, the chapter provides a comprehensive guideline for practicing engineers on the detrimental effects of water, and methods to estimate the remaining life of insulation as a function of the water content of paper. Several case studies on measuring water in field transformers are demonstrated, with explanations of the optimal probe location and how to adjust for hotspot temperatures.

Biodegradable oils are currently being promoted for use in transformers, primarily because of environmental concerns, but also to reduce the fire risk of mineral oil-based transformers. Chapter 8 presents biodegradable oil fundamentals and their impact on paper ageing. Initially, types of biodegradable oils and their difference from mineral oils are explained, along with ageing of biodegradable oils and ageing of biodegradable oil-impregnated paper/pressboard. Then, DGA interpretation schemes for biodegradable oils, (FDS interpretation schemes for transformers impregnated with biodegradable oil, and finally standardizing with currently used condition monitoring techniques are highlighted in this chapter.

There are many offline and online monitoring techniques currently available for transformer condition monitoring. In recent times, we have been working on an intelligent framework for transformer condition monitoring using online sensors to provide total visibility of the transformer and signal processing to extract measured signals from noise, feature extraction to make sense of measurement data, and finally pattern recognition to identify fault types in transformers. Chapter 9 demonstrates the importance of numeric modeling for assisting fault detection in a transformer, statistical learning for dealing with measurement uncertainties, and data and information fusion for transformer condition assessment with the help of signal processing techniques for intelligent monitoring of transformers. Our research in this area is demonstrated, with the introduction of a smart box – a hardware and software platform for implementing a smart transformer condition monitoring system, proposing the concept of a health index and its interpretation.

Chapter 10 summarizes the condition monitoring aspects of transformers and highlights the key components of available tools and techniques, with their limitations and need for future research.

References

1 Schaible M. Electrical insulating papers – an overview. *IEEE Electrical Insulation Magazine* 1987; 3(1): 8–12.

2 Siqueira G, Bras J, Dufresne A. Cellulosic bionanocomposites: A review of preparation, properties and applications. *Polymers* 2010; 2: 728–765.

3 Koestinger P, Aronsen E, Boss P, Rindlisbacher G, Ag M. Practical experience with the drying of power transformers in the field, applying the LFH technology. Paper No. A2-205, CIGRE, Paris, 2004.

4 Rindlisbacher G, von Ah M. Cut and dry: Low frequency drying process in transformer production can cut energy costs in half. ABB Review 3/2005. Available online at: library.e.abb.com/public/22f229958625da3ec125707b004c4466/66-69%203M557_ENG72dpi.pdf.

5 Prevost TA. Thermally upgraded insulation in transformers. Proceedings of Electrical Insulation Conference and Electrical Manufacturing Expo 2005, pp. 120–125.

6 Sharma BK. *Industrial Chemistry Including Chemical Engineering*. Delhi: Goel Publishing; 2006.

7 Haupert T, Hanson D, Savio L. *Insulating Media*. Boca Raton, FL: IEEE Press; 2007.
8 Oommen TV. Vegetable oils for liquid-filled transformers. *IEEE Electrical Insulation Magazine* 2002; 18(1): 6–11.
9 Lundgaard LE, Hansen W, Linhjell D, Painter TJ. Aging of oil-impregnated paper in power transformers. *IEEE Transactions on Power Delivery* 2004; 19(1): 230–239.
10 Emsley AM, Stevens GC. Review of chemical indicators of degradation of cellulosic electrical paper insulation in oil-filled transformers. *IEE Proceedings – Science, Measurement and Technology* 1994; 141(5): 324–334.

Further Reading

Arvind D, Khushdeep S, Deepak K. Condition monitoring of power transformer: A review. Transmission and Distribution Conference and Exposition, Chicago, IL, April 21–24, 2008, pp. 1–6.

Bertrand Y, Hoang LC. Vegetal oils as substitute for mineral oils. Proceedings of 7th International Conference on Properties and Applications of Dielectric Materials, Nagoya, Japan, June 1–5, 2003, vol. 2, pp. 491–494.

CIGRE Brochure 323. Ageing of cellulose in mineral-oil insulated transformers. Task Force D1.01.10, CIGRE, Paris, October, 2007.

Feinberg R (ed). *Modern Power Transformer Practice*. London: Palgrave.

Hohlein I, Heinzig P, Kachler AJ. Transformer materials as a prerequisite for transformer reliability – knowledge, properties and limits. Paper No. D1-204, CIGRE, Paris, 2006.

Hohlein I, Heinzig P, Knorr W, Asano R. New properties of "old" transformer materials. Conference Record of the IEEE International Symposium on Electrical Insulation, Nuremberg, Germany, June 11–14, 2006, pp. 81–84.

IEEE/PES Transformers Committee Meeting, Pittsburgh, PA, October 9, 2003. Available online at: www.transformerscommittee.org/meetings/f2003_pittsburgh/Minutes/F03-MainMinutes.pdf.

Khaparde SA, Kulkarni SV. *Transformer Engineering Design and Practice*. Boca Raton, FL: CRC Press; 2004.

Mirzaie M, Gholamil A, Tayyebi H. Thermal degradation of cellulose paper insulation in power transformers. Proceedings of IEEE International Conference on Solid Dielectrics, ICSD '07, Winchester, UK, July 8–13, 2007, pp. 673–676.

Nunn TR, Cherry DB. Field commissioning and maintenance of small power liquid filled transformers. Conference Record of the 2004 Annual Pulp and Paper Industry Technical Conference, June 27–July 1, 2004, pp. 198–208.

Ohlen M. Estimating moisture in power transformers – how to estimate and what to do. Transformer Life Management Conference, Megger, Sweden.

Simmons M. Determining the "age" of paper insulation. IEE Colloquium on MV Paper Cables: Asset or Liability (Digest No. 1998/290), 1998, pp. 5/1–5/5.

Webb M. Water in oils. IEE Colloquium on an Engineering Review of Liquid Insulation (Digest No. 1997/003), 1997, pp. 2/1–2/6.

2

Overview of Insulation Diagnostics

2.1 Introduction

The oil–paper insulation of a transformer deteriorates over time and the ageing depends primarily on the operating conditions inside the transformer. As mentioned in Chapter 1, different gases can be produced from the oil–paper insulation ageing over time. Such gases are very often used to assess the condition of the transformer. Cellulose insulation (paper/pressboard) in a transformer undergoes chain scission and its molecular weight decreases with increasing thermal ageing in the presence of high moisture and oxygen. Cellulose insulation also produces degraded glucose and furanic compounds in the progression of ageing. A number of chemical-based diagnostic techniques are used extensively by practicing engineers and specialists to assess the insulation condition of aged transformers. Most commonly used chemical-based diagnostics include moisture analysis in transformer oil, dissolved gas analysis (DGA), furan analysis by high-performance liquid chromatography (HPLC), and molecular weight assessment by degree of polymerization (DP) measurement using the viscometry method.

All these techniques have evolved over several decades of research and practical experience within utilities, universities, and research organizations, and have been refined further by practicing engineers through IEEE and IEC standards, CIGRE guidelines/brochures, and standards of other national/international organizations [1–9]. Significant number of papers have been published by IEEE and CIGRE in this area. This chapter highlights some major interpretation schemes based on published papers and international standards.

Among electrical-based diagnostics, insulation resistance, power frequency dissipation factor, polarization index, and partial discharge (PD) measurements have been used widely. In recent years, modern new diagnostic methods either based on time or frequency domain polarization measurements [e.g., recovery voltage/return voltage (RV), polarization–depolarization current (PDC), and frequency domain dielectric spectroscopy (FDS) measurements] have also been recommended to separately assess the condition of oil and paper insulation and to finally relate to moisture content in the transformer. This chapter introduces some theoretical aspects of these chemical and electrical-based traditional techniques and their interpretations. Electrical-based techniques will be expanded further in the following chapters, with the application of these diagnostics.

Transformer Ageing: Monitoring and Estimation Techniques, First Edition.
Edited by Tapan Kumar Saha and Prithwiraj Purkait.

2.2 Dissolved Gas Analysis

2.2.1 Background of DGA

During the operation of a transformer, a number of gases may be formed due to natural ageing of insulation materials and/or from incipient faults due to excessive electrical or thermal stress. Oil ageing can produce different low molecular weight gases. In contrast, fault-related gases can be produced due to arcing, corona discharge, low-energy sparking, severe overloading, cooling pump motor failure, and overheating in the insulation system. These gases can be found dissolved in the insulating oil, in the gas blanket above the oil, or in gas-collecting devices such as the Buchholz relay.

The main gases formed as a result of electrical and thermal faults in transformers are as follows.

Hydrocarbon and hydrogen

- Methane CH_4
- Ethane C_2H_6
- Ethylene C_2H_4
- Acetylene C_2H_2
- Hydrogen H_2

Carbon oxides

- Carbon monoxide CO
- Carbon dioxide CO_2

Non-fault gases

- Nitrogen N_2
- Oxygen O_2

Extensive research and investigations have been carried out over more than five decades to analyze oil samples for dissolved gases and relate gas content to the condition of the equipment, as well documented in IEEE and IEC standards and reports produced by CIGRE. The first step in this analysis is to detect the relative amounts of gases dissolved in the oil; the next step is to diagnose the fault from the gas data; the final step is to take action to avoid further deterioration of the condition of the transformer. One of the main objectives of DGA is to identify whether the incipient fault is due to partial discharge, or because of overheating or from arcing, which can cause severe risk to the operation of a transformer.

An increased level of hydrogen gas is very often related to partial discharge, when hydrocarbon gas production does not concurrently increase. When hydrogen production is significant along with methane, ethane, and ethylene, oil pyrolysis is considered as the cause. High-energy-based arcing is very often related to acetylene. Lower-energy arcing is related to other hydrocarbon gases, along with hydrogen. Since a rise in temperature of 6–10°C doubles the rate of cellulose degradation, it is important to understand the impact of temperature on the fault process. Carbon dioxide is present in the oil due to normal oxidation of solid cellulose-based insulation, while carbon monoxide is produced due to high-temperature-related cellulose involvement in the fault.

According to Emsley and Stevens [10], a healthy transformer should have less than 0.05 ml of combustible gases (hydrogen and short-chain hydrocarbons: methane, ethane, ethylene, acetylene) per 100 ml of oil and insignificant levels of higher hydrocarbon gases. Measurements on free-breathing transformers show average $CO+CO_2$ levels of 0.4 ml/100 ml of oil after 15 years in operation.

Normally, DGA is conducted over a certain period on a regular time interval to understand the rate of variation of the fault gases. The rate of increase of the fault gases can be related to the rate of progression of the fault. DGA has been found to be valuable at all stages in the life of a transformer. Factory acceptance tests on transformers can provide comparison of gas-in-oil analyses before, during, and after a heat run test to identify if any hotspots are present. Many standards related to DGA are available for conducting oil analysis and also for interpretation of DGA data, and some are listed below.

- IEC 60599 Ed. 2.1 (Bilingual 2007) Mineral oil-impregnated electrical equipment in service – Guide to the interpretation of dissolved and free gases analysis.
- IEEE Guide for the Interpretation of Gases Generated in Oil-Immersed Transformers IEEE Std. C57.104-2008 (Revision of IEEE Std. C57.104-1991).
- IEC 60567 Ed. 4.0 (Bilingual 2011) Oil-filled electrical equipment – Sampling of gases and analysis of free and dissolved gases – Guidance.
- ASTM D 923, Standard Practices for Sampling Electrical Insulating Liquids.
- ASTM D 2945, Test Method for Gas Content of Insulating Oils.
- ASTM D 3305, Standard Practice for Sampling Small Gas Volume in a Transformer.
- ASTM D 3612, Test Method for Analysis of Gases Dissolved in Electrical Insulating Oil by Gas Chromatography.
- IEC 60422 (2013) Mineral insulating oils in electrical equipment – Supervision and maintenance guidance.
- IEC 60666 Ed. 2.0 (Bilingual 2010) Detection and determination of specified additives in mineral insulating oils.

2.2.2 Typical Faults, Fault Gases, and their Diagnosis

Three major types of fault – such as partial discharge, thermal overheating, and arcing – are usually diagnosed by the fault gases. These faults are divided into partial discharge, discharges of low energy, discharges of high energy, and thermal faults of three types (where the temperature ranges are $T<300°C$, $300°C<T<700°C$, and $T>700°C$), as shown in Table 2.1 [11].

IEC 60599 has subdivided these major fault classes into several subclasses, as shown in Table 2.2.

Based on IEC 60599 Ed. 2.1 (Bilingual 2007) [11] and IEEE Std. C57.104-2008 [12], fault gases generated inside a transformer can be related to the fault type, which can be summarized as follows.

Table 2.1 Fault abbreviations

PD	Partial discharges
D1	Discharges of low energy
D2	Discharges of high energy
T1	Thermal fault, $T < 300°C$
T2	Thermal fault, $300°C < T < 700°C$
T3	Thermal fault, $T > 700°C$

Gases	Acceptance tests		Service tests	
	Concentrations at 20°C			
	μl/l	μmol/l	μl/l	μmol/l
Hydrogen	2	0.08	5	0.2
Hydrocarbons	0.1	0.004	1	0.042
Carbon monoxide	5.0	0.2	25	1.0
Carbon dioxide	10	0.4	25	1.0
Oxygen	500	21	500	21
Nitrogen	2000	84	2000	84

Source: Ref. [11], Table 1.

- H_2 gas is produced from the corona effect on oil and cellulose.
- CH_4 and C_2H_6 are produced from low-temperature thermal heating of oil.
- C_2H_4 and H_2 are produced due to high-temperature heating of oil, with additional CH_4 and C_2H_6.
- Low-temperature thermal degradation and corona in cellulose produces CO_2.
- High-temperature thermal degradation and corona in cellulose produces CO.
- Low-energy electrical discharges produce H_2 and CH_4, with small quantities of C_2H_6 and C_2H_4.
- Electrical arcing produces large amounts of H_2 and C_2H_2, with minor quantities of CH_4 and C_2H_4.

2.2.3 Diagnostic Methods

The most commonly used gas-in-oil diagnostic methods are as follows.

a) Based on IEEE C57.104-2008 [12]
   - individual and total dissolved combustible gas (TDCG) concentrations
   - key gas method

     ratio-based methods
   - Doernenburg method
   - Rogers method.

Table 2.2 Typical faults in power transformers

Type	Fault	Examples
PD	Partial discharges	Discharges in gas-filled cavities resulting from incomplete impregnation, high humidity in paper, oil supersaturation, or cavitation, and leading to X-wax formation.
D1	Discharges of low energy	Sparking or arcing between bad connections of different or floating potential, from shielding rings, toroids, adjacent disks, or conductors of winding, broken brazing, or closed loops in the core.
		Discharges between clamping parts, bushing and tank, high voltage and ground within windings, on tank walls.
		Tracking in wooden blocks, glue of insulating beam, winding spacers. Breakdown of oil, selector breaking current.
D2	Discharges of high energy	Flashover, tracking, arcing of high local energy, or with power follow-through.
		Short-circuits between low voltage and ground, connectors, windings, bushings and tank, copper bus and tank, windings and core, in oil duct, turret. Closed loops between two adjacent conductors around the main magnetic flux, insulated bolts of core, metal rings holding core legs.
T1	Thermal fault T<300°C	Overloading of the transformer in emergency situations.
		Blocked item restricting oil flow in windings.
		Stray flux in damping beams of yokes.
T2	Thermal fault 300°C<T<700°C	Defective contacts between bolted connections (particularly between aluminum busbar), gliding contacts, contacts within selector switch (pyrolitic carbon formation), connections from cable and draw-rod of bushings.
		Circulating currents between yoke clamps and bolts, clamps and laminations, in ground wiring, defective welds, or clamps in magnetic shields.
		Abraded insulation between adjacent parallel conductors in windings.
T3	Thermal fault T>700°C	Large circulating currents in tank and core.
		Minor currents in tank walls created by a high uncompensated magnetic field.
		Shorting links in core steel laminations.

Source: Ref. [11], Table A.1.

b) Based on IEC 60599-2007 [11]
   - IEC gas ratio method.

c) Duval's triangle.

d) Duval's pentagon.

2.2.3.1 IEEE C57.104-2008 [12]

2.2.3.1.1 *Individual and TDCG Concentrations*

This particular method defines four different conditions, as listed below.

Condition 1: Transformer is operating satisfactorily.
Condition 2: Greater than normal combustible gas level.
Condition 3: High level of decomposition.
Condition 4: Excessive level of decomposition.

For conditions 1–3, IEEE C57.104 recommends that additional investigation should be undertaken when any individual combustible gas exceeds the specified levels. For condition 4, the standard suggests that continued operation could result in failure of the transformer. Table 2.3 shows the value of limits for TDCG in the oil. This table has been constructed based on the experiences of many companies.

2.2.3.1.2 *Key Gas Method*

The KGM actually uses four characteristic charts that represent typical relative gas concentrations for four general fault types: overheating of cellulose (OHC), overheating of oil (OHO), partial discharge (PD), and arcing. The key gases used are carbon monoxide (CO), hydrogen (H_2), methane (CH_4), ethane (C_2H_6), ethylene (C_2H_4), and acetylene (C_2H_2). Figure 2.1 indicates the relative proportions of these "key gases" for the four general fault types [12]. These typical fault signatures can be used for analysis of possible fault conditions.

2.2.3.1.3 *Ratio-Based Methods*

Other methods popularly adopted by utilities for fault diagnosis using DGA are based on a ratio of certain key gases. The use of gas ratios to indicate a possible fault type is based on the investigators correlating the gas analysis results with the fault types.

These ratios are as follows.

Ratio 1 $R1 = \frac{CH_4}{H_2}$
Ratio 2 $R2 = \frac{C_2H_2}{C_2H_4}$
Ratio 3 $R3 = \frac{C_2H_2}{CH_4}$
Ratio 4 $R4 = \frac{C_2H_6}{C_2H_2}$
Ratio 5 $R5 = \frac{C_2H_4}{C_2H_6}$

Doernenburg Ratio Method This method uses the above ratios R1–R4. The values for the concentration of these key gases are firstly compared with the limiting values (L1) as given in Table 2.4, to ascertain whether the transformer is in a normal condition or not [12].

Once the individual gas concentrations exceed the limits prescribed in Table 2.4, the ratios R1–R4 are calculated. These ratios are then compared with the limiting values shown in Table 2.5 to suggest possible fault conditions. To

Table 2.3 Dissolved gas concentration limits as per IEEE[a]

Status	H_2	CH_4	C_2H_2	C_2H_4	C_2H_6	CO	CO_2	TDCG[b]
Condition 1	100	120	35	50	65	350	2500	720
Condition 2	101–700	121–400	36–50	51–100	66–100	351–570	2500–4000	721–1920
Condition 3	701–1800	401–1000	51–80	101–200	101–150	571–1400	4001–10,000	1921–4630
Condition 4	>1800	>1000	>80	>200	>150	>1400	>10,000	>4630

Note 1: The table assumes that no previous tests on the transformer for DGA have been made or that no recent history exists. If a previous analysis exists, it should be reviewed to determine if the situation is stable or unstable.

Note 2: An ASTM round-robin indicated variability in gas analysis between laboratories. This should be considered when having gas analysis carried out by different laboratories.

[a] The numbers shown in the table are in parts of gas per million parts of oil [μl/l (ppm)] volumetrically and are based on a large power transformer with several thousand gallons of oil. With a smaller oil volume, the same volume of gas will give a higher gas concentration. Small distribution transformers and voltage regulators may contain combustible gases because of the operation of internal expulsion fuses or load break switches. The status codes in Ref. [12], Table 1 are also not applicable to other apparatus in which load break switches operate under oil.

[b] The TDCG value does not include CO, which is not a combustible gas.

Source: Ref. [12], Table 1.

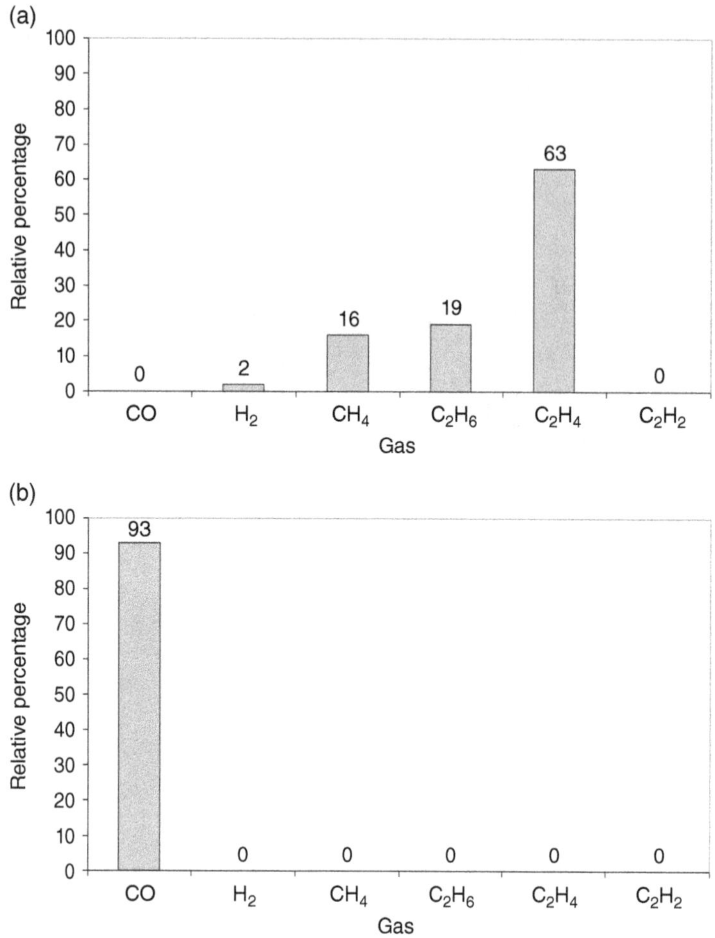

Figure 2.1 (a) Thermal decomposition of oil products include C_2H_4 and CH_4, together with smaller quantities of H_2 and C_2H_6. Traces of C_2H_2 may be formed if the fault is severe or involves electrical contacts. Principal gas: ethylene. (b) Thermal overheating and decomposition of cellulose produces large quantities of CO_2 and CO. Hydrocarbon gases, such as CH_4 and C_2H_2, will be formed if the fault involves an oil-impregnated structure. Principal gas: carbon monoxide. (c) Low-energy electrical discharges – partial discharges produce H_2 and CH_4, with small quantities of C_2H_6 and C_2H_4. Comparable amounts of CO and CO_2 may result from discharges in cellulose. Principal gas: hydrogen. (d) High-energy electrical discharge such as arcing produces large amounts of H_2 and C_2H_2, with minor quantities of CH_4, C_2H_6, and C_2H_4. CO and CO_2 may also be formed if the fault involves cellulose. Principal gas: acetylene. *Source:* Ref. [12].

ensure that the ratio procedure is valid, one has to first ensure that one of the two gases in each ratio exceeds the limit L1 given in Table 2.4. Otherwise, it is recommended to carry out oil resampling and adopt some other procedure for dissolved gas examination.

In case all of the ratios indicate the same kind of fault (as per Table 2.5), the suggested diagnosis is considered to be valid. However, if any one of the ratios is out of the range specified for a certain fault type, then the method is not valid

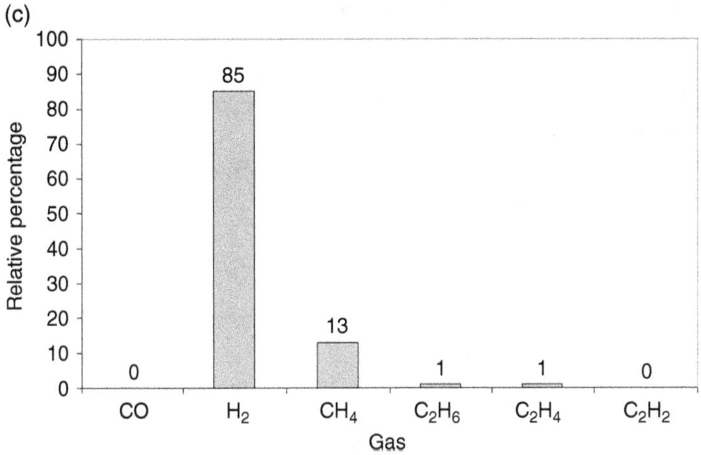

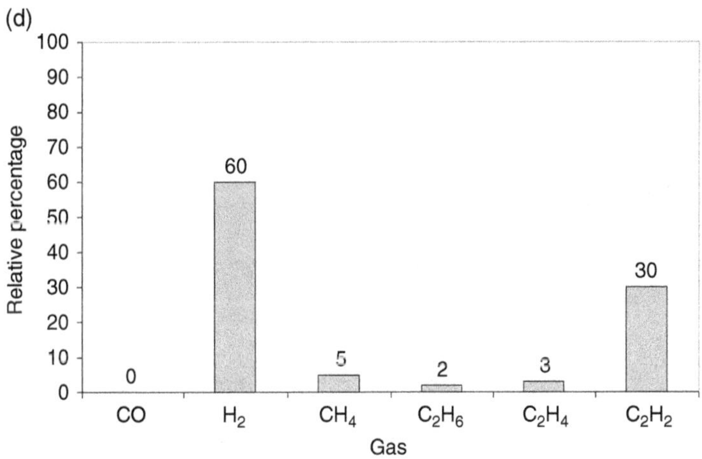

Figure 2.1 (*Continued*)

Table 2.4 Concentration limits of dissolved gas as per Doernenberg (ppm by volume)

Key gas	**Concentration (ppm) (L1)**
H_2	100
CH_4	120
CO	350
C_2H_2	35
C_2H_4	50
C_2H_6	65

Source: Ref. [12].

Table 2.5 Ratio limits of key gases as per Doernenburg

Suggested fault type	R1	R2	R3	R4
Thermal decomposition	>1.0	<0.75	<0.3	<0.4
PD (low-density PD)	<0.1	Not significant	<0.3	>0.4
Arcing (high-density PD)	>0.1, <1.0	>0.75	>0.3	<0.4

Source: Ref. [12], Table 4.

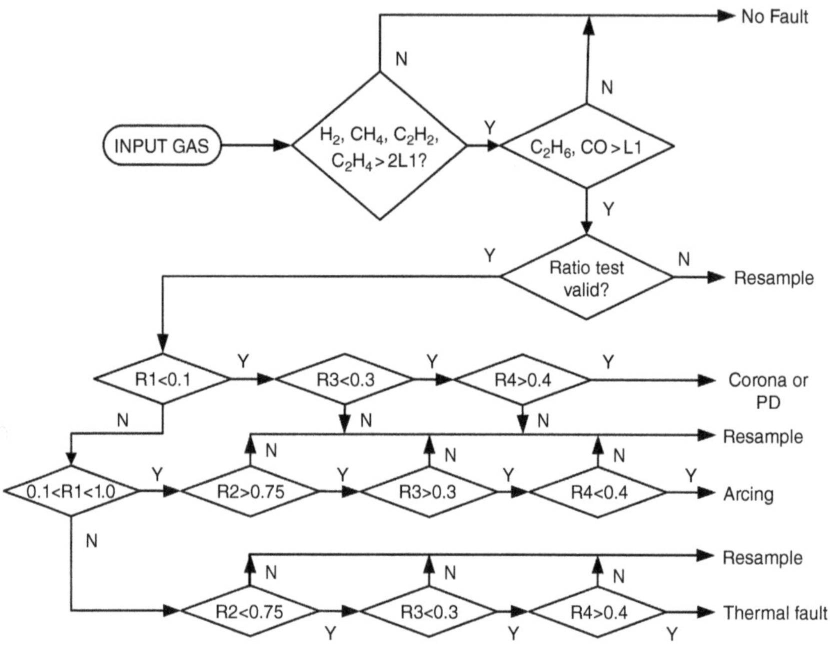

Figure 2.2 Doernenburg's ratio method flow chart. *Source:* Ref. [12], Figure 4.

and a resampling is recommended. The total procedure is shown schematically in the flow chart of Figure 2.2 [12].

For Doernenburg's ratio method to be credible, the levels of gases used in the ratios must be greater than some predetermined levels (in ppm), known as the L1 limits shown in Table 2.5.

Rogers Ratio Method The Rogers ratio method is similar to the Doernenburg ratio method and uses three ratios – R1, R2, and R5 – for fault diagnosis. The Rogers ratio method does not depend on specific gas concentration limiting values to be present in the transformer oil for the diagnosis to be valid. However, it suggests that the method be used only when the normal safe limits of the individual gases have been exceeded. Table 2.6 shows the ratio limits of key gases as per Rogers ratio method for fault diagnosis.

Table 2.6 Ratio limits of key gases as per Rogers ratio method

Fault	$R1 = \frac{CH_4}{H_2}$	$R2 = \frac{C_2H_2}{C_2H_4}$	$R5 = \frac{C_2H_4}{C_2H_6}$
No fault	>0.1, <1.0	<0.1	<0.1
Corona or PD	<0.1	<0.1	<0.1
Arcing	>0.1, <1.0	>0.1, <3.0	>3.0
Low-temperature thermal fault	>0.1	>0.1, <1.0	>1.0, <3.0
Thermal fault <700°C	>1.0	<0.1	>1.0, <3.0
Thermal fault >700°C	>1.0	<0.1	>3.0

Source: Ref. [12], Table 6.

Table 2.7 DGA interpretation table: ratio limits of key gases

Case	Characteristic fault	$\frac{C_2H_2}{C_2H_4}$	$\frac{CH_4}{H_2}$	$\frac{C_2H_4}{C_2H_6}$
PD	Partial discharges (see notes 3 and 4)	NS[1]	<0.1	<0.2
D1	Discharges of low energy	>1	0.1–0.5	>1
D2	Discharges of high energy	0.6–2.5	0.1–1	>2
T1	Thermal fault t<300°C	NS[1]	>1 but NS[1]	<1
T2	Thermal fault 300°C<t<700°C	<0.1	>1	1–4
T3	Thermal fault t>700°C	<0.2[2]	>1	>4

Note 1: In some countries, the ratio C_2H_2/C_2H_6 is used rather than the ratio CH_4/H_2. Also, in some countries, slightly different ratio limits are used.
Note 2: The above ratios are significant and should be calculated only if at least one of the gases is at a concentration and a rate of gas increase above typical values (see clause 9).
Note 3: $CH_4/H_2<0.2$ for partial discharges in instrument transformers. $CH_4/H_2<0.07$ for partial discharges in bushings.
Note 4: Gas decomposition patterns similar to partial discharges have been reported as a result of the decomposition of thin oil film between overheated core laminates at temperatures of 140°C and above (see 4.3 and [1] of annex C).
[1] NS = non-significant whatever the value.
[2] An increasing value of the amount of C_2H_2 may indicate that the hotspot temperature is higher than 1000°C.
Source: Ref. [11], Table 2.

2.2.3.2 IEC 60599-2007 [11]

IEC 60599 recommends three major gas ratios R1, R2, and R5, but with modified code limits. The limiting gas ratios, along with suggested fault types (correlating to Table 2.1), are summarized in Table 2.7. These ratios are significant only when at least one of the gases is above the limit of gas concentration and rate of increase of gas generation.

In IEC 60599-2007, some other gas ratios have also been recommended for specific diagnosis.

- The ratio $CO_2/CO<3.0$ is considered as an indication of cellulose decomposition, with some degree of carbonization.

- The ratio O_2/N_2 is around 0.5 in a healthy transformer, but if this ratio is less than 0.3 it is generally indicative of high oxidation rates.
- On load tap changer (OLTC) operations may produce gases due to discharges of low energy (D1). When there is oil or gas communication between the OLTC compartment and the main tank, the oil in the main tank may be contaminated and hence lead to the wrong diagnosis. This IEC standard suggests that C_2H_2/H_2 ratios higher than 2 to 3 in the main tank could be considered as an indication of OLTC contamination.

Although DGA has been used widely by utilities based on these two standards, some anomalies have also been reported by users. On some occasions, a particular ratio or a combination of ratios have been found to lie outside any of the prescribed limits – or may be in between the two limits. In some cases, this has been related to a transition from one fault type to another, or a combination of more than one fault class. In some cases, this has also been related to the performance of the gas extraction and analysis techniques used by a particular laboratory.

2.2.3.3 Duval's Triangle Method

The assessment of faulty transformers from DGA data can be represented in a more user-friendly graphical form as suggested by Duval and coworkers [13–15]. This triangle representation is shown in Figure 2.3. The nomenclatures used in the triangle correspond to different fault cases described in IEC 60599-2007:

PD = partial discharges
D1 = discharges of low energy
D2 = discharges of high energy

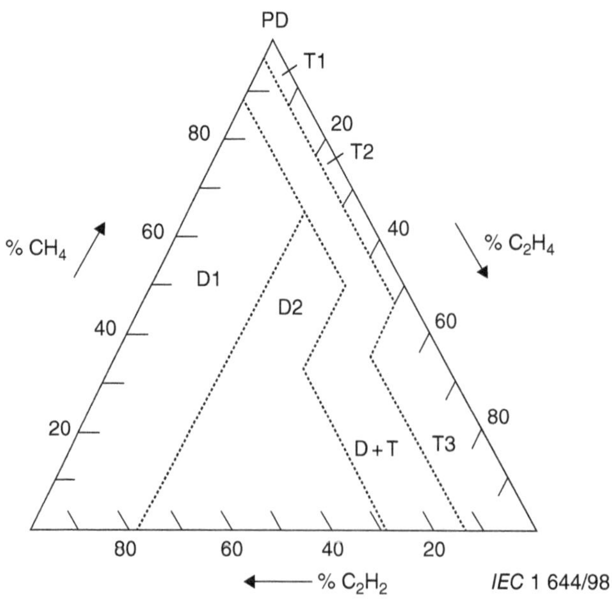

Figure 2.3 Duval's triangle. *Source:* Ref. [11], Figure B.3.

T1 = thermal faults of temperature <300°C
T2 = thermal faults of temperature 300°C<T<700°C
T3 = thermal faults of temperature >700°C
DT = combination of thermal and electrical faults.

The triangle coordinates corresponding to DGA results in ppm can be calculated as

$$\%C_2H_2 = \frac{100x}{x+y+z};\%C_2H_4 = \frac{100y}{x+y+z};\%CH_4 = \frac{100z}{x+y+z}$$

where x=[C_2H_2] (μl/l), y=[C_2H_4] (μl/l), and z=[CH_4] (μl/l). Locating the point on the triangle can be done using a triangular coordinate graph paper. Any unknown fault can thus be identified by finding out the zone where the corresponding point is located in the triangle.

2.2.3.4 Duval's Pentagon Method

Similar to Duval's triangle method, two additional graphical tools (i.e., Duval pentagons of type I and II) have recently been developed as new complementary tools for the interpretation of dissolved gas analysis in mineral oil-filled electrical equipment [16]. Concentrations of five dissolved gases are measured (i.e., %H_2, %CH_4, %C_2H_6, %C_2H_4, and %C_2H_2) and a pentagon coordinate system is adopted, where each axis denotes the relative portion of these gases. The fault types can be determined by computing the geometrical center (gray dot in Figure 2.4) of the polygon (gray lines in Figure 2.4). Each type of Duval pentagon can identify seven potential faults, and the different fault types of these two types of pentagon are summarized in Table 2.8.

The conventional methods of DGA data analysis make decisions based on hard-crisp values of the gas concentration values or the ratio of key gases. Although well received by power utilities, there are several limitations pertaining to the foregoing ratio-based approaches. Uncertainty and ambiguity still exist as to which key gas ratios should be considered and on the credibility of the suggested ratio values, since each scheme has its own recommendation of key gas ratios and their values. A number of modern machine learning techniques have been trialed and tested for various fault diagnoses. These include fuzzy logic-based techniques, several advanced artificial neural network-based techniques, wavelet transform-based methods, genetic algorithms, and a number of other classification tools. A number of expert system techniques have also been trialed by several researchers. This will be discussed in detail in Chapter 4.

2.3 Furan Analysis

2.3.1 Background of Furan Analysis

During the operation of a transformer, when the load increases, the temperature of the windings also increases and as a result, both oil and cellulose-based paper/pressboard insulation undergo a certain degree of degradation.

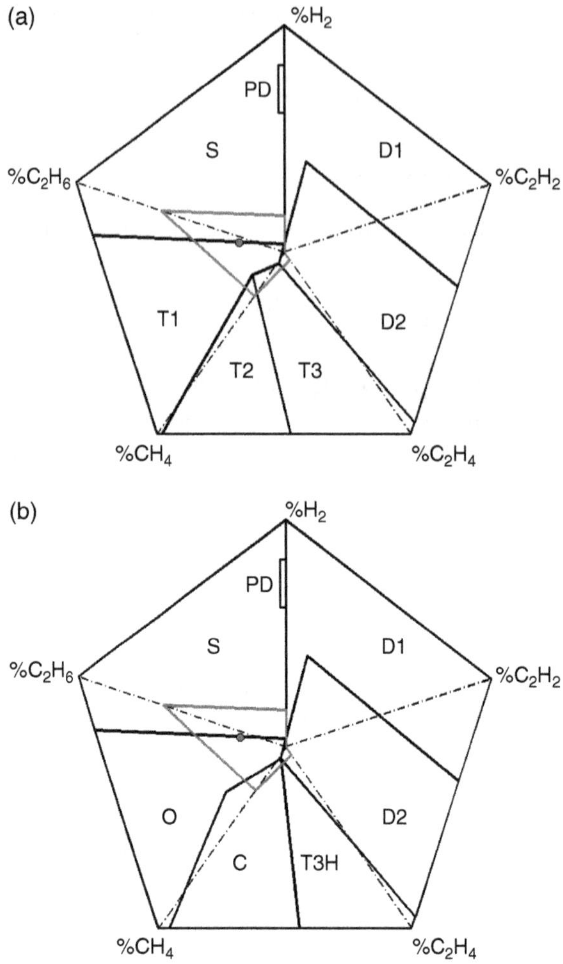

Figure 2.4 Coordinates and fault zones in the Duval pentagon method: (a) Duval pentagon type I; (b) Duval pentagon type II.

Table 2.8 Comparison of fault types of two types of Duval pentagon

Duval pentagon I	Duval pentagon II
PD: corona partial discharges	PD: corona partial discharges
D1: low-energy discharges	D1: low-energy discharges
D2: high-energy discharges	D2: high-energy discharges
T3: thermal faults >700°C	T3-H: thermal faults T3-H in oil only
T2: thermal faults 300–700°C	C: thermal faults T3-C, T2-C, and T1-C with carbonization of paper
T1: thermal faults <300°C	O: overheating T1-O <250°C
S: stray gassing of mineral oil at 120 and 200°C	S: stray gassing of mineral oil at 120 and 200°C

The condition of the oil is normally monitored by chemical analysis of an oil sample. Degradation of transformer paper insulation can be assessed by direct measurement of the paper's DP and tensile strength (TS). Although these are very reliable methods of assessing the degree of ageing of paper insulation and intrinsic properties, they require paper samples to be removed from a transformer. This is possible only while offline during a major refurbishment or overhaul.

At elevated temperatures the cellulose chains in the paper insulation of the windings will cleave. Transformer winding paper begins to degrade noticeably at temperatures of about 100–110°C, and for every 6–10°C rise the degradation rate approximately doubles. Overheating can cause cellulose decomposition and production of CO and CO_2. These two gases can also be produced during thermal decomposition of the oil. Therefore, analysis of CO and CO_2 cannot be used as an unambiguous indication of paper degradation.

Accompanying this paper degradation is the release of a chemical compound furfuraldehyde (FFA) into the oil. FFA is a by-product specific to the thermal degradation of the insulation paper and cannot be produced by the oil. The presence of water and/or oxygen accelerates paper breakdown and accelerates the production of FFA and its derivatives. The monitoring of furanic compounds by oil analysis using HPLC has been used widely by utilities for the last few decades.

FFA concentration has been used as an indirect indication of the condition of the paper and also related to the degree of polymerization. The rate of change of FFA concentration has been used to indicate the rate of ageing of the paper/pressboard insulation. Measuring FFA within oil samples collected periodically from power transformers is usually carried out as a means of transformer condition monitoring. The furfural test is much more convenient than DP or TS measurements, for which the transformer needs to be taken offline and opened to collect paper samples from the windings. The furfural test requires a small amount of oil collected from an operational transformer without interruption of service, and then the measurement of FFAs by HPLC. The HPLC method for determination of FFAs is a well-established technique.

Generally, 2-furfuraldehyde is the major compound found in transformer oil and the rest are present at very low or undetectable levels. The furanic compounds are oil-soluble, so they can easily be sampled and analyzed. Although furan assessment is a reasonably mature condition monitoring tool, the relationship between the generation of these furanic by-products and the condition of in-service paper insulation is not yet well understood.

2.3.2 Paper Degradation and Furan Formation [17]

As mentioned before, paper/pressboard insulation contains about 90% cellulose, 6–7% hemicellulose, and 3–4% lignin. Transformer insulation during operation is affected by thermal and electrical stress, moisture, and oxidation. The insulating material, in particular cellulose-based paper/pressboard, gradually degrades and breaks the hydroxyl bonds. High temperature, oxygen, and moisture are the three main factors responsible for ageing. It has been reported that heating cellulose to a maximum of 200°C in the absence of oxidizing agents and moisture tends to break the glycosidic bonds and open the glucose rings. This chemical reaction further produces free glucose molecules, water, carbon oxides, and organic acids. Cellulose is vulnerable to oxidation and, in the absence of an oxidizing agent, the hydroxyl groups are oxidized to carbonyl and carboxyl (acidic) groups. This can also lead to chain scission and, during the oxidation process, water is produced. As seen above, water and acids are produced from thermal and oxidative degradation, which can further break the glycosidic bond, producing free glucose, and hydrolytic degradation is further accelerated. Moisture has been found to be the main ageing agent of paper/pressboard insulation systems.

So, all three main types of degradation (thermal, hydrolytic, and oxidation) promote chain scission and hence produce glucose and degraded forms of glucose. In the presence of moisture and acidic products, glucose further degrades and produces furanic compounds such as 5-hydroxymethyl-2-furaldehyde, which is an unstable compound and can further decompose to other furfuraldehydes. Six furans are reported to be found in the oil, as first suggested by Burton *et al.* [18]:

- 2-furfural (2-FAL)
- 2-acetylfuran (2-ACF)
- 2-furoic acid
- 5-methyl-2-furfural (5-MEF)
- 2-furfurylalcohol (2-FOL)
- 5-hydroxymethyl-2-furfural (5-HMF).

One of the most probable mechanisms of formation of 2-FAL is shown in Figure 2.5.

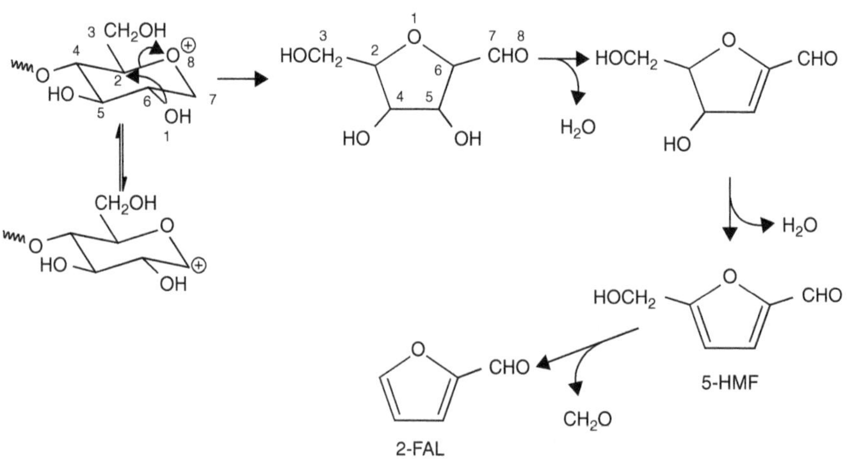

Figure 2.5 One of the most probable cellulose thermal degradation mechanisms. *Source:* Ref. [19], Figure 2.

2.3.3 Furan Analysis for Insulation Condition Monitoring

CIGRE Brochure 494 [19] provides a comprehensive analysis of furanic compounds for thermal degradation of paper/pressboard insulation in a transformer. Extensive literature on this topic is also available from the IEEE, CIGRE, and other journals, as highlighted in this brochure. The brochure also highlights that "the major furanic derivative, formed from thermally non stabilized Kraft paper is 2-furfuraldehyde (2-FAL). A lot of research around the world is devoted to the correlation between 2-FAL concentration and ageing of the solid insulation in terms of remaining life evaluation." Four generator step-up transformers in similar service conditions were monitored for seven years, and the 2-FAL level in one transformer showed a significantly higher level of thermal irregularity, as shown in Figure 2.6.

Practical experience shows that significant 2-FAL concentrations in oil can be generated at service temperatures in wet transformer insulation, as shown in Figure 2.7 [19].

Oil maintenance also has a strong influence on the 2-FAL formation and retention in the oil, as discussed extensively in this brochure. "Although 2-FAL seems to be the most popular thermal ageing marker, the other related furanic compounds mentioned before can also be detected in operating transformers." The brochure reported that "this may lead to a new and reliable way to use the other related furanic compounds." The proposed diagnosis model is presented in Table 2.9.

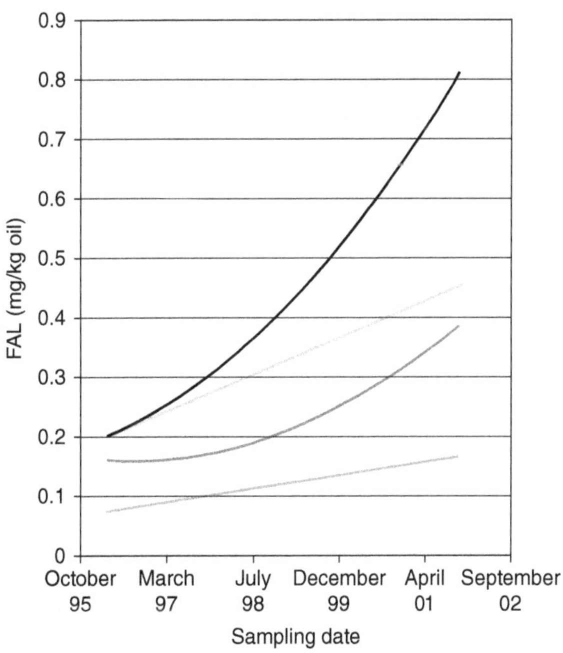

Figure 2.6 2-FAL analysis of four GSU transformers, commissioned in 1992. *Source:* Ref. [19], Figure 7.

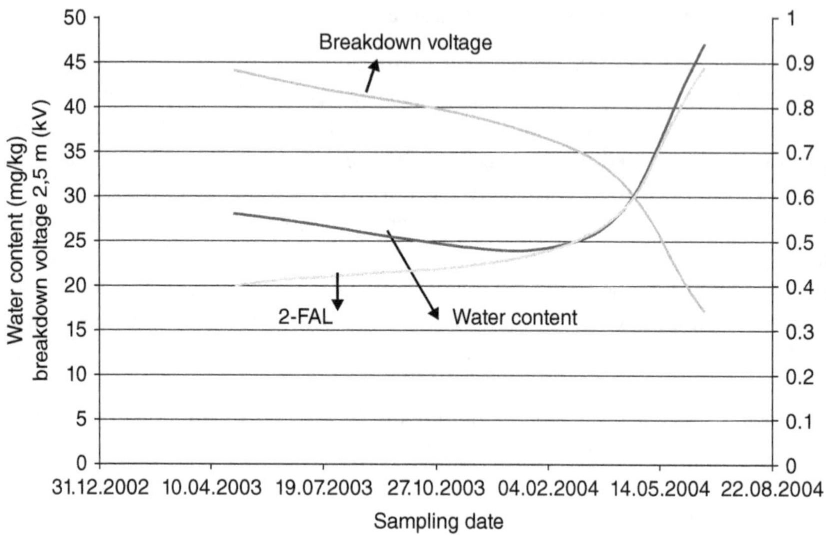

Figure 2.7 Variation of the breakdown voltage, humidity, and 2-FAL in a 75 MVA transformer manufactured in 1981. *Source:* Ref. [19], Figure 8.

Table 2.9 Possible causes of specific furanic compound presence

Compound	Diagnosis proposed
5-HMF	Oxidation
2-FOL	High moisture
2-FAL	General overheating or normal ageing
2-ACF	Rare, causes not fully defined
5-MEF	High temperatures

Source: Ref. [19], Table 5.

In general, most of the authors agree that there is a single critical threshold for equipment of about a few hundred ppb for 2-furfural. According to a U.S. consulting firm, for a representative population of U.S. transformers 90% have less than 0.1 mg/kg of 2-furfural. For European transformers, the 90% value calculated based on a statistical survey including more than 5000 transformers was about 5 mg/kg (only around 60% have less than 0.1 mg/kg of 2-furfural) which is very different. [19]

It should be noted that when European statistics are compared with North American figures, the levels of furanic compounds that are generated are greater in Europe. This fact can be related to the introduction of thermally upgraded paper in North America in the 60s. [19]

The other factor is inhibited oils, and it has been reported that:

> In fact, it seems that for inhibited oils, the concentration of 2-FAL is much lower than for uninhibited oils, if we compare transformers of the same age. Inhibitors may influence the formation of 2-FAL or may have an effect on its migration from the paper to the oil. This phenomenon is still not really well understood by the scientific community. [19]
>
> There is considerable evidence from laboratory studies and practice that clearly demonstrates that the oil type has a major influence on the 2-FAL generation. The 2-FAL content of inhibited and passivated oils is significantly lower (nearly 10 times) than the 2-FAL content of non-inhibited oils with the same degree of polymerisation (DP) of paper. Nevertheless, it is not clear if this is due to the presence of a lower order kinetic of formation of 2-FAL or just another effect of partition. [19]
>
> The presence of air promotes the formation of furanic derivatives from cellulose. Therefore, the 2-FAL analysis may be a better tool for ageing evaluation of free-breathing transformers than that of closed-type transformers. Generally, lower furanic concentrations are to be expected in the case of a transformer with a rubber bag or nitrogen cushion. [19]
>
> Investigations have shown that thermally stabilized Kraft paper develops much less furanic compounds in oil than plain Kraft paper. [19]

PD IEC/TR 62874:2015: Guidance on the interpretation of carbon dioxide and 2-furfuraldehyde as markers of paper thermal degradation in insulating mineral oil is a BSI Standards publication [20], which provides the use of ageing by-products (e.g., carbon dioxide and furanic compounds) to estimate the progress of paper thermal degradation in transformers. This report provides several tables for 2-FAL typical values in Annex A, which are reproduced here in Table 2.10. Note that values reported as N.A. (not available) were not calculated, because the number of cases in the population was too exiguous.

A number of researchers have attempted to correlate 2-FAL production with the other intrinsic properties of paper/pressboard insulation. This includes investigation of 2-FAL with the degree of polymerization, and several empirical relationships have been proposed. This will be discussed after the degree of polymerization is introduced in the following section.

2.4 DP Measurements

2.4.1 Background of DP Measurement

Kraft paper is the best known naturally available insulation material for high-voltage electrical applications such as power transformers, cables, and capacitors. The average number of glucose monomers in the cellulose is referred to as the degree of polymerization. In a transformer, cellulose material is exposed in a complex environment and its deterioration is caused by various ageing factors during normal operation. Paper/pressboard insulation deteriorates with age in the presence of higher water content and excess thermal heat and oxygen. They all

Table 2.10 Typical values of 2-FAL concentration (Conc., expressed in mg/kg) and rate of increase (RoI, expressed in mg/kg/y) for different families of equipment and oil type

A.2.2 Family: GSU (generation step-up units)

NOTE In A.2.2, no distinction is made between units with open conservator and units with sealed conservator.

A.2.2.1 Sub-family: GSU impregnated with uninhibited oil

Table A.1 2-FAL typical values for GSU transformers, filled with uninhibited mineral oil (based on a population of 1860 units)

Age →	<1 Y		1–10 Y		10–30 Y		>30 Y	
Percentile ↓	Conc.	RoI	Conc.	RoI	Conc.	RoI	Conc.	RoI
90°	<0.05	<0.01	0.3	0.04	2.0	0.30	3.0	N.A.
98°	<0.05	<0.01	1.0	0.10	4.0	0.70	6.0	N.A.

A.2.2.2 Sub-family: GSU impregnated with inhibited oil

Table A.2 2-FAL typical values for GSU transformers, filled with inhibited mineral oil (based on a population of 176 units)

Age →	<1 Y		1–10 Y		10–30 Y		>30 Y	
Percentile ↓	Conc.	RoI	Conc.	RoI	Conc.	RoI	Conc.	RoI
90°	N.A.	N.A.	0.1	N.A.	0.80	0.25	N.A.	N.A.
98°	N.A.	N.A.	0.15	N.A.	1.5	0.60	N.A.	N.A.

A.2.3 Family: network transmission units

Table A.3 2-FAL typical values for network transmission transformers, filled with uninhibited mineral oil (based on a population of 2845 units)

Age →	<1 Y		1–10 Y		10–30 Y		>30 Y	
Percentile ↓	Conc.	RoI	Conc.	RoI	Conc.	RoI	Conc.	RoI
90°	0.1	0.01	0.3	0.06	1.6	0.25	2.0	0.3
98°	0.2	0.02	1.1	0.6	3.5	0.80	4.5	1.1

A.2.4 Family: large distribution units

A.2.4.1 Sub-family: large distribution units with open conservator

Table A.4 2-FAL typical values for large distribution transformers, with open breathing conservator, filled with uninhibited mineral oil (based on a population of 7107 units)

Age →	<1 Y		1–10 Y		10–30 Y		>30 Y	
Percentile ↓	Conc.	RoI	Conc.	RoI	Conc.	RoI	Conc.	RoI
90°	<0.05	<0.01	0.7	0.20	1.5	0.30	N.A.	N.A.
98°	0.2	0.03	3.0	0.50	4.5	0.80	N.A.	N.A.

Table 2.10 (Continued)

A.2.4.2 Sub-family: large distribution units with sealed conservator

Table A.5 2-FAL typical values for large distribution transformers, with sealed conservator, filled with uninhibited mineral oil (based on a population of 288 units)

Age →	<1 Y		1–10 Y		10–30 Y		>30 Y	
Percentile ↓	Conc.	RoI	Conc.	RoI	Conc.	RoI	Conc.	RoI
90°	<0.05	N.A.	0.15	0.02	1.6	0.40	N.A.	N.A.
98°	0.15	N.A.	0.85	0.04	5.0	0.70	N.A.	N.A.

Source: Ref. [20].

contribute to the degradation process and subsequently reduce the DP of the paper/pressboard, which subsequently reduces the mechanical strength. Under elevated ageing conditions paper/pressboard undergoes chain scission, which in turn reduces the tensile strength and DP of the paper. Degree of polymerization has been used as the most important parameter for assessing the ageing, or the progress of ageing, of paper insulation in a transformer. The DP value of the paper insulation decreases as the ageing period increases.

2.4.2 Paper Ageing and DP [21]

The solid insulation (paper, pressboard, transformerboard) used in transformers is a sheet of material made from vegetable cellulose. The main source of cellulose fiber is wood. In dry conditions, wood contains 40–50% cellulose, 20–30% lignin, and 10–30% hemicellulose and polysaccharides. The treated paper contains about 90% cellulose, 6–7% hemicellulose, and 3–4% lignin. Cellulose is a linear polymer composed of individual anhydrous glucose units linked at the first and fourth carbon atoms through a glucosidic bond. The structure of glucose and cellulose is shown in Figure 2.8.

The good mechanical properties of cellulose and its derivatives, on which their applicability depends, are due to their polymeric and fibrous nature. When mathematically expressed, cellulose has the molecular formula $(C_6H_{10}O_5)_n$ (the end units are slightly different), where $(C_6H_{10}O_5)$ denotes an anhydroglucose unit and *n* is the DP. Paper derives its mechanical strength from its fibrous nature. Very

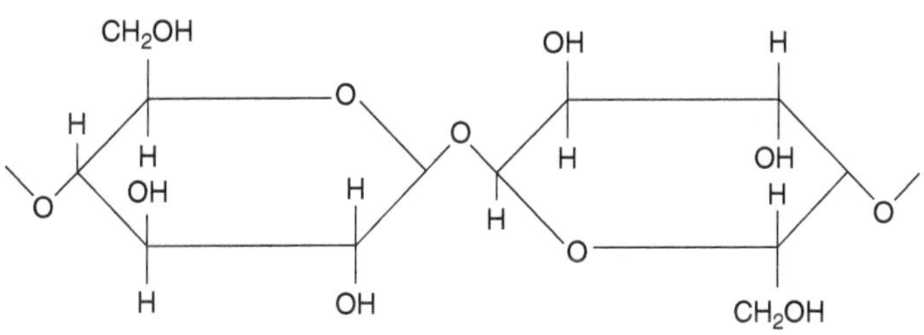

Figure 2.8 Structure of cellulose.

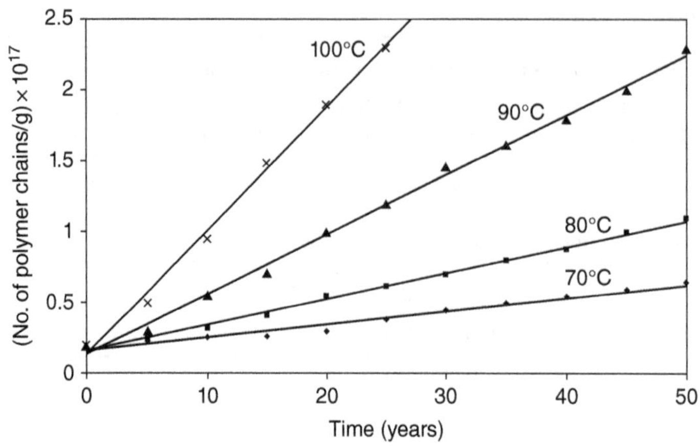

Figure 2.9 Number of polymer chains versus ageing time. *Source:* Ref. [22].

often, the quality of the cellulose is measured in terms of its DP by the average viscometric method. The length of the cellulose chain will thus be denoted DP_v, or more generally DP. Cellulose-based paper and pressboard deteriorates slowly and the process is accelerated by heat, moisture, and oxygen, which are all present in an operating transformer. The amount of chain scission with respect to age can be demonstrated as in Figure 2.9 [22], where the increase in individual polymer chains with increasing age of the transformer can be observed.

Measurement of the DP of cellulose is considered to provide information on the integrity of the polymer chains, and the changing of molecular level (such as chain lengths) determines the macroscopic properties of cellulose insulation. The strength of paper is critically dependent on the DP of the cellulose. During paper manufacture, the pulping process and subsequent washing and drying reduce the DP of the cellulose in the paper to the range 1000 to 1500. After a long period of service at high temperature with high content of water and oxygen, the paper becomes brittle, changes color to dark brown, and its DP drops to the range 200 to 250.

Sometimes, cotton is used as an insulating material. Cotton fiber lengths are greater than those of unbleached softwood sulfate cellulose, but their diameter is smaller. The average DP for new cotton is higher than that of Kraft paper.

At a DP of 150–200, the strength of paper is reduced to 20% of its initial value and below 150, paper has no strength at all. This point is often regarded as the end-of-life criterion for transformer paper/pressboard insulation, because at DP lower than 200, the paper loses all its mechanical strength, although its dielectric constant is not greatly affected [10]. According to some authors, at a DP of 900 to 500, the strength of the paper is virtually constant, but in the range 500 to 200, it decreases in direct proportion to the DP. Comprehensive literature is now available based on accelerated ageing of oil–paper insulation. Most of these publications take the paper end-of-life criterion as 50% of its original tensile strength, or 200–250 as DP value [23–27]. McNutt described the importance of insulation end-of-life criterion [28]. He pointed out that different investigators tend to choose different endpoint DP levels. Bozzini suggested 100–150 [24], Lampe *et al.* used 200 [25], Fabre and Pichon proposed 100–200 [26].

2.4.3 Measurement of DP

The molecular weight of polymers can be measured experimentally by a wide variety of methods. The methods based on solution viscosity have attracted the most attention for measurement of polymer molecular weight in industry generally, as well as in the electricity industry for measurements of the molecular weight of cellulose in insulation paper. IEC 60450:2004 [1] and ASTM D4243 [2] provide comprehensive guidance on DP measurement.

The viscosity measurements only provide the so-called viscosity average molecular weight or average degree of polymerization (DP_v) of the cellulose paper, and the viscosity measurements also require calibration. Because the method only provides an estimate of the average molecular weight, it gives no measure of the molecular weight distribution of the cellulose chains, which can play an important role in determining other polymer properties such as mechanical strength. In other words, the viscosity measurements do not provide a complete picture of the molecular weight changes which take place from degradation of the cellulose, and hence it does not provide complete information for estimation of the condition of the insulation in a transformer. In addition, the viscosity measurements require a significant amount of paper. However, in an operational transformer, sampling the insulation paper can only be made from areas where the insulation is relatively unimportant, such as the top of the winding near the terminal junction. The samples from these remote areas of the transformer generally provide a poor representation of the material in other sections of the transformer where the temperature is higher and the degradation rate is probably also higher [22].

The gel permeation chromatography (GPC) method can overcome the sampling problems as found in the viscometric method, since it is based on measurements of small samples of the insulation paper (as low as 1–2 mg) and can provide the full molecular weight distribution for the cellulose in the insulation paper. Therefore, studies of cellulosic insulation materials based on the molecular weight distribution obtained by GPC have been undertaken recently.

Figure 2.10 [22] shows typical distribution results of GPC analysis performed on a new paper sample and also an aged paper sample. The chromatogram of the new

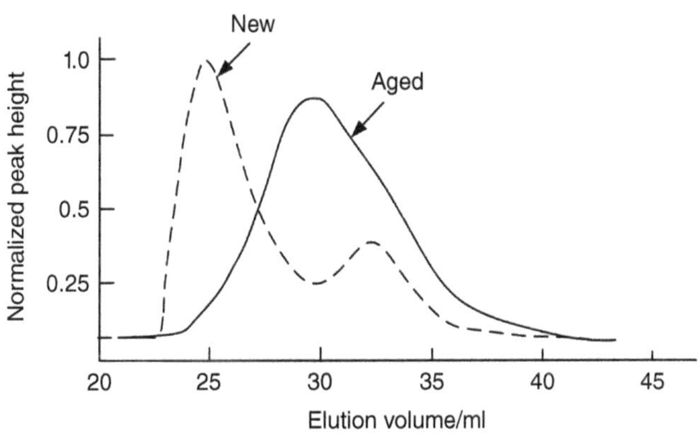

Figure 2.10 Gel permeation chromatograms of new and aged insulation paper [22, 27].

cellulose insulation paper shows the presence of two components. The component at the lower elution volume or high molecular weight is due to cellulose, while the smaller, lower molecular weight component is due to hemicellulose. The GPC profile for the cellulose from the used paper has a clearly different shape from that obtained for the new paper, indicating a change in the molecular weight distribution of the cellulosic components of the paper. In the chromatogram of the cellulose insulation paper taken from an aged transformer, the molecular weight of the cellulose component has decreased significantly. The curve is shifted toward a lower molecular weight (higher elution volume). The molecular weight distribution of the cellulose has also broadened considerably and the peak of hemicellulose has become almost indiscernible, suggesting that the hemicellulose component of the paper may have been largely degraded and hemicellulose may not in fact be present in the aged paper [22].

Hill and coworkers [22, 27] observed that chromatograms of aged paper show that the molecular weight at the peak of each of the components in the chromatograph decreases as the degradation time increases, which is consistent with the polymer chains being subject to scission. Scission of a cellulose chain will result in the loss of one chain of a particular molecular weight and the formation of two new chains of a lower molecular weight. Thus, as degradation proceeds, the number of cellulose chains in a given mass of paper increases and the average molecular weight of the cellulose decreases.

Molecular weight studies by GPC have been found to be more useful when dealing with Kraft paper, particularly in cases where the molecular weight distribution of the paper changes significantly as the degradation proceeds. GPC provides the most convenient way to obtain a detailed molecular weight distribution for the polymer. As GPC gives the entire distribution for the polymer, any small change in molecular weight during the ageing process of an insulation material is easily observable through the chromatogram. GPC measurement provides information about the change in molecular weight and molecular weight distribution.

2.4.4 Insulation Condition Monitoring by DP Measurements and their Relations with Other Properties

McShane *et al.* [29] show the nature of variation of tensile strength and DP with ageing of thermally upgraded paper at three different temperatures (130°C, 150°C, and 170°C), as in Figure 2.11. Hill *et al.* [30] comprehensively studied artificially aged insulation paper tensile strength and correlated it with weight average degree of polymerization (DP_w), as shown in Figure 2.12.

Water content in oil–paper insulation is considered one of the most influential factors affecting ageing. Water molecules are produced during chain scission, oxidation, and glucose degradation. The speed of water forming is different for different types of degradation, and is influenced by a number of conditions, mainly by temperature, amount of oxygen, and water content.

Thus, it is evident that the amount of water cannot be directly proportional to just the quantity of broken bonds and hence DP. The moisture at a given place is not proportional only to the degree of local ageing, but is spread into the surrounding space under the influence of diffusion. The hottest place will probably have the

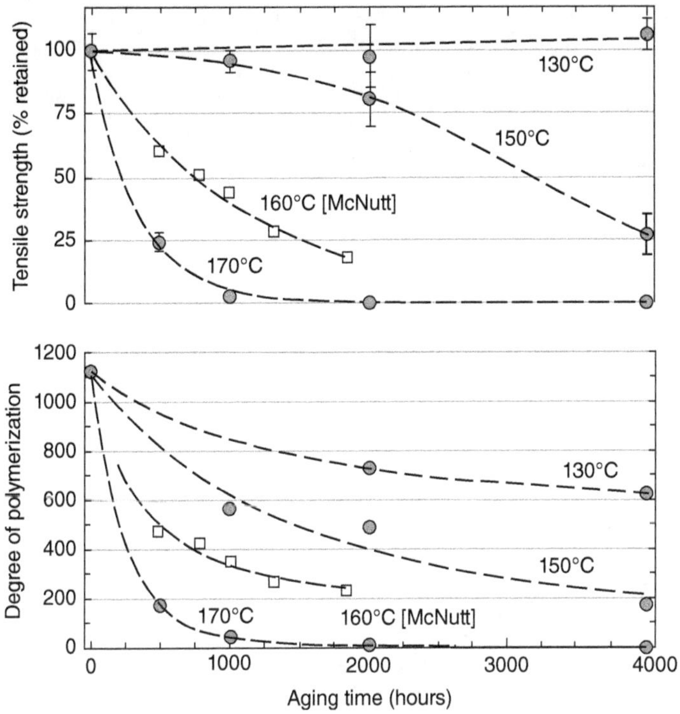

Figure 2.11 Tensile strength and DP of thermally upgraded paper in mineral oil (sealed vessel at 130°C, 150°C, and 170°C). Data at 160°C is included for comparison with [28]. *Source:* Ref. [29], Figure 2.

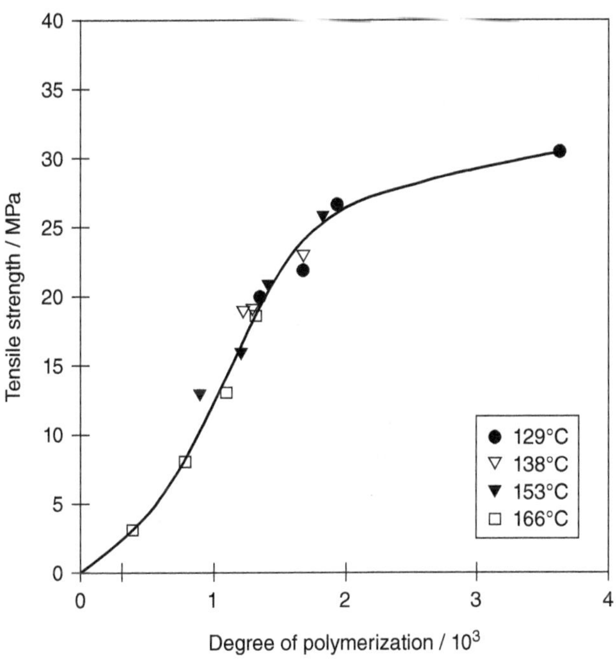

Figure 2.12 Relationship between tensile strength and DP_w for artificially aged insulation paper. *Source:* Ref. [30], Figure 5.

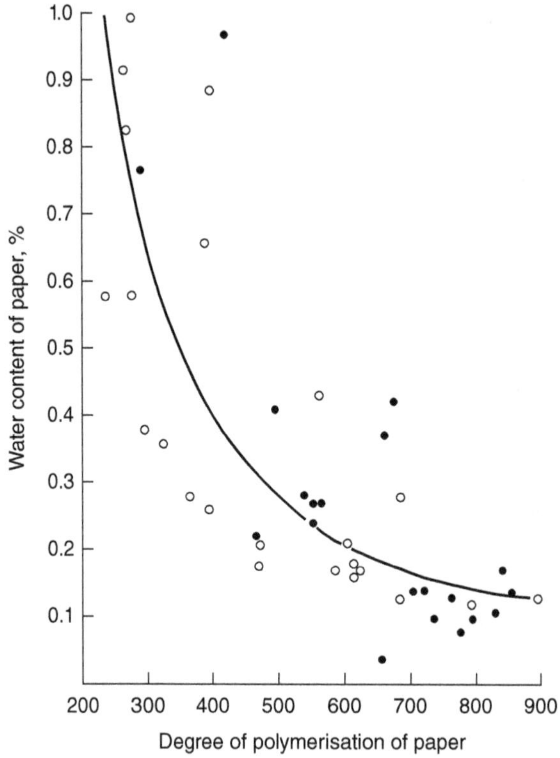

Figure 2.13 Water evolved during ageing of paper. *Source:* Ref. [31], Figure 5.

lowest DP, but it does not necessarily have the highest water content. Laboratory experimental results showing variation of DP with paper moisture content are shown in Figure 2.13 [31]. Moisture developed during ageing will stay in the volume of insulation more or less uniformly by influence of temperature gradients and type of materials (transformerboard or paper on conductor), but the majority of moisture formed by degradation will stay accumulated in the paper.

It is evident from Figure 2.13 that the relative moisture content (H) of paper can be empirically related to the DP by the following relationship on the condition that moisture and oxygen content are adequately low:

$$H = K_1.\exp(-\mathrm{DP}_t.K_2) \tag{2.1}$$

Here, H is relative water content (%), DP_t is the degree of polymerization at time t, and K_1, K_2 are constants to be calculated from experimentally obtained data. It has been reported that with the moisture level in the paper increasing by 0.5%, the DP of the cellulose is halved, resulting in a decrease in electrical strength. This gives rise to another relationship between water concentration in the paper and DP:

$$H = \frac{0.5\log\frac{\mathrm{DP}_0}{\mathrm{DP}_t}}{\log 2} \tag{2.2}$$

Here, DP_0 is the initial value of DP.

2.4.5 Insulation Life Prediction from DP Measurements

DP has been widely used for transformer life prediction, as the life of a transformer is mainly dependent on the life of its paper/pressboard insulation. This is due to the fact that end of life is predominantly determined by the loss of tensile strength, which has been related directly to polymer chain scission. This chain scission results in a lowering of the DP and a decrease in tensile strength.

Emsley and Stevens [10, 32] showed that the reaction rate at any time could be assumed to be proportional to the number of unbroken chain bonds available and their analysis revealed an equation of the form

$$\frac{1}{\mathrm{DP}_t} - \frac{1}{\mathrm{DP}_0} = k \times t \tag{2.3}$$

where $\mathrm{DP}_t = \mathrm{DP}$ at time t, $\mathrm{DP}_0 =$ initial DP, and k is a constant dependent on temperature and other operating conditions:

$$k = A.\exp\left(\frac{-E}{R(T+273)}\right) \tag{2.4}$$

T = temperature (°C)
R = gas constant = 8.314 J/mol/K
E = activation energy = 113 kJ/mol.

Emsley and Stevens [10] related the slope of the kinetic plot (that is, the reaction rate constant k) to temperature using the Arrhenius relationship, plotting log(k) against the reciprocal of absolute temperature. They also showed that the slopes of the plots, and hence the activation energy of the reaction, remain constant under a wide variety of experimental conditions. Substituting values of initial and final DP of 1000 and 200, respectively, and combining with equation (2.4), the equation for transformer remaining life can be expressed as in

$$\text{life} = \frac{0.004}{A}.\exp\left(\frac{13{,}600}{T+273}\right)\text{hours} \tag{2.5}$$

This equation can be rearranged in the following form for transformer life estimation:

$$\frac{1}{\mathrm{DP}_{\mathrm{final}}} - \frac{1}{\mathrm{DP}_{\mathrm{initial}}} = k \times \text{life}$$

The data can be extrapolated to transformer operating temperatures using the above equations. Figure 2.14 can be used for prediction of transformer life against the final DP value. This plot was obtained with the initial unaged DP value estimated as 1100 and data taken from Ref. [32]. Figure 2.15 shows the variation of predicted insulation life with the choice of initial and final DP values for dry Kraft paper in dry oil, also from Ref. [32].

The profile of DP values over the ageing period is, however, strongly dependent on temperature. A sample plot of this relationship is shown in Figure 2.16.

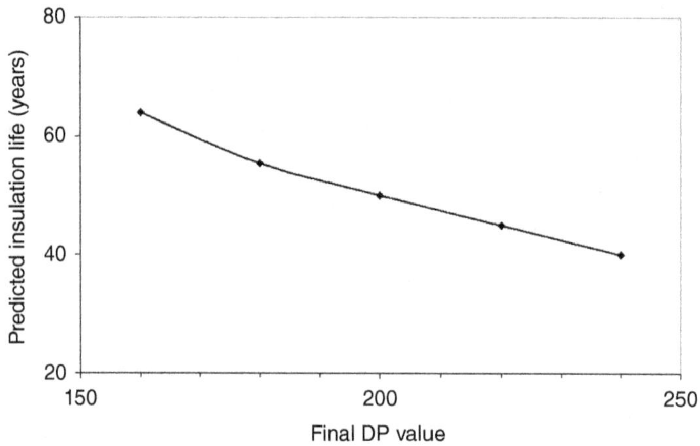

Figure 2.14 Variation of predicted insulation life with choice of final DP values for oil-impregnated paper.

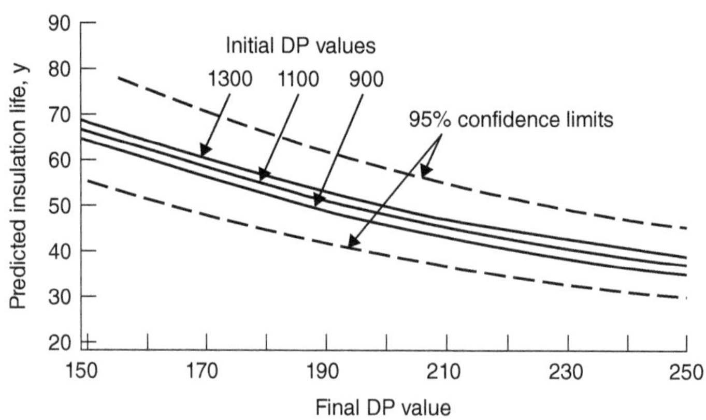

Figure 2.15 Variation of predicted insulation life with the choice of initial and final DP values for dry Kraft paper in dry oil. *Source:* Ref. [32], Figure 5.

The advantages of using DP measurements to predict insulation remaining life are: DP measurements are relatively easy; DP can be empirically related to insulation condition. Life estimates using DP values are, however, subject to the following shortcomings.

- They assume that the kinetics and mechanisms of degradation are the same at transformer service temperatures. This is not necessarily valid, and detailed studies are still required to elucidate the chemistry of degradation. The degradation factor in reality does not vary linearly with time and temperature. In some cases, an initial period of rapid degradation is usually observed before the linear relationship settles down. The rate of degradation is also dependent on the age of the paper.

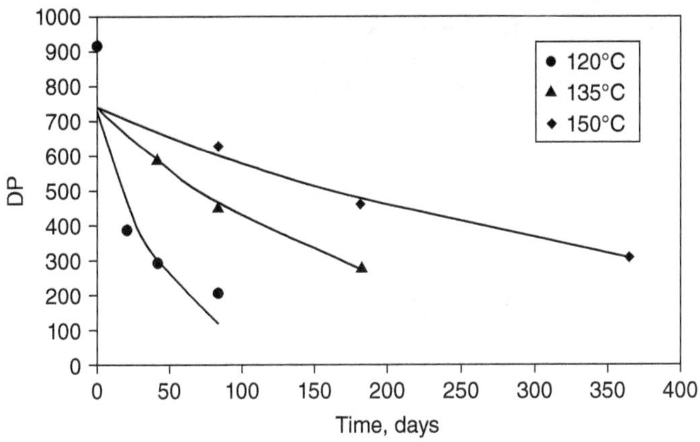

Figure 2.16 Change of DP with time and temperature. *Source:* Ref. [33], Figure 2.

- It is necessary to have available detailed information about past and present operating conditions if a realistic assessment is to be made.
- The initial paper DP needs to be known, which is often quite a difficult piece of information to gather.
- There is still no consensus regarding the end-of-life DP value. The predicted life will obviously vary with different chosen values of end-of-life DP.
- The mechanisms and kinetics of the process are not properly defined, because a number of different reactions could be operating simultaneously.
- The rates of hydrolytic and oxidative degradation of paper depend on the accessibility of the reactants to the interior of the insulation. It has been demonstrated that the rate of degradation varies across the thickness, and is greatest in the inner and outer layers.

The impact of oxygen and moisture also needs to be considered for life prediction based on DP measurements, as will be better explained in Chapter 7. Previous work by Heywood and coworkers [34] supported this from their findings. They reported that "water level in transformer paper at the end of life at 80°C could be 5%, with about 0.1% in the oil." The rate of degradation at normal service temperatures and 4% moisture in the paper has been shown to be 20 times greater than at 0.5% (dried condition), so moisture accumulation has a significant effect on insulation life. Figure 2.17 shows the effects of oxygen and water on DP values.

2.4.6 Relationship of DP with Other Chemical-Based Condition Monitoring Indicators

As mentioned before, DP has been used widely to predict the remaining life of aged power transformers. However, DP measurements require the transformer to be de-energized and paper samples collected from the transformer, which is cumbersome and time-consuming. Many researchers have attempted to correlate DP with furfural content, tensile strength, and dissolved gases. There has been a

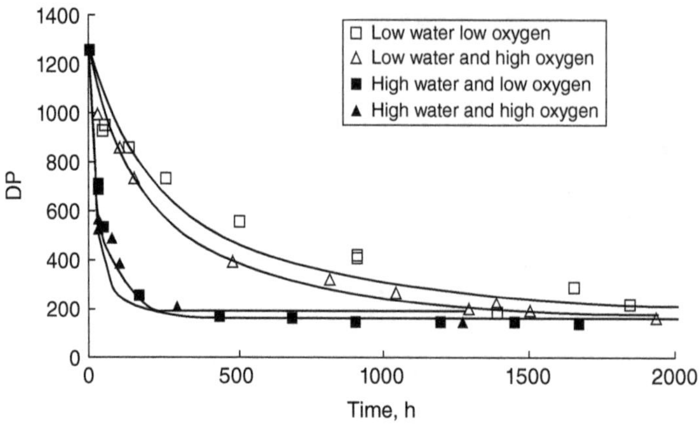

Figure 2.17 Effects of water and oxygen on DP change during ageing of Kraft paper in oil at 140°C. *Source:* Ref. [34], Figure 1.

substantial effort toward correlating furan and DP analysis results for transformer insulation condition monitoring.

The first attempt to correlate the amount of furfural dissolved in the oil and the DP of the paper was based on empirical formulas. These have been derived from both laboratory experiments and field measurements. A statistical relationship between furfural concentration in oil and the DP of solid insulation is investigated, with some scatter plotting as shown in Figure 2.18.

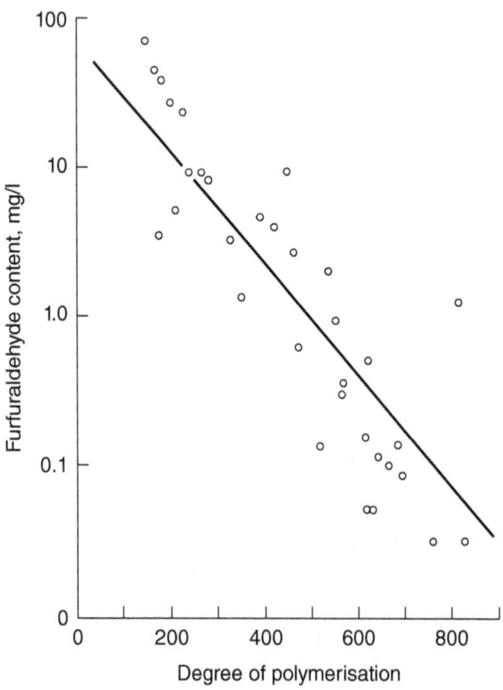

Figure 2.18 Scatter plot for DP versus total furan concentration in oil. *Source:* Ref. [31], Figure 13.

Among the total furans, however, 2-FAL is the most prominent. Efforts have thus been made to correlate DP with 2-FAL, as shown in Figure 2.18.

The linear regression equation is

$$DP = 469 - 370\log_{10}[2FAL\ (mg/l)] \tag{2.6}$$

where the correlation coefficient is −0.849, which is quite satisfactory for operating transformers.

There are other approaches for correlating DP with the furfural content in oil, based on experimentation and empirical relations, though expressed in a different way from the above.

It has been shown that the amount of furfural dissolved in oil is defined by the formula

$$2FAL(\mu g/g paper) = \frac{10^6}{162 \times DP_0} \times N \times 96 \times 0.3 = 222 \times N \tag{2.7}$$

where DP_0 is the initial DP (800), 162 is the molecular weight of the glucose units constituting the cellulose, N is the number of chain scissions, 96 is the molecular weight of furfural, and 0.3 is the reaction yield. Assuming an oil-to-paper ratio (R) of 25, equation (2.7) can be rewritten as

$$2FAL(mg/kg\ oil) = \frac{222 \times N}{R} = 8.88 \times N \tag{2.8}$$

Here N, the average number of bond scissions, is defined by the formula

$$N = \frac{DP_0}{DP} - 1 \tag{2.9}$$

Combination of formulas (2.7) and (2.8) gives

$$DP = \frac{8.88 \times DP_0}{8.88 + 2FAL} = \frac{7100}{8.88 + 2FAL} \tag{2.10}$$

Equation (2.10) is a formula which allows correlation of the amount of furfural (2-FAL), expressed in mg/kg of oil, analyzed in a transformer in service with the DP of the cellulosic paper in the windings of the same transformer.

Hill *et al.* [35] correlated total furan concentration versus DP calculated by average molecular weight (DP_n), as shown in Figure 2.19.

It has been shown that the minimum value of DP which cellulose may reach is about 180 to 200 units. Then, the maximum amount of furfural which may be found dissolved in the oil, calculated by formula (2.10), is about 30 mg/kg, which is quite consistent with the data reported for real transformers. However, it is unlikely that all winding paper degrades to the same extent, as not all the paper is subjected to the same conditions (i.e., the innermost layers of paper around the conductors are subjected to higher temperatures than the outermost, the transformer may have a localized hotspot, etc.). Then, formula (2.8) should be modified on the basis of the transformer operating conditions. For instance, in a normal

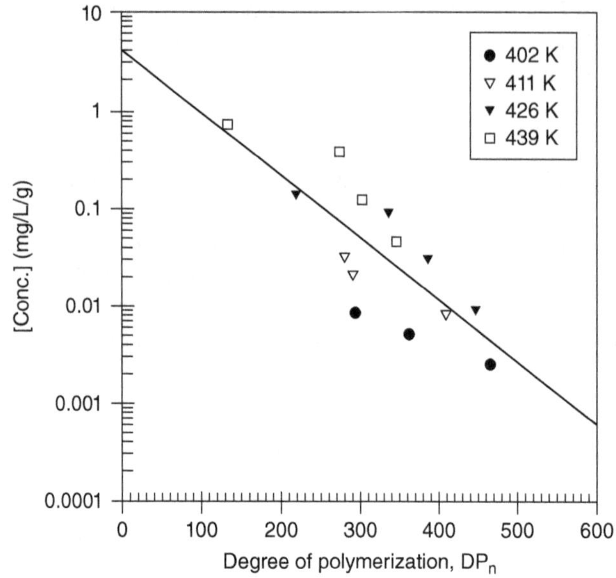

Figure 2.19 Total furan concentration versus DP calculated by number average molecular weight (DP_n). *Source:* Ref. [35], Figure 4.

operating transformer it can be assumed that 20% of the winding conductor insulation (the innermost layers) degrades twice as fast as the rest of the conductor insulation, thus accounting for 33% of total furfural. Then, the maximum level of degradation should be

$$N_{max} = \frac{2FAL \times 0.33}{8.88 \times 0.2} = 0.186 \times 2FAL \tag{2.11}$$

and

$$DP_{min} = \frac{800}{(0.186 \times 2FAL) + 1} \tag{2.12}$$

Also, the range of degradation of the remaining winding insulation, assuming a variation of 50% from the average, should be

$$N = 0.5 \times \frac{0.67}{0.8} \times \frac{2FAL}{8.88} \text{ to } N = 1.5 \times \frac{0.67}{0.8} \times \frac{2FAL}{8.88} \tag{2.13}$$

and

$$DP = \frac{800}{0.141 \times 2FAL + 1} \text{ to } DP = \frac{800}{0.047 \times 2FAL + 1} \tag{2.14}$$

Insulation ageing estimations based on DP and furan have also been used extensively, and Lundgaard *et al.* [36] presented results for up to 10,000 hours of ageing

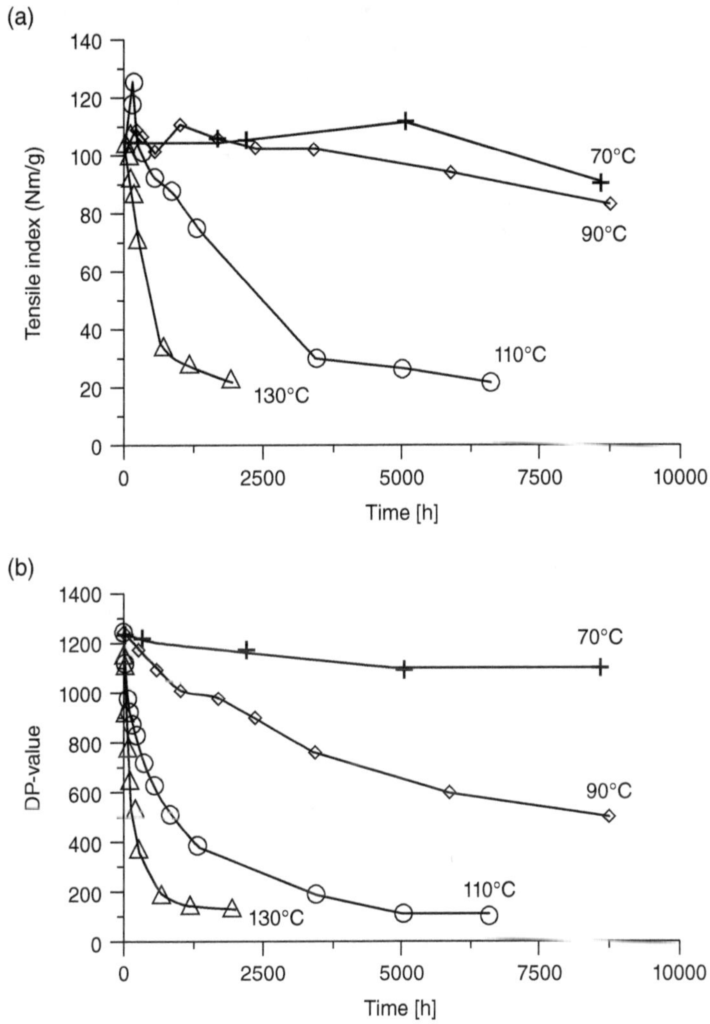

Figure 2.20 Tensile index and DP plot for Kraft paper versus time at 3% moisture. *Source:* Ref. [36], Figure 1.

at 3% added moisture, as shown in Figure 2.20. Allan [37] correlated the age calculated from DP measurements and also from furan measurements, as shown in Figure 2.21. Allan also studied the correlation between insulation ageing determined by CO_2 and DP measurements, as shown in Figure 2.22.

Figure 2.23 shows the correlation between tensile strength and DP value for non-upgraded Kraft paper [38]. Another correlation between tensile strength and DP is shown in Figure 2.24, which is based on work by Emsley *et al.* [34].

Chendong *et al.* [39] found a linear relationship between the furfural concentration in logarithmic scale and the degree of polymerization. This was

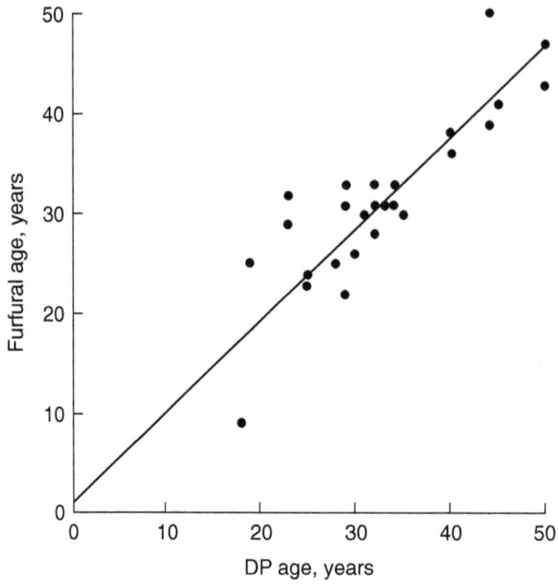

Figure 2.21 Correlation between insulation ageing determined by furfural and DP measurements. *Source:* Ref. [37], Figure 4.

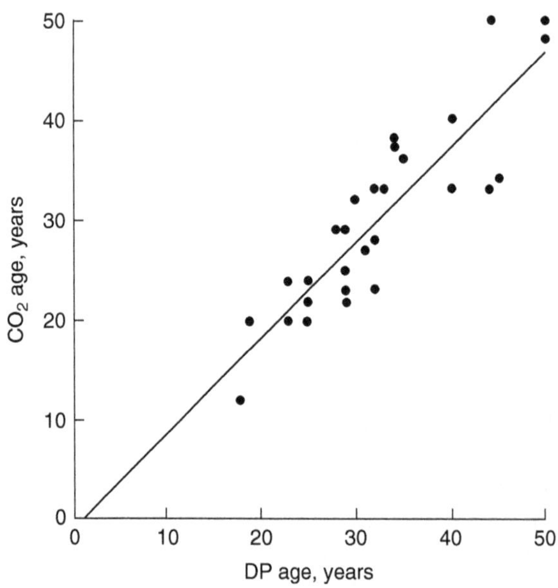

Figure 2.22 Correlation between insulation ageing determined by CO_2 and DP measurements *Source:* Ref. [37], Figure 4.

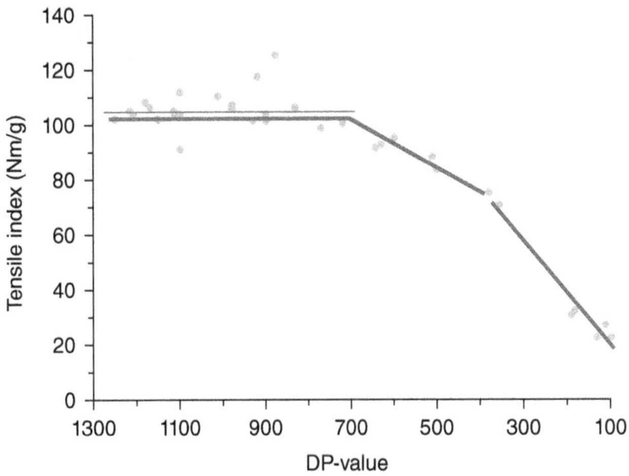

Figure 2.23 Correlation between tensile strength and DP value for non-upgraded Kraft paper. *Source:* Ref. [38], Figure 32.

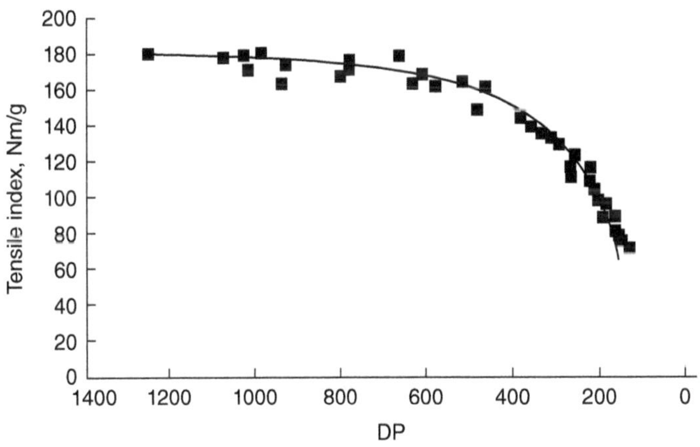

Figure 2.24 Correlation between tensile strength and DP. *Source:* Ref. [19], Figure 20.

represented by the formula $\log(F) = 1.5 - 0.0035(\mathrm{DP})$, where F is the furfuraldehyde concentration (mg/l). This is shown in Figure 2.25.

CIGRE Brochure 494 [19] also provided more information about the correlation of DP and furan, as below.

In the past, CIGRE WG15.01 (TF03) had shown the correlation between the DP and the furanic compounds evolution obtained in laboratory ageing tests. This correlation is shown in Figure 2.26 [19]. As can be seen above, extensive literature exists on the correlation between DP, furan, and tensile strength. However, some of the literature is based on ageing experiments conducted in laboratories and some is from field measurements. Care should be taken in using the results for field transformers.

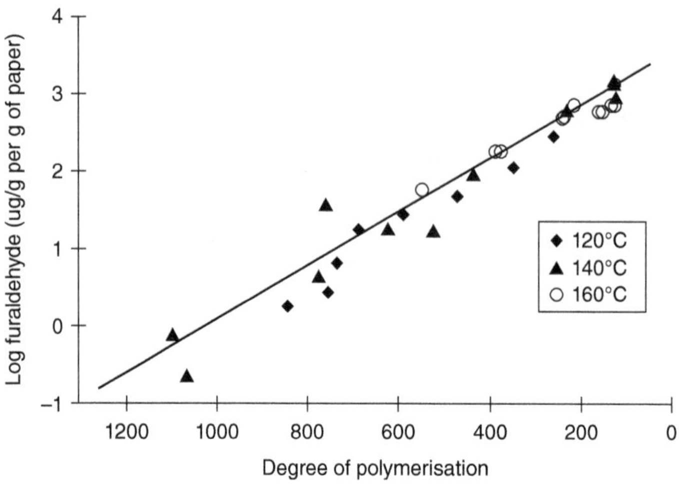

Figure 2.25 Residual life estimation – laboratory experiment, Chendong equation. *Source:* Ref. [19], Figure 21.

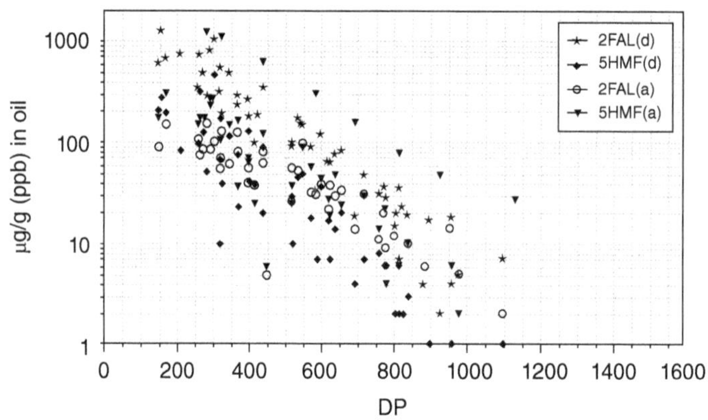

Figure 2.26 2-FAL and other related furanic compounds in oil versus DP: collected laboratory data. *Source:* Ref. [19], Figure 17.

2.5 Traditional Electrical Techniques of Insulation Diagnosis

2.5.1 Introduction

It has been reported that when the cellulose insulation ages in a transformer, the dielectric properties (dissipation factor at power frequency and breakdown strengths at both power frequency and lightning impulse) do not change drastically. Based on this general assumption, very little systematic research had been undertaken in the area of electrical-based traditional diagnostic techniques for condition monitoring or for studying the degradation of electrical insulation in power transformers until 1990.

Insulation resistance and polarization index have long been used by the electricity utilities to ascertain the moisture condition of transformers. Partial discharge is an electrical phenomenon that occurs inside a transformer, and the magnitude of such discharges can cause progressive deterioration and sometimes may lead to insulation failure. There are vast numbers of papers available on PD processes, PD patterns, and fault identification, which will be presented in Chapter 5. A number of researchers have worked on the measurement of dielectric strength of pressboard and paper with different wave shapes (power frequency, lightning impulse, switching impulse, or combinations of these). However, in recent years new diagnostic methods have been promoted that are complementary to the classical insulation resistance, power frequency dissipation factor, and polarization index measurements. These polarization-based diagnostic methods will be presented in Chapter 3 [40].

2.5.2 Insulation Resistance [41]

The measurement of insulation resistance (IR) has been applied to the assessment of transformer condition since the early twentieth century. The method relies simply on the application of a known DC voltage, the measurement of the resulting current, and then the calculation of the IR from Ohm's Law. The absolute value and the shape of the IR curve versus time are both used to assess the condition of the insulation. The shape of the IR versus time curve is expressed by calculating the ratio of the IR measured after 10 minutes to the 1-minute value. This is known as the polarization index, and provides information on the concentration of contaminants, including moisture in the insulating system.

For transformers, insulation resistance measurements are conducted to test the moisture content (or dryness) of the insulation. Insulation resistance can also provide information about unknown damage to bushings. Normally, insulation resistance values are compared over a time period to find the change in moisture conditions of a transformer. Any single-value test may not be very useful.

When a DC voltage is applied to an insulation system, the current has three components, which can be described as below.

- *Capacitance charging current.* This is due to the geometric capacitance of the insulation system. Depending on the insulation capacitance, the capacitance charging current can be reasonably large, but this will decay to zero over time when the insulation is totally charged.
- *Absorption current.* This is normally due to molecular changes within the insulation system, which could exist typically in the range of several seconds to several minutes for transformers.
- *Leakage current.* This current is calculated from the applied voltage and insulation resistance. It is recommended that the leakage current should be taken when the capacitance charging current and the absorption current have become negligible.

Insulation resistance measurements on transformers are normally performed at DC voltages up to 5000 V. There is a wide variety of equipment available for insulation resistance measurements, and the applied voltage magnitude depends on the size and condition of insulation resistance to be measured. During insulation resistance measurement of a transformer, it is recommended that the tank and core should be grounded and the windings should be short-circuited, the windings not being tested should also be grounded. The temperature of the insulation system has an impact on the measurement, and hence insulation resistance is normally measured at a known temperature and very often corrected to a standard temperature for comparison purposes. One-point insulation resistance or insulation current measurement has been in use for decades. However, its usefulness has recently been questioned by the advent of more sophisticated PDC measurements, which can measure insulation resistance and also provide comprehensive ageing and moisture information about the oil–paper insulation system, as discussed in detail in Chapter 3.

2.5.3 Polarization Index Test

The polarization index test is a ratio-based measurement, which is used to predict insulation performance even if the polarization currents have not diminished to a lower value. A very important aspect of this test is its temperature insensitivity. Since the test is ratio-metric in nature, results do not need to be temperature corrected.

For the polarization index test, an insulation resistance is recorded after 1 minute, then again after 10 minutes. The polarization index is given as

$$\mathrm{PI} = (R_{10}/R_1) \tag{2.15}$$

where PI is the polarization index, and R_1, R_{10} are the resistances measured after 1 and 10 minutes, respectively.

According to IEEE Std. 62-1995 (Revision of IEEE Std. 62-1978) [40]:

> For small transformers the polarization index will be equal to 1 or slightly higher. Larger transformers may exhibit a polarization index of 1.1–1.3. In general, a high value of polarization index indicates that the insulation system is in good condition. A polarization index of <1 indicates that immediate corrective action is required.

2.6 DDF, Capacitance, and Power Factor

The capacitance of a transformer depends on the characteristics of the dielectric material and the physical size of the insulation system. The capacitance value can change by degradation of the insulation, contamination, and physical damage.

2.6.1 Dissipation Factor, Loss Factor, or tan δ

The dielectric loss is the power dissipated by the insulation system when an AC voltage is applied. The dielectric loss of a transformer is an important parameter,

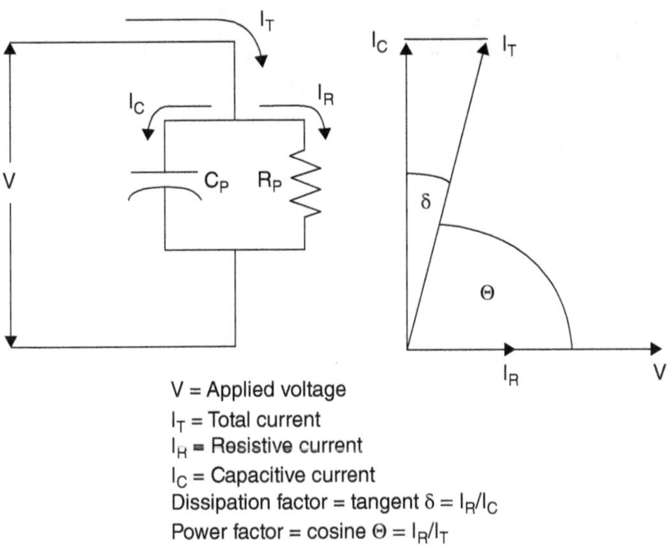

Figure 2.27 Simple vector diagram for loss factor test. *Source:* Ref. [41], Figure 7.

which is dependent on the condition of the insulation. When a transformer has a low loss, it is normally assumed to have a good insulation system (dry and not degraded). A high loss factor indicates ageing and/or moist conditions in the insulation system. The loss factor has been used widely as an indication of the condition of transformer insulation. It is expressed in different terminologies – for example, dissipation factor ($\tan\delta$) or power factor.

When an AC voltage is applied to any insulation system, the total current has resistive and capacitive components. The dissipation factor ($\tan\delta$) is the ratio of resistive current to capacitive current, while the power factor is calculated from the ratio of resistive current to total current flowing through the insulation.

Based on ASTM D D150 – 11, an insulation system can be modeled either as series or parallel combinations of resistance and capacitance. A parallel combination of R and C (as shown in Figure 2.27) is very commonly used for modeling an insulation system at power frequency [40].

DDF measurement is normally carried out at power frequency using a voltage level that is high enough to overcome any local interference at the test site (typically 10–30 kV). The DDF also represents the ratio between the real and imaginary parts of the permittivity. Testing is often performed at different temperatures. As the presence of contaminants – including moisture – influences the DDF at different temperatures, this technique can be used to identify the concentration of these contaminants. Many instruments are available for accurate measurement by experienced staff at high voltages. Measurement equipment (e.g., Schering bridge) has been used widely to measure the dissipation factor at power frequency. Today, easily operated ratio-arm bridges are available, which allow staff with little advanced testing experience to assess the condition of the insulation. These typically operate at power frequency (50 or 60 Hz), up to

20 kV. International standards have laid out guidelines with temperature correction data to use the DDF as an acceptance criterion. Some organizations carry out DDF testing at different temperatures as an acceptance test for transformers. Many of the limitations of power-frequency DDF measurement are expected to be better understood and even resolved through low-frequency dielectric spectroscopy. In recent years, dissipation factor measurement at a variable frequency, commonly known as frequency domain dielectric spectroscopy (FDS) measurement (with varying frequencies from 0.1 mHz to 1 kHz) is recommended to get a comprehensive status of the oil–paper insulation system in a transformer. This will be discussed in detail in Chapter 3.

2.7 Partial Discharge Measurements

IEC 60270:2000 [42] provides comprehensive specifications for measurement of partial discharges, it defines PD as a "localized electrical discharge that only partially bridges the insulation between conductors and which can or cannot occur adjacent to a conductor."

IEEE Std. C37.301-2009 [43] defines PD as "a localized dielectric breakdown of a section of dielectrically stressed insulation path that occurs generally in voids, cracks, or interfaces within that insulating system or from the sharp edges of energized apparatus parts. These discharges may or may not exhibit a glow discharge, based on location and the intensity of these discharges."

To evaluate the dynamic behavior of a power transformer in the electricity system, appropriate online sensor-based measurement techniques need to be developed, which can readily be used for transformer condition monitoring without disrupting the normal operation of the transformer. PD measurement is one such technique for monitoring the dynamic behavior of a transformer's insulation system and assessing the quality and performance characteristics of the transformer [44].

PD can cause progressive deterioration of transformer insulation and may eventually lead to catastrophic failure of the whole unit. Inside a transformer's tank, PD can be generated by a variety of sources – for example, by cavities in solid insulation, by protrusions, by bubbles in the insulating liquid, by drifting metal particles, and due to an electric field tangential to the surface of solid insulation [45]. The ultimate objectives of PD measurement of a transformer are to identify the PD source that causes discharge, quantify the discharge severity of the PD source, and localize the PD source inside the transformer tank.

Over the last three decades, PD phenomena in high-voltage apparatus (including power transformers) has attracted extensive investigation related to mechanisms, physical and chemical effects, detection and measurement techniques, and PD source recognition and localization [45]. However, PD measurement – including detection, signal de-noising, PD source recognition, and localization in a transformer – is considerably more complicated as a result of the complex transmission line behavior of transformer windings and coupling and resonance effects between the windings. Properly performing online PD measurements of a transformer, effectively analyzing the measured PD signals and subsequently making an informed condition assessment on the transformer's insulation system are still challenging.

PD in a transformer is a stochastic phenomenon. PD pulse height or phase distributions have a dependence on time [45, 46]. The variation of PD pulses with respect to the applied sinusoidal voltage is decided by a statistical time lag, which can influence the rise time, magnitude, and shape of the detected PD pulses. Moreover, the memory effects, which are associated with residuals produced by previous discharges, can largely affect the behavior of PD pulses. Furthermore, discharge pulses generated inside a transformer need to propagate along the transformer winding before reaching the PD detector. This will cause both attenuation and distortion in the high-frequency component of the PD signal. The above stochastic PD behavior and PD pulse propagation inside a transformer winding may result in considerable difficulties in interpreting PD measurement data.

PD pulses are impulsive signals and their durations range from several to tens of nanoseconds with a frequency band ranging from several hundred kilohertz to several hundred megahertz [30–33]. There exist different types of PD signal acquisition methods, such as capacitive measurement, inductive measurement, and acoustic measurement systems [44, 47]. Therefore, a number of important issues – such as determination of the signal acquisition method, selection of frequency bandwidth, and sampling rate of the measurement system – need to be considered to develop a PD measurement system for online monitoring of transformers.

PD signals acquired from measurements on field transformers at substation environments are always coupled with extensive noise, which exhibits different distribution properties. Therefore, PD signal de-noising is an essential process for accurately extracting PD signals from the acquired noise-corrupted signals before further analysis. It is necessary to investigate the types and characteristics of noise at the substation, and to develop signal processing techniques to separate PD signals from noise either in the time domain or the frequency domain or joint time–frequency domain [48].

In practice, it is not uncommon that more than one PD source can co-exist in a transformer and discharge simultaneously. As such, before identifying each PD source, there is a need to separate multiple PD sources in a PD signal. For an effective multi-PD source separation, an accurate representation of individual PD pulses is important. This can be achieved by decomposing individual PD pulses into time and frequency domains at multi-resolutions, and then using the components from the decomposed PD pulse (signal) to represent the original signal. These representative components (features) can be used for both PD source separation and subsequent PD source classification [49, 50].

After PD source separation, the remaining task of PD measurement for transformer condition monitoring is to identify the types of PD source (insulation defects) that cause discharges inside the transformer. Various pattern recognition algorithms – such as artificial neural networks (ANNs) and support vector machines (SVMs) – can be applied for PD source recognition [51].

Another important task of PD monitoring of transformers is PD source localization, which helps identify the location of the fault that causes the discharge in the transformer. Normally, a number of acoustic emission (AE) sensors are attached to the transformer's tank to obtain the acoustic signal generated by PD events inside the transformer tank. A spatial intersection method can then be employed to localize the PD source. In this process, the difficulty is the identification of the critical points of the arriving sound waveforms measured by AE sensors [52].

Over the past four years, the authors of this book have investigated the above key issues involved in conducting online PD measurement of transformers and interpreting PD measurement results. On the basis of such investigations, a number of online PD measurement systems and a variety of advanced signal processing and pattern recognition techniques have been developed [48–55]. These techniques will be presented in Chapter 5.

2.8 Conclusion

Though DGA, furan, tensile strength, and DP measurements are still being utilized extensively by the utilities, in recent years new chemical-based diagnostic methods such as GPC, UV spectroscopy and Fourier transform infrared spectroscopy methods have been promoted. Similarly, classical insulation resistance, power frequency dissipation factor, and polarization index measurements are still used by the utilities; new dielectric response-based diagnoses are promoted widely due to the availability of modern computer-controlled instrumentation and better data-interpretation methodologies. These new methods are based on either time or frequency domain polarization measurements and will be analyzed extensively in Chapter 3.

References

1 IEC 60450:2004. Measurement of the average viscometric degree of polymerization of new and aged cellulosic electrically insulating materials.
2 ASTM D 4243. Standard Test Method for Measurement of Average Viscometric Degree of Polymerization of New and Aged Electrical Papers and Boards.
3 IEC 60567 Ed. 4.0 (Bilingual 2011). Oil-filled electrical equipment – Sampling of gases and analysis of free and dissolved gases – Guidance.
4 ASTM D 923. Standard Practices for Sampling Electrical Insulating Liquids.
5 ASTM D 2945. Test Method for Gas Content of Insulating Oils.
6 ASTM D 3305. Standard Practice for Sampling Small Gas Volume in a Transformer.
7 ASTM D 3612. Test Method for Analysis of Gases Dissolved in Electrical Insulating Oil by Gas Chromatography.
8 IEC 60422:2013. Mineral insulating oils in electrical equipment – Supervision and maintenance guidance.
9 IEC 60666 Ed. 2.0 (Bilingual 2010). Detection and determination of specified additives in mineral insulating oils.

10 Emsley AM, Stevens GC. Review of chemical indicators of degradation of cellulosic electrical paper insulation in oil-filled transformers. *IEE Proceedings A: Science, Measurement and Technology* 1994; 141: 324–334.
11 IEC 60599 Ed. 2.1 (Bilingual 2007). Mineral oil-impregnated electrical equipment in service – Guide to the interpretation of dissolved and free gases analysis.
12 IEEE Std. C57.104-2008 (Revision of IEEE Std. C57.104-1991). IEEE Guide for the Interpretation of Gases Generated in Oil-Immersed Transformers.
13 Duval M, Dukarm J. Improving the reliability of transformer gas-in-oil diagnosis. *Electrical Insulation Magazine* 2005; 21(4): 21–27.
14 Duval M. A review of faults detectable by gas-in-oil analysis in transformers. *Electrical Insulation Magazine* 2002; 18(3): 8–17.
15 Duval M, dePablo A. Interpretation of gas-in-oil analysis using new IEC publication 60599 and IEC TC 10 databases. *Electrical Insulation Magazine* 2001; 17(2): 31–41.
16 Duval M, Lamarre L. The Duval pentagon – a new complementary tool for the interpretation of dissolved gas analysis in transformers. *IEEE Electrical Insulation Magazine* 2014; 30(6): 9–12.
17 Unsworth J, Mitchell F. Degradation of electrical insulating paper monitored with high performance liquid chromatography. *IEEE Transactions on Electrical Insulation* 1990; 25(4): 737–746.
18 Burton PJ *et al.* Recent developments by CEGB to improve the prediction and monitoring of transformers. CIGRE Paper No. 12-09, 1984.
19 CIGRE Brochure 494. Furanic Compounds for Diagnosis Working Group (D1.01, TF13), April 2012.
20 PD IEC/TR 62874:2015. Guidance on the interpretation of carbon dioxide and 2-furfuraldehyde as markers of paper thermal degradation in insulating mineral oil (this published document is the UK implementation of IEC/TR 62874:2015). BSI Standards Ltd., 2015.
21 Saha TK. Review of modern diagnostic techniques for assessing insulation condition in aged transformers. *IEEE Transactions on Dielectrics and Electrical Insulation* 2003; 10(5): 903–917.
22 Darveniza M, Hill DJT, Le TT, Saha TK, Williams B. Chemical degradation of cellulosic insulation paper for power transformers. Proceedings of the 4th International Conference on Properties and Applications of Dielectric Materials, 1994, pp. 780–783.
23 Pahlavanpour B, Martins MA, DePablo A. Experimental investigation into the thermal-ageing of Kraft paper and mineral insulating oil. IEEE International Symposium on Electrical Insulation, Boston, MA, 2002, pp. 341–345.
24 Bozzini CA. Transformer ageing diagnosis by means of measurements of the degree of polymerization – results of new experiments. Paper No. 12-08, CIGRE, Paris, 1968.
25 Lampe W, Spicar E, Carrander K. Continuous purification and supervision of transformer insulation system in service. IEEE Winter Power Meeting, 1978, pp. 111–117.
26 Fabre J, Pichon A. Deteriorating Processes and Products of Paper in Oil – Application to Transformers. Report No. 137, CIGRE, Paris, 1960.

27 Hill DJT, Le TT, Darveniza M, Saha TK. A study of degradation of cellulosic insulation materials in a power transformer. Part 1: Molecular weight study of cellulose insulation paper. *Polymer Degradation and Stability* 1995; 48(1): 79–87.
28 McNutt WJ. Insulation thermal life considerations for transformer loading guides. *IEEE Transactions on Power Delivery* 1992; 7: 392–401.
29 McShane CP, Rapp KJ, Corkran JL, Gauger GA, Luksich J. Aging of paper insulation in natural ester dielectric fluid. Transmission and Distribution Conference and Exposition, IEEE/PES, Vol. 2, 2001, pp. 675–679.
30 Hill DJT, Le TT, Darveniza M, Saha TK. A study of degradation of cellulosic insulation materials in a power transformer. Part 2: Tensile strength of cellulose insulation paper. *Polymer Degradation and Stability* 1995; 49(3): 429–435.
31 Shroff DH, Stannett AW. A review of paper aging in power transformers. *IEE Proceedings C: Generation, Transmission and Distribution* 1985; 132(6): 312–319.
32 Emsley AM, Stevens GC. A reassessment of the low temperature thermal degradation of cellulose. Sixth International Conference on Dielectric Materials, Measurements and Applications, 1992, pp. 229–232.
33 Gasser HP, Huser J, Krause C, Dahinden V, Emsley AM. Determining the ageing parameters of cellulosic insulation in a transformer. Eleventh International Symposium on High Voltage Engineering 1999, Vol. 4, pp. 143–147.
34 Emsley AM, Xiao X, Heywood RJ, Ali M. Degradation of cellulosic insulation in power transformers. Part 3: Effects of oxygen and water on ageing in oil. *IEE Proceedings A: Science, Measurement and Technology* 2000; 147(3): 115–119.
35 Hill DJT, Le TT, Darveniza M, Saha TK. A study of degradation in a power transformer. Part 3: Degradation products of cellulose paper insulation. *Polymer Degradation and Stability* 1996; 51: 211–218.
36 Lundgaard LE, Hansen W, Linhjell D, Painter TJ. Aging of oil-impregnated paper in power transformers. *IEEE Transactions on Power Delivery* 2004; 19(1): 230–239.
37 Allan DM. Practical life-assessment technique for aged transformer insulation. *IEE Proceedings A: Science, Measurement and Technology* 1993; 140(5): 404–408.
38 CIGRE Brochure 323. Ageing of Cellulose in Mineral Oil Insulated Transformers (D1.01, TF10), October 2007.
39 Chendong X, Qiming F, Shiheng X. To estimate the ageing status of transformers by furfural concentration in the oil. CIGRE Study Committee 33 Colloquium, Leningrad, Moscow, 1991.
40 IEEE Std. 62-1995 (Revision of IEEE Std. 62-1978). IEEE Guide for Diagnostic Field Testing of Electric Power Apparatus Part 1: Oil Filled Power Transformers, Regulators, and Reactors.
41 ASTM D D150 – 11. Standard Test Methods for AC Loss Characteristics and Permittivity (Dielectric Constant) of Solid Electrical Insulation.
42 IEC IS 60270. High-Voltage Test Techniques – Partial Discharge Measurements, 3rd edn, 2000.
43 IEEE Std. C37.301-2009. IEEE Standard for High-Voltage Switchgear (Above 1000 V) Tests – Partial Discharge Measurements Defines Partial Discharge.
44 IEEE C57.113-1991 (R2002). IEEE Guide for Partial Discharge Measurement in Liquid-Filled Power Transformers and Shunt Reactors.

45 Bartnikas R. Partial discharges: Their mechanism, detection and measurement. *IEEE Transactions on Dielectrics and Electrical Insulation* 2002; 9(5): 763–808.

46 Kreuger FH. *Partial Discharge Detection in HV Equipment*. London: Butterworth, 1989.

47 IEC 62478/TS/Ed. 1. High Voltage Test Techniques – Measurement of Partial Discharges by Electromagnetic and Acoustic Methods, 2012.

48 Seo J, Ma H, Saha T. Probabilistic wavelet transform for partial discharge measurement of transformer. *IEEE Transactions on Dielectrics and Electrical Insulation* 2015; 22(2): 1105–1117.

49 Chan J, Ma H, Saha T. Automatic partial discharge sources separation for power transformer condition assessment. *IEEE Transactions on Dielectrics and Electrical Insulation* 2015; 22(4): 2271–2283.

50 Chan J, Ma H, Saha T. A hybrid method on signal de-noising and representation for online partial discharge monitoring of power transformers at substations. *IET Science, Measurement & Technology* 2015; 9(7): 890–899.

51 Ma H, Chan JC, Saha T, Ekanayake C. Pattern recognition techniques and their applications for automatic partial discharge source classification. *IEEE Transactions on Dielectrics and Electrical Insulation* 2013; 20(2): 468–478.

52 Seo J. Advanced signal processing techniques for online partial discharge monitoring and assessment of power transformers. MPhil thesis, University of Queensland, Australia, 2014.

53 Chan J, Ma H, Saha T. Automatic blind equalization and morphological thresholding for partial discharge signal de-noising. *IEEE Transactions on Power Delivery* 2014; 29(4): 1927–1938.

54 Chan JC, Ma H, Saha T, Ekanayake C. Self-adaptive partial discharge signal extraction based on ensemble empirical mode decomposition and optimal morphological thresholding. *IEEE Transactions on Dielectrics and Electrical Insulation* 2014; 21(1): 294–303.

55 Seo J, Ma H, Saha T. A novel signal extraction technique for online partial discharge (PD) measurement of transformers. *International Transactions on Electrical Energy Systems* 2016; 26(5): 1032–1048.

3

Dielectric Response Measurements

Interest is increasing in replacing the traditional way of determining moisture content in solid insulation of power transformers based on chemical analysis of oil by new methods utilizing dielectric response measurements. Chemical-based measurements of moisture in oil and then using equilibrium curves for estimating moisture in solid insulation has lower accuracy at lower temperatures and lower moisture levels. Evaluation of moisture content in oil–paper-insulated power transformers is of vital importance, since water significantly accelerates the ageing of solid insulation. Moisture content above 3% has been considered by many users as a warning level. Oil analysis by means of Karl Fischer titration (KFT) has traditionally been used widely for moisture analysis. The evaluation of moisture content in solid insulation assumes that the transformer is in an equilibrium condition with respect to the distribution of moisture between oil and the paper/pressboard insulation system. Unfortunately, a transformer in operation is not in a state of equilibrium due to the changes in loading and subsequently the temperature. This is one of the reasons why the moisture estimations are not always very accurate. In addition, the equilibrium curves are not very accurate for the oil sampled at low temperature, and this further introduces uncertainty in the moisture estimation process of solid insulation.

In recent years, polarization-based measurements either in the time or the frequency domain have attracted a lot of attention from utility engineers and scientific communities. The main reason for this interest is due to the availability of a number of commercial pieces of equipment, as well as a better understanding of the measurement process and its interpretation. Owing to the thermal and moisture-related ageing process, the dielectric properties of insulation changes. Measurement of dielectric dissipation factor or loss tangent at power frequency for any insulation system has been in use for many decades. When the measurement of dielectric dissipation factor (DDF) and loss tangent can be performed in a wide frequency range, the method is commonly known as frequency domain spectroscopy (FDS) and the measurement is performed by applying a variable-frequency sinusoidal voltage source across the insulation. Similarly, insulation resistance has been in use for many decades to assess the condition of an insulation system. This has been performed in the time domain by applying a

Transformer Ageing: Monitoring and Estimation Techniques, First Edition.
Edited by Tapan Kumar Saha and Prithwiraj Purkait.

DC voltage across the insulation. In recent times, PDC measurements have been found to be a useful tool to quantify moisture and ageing information on the oil–paper insulation system of a transformer in the time domain. This technique is simple, and the methodology is based on linear polarization phenomena, which will be described in detail in later sections of this chapter. Both time domain and frequency domain measurements require a comprehensive understanding of the polarization processes and the impact of different ageing parameters on their measured values. This will be the focus of this chapter.

Part A: Time Domain Polarization-Based Dielectric Response Measurements and Interpretations

3A.1 Basic PDC Measurement Procedure [1]

PDC analysis measures the polarization current during the application of a DC voltage (as shown in Figure 3A.1) and the depolarization current after the DC voltage is removed from the insulation system. The insulation system in this case can be the insulation between windings of a transformer. These currents will provide the condition of oil in the main duct between the windings and also the condition of the oil-impregnated solid insulation system. The PDC method has been used to provide the moisture in transformer solid insulation as a percentage. We have conducted hundreds of measurements and compared the moisture analysis with the results from Karl Fischer titration measurements. PDC measurement can provide the conductivities of oil and paper separately. PDC data can also be used to calculate insulation resistance and polarization index. PDC data can be used to estimate complex capacitance and dissipation factors in the frequency domain by mathematical models.

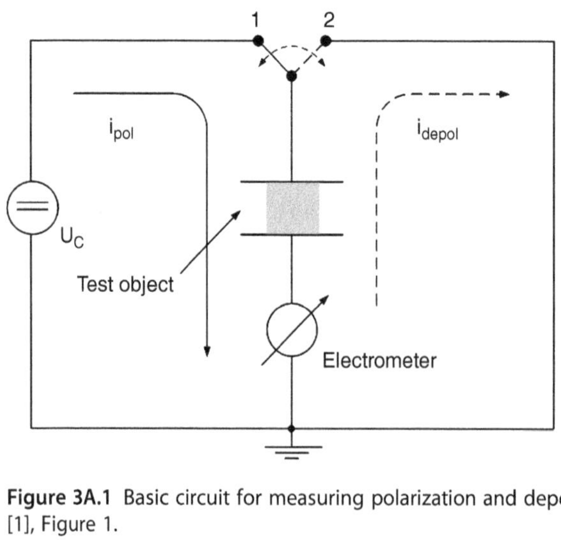

Figure 3A.1 Basic circuit for measuring polarization and depolarization currents. *Source:* Ref. [1], Figure 1.

Figure 3A.1 shows the polarization and depolarization measurement methodology. When the switch is in the left position (position 1), the voltage source is connected to the test object (e.g., transformer with primary winding terminals short-circuited and secondary winding terminals short-circuited to form a two-terminal dielectric system). The polarization current flows and is measured through an electrometer or by any low-current measurement device. After the charging time t_p, the switch goes to the right position (position 2). The test object is short-circuited. The depolarization current i_{depol} flows though the electrometer. The electrometer measures the charging and discharging (polarization and depolarization) currents.

The working principle of the PDC method is shown in Figure 3A.2. During the polarization time, the current arises from the activation of the polarization process with different time constants corresponding to different insulation materials and due to conductivity of the test object. Then, when the voltage is removed and the object is short-circuited, the previously activated polarization process gives rise to the depolarization or discharging current in the opposite direction, where no contribution of the conductivity is present. Figure 3A.2 shows the typical nature of these currents due to DC charging voltage U_0. Polarization and depolarization currents are influenced by the properties of the insulating materials, as well as by the geometric structure of the insulating system. When a DC charging voltage is suddenly applied to the previously uncharged test object, there is a motion of charges due to the interaction of the electrical field with the free and different kinds of bound charges within the dielectric. This so-called polarization current i_p is a pulse-like current at time $t = 0$, then decreases during the polarization time t_p down to a certain value related to the insulation intrinsic

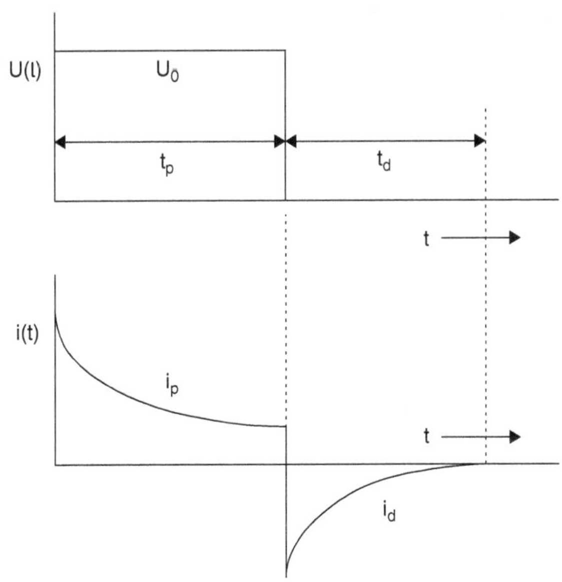

Figure 3A.2 Polarization and depolarization currents. *Source:* Ref. [1], Figure 2.

conductivity. The polarization current measurement can be stopped when it becomes stable or very low. If the test object is then suddenly short-circuited, a discharging current i_d (depolarization current) jumps to a negative value that gradually decreases down to zero. If t_p is large, then i_d becomes directly proportional to the dielectric response function.

While measuring PDC for a transformer, all the HV bushings are connected together to form one electrode and all the LV short-circuited bushings constitute the second electrode. In this arrangement, polarization and depolarization currents are measured between the LV and HV windings. During the PDC measurement, it is recommended that the transformer tank be grounded. In this measurement, the insulation between the two electrodes is charged by a DC voltage between 500 V and 1000 V for a long period, where a typically recommended period is 10,000 s. A long charging/polarization time is required to assess the solid insulation DC conductivity. The charging current or polarization current measured during this period is the combination of absorption current and conduction current. At the end of the preset charging time, the test object is switched over to disconnect the insulation from the DC source voltage and the test object is short-circuited for another long period, typically 10,000 s. The depolarization (discharging) current measured during this duration consists only of the absorption current, as the conduction current only exists during the DC voltage application. The depolarization current decays much faster and should be very small – theoretically zero for very long discharging time.

The basic theory of the dielectric response measurements will be given in the following section, with special emphasis on the PDC technique and the different insulation condition assessment parameters that can be obtained from the PDC measurement.

3A.2 Basic Theory of Dielectric Response [1–4]

When a homogeneous electric field $E(t)$ is applied to dielectric material, the current density through the surface of the material can be written as

$$J(t) = \sigma E(t) + \frac{dD(t)}{dt} \tag{3A.1}$$

The current density $J(t)$ is the sum of the conduction current and the displacement current, where σ is the DC conductivity and $D(t)$ is the electric displacement:

$$D(t) = \varepsilon_0 \varepsilon_r E(t) + \Delta P(t) \tag{3A.2}$$

where ε_0 is the vacuum permittivity and ε_r is the relative permittivity of the insulation at power frequency (typically 4.5 for cellulose paper/pressboard and 2.2 for transformer hydrocarbon oil). $P(t)$ is the dielectric polarization and is related to the dielectric response function $f(t)$ of the material by the relationship

$$\Delta P(t) = \varepsilon_0 \int_0^{\infty} f(t-\tau).E(\tau)d\tau \tag{3A.3}$$

The response function $f(t)$ describes the fundamental memory property of the dielectric system and can provide significant information about the insulating material. The function $f(t)$ can be determined experimentally as the response of the dielectric to a step-function charging field.

3A.2.1 Polarization and Depolarization Currents

Combining equations (3A.2) and (3A.3), equation (3A.1) can be rewritten as

$$J(t) = \sigma E(t) + \varepsilon_0 \varepsilon_r \frac{dE(t)}{dt} + \varepsilon_0 \frac{d}{dt}\int_0^t f(t-\tau).E(\tau)d\tau \tag{3A.4}$$

For a homogeneous material, the field strength $E(t)$ can be considered as generated by an external voltage $U(t)$. The current through a test object with geometric capacitance C_0 (measured capacitance at or near power frequency, divided by ε_r) can be written as

$$i(t) = C_0\left[\frac{\sigma}{\varepsilon_0}U(t) + \varepsilon_r \frac{dU(t)}{dt} + \frac{d}{dt}\int_0^t f(t-\tau).U(\tau)d\tau\right] \tag{3A.5}$$

The test object can be a single dielectric material or an arrangement of several dielectric materials in series or parallel. For more than one dielectric material, σ, ε_r, and $f(t)$ represent, respectively, composite conductivity, relative permittivity, and dielectric response function of this heterogeneous test object. Assuming that the test object is totally discharged and that a step voltage is applied with the following characteristics:

$$U(t) = \begin{cases} 0 & t<0 \\ U_0 & 0 \le t \le t_p \\ 0 & t > t_p \end{cases} \tag{3A.6}$$

This will give zero current for times before $t = 0$ and so-called polarization currents for times $0 \le t \le t_p$. The polarization current is built up in two parts – one related to the conductivity of the test object and the other related to the activation of the different polarization processes within the test object. The polarization (charging) current through the object can thus be expressed as

$$i_p(t) = C_0 U_0\left[\frac{\sigma}{\varepsilon_0} + f(t)\right] \tag{3A.7}$$

Once the step voltage is replaced by a short circuit, a depolarization current is built up. The magnitude of the depolarization current is expressed as

$$i_d(t) = -C_0 U_0\left[f(t) - f(t+t_p)\right] \tag{3A.8}$$

where t_p is the time during which the voltage has been applied to the test object.

3A.2.2 Dielectric Response Function Estimation

It has been shown that, for oil/cellulose insulation systems, the "general response function" can be expressed in parametric form as:

$$f(t) = \frac{A}{\left(\frac{t}{t_0}\right)^n + \left(\frac{t}{t_0}\right)^m} \tag{3A.9}$$

with A, $t_0 > 0, m > n > 0$, and $m > 1$.

In order to estimate the dielectric response function $f(t)$ from a depolarization current measurement, it is assumed that the dielectric response function is a continuously decreasing function in time, then if the polarization period is sufficiently long, so that $f\left(t + t_\mathrm{p}\right) \cong 0$, the dielectric response function $f(t)$ is proportional to the depolarization current. Thus, from equation (3A.8)

$$f(t) \approx \frac{-i_\mathrm{d}(t)}{C_0 U_0} \tag{3A.10}$$

The parameters of $f(t)$ are obtained from a nonlinear least-squares fit to equation (3A.10).

3A.2.3 Estimation of the Conductivity

From the measurements of polarization and depolarization currents, it is possible to estimate the DC conductivity σ of the test object. If the test object is charged for a sufficiently long time so that $f\left(t + t_\mathrm{p}\right) \cong 0$, then equations (3A.7) and (3A.8) can be combined to express the DC conductivity of the composite dielectric as

$$\sigma \approx \frac{\varepsilon_0}{C_0 U_0} \left(i_\mathrm{p}(t) - i_\mathrm{d}(t)\right) \tag{3A.11}$$

The average conductivity for a given insulation system is thus found to be dependent on the difference between the polarization and depolarization current values.

3A.2.4 Modeling of the Response Function *f*(*t*) and Estimation of Conductivity

The first step is to fit the measured depolarization current in parametric form to obtain the dielectric response function $f(t)$ from equations (3A.9) and (3A.10). The nature of the depolarization current is dependent on several factors, including geometry, moisture content, conductivity, and other ageing conditions of the oil–paper composite insulation system. Thus the response function $f(t)$, when fitted to the depolarization current, is supposed to characterize all these physical properties and conditions of the composite insulation system.

The next step is to compute the conductivities of oil and paper insulation utilizing equation (3A.11). The quantity C_0 in equation (3A.11) is the geometric capacitance of the insulation given by

$$C_0 = \frac{C_\mathrm{m}}{\varepsilon_\mathrm{r}} \tag{3A.12}$$

where C_m is the capacitance measured between the two terminals of the insulation system under test. It can be measured with any capacitance-measuring AC bridge at or around the power frequency. ε_r is the effective relative permittivity of the composite oil–paper insulation system. For accurate estimation of this effective permittivity, it is desirable to have correct and complete knowledge of the design and exact composition of the oil–paper insulation system of the transformer. In most cases, however, information about the exact design and composition of the transformer insulation system is not readily available from the utilities. For modeling purposes (and most practical cases), it is sufficient to assume a series arrangement of paper/pressboard and oil duct of the transformer insulation system. In such a case, Gäfvert *et al.* [2] suggested as the expression for the effective permittivity

$$\varepsilon_r = \frac{\varepsilon_p.\varepsilon_d}{\varepsilon_p.(1-X) + \varepsilon_d.X} \tag{3A.13}$$

where ε_p and ε_d are the relative permittivities of paper and oil, respectively. X is the relative amount of paper in the composite system. The range of X is typically 20–50% [2] for a transformer. Similarly, the effective conductivity may be written as

$$\sigma_r = \frac{\sigma_{paper}.\sigma_{oil}}{\sigma_{paper}.(1-X) + \sigma_{oil}.X} \tag{3A.14}$$

where σ_{paper} and σ_{oil} are paper and oil conductivities, respectively.

Once the values of the effective permittivity ε_r and hence C_0 are estimated, the DC conductivity σ can be determined using equation (3A.11) from the difference between the polarization and depolarization currents. For a linear dielectric, one would expect that the difference between the polarization and depolarization currents is constant throughout and equal to the conduction current. This condition is only true for a linear dielectric and when the charging time is sufficiently long. But in practical dielectric systems, like in the complex oil–paper insulation system of a transformer, the assumption of linearity is not always true. Moreover, due to time constraints in actual field testing it is often not possible to charge the insulation for periods longer than 10,000 s. The conduction current (difference between polarization and depolarization currents) is thus found to be time-varying.

Earlier researchers [2, 4–6] have reported that the initial parts of the polarization and depolarization currents are sensitive to the condition of the oil, whereas the final long-term values of these currents are influenced by the condition of the solid insulation (paper). In other words, the initial higher magnitude of the conduction current is due to the higher mobility of the charge carriers in the liquid dielectric (oil). This initial part of the response is found to die down after some time, and the steady conduction current due to the less mobile charge carriers in the solid insulation (paper/pressboard) becomes predominant. The initial polarization current (after the first transient, which is normally not recorded) can be written as:

$$i_p(+0) = C_0 U_0 \frac{\sigma_{oil}}{\varepsilon_0}.\frac{\varepsilon_r}{\varepsilon_d} \tag{3A.15}$$

such that

$$\sigma_{\text{oil}} = \frac{\varepsilon_0.\varepsilon_{\text{d}}}{\varepsilon_{\text{r}}.C_0.U_0}.i_{\text{p}}(+0) \tag{3A.16}$$

In contrast, the long-time polarization current (steady DC value i_{dc}) can be related to the paper conductivity by

$$i_{\text{dc}} = C_0.U_0\frac{\sigma_{\text{r}}}{\varepsilon_0} \tag{3A.17}$$

If $\sigma_{\text{oil}} \gg \sigma_{\text{paper}}$, then from equation (3A.14) we get

$$\sigma_{\text{r}} = \frac{\sigma_{\text{paper}}}{X} \tag{3A.18}$$

Combining equations (3A.17) and (3A.18), we get

$$i_{\text{dc}} = C_0.U_0.\frac{\sigma_{\text{paper}}}{\varepsilon_0.X}$$

or

$$\sigma_{\text{paper}} = \frac{\varepsilon_0.X}{C_0.U_0}.i_{\text{dc}} \tag{3A.19}$$

An average value of $X = 0.35$ (if actual design parameters are not available) can be chosen for the calculation of paper conductivity.

3A.3 Practical PDC Measurement Issues [7]

For PDC measurement in a substation, certain practical issues need to be addressed. These include the measuring circuit requirements, transformer connections, nature of the excitation voltage, charging and discharging periods, and environmental influences such as rain, temperature variations, and the effects of noise and interference.

3A.3.1 Terminal Connections

The basic circuit arrangement for the measurement of PDC of insulation is shown in Figure 3A.1. It is recommended to use the "two active electrodes" technique for the measurement. According to this technique, the insulation to be analyzed is connected between two electrodes, thus forming a capacitor with the insulation as the dielectric. One of the electrodes is marked as the "excitation electrode" and the test voltage is applied to it, referenced to ground, while the current is measured along a line connected between the second "sensing electrode" and ground. With such a "two active electrodes" arrangement, the effects of stray capacitances of the insulation to ground and also cable capacitances can be reduced to a minimum.

Transformers can have different connections, including single-phase, three-phase, three-phase star, three-phase delta, with or without tertiary winding, with or without tap winding, and even three-phase and single-phase autotransformers.

The basic idea is to assess the condition of the bulk insulation consisting of paper and oil between the windings. Thus, for all types of transformer configuration, the insulation to be diagnosed needs to be connected as the dielectric between the two electrodes of the measuring unit. Figure 3A.3 demonstrates the recommended connections at the transformer terminals for different types of transformer.

Figure 3A.3 includes most of the commonly available types of transformer winding arrangement. However, there may be some more arrangements possible which are nothing but different combinations of those shown in Figure 3A.3. In all of these cases, the basic idea is to include the bulk insulation (oil and paper) between the two electrodes of the measuring system. Adequate care must be taken while connecting the transformer terminals to the measuring equipment, such that there is no possible short-circuit path between the two electrodes – or the measuring and exciting equipment may get damaged.

3A.3.2 High-Voltage Source

The high-voltage source to be used for exciting the insulation during a PDC test should be free from drift and disturbances. The voltage source must also be stable during its transient switching to the capacitive load of the insulation. During the PDC tests, it is often very convenient to make the assumption of linearity of the dielectric in order to simplify the analysis of the measurement results. However, it remains an experimental fact that very few systems are truly linear in the presence of a high electric field. At higher charging voltages (high field), charge movements may take place on a significant scale. Figure 3A.4 shows the influence of an increase in charging voltage from 500 V to 1000 V on a 100 kVA transformer. To simplify the comparison, currents obtained with 500 V charging voltage are multiplied by 2. The results show that the actual amplitude of depolarization current for 1000 V is lower. This nonlinearity is caused by the high excitation voltage. This will, in turn, give rise to error in the interpretation of results.

If the amplitude of the field is raised sufficiently, a complete crossing over of the polarization and depolarization currents at short times may be observed. The phenomenon of crossing over is due to the injection of space charge into the dielectric during the charging process, leading to gradual accumulation of excess charge near one or both electrodes, according to the nature of the interface. When the field is removed abruptly at the commencement of the discharge process, the initial depolarization current consists of the true depolarization current along with the current due to the withdrawal of the previously injected charges superimposed. Figure 3A.5 depicts such a case corresponding to a 30 MVA transformer where the crossing over of the depolarization and polarization currents at a small time is observed.

Therefore, it is important to stay within a linear dielectric domain by performing the polarization/depolarization current measurements with not too high charging voltages.

3A.3.3 Charging and Discharging Periods

In order to study the discharge characteristics of a material with a view to accurately determine its dielectric response function, it is necessary to charge the

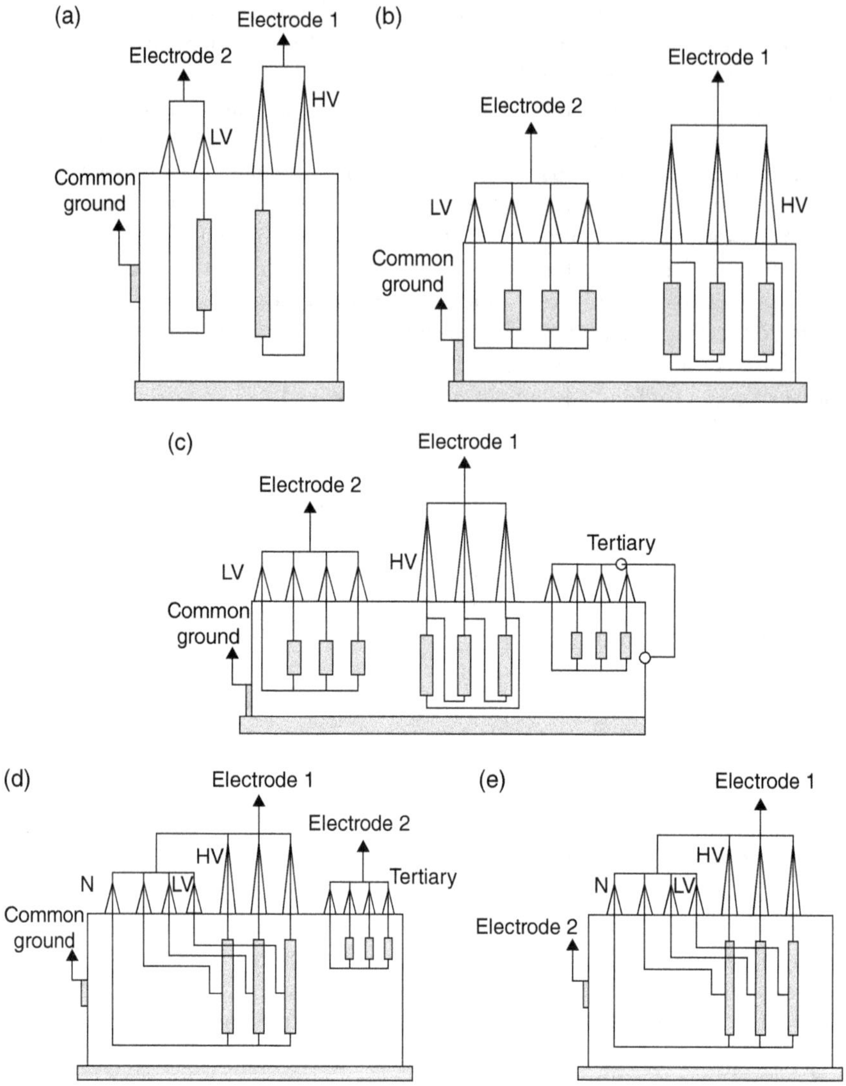

Figure 3A.3 Different transformer terminal connections for PDC measurements: (a) single-phase two-winding transformer; (b) three-phase transformer with star-connected LV winding with neutral and delta-connected HV winding; (c) three-phase transformer with star-connected tertiary winding; (d) three-phase autotransformer with tertiary winding; (e) three-phase autotransformer without tertiary winding.

material prior to the beginning of the discharge process for at least ten times longer than the maximum desired time of the discharge measurement. The depolarization current at times that are short compared with the preceding charging time follow the genuine characteristics expected of the material in question, but at times long compared with the charging time, the slope of the logarithmic plot of the depolarization current becomes much steeper than the genuine

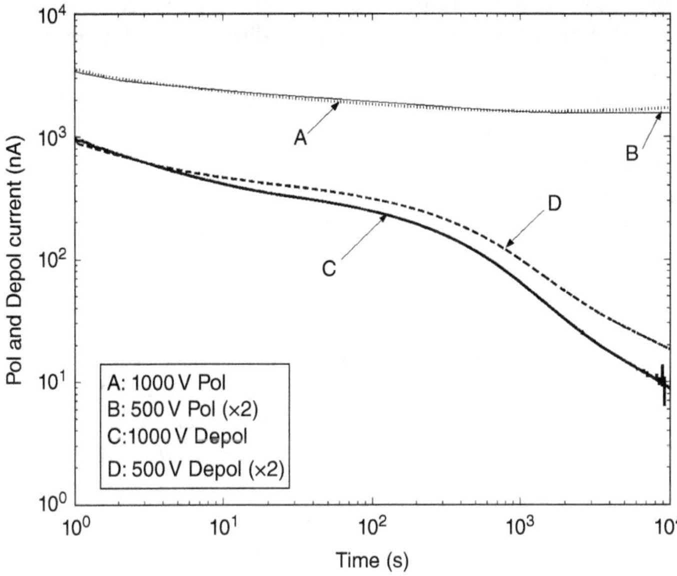

Figure 3A.4 Relaxation currents at different excitation voltages (scaled to the same level for comparison).

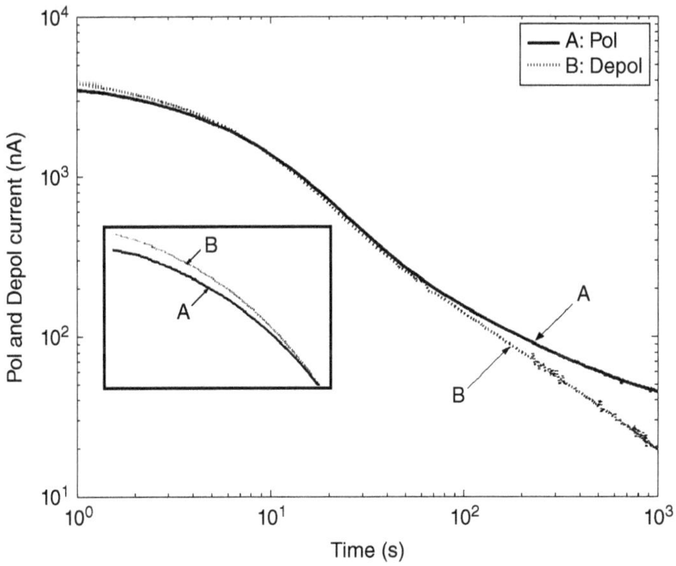

Figure 3A.5 Crossing over of depolarization and polarization currents; inset shows the initial period on an enlarged scale.

characteristics. Figure 3A.6 is a plot of the two dielectric response functions $f(t)$ calculated for a 70 MVA transformer – at charging to discharging time ratios of 10:1 and 1:1, respectively.

Apart from the serious errors that may arise if this condition is not fulfilled, these requirements mean that the time taken for an experiment is much longer than the actual measuring time itself.

The same conditions apply to the opposite process of polarization current measurement: the material must be thoroughly discharged for at least ten times as long as it is intended to measure the polarization current. If the material is not fully discharged before starting a new cycle of measurement, the charges remaining from the previous measurement may introduce errors into the current measurement. This effect is more severe in new transformers, where the magnitudes of the relaxation currents are quite low themselves. This is often referred to as the "memory effect." Figure 3A.7(a) shows such a condition when the transformer is charged again without discharging for a sufficiently long time. The apparently higher magnitude of the polarization and depolarization current is due to the "memory effect" – that is, the influence of previous measurements. As observed in Figure 3A.7(a), the difference in magnitude of the two depolarization currents is more prominent than the corresponding difference in polarization currents. This may result in erroneous calculations of response function and conductivities. A similar nature of polarization and depolarization currents with varying charging times is reported by Zaengl [8], as shown in Figure 3A.7(b).

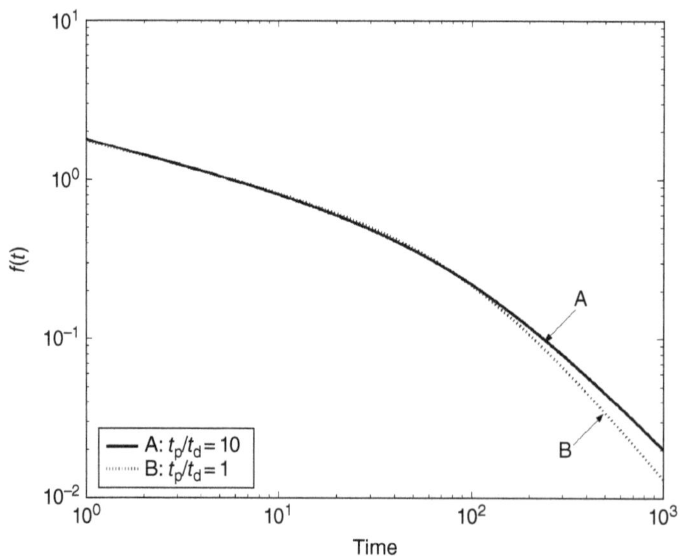

Figure 3A.6 Dielectric response function $f(t)$ at different ratios of charging and discharging times t_p and t_d.

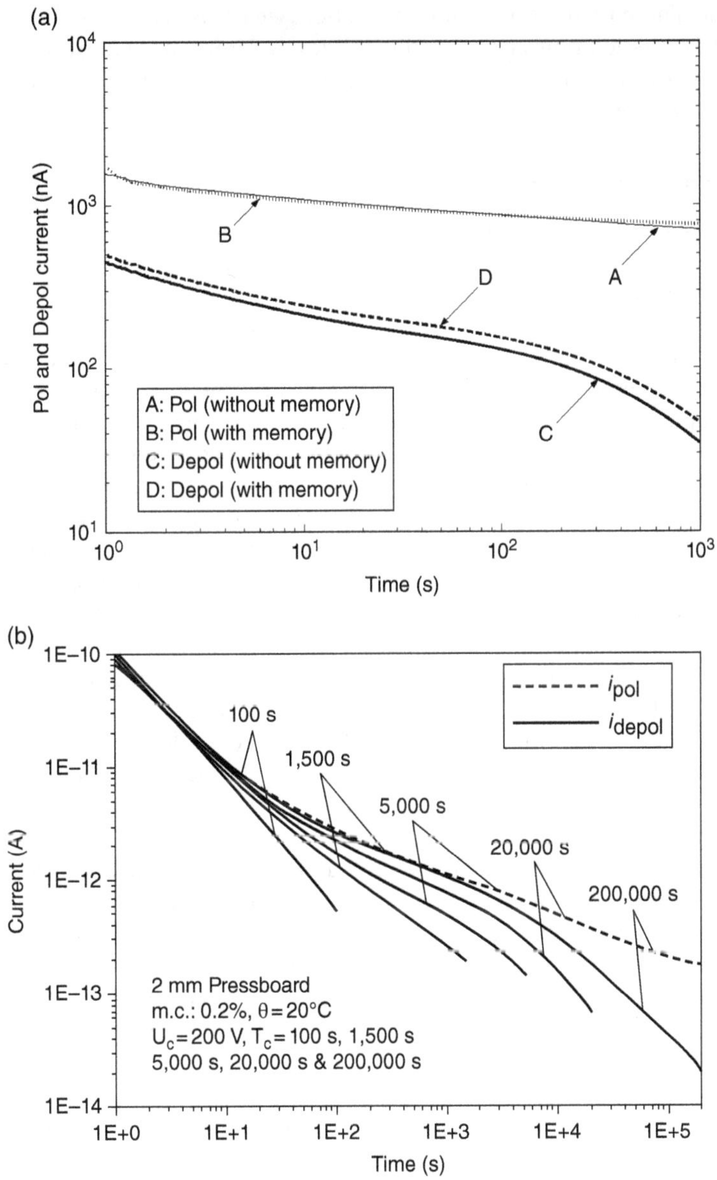

Figure 3A.7 (a) Polarization and depolarization currents with and without adequate pre-discharging. (b) Polarization and depolarization currents of a dry paper sample as a function of charging time. *Source:* Ref. [8], Figure 5.

It is thus evident that in order to obtain the correct estimation of the dielectric response or conductivities from the polarization and depolarization currents, it is essential to short the transformer to ground for a sufficiently long time before conducting the test. It is also recommended that the charging time be at least ten times longer than the intended discharging time in one cycle of measurement. These requirements, however, often need to be sacrificed due to the constraint of testing times given by the utilities. A compromise needs to be made between the intended charging/discharging periods and the actually available total testing time.

3A.3.4 Noise and Interference

Owing to the inherently high values of insulation resistance of oil and paper, polarization and depolarization currents in normal transformers are very small – in the range of nanoamperes. The measurement of these small currents can be affected by induced AC currents, electromagnetic interference, and electrostatic induction from nearby high-voltage installations. In actual field test conditions, these interfering objects are unavoidable. It has been observed that other nearby energized equipments, especially other power transformers, cause coupling of AC currents in the measurement circuits. This current superimposes the desired DC current and often exceeds the DC current by a factor of 10^3 to 10^4. Usually, other equipment cannot be switched off due to operational requirements. These effects are also aggravated by wind and humidity in open substations. Long testing leads connecting the bushings to the testing instrument also add to the problem of interference. It has also been observed during onsite measurements that too much movement of people close to the measurement setup may introduce noise in the measurement. It is also recommended that the transformer under test should in no way be touched from outside while the testing is going on, otherwise spurious signals may get injected into the measurement. Figure 3A.8 displays the polarization and depolarization currents of a 7 MVA transformer tested in the field under extreme windy conditions and in close proximity to live overhead busbars and other live transformers.

Apart from the obvious and practicable precautions to reduce interference in the field, the situation can be improved by the use of filters. These filters can be either analogue filters integrated with the measurement system or software-implemented digital filters. Analogue filtering, however, should be kept to a minimum as the capacitors needed for the filters will exhibit dielectric relaxation phenomena much like the insulation to be diagnosed – thereby introducing unwanted errors. Digital filtering consists of notch filtering at mains supply frequency and low-pass filtering. Digital filtering has the advantage for this application in that the filter quality is independent of the signal amplitude and the filter properties can easily be adjusted by the software.

In the PDC measuring equipment developed by the authors of this book, software-based digital filters have been implemented to reduce the noise from the measured relaxation current signals. Figure 3A.9 shows the polarization and depolarization currents of the same transformer as in Figure 3A.8, but after the signals have been processed by a digital filter.

3A.3.4.1 Rain/Excessively Humid Conditions

Rain makes onsite measurement not only uncomfortable but often impossible. Rain seems to generate leakage DC currents which superimpose the desired measurement currents. The magnitude of these leakage currents depends on the rain

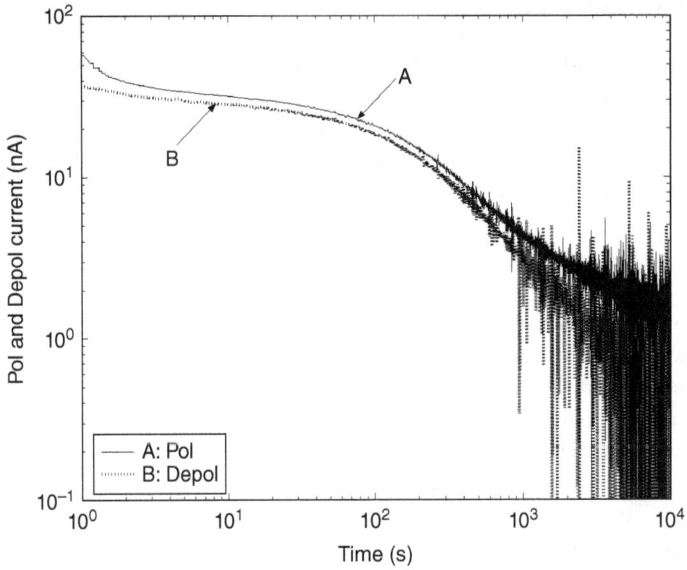

Figure 3A.8 Polarization and depolarization currents in the presence of noise.

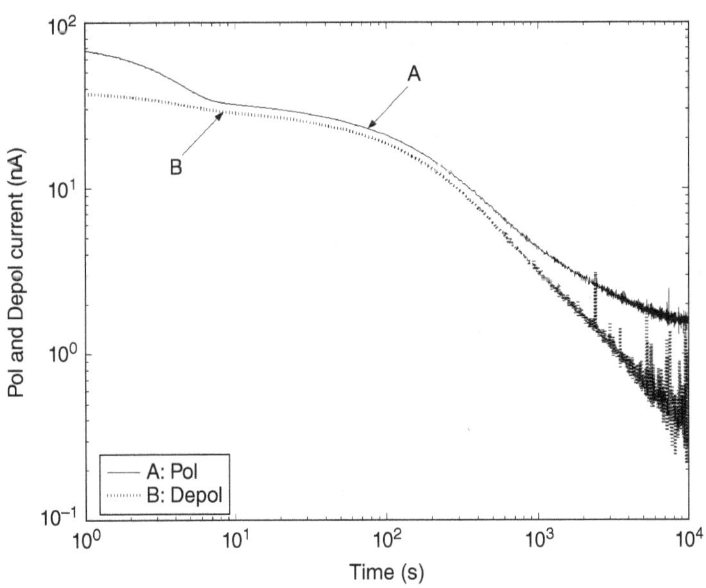

Figure 3A.9 Polarization and depolarization currents after filtering out noise.

intensity and sometimes exceeds 100 nA, which makes measurement difficult to almost impossible. As there seems to be no chance of distinguishing between currents through the oil–paper insulation system and leakage currents caused by rain, the effect has not yet been investigated further.

3A.3.5 Temperature Instability

Several researches and field measurements have reported that relaxation currents are very sensitive to temperature. This phenomenon has been attributed to the increased mobility of charge carriers during the polarization/depolarization process at elevated temperatures. This dependence of the relaxation currents may seriously affect the PDC measurement results during field testing in more than one way.

It is often experienced in actual field testing that the transformer to be tested was previously connected to the grid and in operating condition. During normal operating conditions, the temperature inside a transformer may reach a value much higher than ambient, depending on the loading condition. For PDC measurement purposes, if a transformer is taken out of service it must be given adequate time for the temperature to settle down to ambient conditions before commencing the actual PDC testing.

During thermal transition, complex dynamic processes occur as temperature gradients develop. Temperature transitions disturb the moisture equilibrium of the system, causing the initiation of moisture mass transfer processes. Transformer oil and paper/pressboard are very dissimilar materials, in that the former is hydrophobic and the latter is hydrophilic. As a consequence, almost all of the moisture present in the system resides in the pressboard. As the temperature changes, moisture will move into or out of the paper/pressboard via diffusion. It is thus evident that unless the temperature is stable and the oil–paper has attained equilibrium, any dielectric test done in transition will not reflect the true condition of the insulation.

Figure 3A.10 shows a plot of PDC test on a transformer, where PDC measurements were carried out while the transformer was in the process of cooling down. Before start of the test, the transformer was running at a temperature of 60°C. The transformer was then switched off from the source and allowed to cool down to the ambient temperature of 23°C. The PDC measurement was done during this cooling process. The plot of Figure 3A.10 also contains the polarization and depolarization currents conducted for the same transformer at the normal ambient condition of 23°C, for comparison.

If a thermal step is applied to the sample, there is a thermal wave diffusion into the material. This generates a slight displacement of charges placed in the bulk of the material. This causes the appearance of a current in the external circuit. This current may interfere with the polarization and depolarization currents under measurement, giving rise to unwanted errors. This observation is well supported by the plots of Figure 3A.10. The polarization and depolarization currents corresponding to the case when the transformer temperature was under transition are higher than their ambient-temperature counterparts. This may result in erroneous calculations of the response function and conductivities.

In Figure 3A.10, the depolarization current values corresponding to the two separate cases (plot B and plot D) almost overlap after about 1000 s discharging. This is due to the fact that, during this time, the temperature had already come down close to ambient from its initial higher value. Thus, it is essential to allow

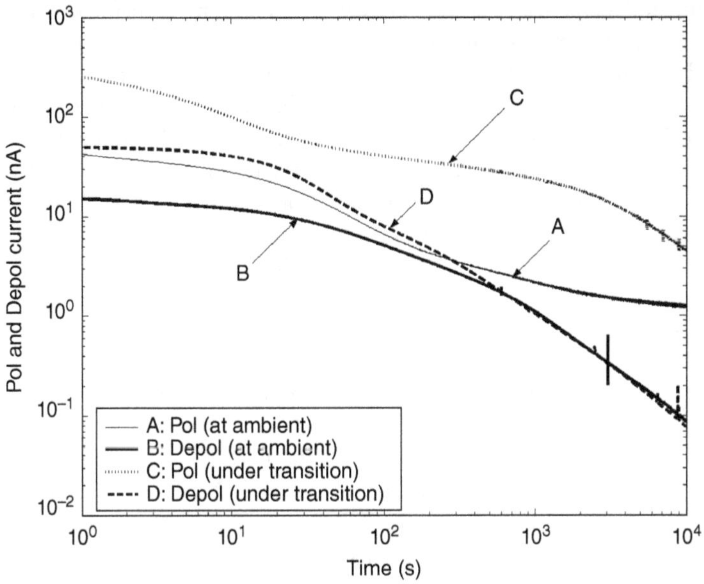

Figure 3A.10 Polarization and depolarization current plots for transformer under downward transition of temperature.

sufficient time for the transformer temperature to settle down to ambient before commencing the actual PDC test.

An opposite kind of temperature transition effect on the PDC measurement can be observed while field testing in open substations. Especially in tropical cities like Sydney, Melbourne, or Brisbane and other outback towns in Australia and similar places, the ambient temperature varies widely during the day. The ambient temperature may start at 18–20°C in the morning and shoot up to 30–35°C around midday. In effect, the temperature inside the transformer tank will also go up depending on the volume and quality of oil, paper, and materials involved in the heat conduction process. This may have a noticeable impact on the PDC measurement. In a normal case, the PDC test takes around six hours to complete with 10,000 s charging and 10,000 s discharging periods. This means that if the polarization current measurement starts in the morning with a lower ambient temperature, then the depolarization current measurement is expected to be at a much higher ambient temperature as it goes into the day. Both the polarization and depolarization currents being very sensitive to temperature variations, these kinds of temperature variation during the measurement period itself may introduce unwanted variations in current. An experiment was performed in the laboratory with the test transformer being put inside a temperature-controlled chamber. The PDC test was started at 20°C and then the chamber temperature was increased at a steady rate of 2°C per hour up to 30°C while the test was going on. Figure 3A.11 shows the polarization and depolarization currents in the case with the temperature around the transformer tank rising steadily from 20°C to 30°C during the test. This figure also includes the polarization and depolarization currents when the same transformer was tested at a constant temperature of 20°C.

It is observed in Figure 3A.11 that – as expected – the two polarization currents corresponding to that at 20°C and the other under transition from 20°C to 30°C overlap each other during the initial period. However, as the temperature increases gradually from 20°C toward 30°C, the polarization plot under transition (plot A) shoots above the corresponding polarization current at 20°C (plot C). The depolarization current under transition (plot B) is substantially higher than the corresponding depolarization current at 20°C (plot D) for most of the time, only coming close to each other at large times when the current magnitudes themselves are too small to distinguish. It is thus clear from Figure 3A.11 that any dielectric response function and conductivity computed from the polarization and depolarization currents obtained under such temperature transitions during the PDC measurement will be erroneous.

This effect of an increasing day temperature on the PDC measurement makes the interpretation of results even more confusing in cases where the oil condition is good (i.e., when the difference between the polarization and depolarization currents during the initial part is small). In such cases, the increased magnitude of the depolarization current due to increased temperature may even overshoot the magnitude of the corresponding polarization current during the initial period – making computation of conductivity impossible. One such instance is shown in Figure 3A.12, which was performed on a field transformer at an outdoor substation. The air temperature during the measurement was recorded to vary from

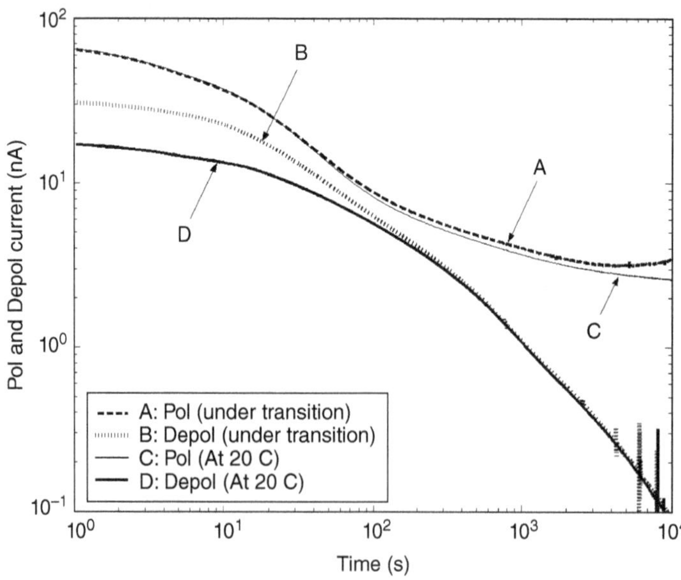

Figure 3A.11 Polarization and depolarization current plots for transformer under upward transition of temperature.

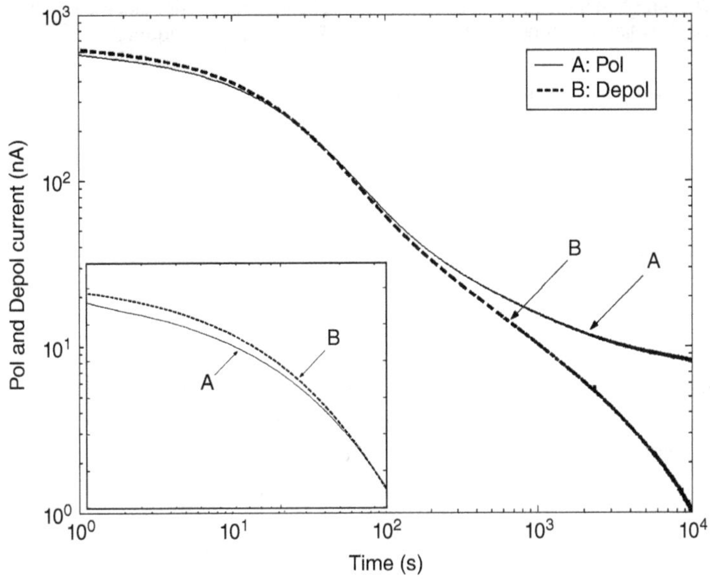

Figure 3A.12 Higher magnitude of depolarization current over polarization current due to temperature transient; inset shows the initial period on an enlarged scale.

18°C at the start to 34°C till the test was over. The resulting depolarization current was found to have higher values during the initial period than the polarization current.

It is thus apparent that the interpretation of the PDC test results should be done with sufficient attention being given to such temperature variations during the measurement. This phenomenon will be discussed further later in this chapter, with appropriate correction factors for such a variation of temperature during the PDC measurements.

3A.3.6 Signal/Supply/Measurement Cable Length

Another issue that needs to be addressed in relation to the measuring setup is the cables that are used to connect the transformer terminals with the low-current measurement device (e.g., electrometer). These signal cables need to be shielded and the shield connected to the instrument common ground point to minimize the effects of any interference. It is also recommended to keep the cable length as short as possible to avoid the effects of stray capacitance and resistance of the signal cable from affecting the test results. This has been well explained by Homagk and Leibfried [9].

For large high-voltage power transformers with long bushings, cable quality has to be taken into consideration.

> To keep distances short or due to local conditions, measurement cables often hang down from bushings directly to the ground. Sometimes this causes cable oscillation excited even by light wind. If during measurement

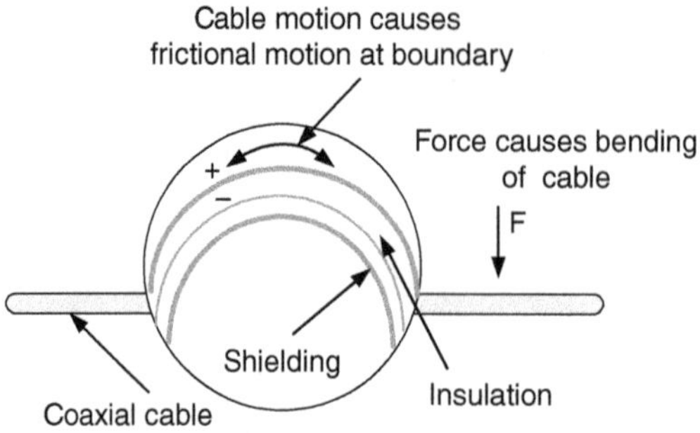

Figure 3A.13 Triboelectric effect. *Source:* Ref. [9], Figure 3.

> the cables are moved, there can be a significant noise current caused by the triboelectric effect. Triboelectric currents are generated by charges created at the interface between a conductor and an insulator due to friction [see Figure 3A.13]. Free electrons rub off the conductor and create a charge imbalance that causes a current flow. There are special low noise cables available with semi conducting layers between insulators and conductors which significantly decrease triboelectric noise. It is also important to ensure that cables are not moved during measurement. [9]

A similar imminent problem of PDC measurements is the creation of a more or less cabling loop. Following the Maxwell equations, this creates a plane where parasitic AC currents can be induced (e.g., from ground currents in power plants or in switchyards). For good measurement results, this plane should be reduced to a minimum, depending on the local situation.

3A.4 Interpretation of PDC Measurement Results

3A.4.1 Laboratory Experiments

Transformer insulation structure often comprises of paper or cotton insulation over conductors and pressboard spacers placed in the oil gap between windings. Under operating stresses and ambient conditions, this insulation undergoes different degrees of ageing over the years and absorbs different degrees of moisture. Moisture not only advances the ageing rate, but also influences the characteristics of insulation – such as permittivity, conductivity, and dielectric loss.

Thus, dielectric response measurements are strongly influenced by insulation ageing and moisture content, in addition to operating temperature and geometrical structure. Tests are then conducted under controlled laboratory conditions to understand such effects on dielectric response measurements before the method can be used for assessment of insulation in field transformers.

Several researchers have developed different ageing and moisture conditioning experimental setups in a laboratory environment to study these effects on dielectric response measurements.

Saha and Purkait [10] developed a moisture conditioning method in which a set value of paper moisture level was achieved by controlling the pressure of water vapor and temperature inside a closed container for a long period of time. The test samples used for the purpose were a pair of paper-wrapped copper conductors impregnated with transformer oil. The starting moisture content of both the oil and paper samples was measured using Karl Fischer titration with a 737 KF coulometer from Metrohm. According to ASTM D 3277-73, a methanol/chloroform mixture was used to extract water from the paper samples for measurement. Dry and degassed transformer oil (Shell Diala B) was then added to the container, so that the conductors were completely immersed in the oil. Figure 3A.14 shows the moisture conditioning setup developed at the University of Queensland.

To perform ageing experiments, moisture-conditioned conductor samples and oil were transferred from the conditioning vessel to the ageing ampoules after oil/paper equilibrium was achieved. The ampoules were subsequently placed in a controlled-temperature oven for various ageing periods: 120 days, 240 days, and 360 days. The temperature of the oven was set to 95°C during the ageing period. The ageing ampoules contained the moisture-conditioned paper-wrapped conductor, oil, and air in the remaining space. The ageing ampoules were sealed with vacuum grease so that there was no contact with the outside atmospheric conditions.

After completion of ageing according to the set period, the ampoules were taken out of the oven and placed in a humidity-controlled chamber. After cooling down,

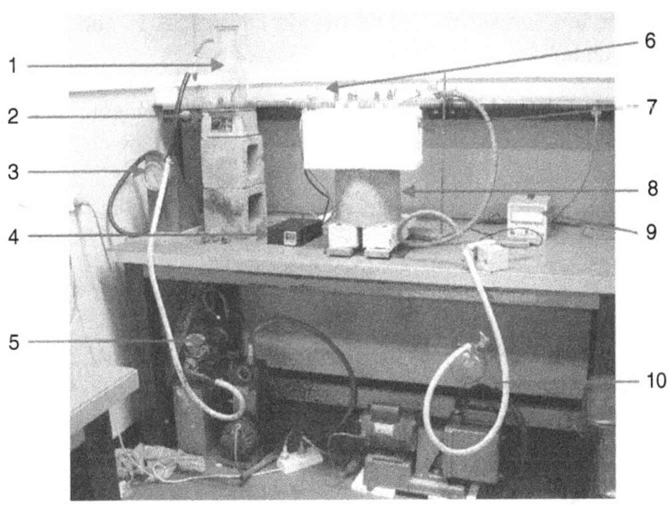

Figure 3A.14 Paper moisture conditioning setup: 1, oil container; 2, heater and stirrer for oil; 3, pressure gauge for oil; 4, temperature controller for the conditioning vessel; 5, vacuum pump for oil; 6, thermocouple probe for conditioning vessel temperature control; 7, conditioning vessel with heat-reserving jacket; 8, heating block for conditioning vessel; 9, pressure gauge for conditioning vessel; 10, vacuum pump for conditioning vessel. *Source:* Ref. [10], Figure 8.

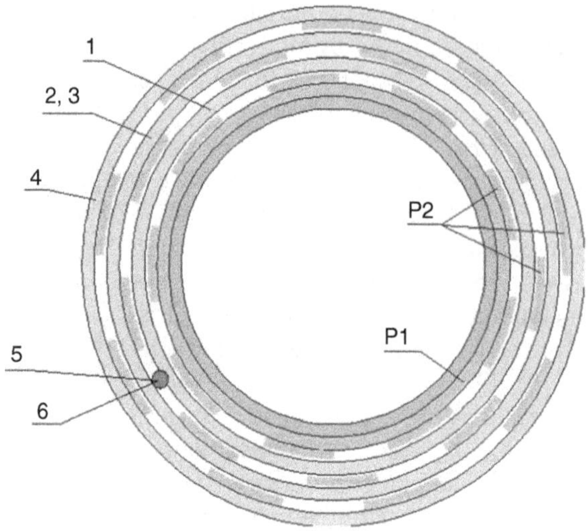

Figure 3A.15 Internal winding arrangement of the "pancake" model.

the aged samples and oil were transferred into the test cells for dielectric response measurements. The moisture contents of paper and oil after ageing were measured by the Karl Fischer titration method.

In Refs [11–15] a test object modeling power transformer insulation systems was designed, constructed, and examined. The design and construction of the test object was published in detail in Refs [11–14]. The winding arrangement and constructional details of the test transformer are shown in Figures 3A.15 and 3A.16. The model can be characterized as follows:

- A metal tank with bushings to connect to the flat "pancake" electrodes, between which different insulation configurations could be sandwiched. The electrodes consisted of shell-type coils.
- Each insulation configuration between the electrodes consisted of flat plates of pressboard ("Kraft Thermo 70") with or without oil gaps between them. If oil gaps were present, the pressboard plates were fixed by means of pressboard spacers.

The insulation system of the model consists of three windings insulated with oil and cellulose, as in a real power transformer, with geometric properties as given in Table 3A.1. A temperature sensor of type Pt 100 was inserted into the tank to measure the actual temperature inside the transformer tank. The amount of solid insulation in the model is approximately 1445 g and of oil 8400 g; the ratio of oil/solid insulation is 5.8. The ratio of oil/cellulose material is about 10:1 to 6:1 in a real transformer.

A photograph of the "pancake" model is shown in Figure 3A.17.

A number of other laboratory test cells were manufactured by other researchers to perform accelerated ageing experiments. See Figures 3A.18–3A.20.

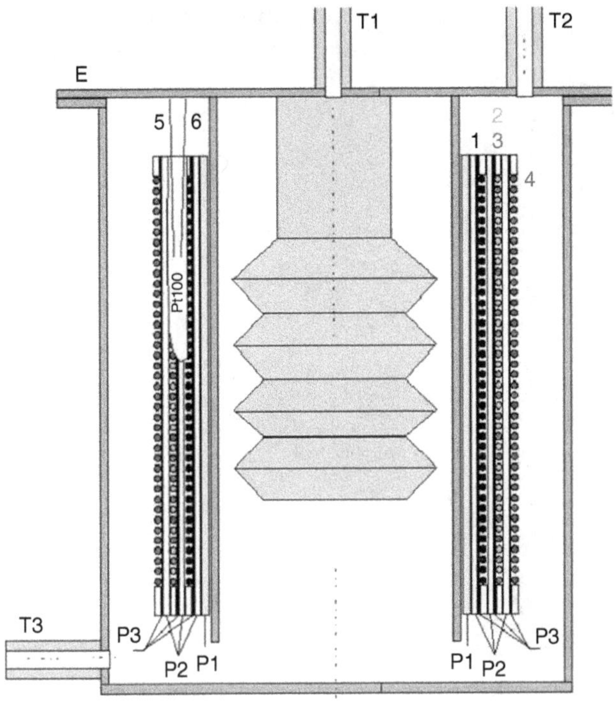

Figure 3A.16 Internal construction details of the "pancake" model.

Table 3A.1 Geometric details of test transformer model

Label	Description
1	Inner coil, made from copper conductor with diameter of 1.5 mm insulated with two layers of paper, with 146 turns in one layer.
2, 3	Middle coil bifilar (double) wound, made from copper conductor with diameter of 1.5 mm insulated with two layers of paper, with 73 turns in one layer.
4	Outer coil, made from copper conductor with diameter of 1.5 mm insulated with two layers of paper, with 146 turns in one layer.
T1	"Breathing" tap (only for inner equalizing).
T2	Oil fill tap.
T3	Oil drain tap.
P1	Presspan tube of 3 mm thickness, support for the whole windings (inner/outer diameter 135/141 mm, length 300 mm).
P2	18 pieces of presspan "spacer" band of 2 mm thickness/10 mm width/300 length on presspan holding band of 1 mm forming oil duct.
P3	Presspan band closing end of winding, upper part of 25 mm, lower part of 12 mm, 1.7 mm height.
E	Grounding point on the tank.

Figure 3A.17 "Pancake" model.

3A.4.2 Effects of Paper/Pressboard Moisture Content on PDC

Figure 3A.21 shows the results of PDC measurements performed by Zaengl [8] with new 2 mm samples of impregnated pressboard (Weidmann Transformerboard Type T IV, impregnated with Technol US3000 mineral oil) with 0.2%, 1%, 2.5%, and 4% moisture contents. These polarization and depolarization currents were measured by applying a charging voltage of 200 V for a very long time. The magnitude and nature of these currents are found to be very sensitive to moisture content.

Der Houhanessian and Zaengl [19] have demonstrated that different pressboard samples, even with the same moisture content, can have different dielectric responses. Figure 3A.22 displays the variations of relaxation currents for two different pressboard samples, each with the same moisture content of 2.5% and 5.0%. It is important to note here that the samples of the same moisture content have been prepared at the same time and have been stored together for several months in hermetically closed vessels.

From Figure 3A.22 it can be observed that the dielectric response of several samples of new pressboard is not unique and may show differences in dependence on local chemical composition and local physical properties, such as specific density, fiber orientation, or surface roughness. The manner of preparation of new samples, the procedure to set the moisture content, and particulars of the impregnation process could also change the time dependence and amplitude of relaxation currents. Frimpong *et al.* [20] reported laboratory experimental results demonstrating variations in relaxation currents with paper condition. It has previously been pointed out [20] that the final parts of polarization and

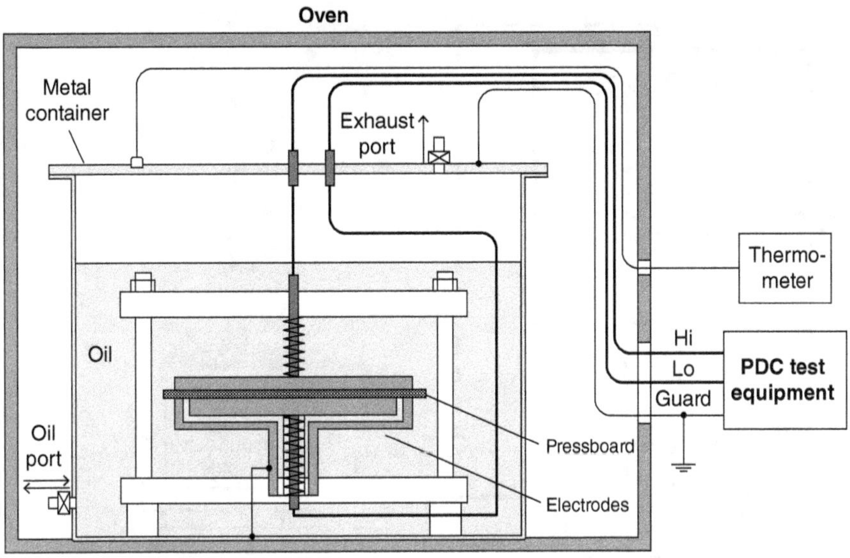

Figure 3A.18 Structure schematic of ageing test device. *Source:* Ref. [16], Figure 5.

Figure 3A.19 Measurement cell in climatic chamber. *Source:* Ref. [17], Figure 3.

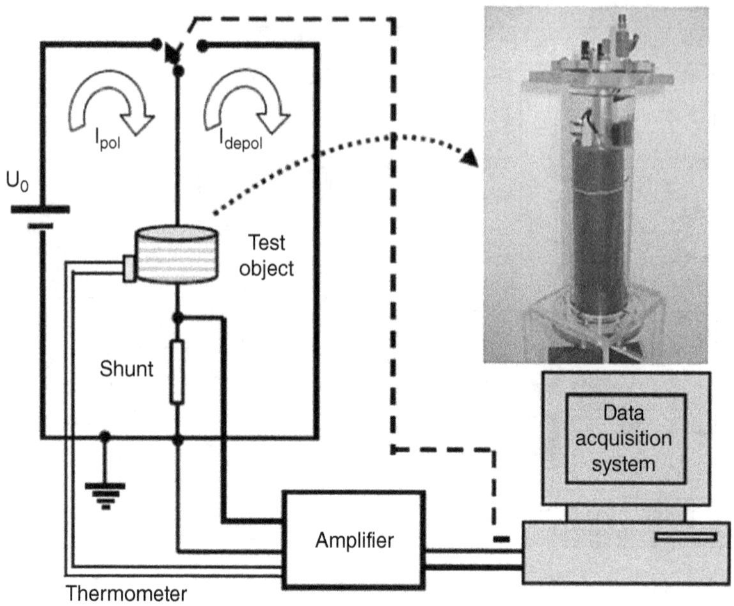

Figure 3A.20 Test arrangement for the "PDC" measuring technique. *Source:* Ref. [18], Figure 3.

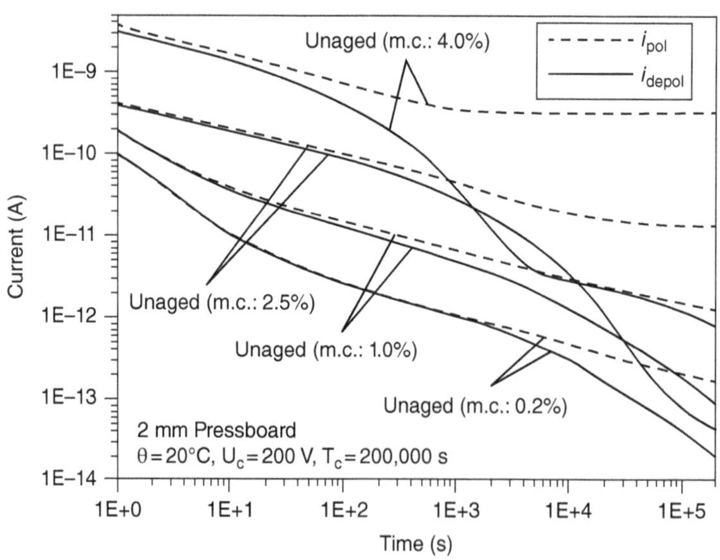

Figure 3A.21 Polarization and depolarization currents of unaged oil-impregnated pressboard samples with different moisture content (m.c.). *Source:* Ref. [8], Figure 4.

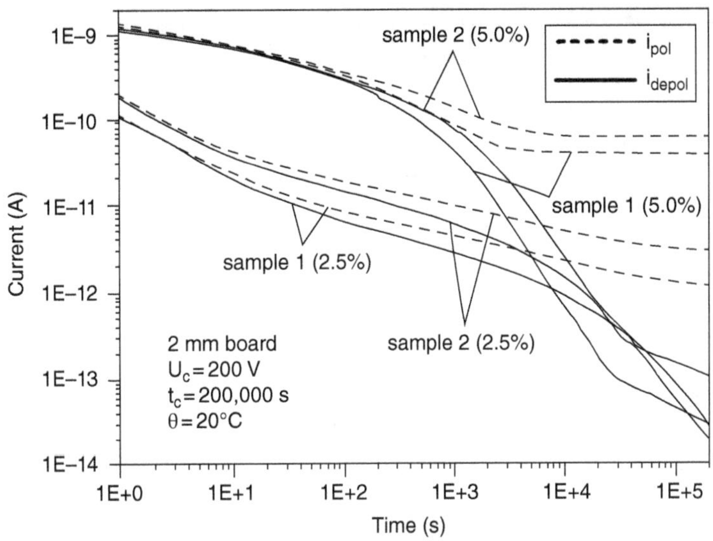

Figure 3A.22 Scatter in relaxation current measurements for different pressboard samples with same moisture content. *Source:* Ref. [19], Figure 4.

depolarization currents are controlled by the conductivity of the solid insulation. Figure 3A.21 shows the effect of change of oil conductivity on the shape of the polarization and depolarization currents. The initial current is changed due to the change in conductivity of the oil. Figure 3A.22 shows the change in polarization current due to a change in moisture. Figure 3A.23 shows the variation of depolarization current due to paper conductivity.

Figure 3A.24 shows the variation of depolarization current with oil conductivity. In general, moisture and contamination tend to increase the conductivity of both the solid and liquid dielectric in the transformer. As shown here, changes in paper and oil conductivity affect both the polarization and depolarization currents. Figure 3A.25 shows the variation of polarization current with board moisture. As shown in Figures 3A.24, 3A.26, and 3A.27, changes in paper and oil conductivity affect both the polarization and depolarization currents.

Experimental results were reported in Ref. [21], where PDC measurements were performed on paper samples with different moisture contents, but with the same oil sample. Figures 3A.28 and 3A.29 show polarization and depolarization currents with the following test samples:

- A – reference, dry and unaged paper sample
- B2 – 3% paper moisture content
- C2 – 3.3% paper moisture content
- D2 – 3.2% paper moisture content.

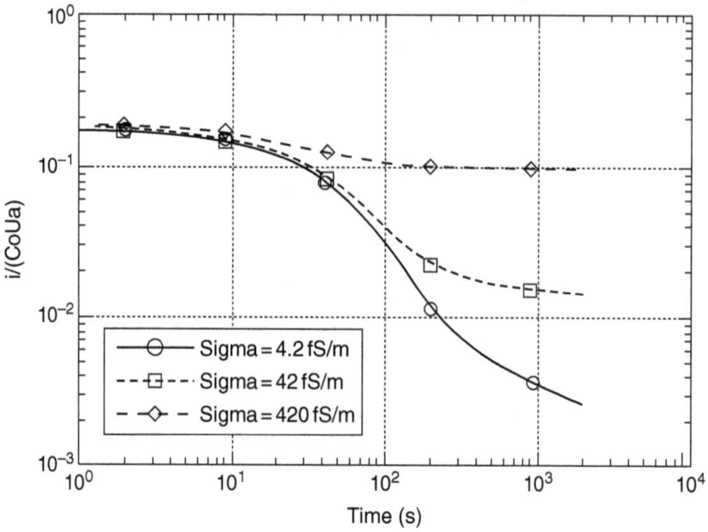

Figure 3A.23 Variation of polarization currents with paper conductivity. *Source:* Ref. [20], Figure 4.

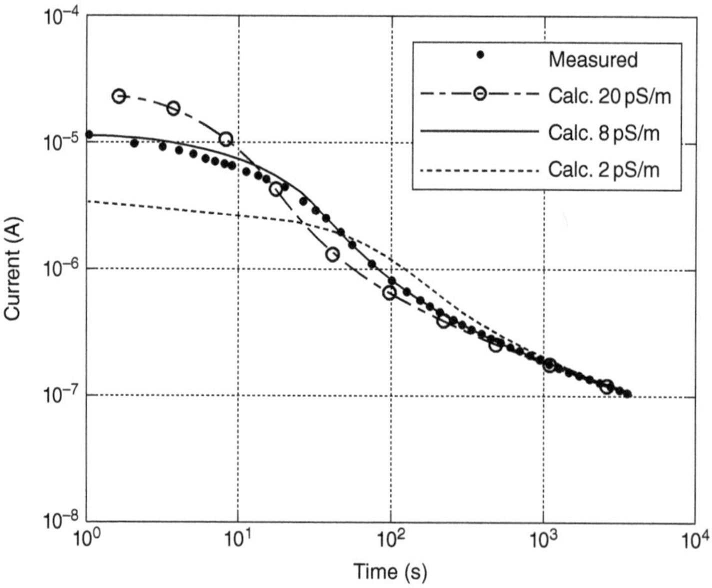

Figure 3A.24 Variation of polarization currents with oil conductivity. *Source:* Ref. [2], Figure 8.

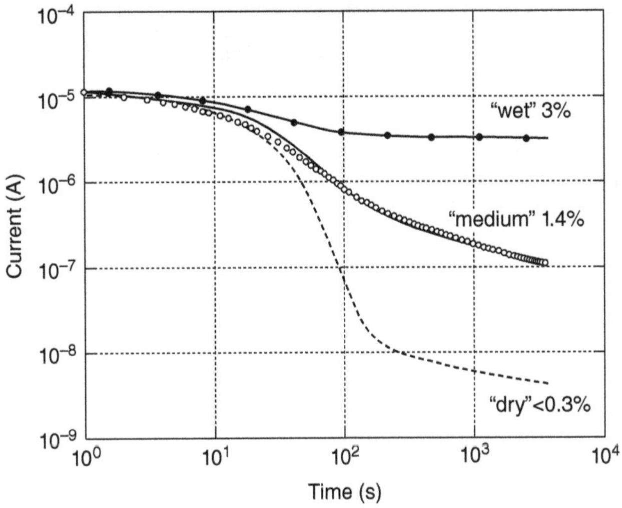

Figure 3A.25 Influence of board moisture on the polarization current. *Source:* Ref. [2], Figure 9.

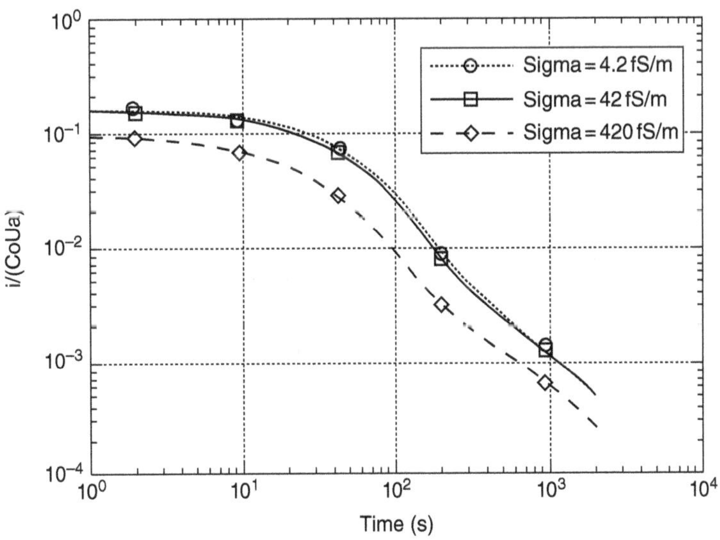

Figure 3A.26 Variation of depolarization currents with paper conductivity. *Source:* Ref. [20], Figure 5.

As seen in Figure 3A.28, the polarization currents for aged and moist paper samples B2, C2, and D2 are higher than the dry and unaged reference sample A. The relative positions of the polarization currents for B2, C2, and D2 are dependent on both the moisture and ageing. The higher the ageing and/or moisture, the higher will be the value of the polarization current. Since the oil condition is the same in all three cases, as suggested by other authors [2, 20], the initial

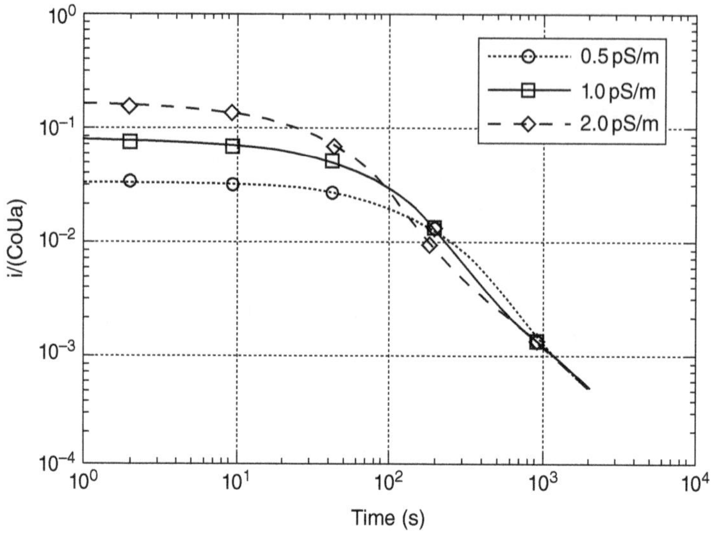

Figure 3A.27 Variation of depolarization currents with oil conductivity. *Source:* Ref. [20], Figure 8.

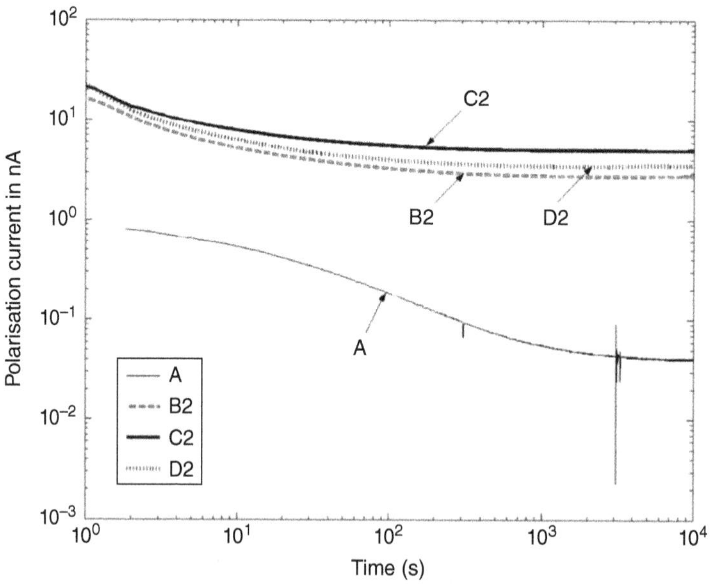

Figure 3A.28 Polarization currents for different paper ageing and moisture but with the same oil. *Source:* Ref. [21], Figure 3.

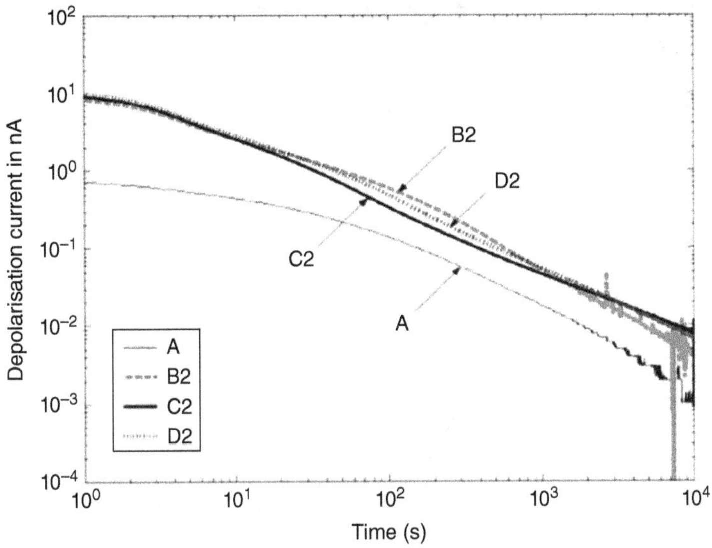

Figure 3A.29 Depolarization currents for different paper ageing and moisture but with the same oil. *Source:* Ref. [21], Figure 4.

values of the currents corresponding to B2, C2, and D2 are very close to each other. The difference is mainly after 100 s, when the effect of paper ageing and/or moisture shows up in the values of the currents. C2, having the highest amount of moisture in the paper, lies on top. B2, corresponding to the unaged paper sample, lies at the bottom of the three. The D2 sample, though more aged than B2 and C2, lies between B2 and C2 since its paper moisture content is lower than that of C2. The differences in the current values are, however, not too much since all these samples B2, C2, and D2 have paper moisture content values quite close to each other.

As expected, the depolarization currents (Figure 3A.29) for the aged and/or moist samples B2, C2, and D2 lie above the reference sample A. The difference between the depolarization currents for B2, C2, and D2 are, however, not significant since their moisture contents are quite close. Only a very close look at Figure 3A.29 will reveal the difference between the depolarization currents of B2, C2, and D2 at the final stage.

Figures 3A.30 and 3A.31 [17] show the polarization and depolarization currents under the influence of moisture in pressboard at an insulation temperature of 23°C. A clear dependence on moisture content is visible; the amplitude increases with moisture and the currents reach their final value faster. Depolarization currents show the same initial amplitude as polarization currents. They decrease faster at samples with a higher moisture content.

Saha and Purkait [1] reported the same from their simulation as well. Figures 3A.32 and 3A.33 show the effect of change of paper conductivity on the shape of the polarization and depolarization currents.

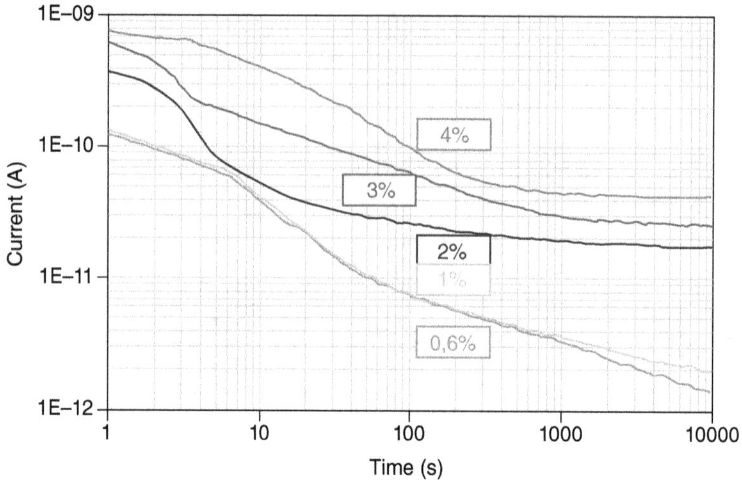

Figure 3A.30 Polarization currents under the influence of moisture. *Source:* Ref. [17], Figure 5.

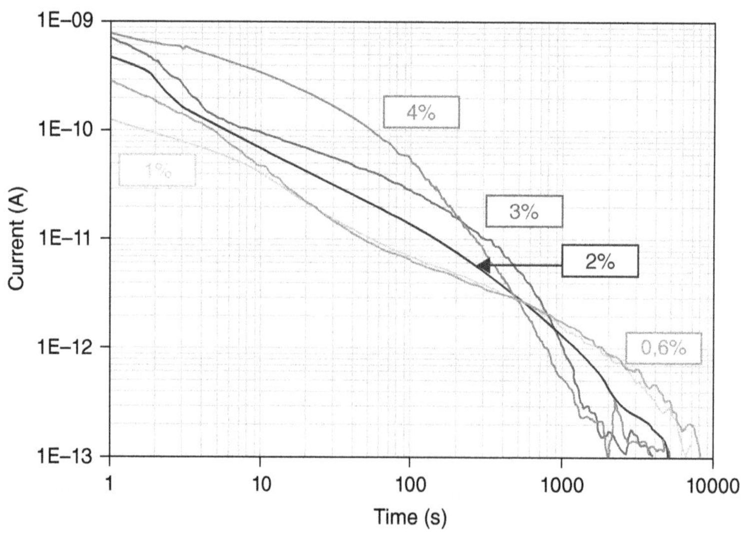

Figure 3A.31 Depolarization currents under the influence of moisture. *Source:* Ref. [17], Figure 6.

3A.4.3 Effects of Oil Moisture Content on PDC

Der Houhanessian and Zaengl [19] performed many measurements on multilayer arrangements (pressboard in series with oil gaps, a configuration relevant for power transformers) that have demonstrated the dielectric response in the

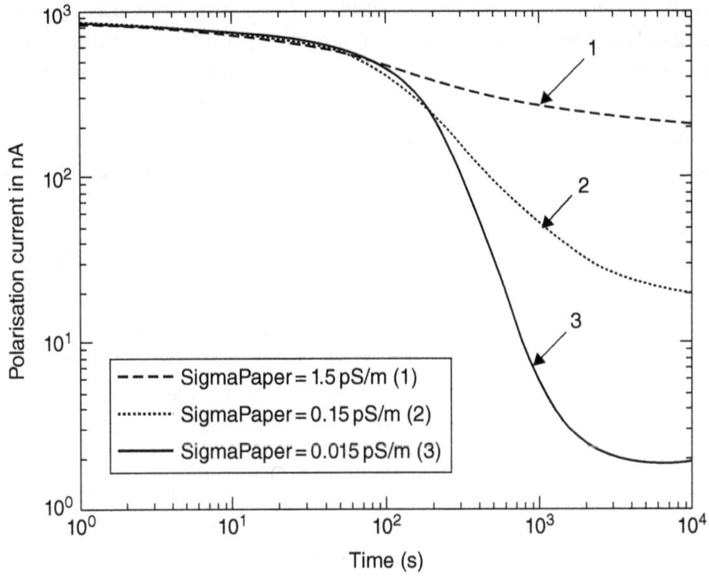

Figure 3A.32 Variation of polarization currents with paper conductivity: (1) $\sigma_{paper} = 1.5$ pS/m; (2) $\sigma_{paper} = 0.15$ pS/m; (3) $\sigma_{paper} = 0.015$ pS/m. *Source:* Ref. [1], Figure 5.

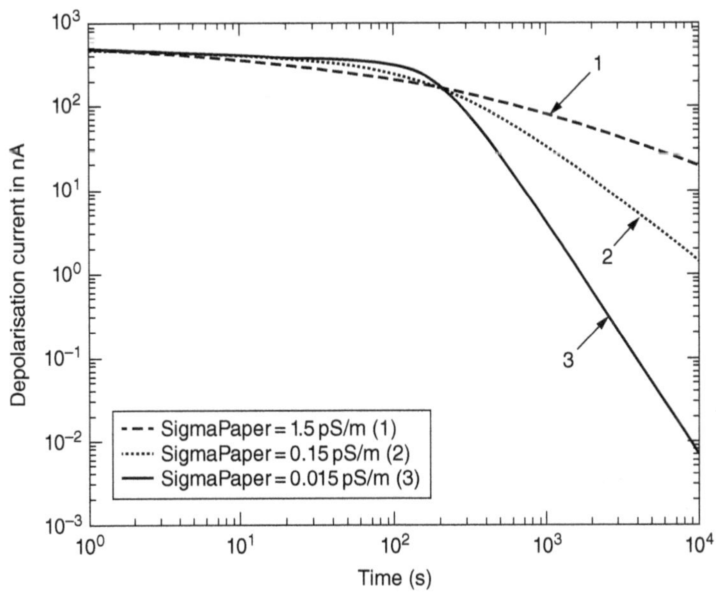

Figure 3A.33 Variation of depolarization currents with paper conductivity: (1) $\sigma_{paper} = 1.5$ pS/m; (2) $\sigma_{paper} = 0.15$ pS/m; (3) $\sigma_{paper} = 0.015$ pS/m. *Source:* Ref. [1], Figure 6.

time domain is dependent on the material condition of each component. The following results of relaxation current measurements are taken from samples consisting of a three-layer arrangement, in which two 2 mm pressboard disks are placed on the surface of parallel-plate electrodes and are separated by a central oil duct of variable thickness. The current measurements are performed with a voltage to electrode spacing ratio of 3 V/mm and the charging times have been limited to 5000 s.

Figure 3A.34 shows the results of these measurements. The relaxation currents (a) to (c) are measured with an unaged system characterized by a board moisture content of 0.2%. For system (a), the oil moisture content was 5 ppm and the oil conductivity was quantified to 0.5 pS/m. The relaxation currents for system (b) have been measured after replacing the oil of system (a), decreasing the oil conductivity from 0.5 pS/m to 0.3 pS/m. Relaxation currents for system (c) belong to an aged system characterized by a board moisture content of 2.0%, an oil moisture content of 19 ppm, and an oil conductivity of 0.1 pS/m. The reason for the low conductivity of this aged oil is assumed to be the "purification" phenomenon caused by the preceding relaxation current measurements which had long charging times. These measurements show clearly that the predominantly exponential decay of relaxation currents, which is due to the interfacial polarization, is essentially controlled by the change of oil conductivity. The dielectric response of the board appears mainly after long times.

To investigate the influence of the oil and its conductivity on the different responses, Zaengl [14] reported experimental results on oil–paper multi-layer insulation samples with two different oils. Firstly with new insulation oil

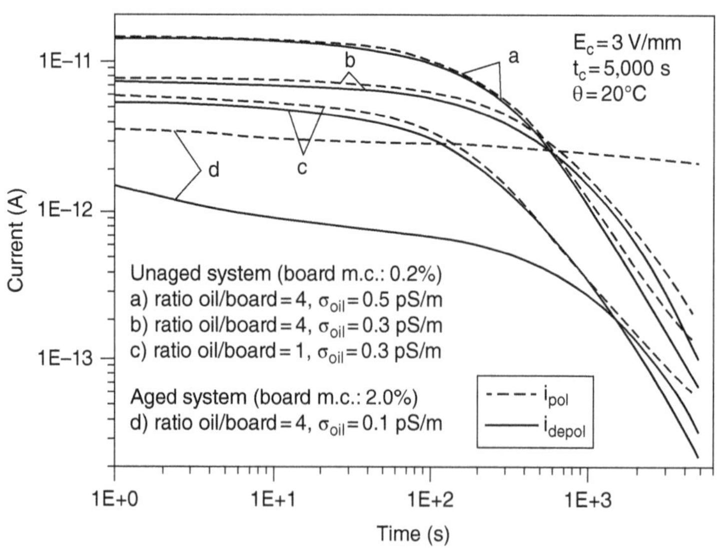

Figure 3A.34 Relaxation currents of different multi-layer samples: (a) $\sigma_{oil} = 0.5$ pS/m, (b, c) $\sigma_{oil} = 0.3$ pS/m, (d) $\sigma_{oil} = 0.1$ pS/m. *Source:* Ref. [19], Figure 5.

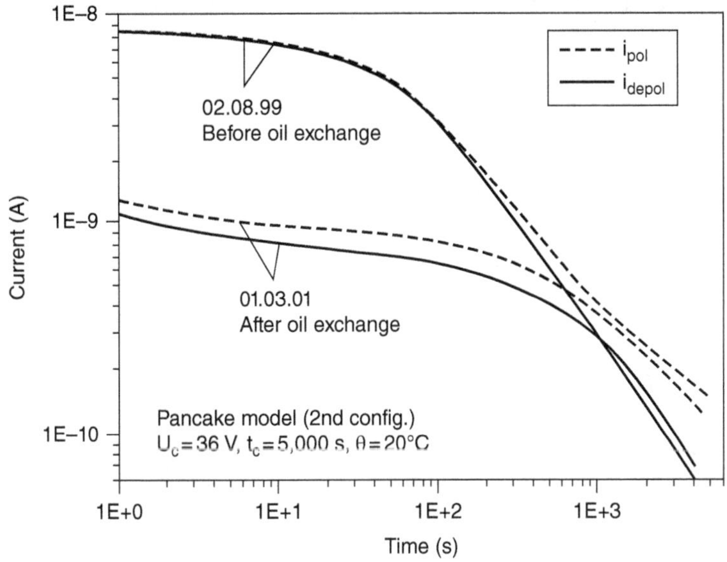

Figure 3A.35 Measured relaxation currents before and after oil exchange. *Source:* Ref. [14], Figure 4.

(Nynas Nytro 10GBN), the conductivity σ of which was about 2 pS/m (20°C), then replaced with oil (Shell Diala D) having a significantly lower conductivity $\sigma = 0.24$ pS/m (21°C). Figure 3A.35 shows the measured polarization and depolarization currents for the two oils.

The initial amplitudes of the currents dropped by a factor of about nine, and the duration of the nearly exponential decay of the currents was significantly prolonged. The decrease in the initial current amplitudes and the increase in the time constant dominating the macroscopic interfacial polarization is thus caused solely by the decrease in oil conductivity. The result of the PDC measurement of Figure 3A.35 confirms that the initial part of the currents is influenced by the conductivity of the oil.

Figures 3A.36 and 3A.37 show investigation results of the effect of the variation of oil conductivity on the polarization and depolarization currents.

In general, moisture and contamination tend to increase the conductivity of both the solid and liquid dielectric in the transformer. Figures 3A.36 and 3A.37 once again confirm the fact that the initial part of the dielectric responses is primarily controlled by the parameters of the liquid dielectric. It is also observed that higher values of conductivity tend to increase the magnitude of both the polarization and depolarization currents.

Saha and Purkait [22] presented experimental investigation reports with two oil–paper samples B1 and B2 that have papers of the same ageing condition (no ageing) and whose paper moisture values are also quite close. The oils in those samples are, however, of different ageing and moisture conditions. The polarization and depolarization currents for the set of samples B1 and B2 are plotted in

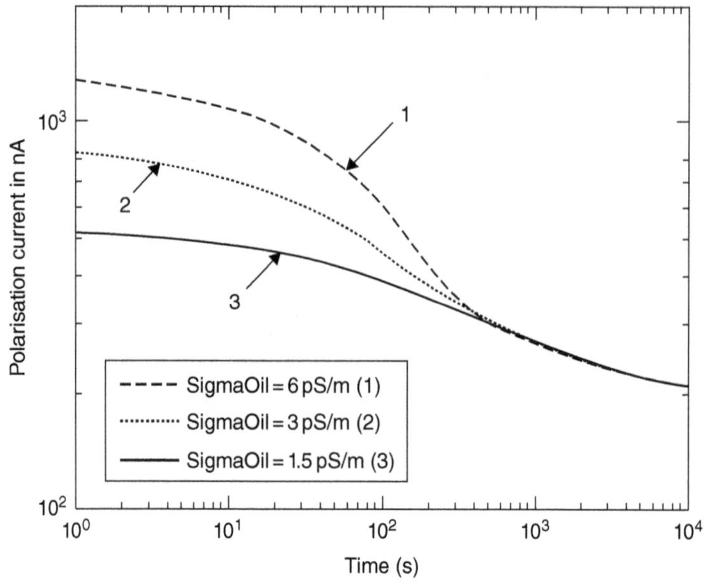

Figure 3A.36 Variation of polarization currents with oil conductivity. *Source:* Ref. [1], Figure 7.

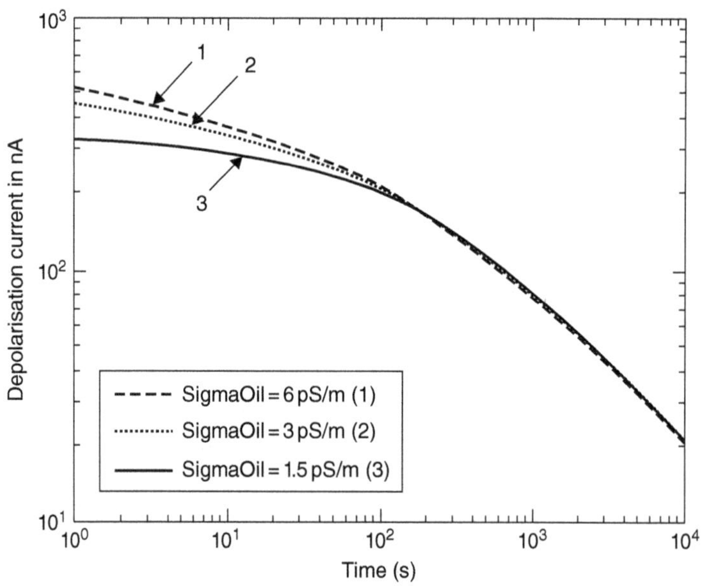

Figure 3A.37 Variation of depolarization currents with oil conductivity. *Source:* Ref. [1], Figure 8.

Figures 3A.38 and 3A.39, respectively. The currents for a reference dry sample with new oil (sample A) are also included for comparison.

As seen in these figures, sample A having the least value of oil ageing and moisture lies at the bottom, followed by samples B1 and B2, respectively. Magnitudes

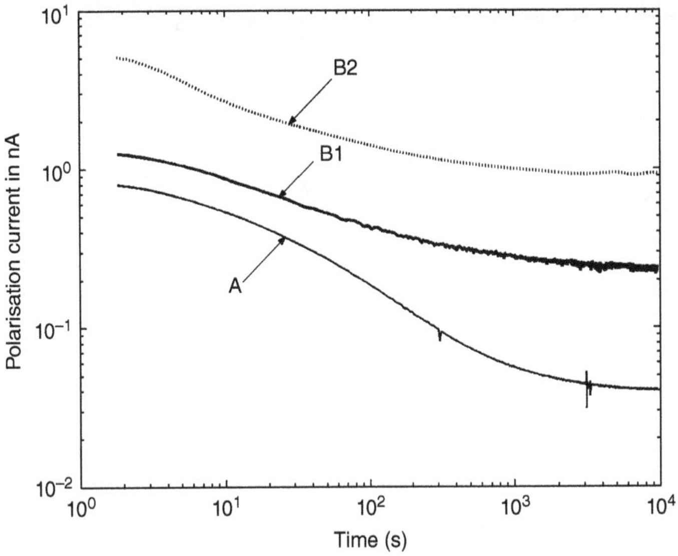

Figure 3A.38 Variation of polarization currents with oil condition. *Source:* Ref. [22], Figure 3.

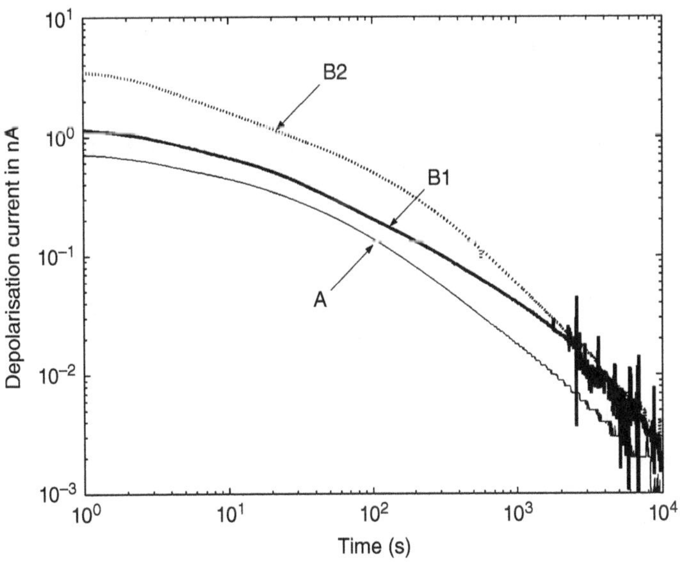

Figure 3A.39 Variation of depolarization currents with oil condition. *Source:* Ref. [22], Figure 4.

of these currents are related to the oil/paper moisture contents. Higher the moisture content, higher will be the current levels.

Similar experiments were performed with two aged paper samples having the same ageing condition and very close moisture contents. The oil samples used

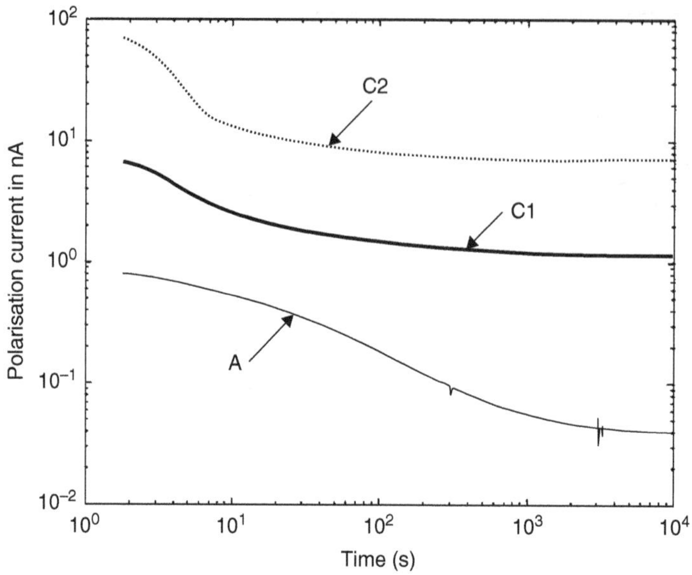

Figure 3A.40 Polarization currents for different oil ageing and moisture. *Source:* Ref. [22], Figure 6.

in those two cases, however, had different moisture contents of 25 ppm (sample C1) and 31 ppm (sample C2), respectively. Similar plots for samples C1 and C2 are shown in Figures 3A.40 and 3A.41, respectively.

Similar conclusions can be drawn for samples C1 and C2.

Oil conductivity values obtained from the initial values of the polarization and depolarization currents using the method described in Ref. [1] are presented in Table 3A.2.

As seen from Table 3A.2, B2 has more severe oil ageing and moisture contents among A, B1, and B2. Similarly, C2 has worse condition of oil ageing and moisture condition compared with C1 and A. PDC measurement results and their analyses was thus found to provide salient information about the conditions of oil [22].

3A.4.4 Effects of Ageing and Related Impurities on PDC

Der Houhanessian and Zaengl [23] have performed PDC measurements on oil-impregnated pressboard samples (2 mm thick; Weidmann Transformerboard Type TIV, impregnated with Technol US 3000 mineral oil). The measurements were done in an oil-filled glass vessel in a grounded metallic oven to control temperature and other environmental conditions. The investigation considered aged board and virgin board with a moisture content of 1% or 0.2%, respectively, impregnated with degassed and dry oil with a moisture content of 5 ppm. They performed ageing for 90 days in a hermetically closed stainless steel vessel at a constant temperature of 120°C (weight ratio of oil to pressboard: 8.4). The ageing process increased the moisture content of the pressboard from 0.2% to 0.5% for virgin boards and from 1.0% to 2.0% for moist boards. The moisture content of the

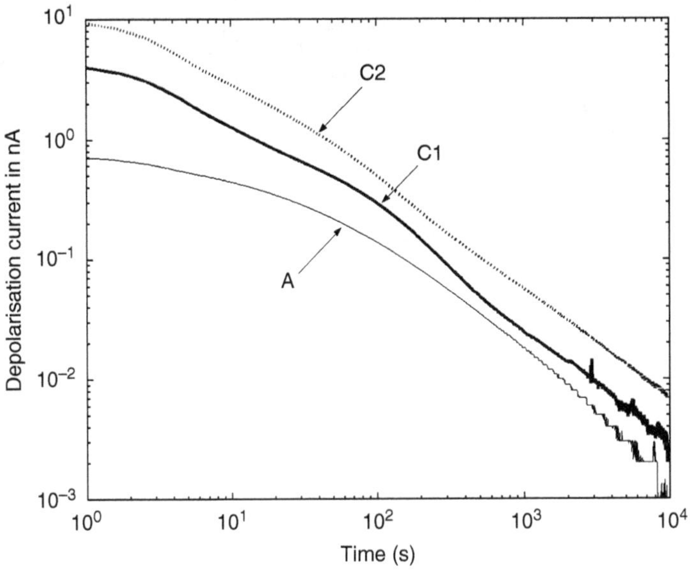

Figure 3A.41 Depolarization currents for different oil ageing and moisture. *Source:* Ref. [22], Figure 7.

Table 3A.2 Oil conductivity [22]

Sample	Oil conductivity (pS/m)
A	0.165
B1	1.17
B2	11.63
C1	3.2
C2	12.38

oil also increased to 9 ppm and 19 ppm, respectively. Moisture measurements were performed by Karl Fischer titration. The DP value during ageing changed from an initial value of 950 to 479 and 195, respectively. PDC measurements were performed at 20–21°C.

Figures 3A.42 and 3A.43 show the polarization (i_{pol}) and depolarization (i_{depol}) currents, where the charging voltage was 2.0 kV DC and charging and discharging durations were 200,000 s for the impregnated board with low (Figure 3A.42) and higher (Figure 3A.43) moisture content, before and after ageing conditions, respectively. The relaxation currents of the pressboard with low moisture content are obviously less affected by the ageing process than those with high moisture content. The results demonstrate the sensitivity of relaxation currents to the change of material properties.

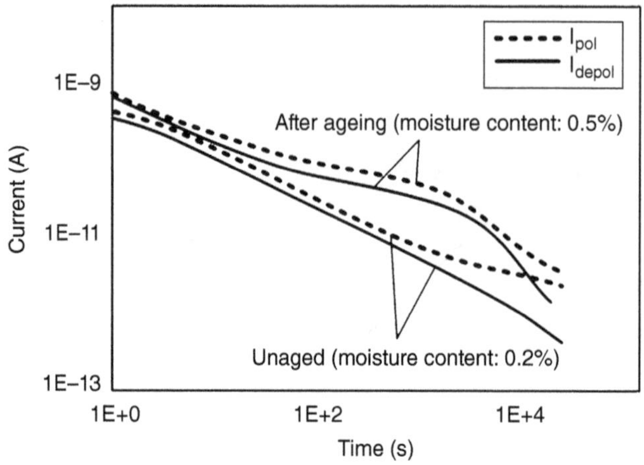

Figure 3A.42 Relaxation currents of a pressboard sample with an initial moisture content of 0.2%, before and after ageing. *Source:* Ref. [23], Figure 2.

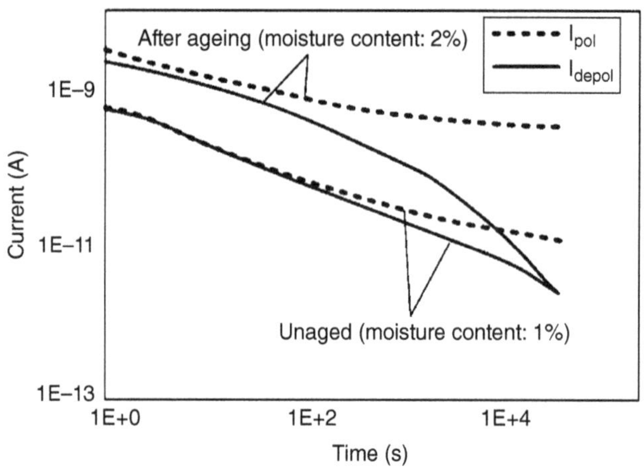

Figure 3A.43 Relaxation currents of a pressboard sample with an initial moisture content of 1.0%, before and after ageing. *Source:* Ref. [23], Figure 3.

Noirhommel *et al.* [24] presented experimental results of transformer models built around a 100 kVA distribution transformer tank and core using Kraft paper-wrapped copper wire as high-voltage and low-voltage windings, while Voltesso 35 was used as insulating oil. These transformers were aged at 135°C for different periods of time. Dielectric measurements were taken at different stages of ageing at a constant temperature of 25°C. Figures 3A.44 and 3A.45 show the variation of polarization currents at different ageing conditions. It is found that the polarization current occupies a higher position in case the insulation is more aged.

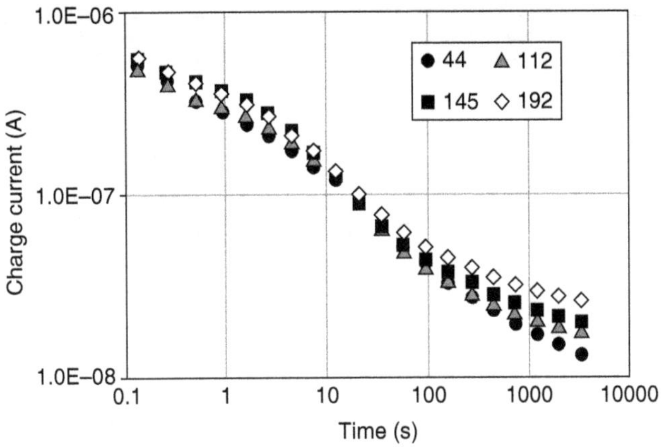

Figure 3A.44 Polarization current at different ageing conditions (with 1.2% moisture content). *Source:* Ref. [24], Figure 2(a).

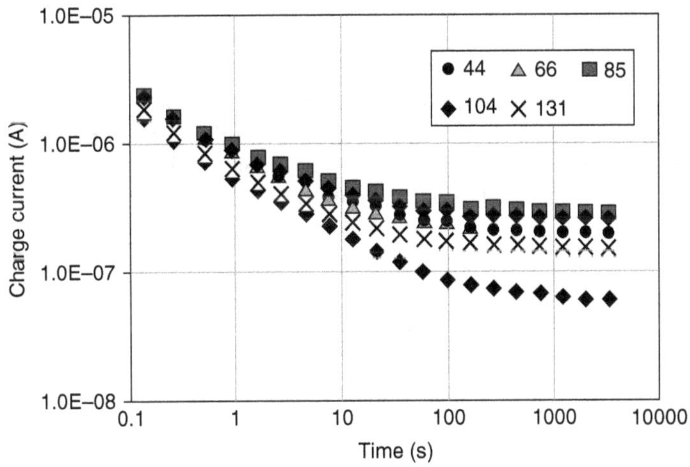

Figure 3A.45 Polarization current at different ageing conditions (with 3.48% moisture content). *Source:* Ref. [24], Figure 3(a).

Following the work of Koch *et al.* [17], Figure 3A.46 compares the PDC of new and aged pressboard. The initial and final current magnitudes are higher for aged materials compared with new materials. This variation could be due to ageing by-products in cellulose, which might have increased the conductivity in addition to an increase in moisture.

3A.4.5 Effects of Temperature on PDC

PDC results are strongly influenced by operating temperature inside a transformer. This is due to the fact that dielectric behavior of the insulation system

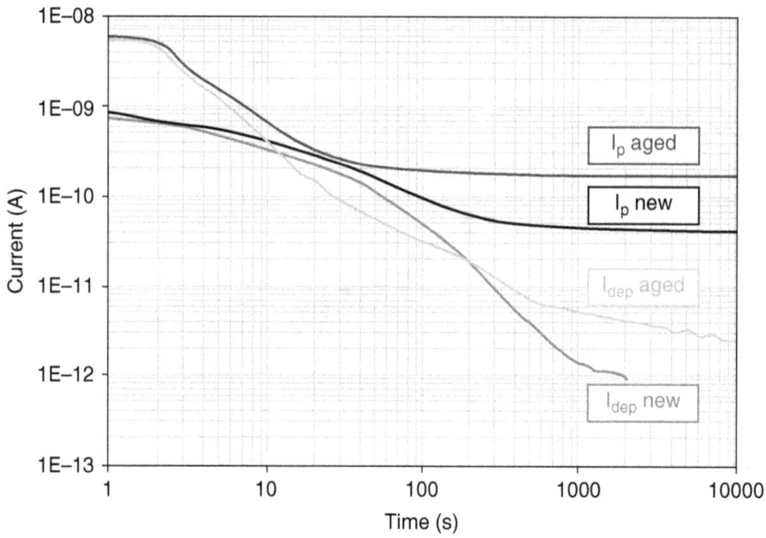

Figure 3A.46 Comparison of PDC for new and aged pressboard with 4% moisture content. *Source:* Ref. [17], Figure 4.

is influenced by moisture content of oil and paper. In a transformer, dissolved water is distributed between paper and oil, which is dependent on temperature. When the temperature rises, water migrates from paper to oil and vice versa. Hence, any significant change in temperature modifies the relative water content of the oil and paper. It is therefore essential to study the effect of temperature and subsequently the impact of moisture distribution on the dielectric behavior of oil and paper separately. Saha and Purkait [25] studied extensively the variation of PDC measurement results with temperature. This is needed specifically to understand the variations of PDC measurement results with temperature. A number of laboratory experiments were conducted in which PDC measurements were performed on a transformer with controlled variations of temperature.

Der Houhanessian and Zaengl [26] also showed that relaxation currents are significantly influenced by temperature. Figure 3A.47 shows the temperature dependence of relaxation currents for a 2 mm impregnated pressboard sample with a moisture content of 0.2%. The measurement conditions are shown in the figure.

They [26] reported relaxation current measurements on pressboard samples and oil forming a multi-layer arrangement at different temperatures. Figure 3A.48 shows temperature-dependent currents for a composite insulation, built from two 2 mm pressboard disks separated by a central oil duct of 16 mm. The moisture content of the pressboard was 0.2% and that of the oil <5 ppm. The measurements were performed at temperatures of 5°C, 20°C, and 35°C at a DC voltage of 60 V for a charging period of 5000 s. The initial current is typically related to the interfacial polarization. When the temperature is increased, the increase in current is due to the decrease in time constant of interfacial

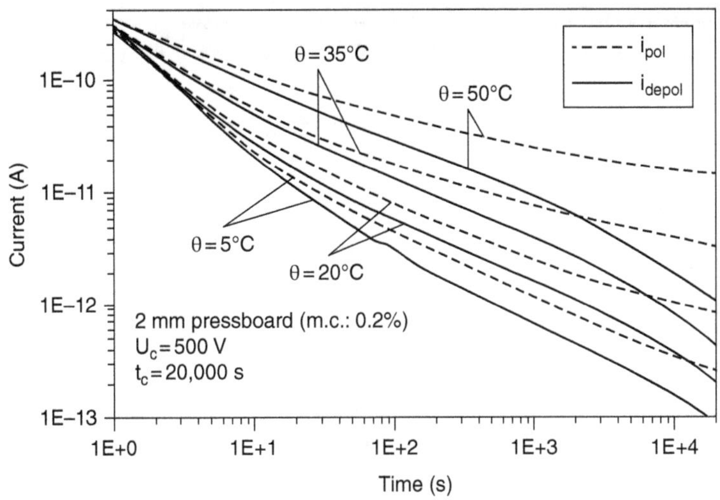

Figure 3A.47 Relaxation currents of 2 mm pressboard with 0.2% moisture content measured at different temperatures. *Source:* Ref. [26], Figure 5.

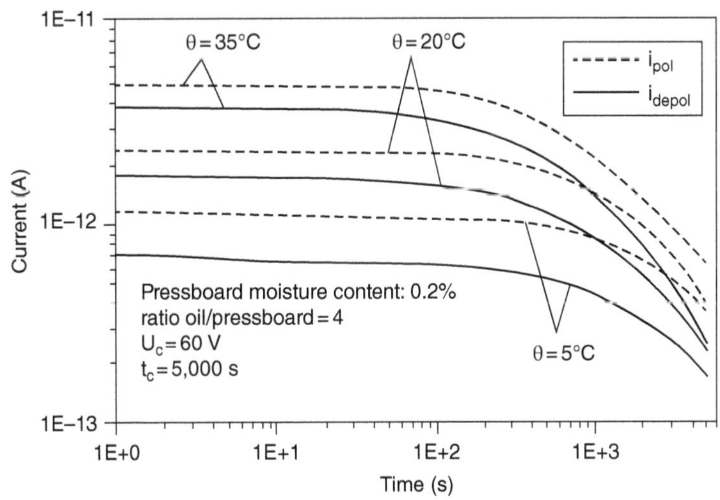

Figure 3A.48 Relaxation currents of a multi-layer sample in dependence on temperature. *Source:* Ref. [26], Figure 7.

polarization. The changes in final current or DC components are due to the shift in dielectric response of the pressboard.

Similar findings were reported by Saha and Purkait [25], where the experiments were conducted on a "pancake" model transformer as shown in Figures 3A.15–3A.17. Figures 3A.49 and 3A.50, respectively, show the polarization and depolarization currents obtained at different temperatures. In each case, the transformer was charged (polarized) with 500 V for 10,000 s and then discharged (depolarized)

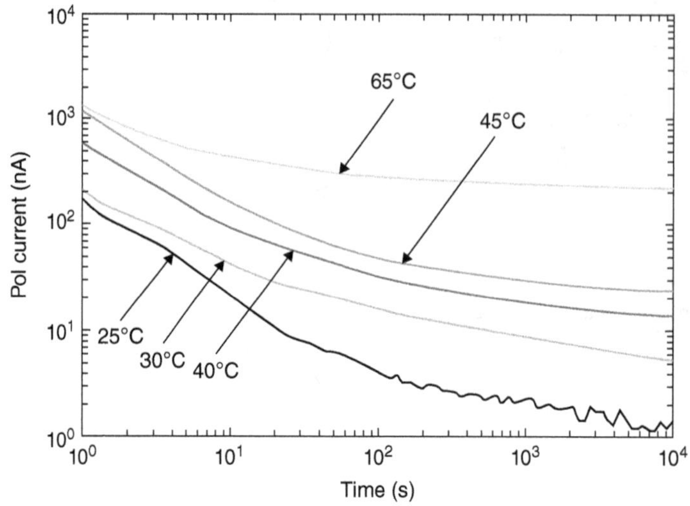

Figure 3A.49 Variation of polarization current with temperature. *Source:* Ref. [25], Figure 9.

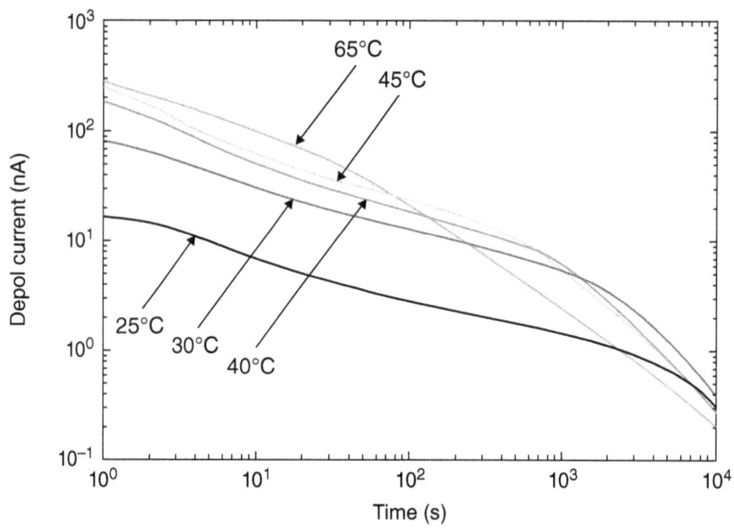

Figure 3A.50 Variation of depolarization current with temperature. *Source:* Ref. [25], Figure 10.

for 10,000 s. It can be seen that the variation in temperature causes significant displacement of both the polarization and depolarization currents. It can also be seen that the magnitude of the polarization and depolarization currents tends to shift to higher values with rising temperature.

Previously, it was shown that

> the conductivity for a given insulation system thus is found to be dependent upon the difference between the polarization and depolarization current

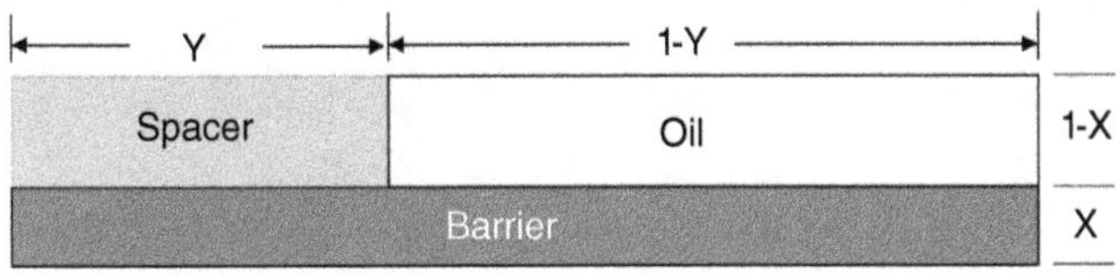

Figure 3A.51 *X–Y* arrangement structure of oil and paper [27].

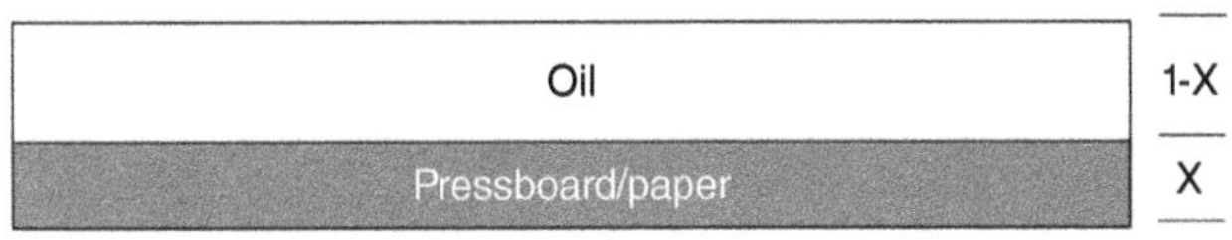

Figure 3A.52 Series arrangement structure of oil and paper [27].

> values. This conductivity, in practice, is the convolution of the conductivities of the oil and the paper that make up the insulation structure. For modeling purposes, it is very common to represent the insulation structure by the relative amount of spacers and barriers in the duct, commonly known as the *X–Y* model, as shown in [Figure 3A.51] [27]. In this model, a parameter *X* is defined as the ratio of the sum of all the thicknesses of all the barriers in the duct, lumped together and divided by the duct width. A spacer coverage *Y* is defined as the total width of all the spacers divided by the total length of the periphery of the duct [27]. However, when sufficient information about insulation is not available for estimation of the oil and paper conductivities separately, it is very often enough to consider the insulation system to be composed of a simple series arrangement of oil duct and paper insulation as shown in [Figure 3A.52], and this does not significantly lose its accuracy. Each material is characterized by its conductivity and permittivity, along with the composite dielectric response function $f(t)$. [25]

The range of X is typically 20–50% and that of Y typically 10–30% [25] for a transformer. The value of X and Y can be calculated more precisely only when the exact structure of the insulation system and all its design parameters are available. For a series structure of oil and paper, model oil/paper conductivity calculations were shown previously in equations (3A.12)–(3A.19).

For the $X–Y$ model (as shown in Figure 3A.51), the composite conductivity may be written in terms of paper and oil conductivities (σ_{paper} and σ_{oil}, respectively) as

$$\sigma_{\text{r}} = \frac{Y}{\dfrac{1-X}{\sigma_{\text{spacer}}} + \dfrac{X}{\sigma_{\text{barrier}}}} + \frac{1-Y}{\dfrac{1-X}{\sigma_{\text{oil}}} + \dfrac{X}{\sigma_{\text{barrier}}}} \tag{3A.20}$$

If we take $\sigma_{\text{spacer}} = \sigma_{\text{barrier}} = \sigma_{\text{paper}}$, then

$$\sigma_r = \frac{Y}{\dfrac{1}{\sigma_{\text{paper}}}} + \frac{1-Y}{\dfrac{1-X}{\sigma_{\text{oil}}} + \dfrac{X}{\sigma_{\text{paper}}}} = \frac{Y}{\dfrac{1}{\sigma_{\text{paper}}}} + \frac{1-Y}{\dfrac{\sigma_{\text{paper}}(1-X) + X\sigma_{\text{oil}}}{\sigma_{\text{oil}}\sigma_{\text{paper}}}} \tag{3A.21}$$

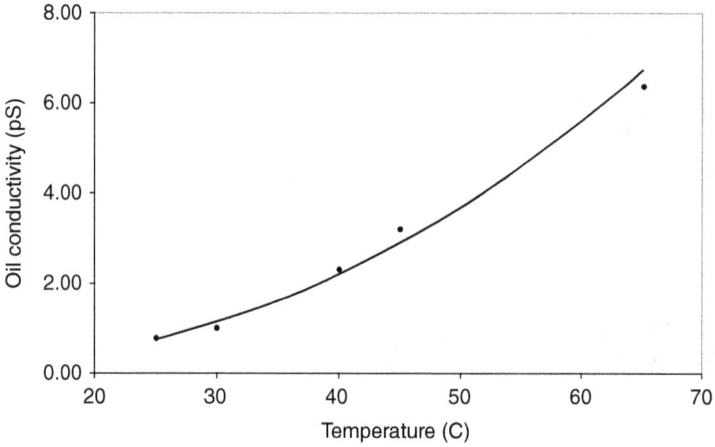

Figure 3A.53 Variation of oil conductivity with temperature. *Source:* Ref. [25], Figure 11.

Considering $\sigma_{\text{oil}} \gg \sigma_{\text{paper}}$

$$\sigma_{\text{r}} \approx Y\sigma_{\text{paper}} + \frac{(1-Y)\sigma_{\text{paper}}}{X}$$

After rearrangement of the above equation:

$$\frac{\sigma_{\text{r}}}{\sigma_{\text{paper}}} \approx \frac{YX + (1-Y)}{X} \tag{3A.22}$$

Figures 3A.53 and 3A.54 show the nature of variation of oil and paper conductivities with temperature (considering the model). Initial currents are considered for the oil conductivity calculation, and hence the initial current will be taken from the measurement conducted at the temperature during the start of

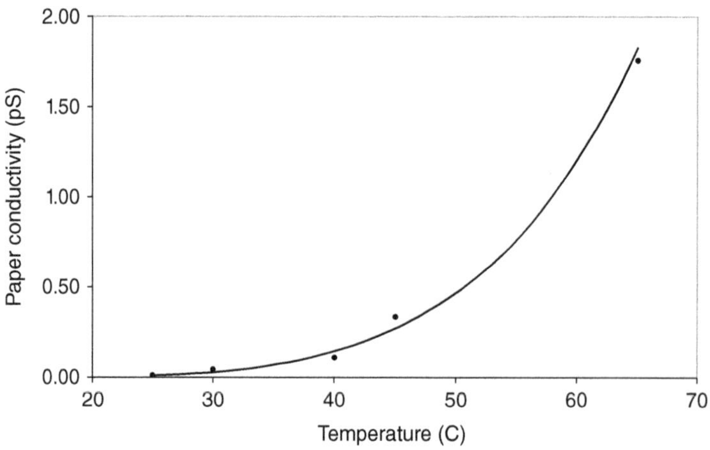

Figure 3A.54 Variation of paper conductivity with temperature. *Source:* Ref. [25], Figure 12.

measurement and corrected accordingly for the ambient temperature (using Figure 3A.53). Similarly, since final currents are considered for the paper conductivity calculation, the final current will be taken from the corresponding measurement at the last temperature and corrected accordingly using Figure 3A.54.

> Both the oil and paper conductivities are found to increase exponentially with temperature. It was also reported by other authors that conductivity follows an exponential law:
>
> $$\sigma(T) \approx A.e^{\left(-\frac{E_{ac}}{kT}\right)} \tag{3A.23}$$
>
> where T is the absolute temperature in Kelvin, A is a constant related to the mobility of ions in the insulation, K is the Boltzmann constant and E_{ac} is the activation energy. Taking the natural logarithm of both sides of [3A.23], it can be shown that the conductivity varies linearly with inverse of the absolute temperature ($1/T$). It is interesting to note that both oil and paper conductivities separately demonstrate linear variations with an inverse of absolute temperature. [25]

Homagk and Leibfried [9] pointed out that PDC measurements require constant insulation temperatures for accurate result interpretation. Figure 3A.55 shows the dependence of the polarization current on temperature. They explain the importance of temperature correction due to the variation of temperature during onsite measurements.

Koch *et al.* [17] also demonstrated the massive influence of temperature on PDC at a moisture content of 4%, as in Figure 3A.56. At higher temperatures,

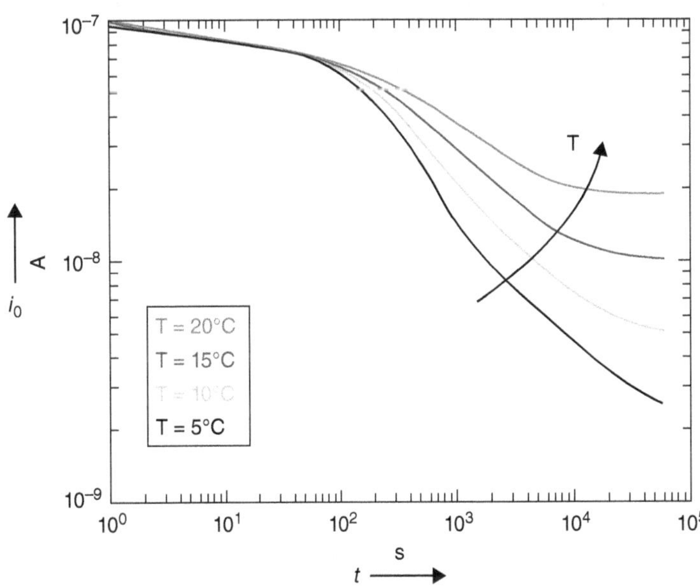

Figure 3A.55 Influence of temperature on polarization current. *Source:* Ref. [9], Figure 4.

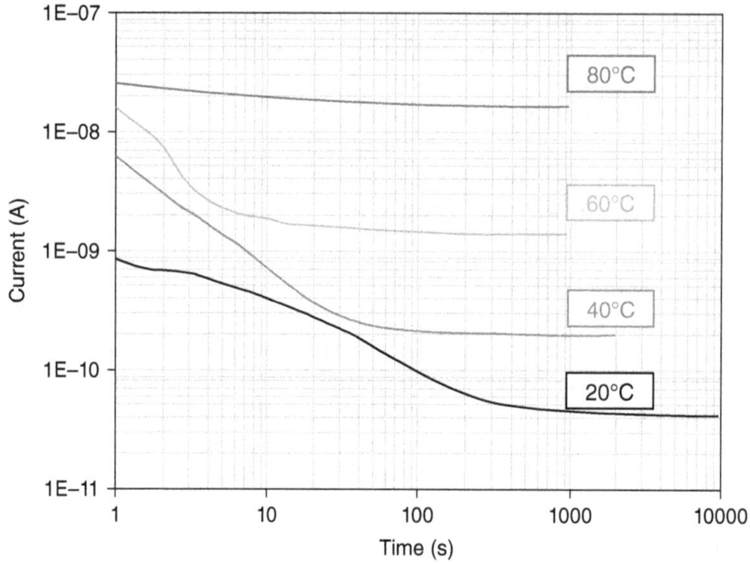

Figure 3A.56 Polarization currents for 4% moisture content at different temperatures. *Source:* Ref. [17], Figure 7.

the polarization processes occur much faster, and the currents are increased, with an increase in conductivity of pressboard and oil.

Because of the strong influence of insulation temperature, reliable temperature compensation is essentially needed to accurately interpret PDC measurement data. Since the influences of moisture and insulation temperature are difficult to distinguish, a reliable temperature measurement is of utmost importance to investigate the variation of temperature on the PDC measurements. Wei *et al.* [16] studied extensively the temperature effect on PDC current measurements. Fofana *et al.* [18] also studied this phenomenon in detail.

3A.4.6 Effects of Charging Voltage on PDC

As described by Der Houhanessian and Zaengl [23], PDC measurement results are also greatly affected by charging voltage magnitude. Figure 3A.57 shows the variation of polarization current for different charging voltages; a charging time of 2000 s was used for these measurements.

However, Saha and Purkait [7] pointed out that during the PDC tests, it is often very convenient to make the assumption of linearity of the dielectric in order to simplify the analysis of the measurement results. However, it remains an experimental fact that very few systems are truly linear in the presence of a high electric field. At higher charging voltages (high field), charge movements may take place on a significant scale that affect the linearity properties. Figure 3A.58 shows the influence of an increase in charging voltage level from 500 V to 1000 V on a 100 kVA transformer. To simplify the comparison, currents obtained with 500 V charging

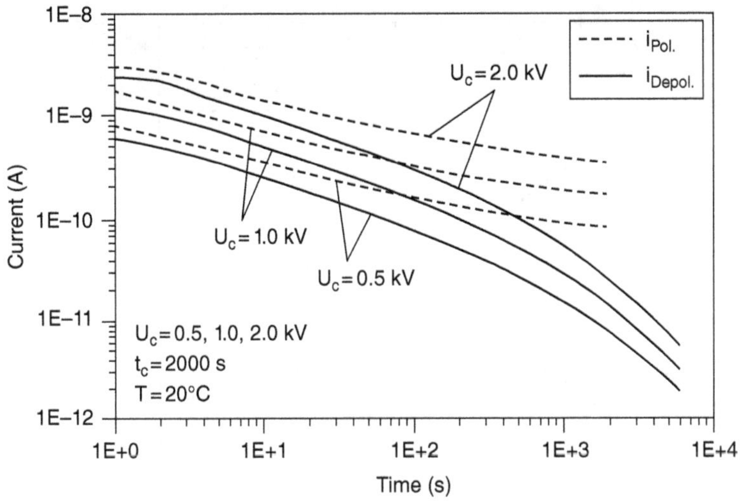

Figure 3A.57 Relaxation currents of an aged pressboard sample as a function of charging voltage.

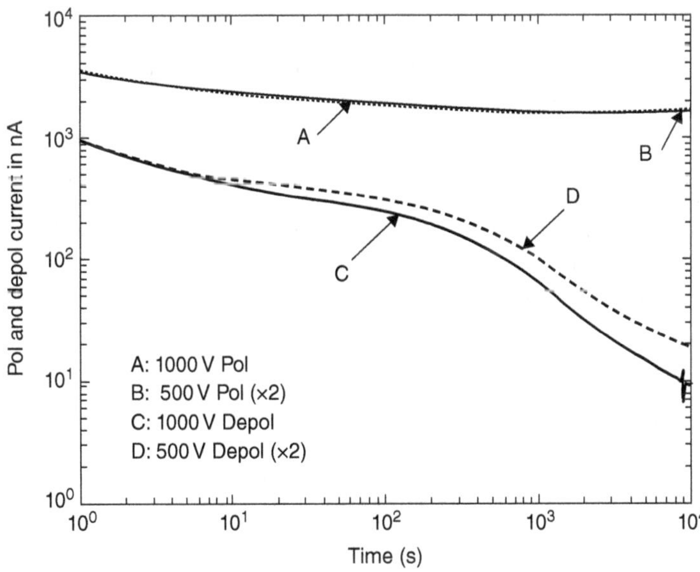

Figure 3A.58 Relaxation currents at different excitation voltages (scaled to the same level for comparison). *Source:* Ref. [7], Figure 5.

voltage are multiplied by 2. The results show that the actual amplitude of the depolarization current for 1000 V is lower.

This nonlinearity is caused by high excitation voltage. This will, in turn, give rise to an error in the computation of the dielectric response function and the

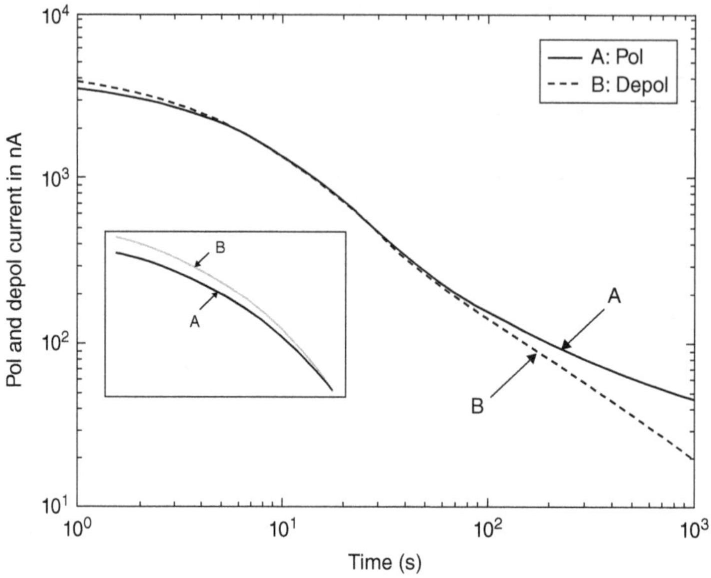

Figure 3A.59 Crossing over of depolarization and polarization currents; inset shows the initial period on an enlarged scale. *Source:* Ref. [7], Figure 6.

conductivities. The polarization currents, however, are close to each other. Similar findings were reported by other researchers as well [19].

According to Jonscher [3], if the amplitude of the field is raised sufficiently, a complete crossing over of the polarization and depolarization currents at short times may be observed due to the injection of excess space charges into the dielectric during the charging process. Figure 3A.59 shows such a case for a 30 MVA transformer where there is a complete crossing over of the depolarization and polarization currents at small time.

3A.4.7 Effects of Charging Time on PDC

Der Houhanessian and Zaengl [23] gave an example of how significantly the relaxation currents will change with charging time $t_{c.}$ Measured values of such currents on a 2 mm thick pressboard sample are displayed in Figure 3A.60. The figure shows that at different charging periods, the discharge currents deviate noticeably from each other. The cause of this deviation, considering the low offset current in the measurement setup (0.3 PA), may be due to the influence of previous measurements or the activation of some electrochemical processes generating a persistent current in the sample (also known as the memory effect).

They have also shown PDC measurement results from a sample consisting of a two-layer arrangement of 2 mm oil gap and 2 mm pressboard placed between parallel-plate electrodes. The pressboard and the oil used are aged ones with final moisture contents of 2.0% and 19 ppm, respectively. Figure 3A.61 shows the measured relaxation currents under the same conditions as used for the results of Figure 3A.60. The deviation of depolarization currents at short times could

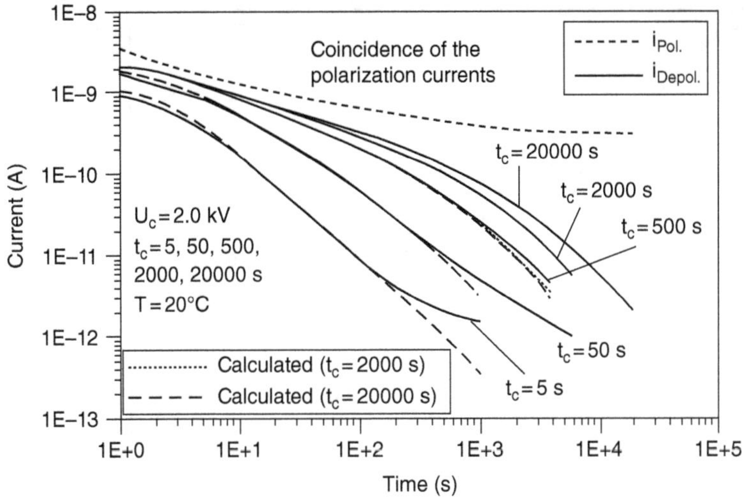

Figure 3A.60 Relaxation currents as a function of charging time of an aged pressboard sample and depolarization currents for 5, 50, and 500 s charging time as calculated from equivalent circuits obtained from relaxation currents at 2000 and 20000 s charging time. *Source:* Ref. [23], Figure 5.

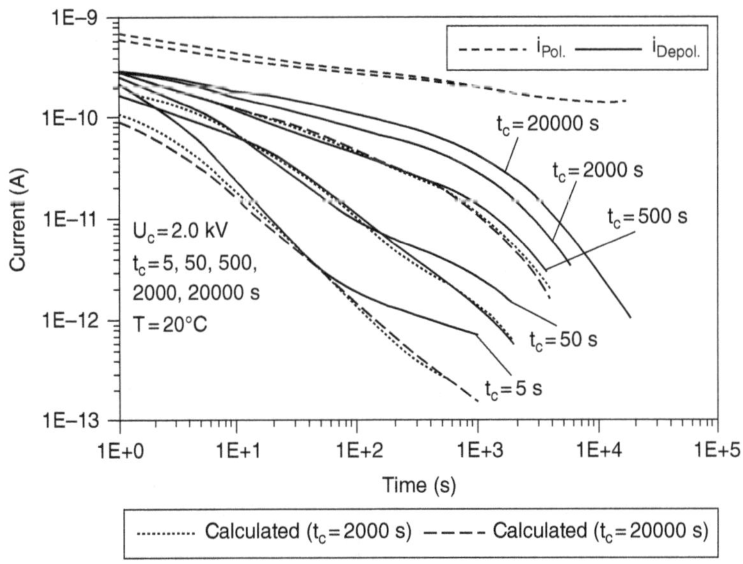

Figure 3A.61 Relaxation currents as a function of charging time of an aged multi-layer sample and calculated depolarization curves for 5, 50, and 500 s charging time from equivalent circuits obtained from relaxation currents at 2000 and 20000 s charging time. *Source:* Ref. [23], Figure 8.

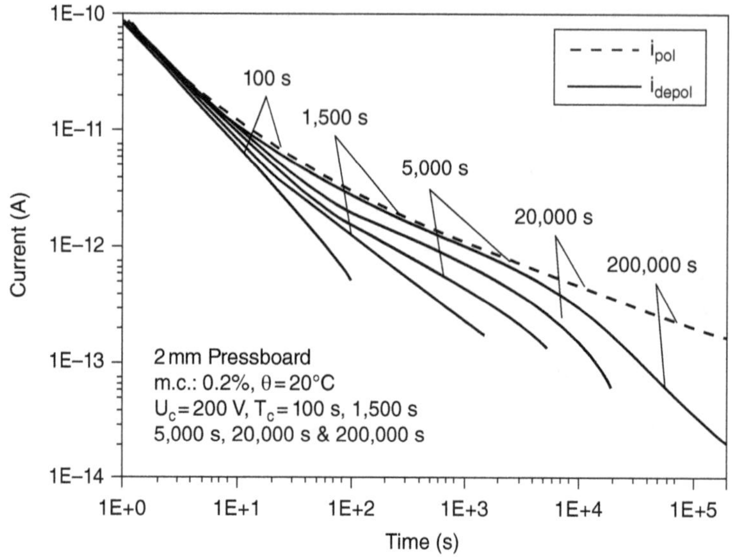

Figure 3A.62 Polarization and depolarization currents of a dry pressboard sample as a function of charging duration. *Source:* Ref. [8], Figure 5.

be explained by the memory effect or a non-restoration of the initial condition of the sample, preceding the measurements of relaxation currents.

Zaengl [8] reports that adequately long charging durations are necessary to complete most of the polarization processes taking place inside the insulation during dielectric response measurements. Otherwise, the "memory effect" may lead to erroneous results. Such a situation of PDC measurements performed with varying charging times is shown in Figure 3A.62.

3A.4.8 Effects of Insulation Geometry on PDC

Der Houhanessian and Zaengl [19] performed many measurements on multi-layer arrangements (pressboard in series with oil gaps, a configuration relevant for power transformers). They demonstrated that the dielectric response in the time domain is dependent on the volume ratios. The samples consist of a three-layer arrangement, in which two 2 mm pressboard disks are placed on the surface of parallel-plate electrodes and are separated by a central oil duct of variable thickness. The current measurements are performed with a voltage to electrode spacing ratio of 3 V/mm and the charging times have been limited to 5000 s. Figure 3A.63 shows measurements changing the ratio of oil to board from 4 to 1 (from 16 mm to 4 mm oil duct). These measurements clearly show that the predominantly exponential decay of relaxation currents, which is due to the interfacial polarization, is essentially controlled by the volume ratio of oil to board along with oil condition – such as moisture content and conductivity.

Zaengl [14] used a pancake model, a test object modeling power transformer insulation systems. Detailed information can be obtained from Ref. [14]. Different

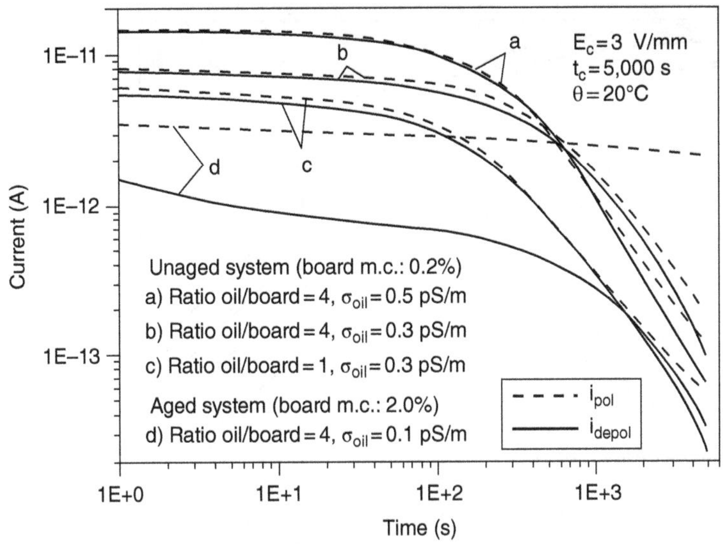

Figure 3A.63 Relaxation currents of different multi-layer samples: (a, b, d) ratio oil/board = 4, (c) ratio oil/board=1. *Source:* Ref. [19], Figure 5.

insulation configurations were studied, and the following codes quantify the dimensions of the whole gap between the electrodes by the thickness (mm) of board and oil:

01 – 10/0
02 – 2/10
03 – 4/10
04 – 10/10.

See Figure 3A.64 [19].

Figure 3A.65 [17] shows the impact of insulation geometry on polarization currents. Three different geometries (100% only pressboard G0, 33% barriers to oil G1, and 20% barriers to oil G2) were studied. It has been observed that with an increasing portion of insulation oil (oil duct), the current remains nearly constant for a longer time and this may have been caused by interfacial polarization at the interfaces between oil and pressboard.

3A.5 PDC Measurement Results on Field Transformers

PDC measurements have been conducted on substation transformers by the authors of this book, as well as by others in the research community. Some works are presented here to highlight better understanding of PDC measurements. Figure 3A.66 [19] presents relaxation currents of two identical single-phase 150 kV, 21 MVA power transformers, manufactured in 1962. The measurements were performed on site, applying a voltage of 2000 V for a duration of 5000 s. The

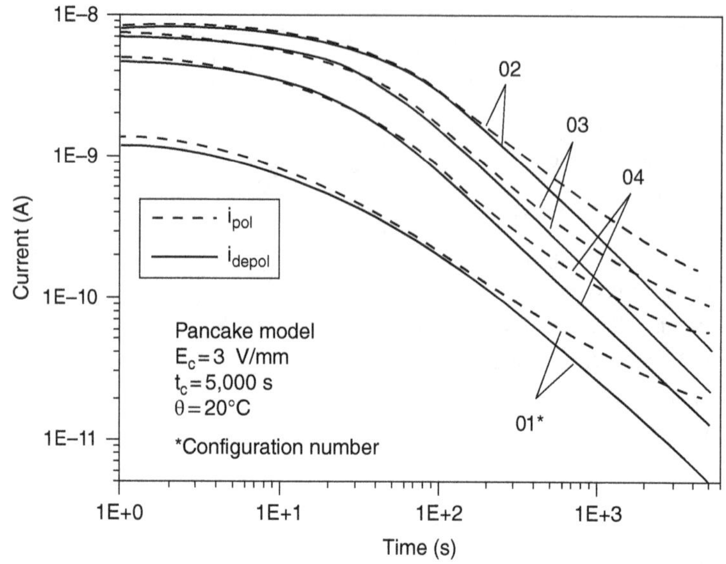

Figure 3A.64 Relaxation currents measured on four different configurations of the "pancake" model. The measurements have been performed for a voltage to electrode spacing ratio of 3 V/mm. *Source:* Ref. [19], Figure 1.

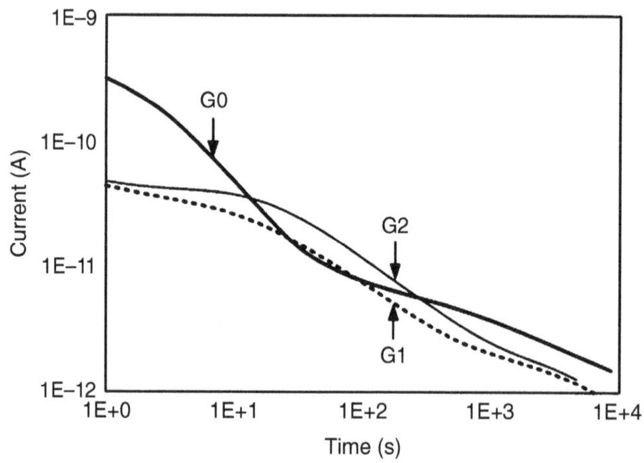

Figure 3A.65 Influence of insulation geometry on polarization currents at 0.6% moisture. *Source:* Ref. [17], Figure 11.

oil temperature of both transformers was 24°C. The difference in initial currents has been related to the difference in oil conductivity, which has been found to be a major factor in interfacial polarization. Conductivity measurements on oil samples from transformer 'a' were 0.9 pS/m, compared with 4.2 pS/m for transformer 'b'.

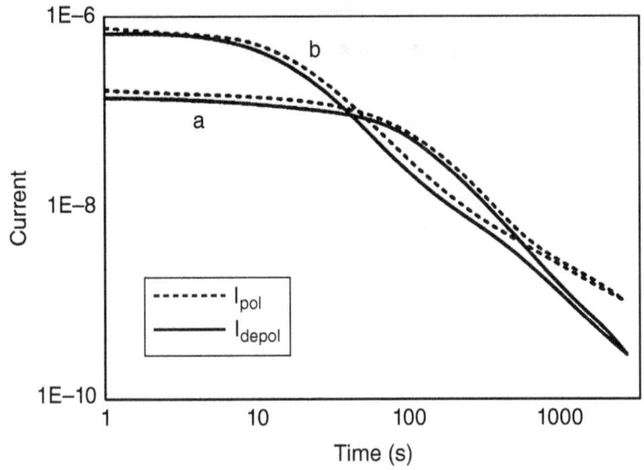

Figure 3A.66 Relaxation currents of two identical single-phase 150 kV, 21 MVA power transformers. *Source:* Ref. [19], Figure 10.

In Ref. [28], dielectric relaxation current measurements were performed on two transformers as characterized below.

1) Used transformer: three-phase 251/15.5 kV 220 MVA, manufactured in 1967 and fully loaded since installation with a scheduled stop every summer.
2) New unit: three-phase 245/16 kV 230 MVA, manufactured in 2000 which replaced the used transformer above.

The results of these measurements are presented in Figure 3A.67, where PDC was performed at 1 kV DC. The polarization/depolarization current measurement duration for the new transformer was 10,000 s and for the used one 6000 s.

> The characteristic, initially predominant exponential shape of the measured currents is due to the exponential time dependence of the interfacial polarization and depolarization currents as caused by the series arrangement of the oil ducts and pressboard barriers. For long times of voltage application, the dielectric response of pressboard barriers becomes more apparent due to the completion of this interfacial polarization. Also the small contribution of the currents generated by the spacers influence mainly the shape of the resulting currents at long times. [28]

Saha and Purkait [1] presented field test results for a number of transformers, as detailed in Table 3A.3.

> These transformers were tested in an open substation environment. The transformers were taken out of service at least one day prior to the testing so that the transformer temperature could settle down to ambient temperature before starting the tests. All the tests were carried out at a charging voltage of 500 V DC and with 10^4 s charging and discharging periods each. For the purpose of better understanding, they have been categorized into

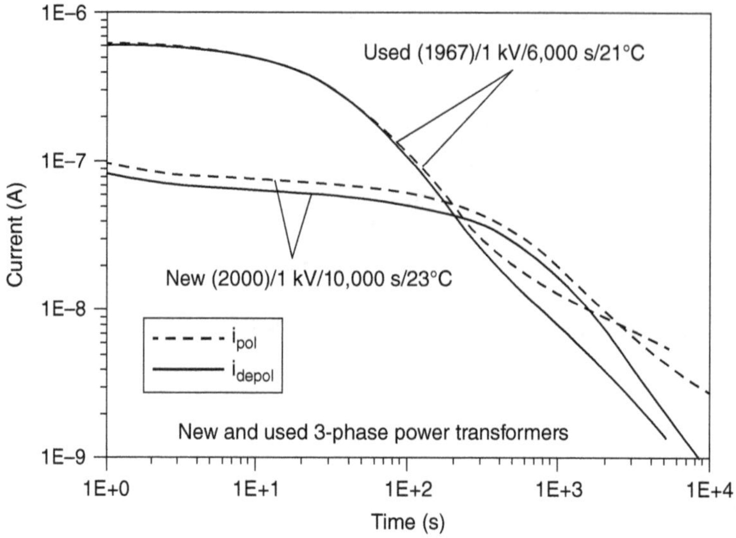

Figure 3A.67 Measured polarization and depolarization currents on a new three-phase 245/16 kV 230 MVA power transformer manufactured in 2000 and a used three-phase 251/15.5 kV 220 MVA unit manufactured in 1967. *Source:* Ref. [28], Figure 2.

Table 3A.3 Transformer details

Transformer	A1	A2	B1	B2	C1	C2
MVA	7	30	100	100	35	45
Voltage (kV)	66/11	132/66/11	330/132/16	330/132/16	66/22	66/22
Connection	Y/Δ	Y/Y/Δ	Y/Y/Δ	Y/Y/Δ	Y/Y	Y/Y
Year of manufacture	1968	1966	1966	1966	1936	1959
Service record	Lightly loaded	Suspected to be very aged	Before oil reclamation	After oil reclamation	Operating, but aged	Operating, but aged

Source: Ref. [1], Table 1.

three groups A, B, and C. The first group (A) consists of two transformers of which one is lightly loaded and the other one is heavily loaded. The second group (B) consists of a single transformer, on which PDC measurements have been carried out before and after oil reclamation and the third group (C) consists of two very old transformers (still in service). The polarization and depolarization currents for the transformers in group A, group B, and group C are plotted in [Figures 3A.68 to 3A.70], respectively. The current plots in each case are divided by the corresponding charging voltage and geometric capacitance values. In this way, they have been normalized to [an] equivalent base of geometry and charging voltage, for easier inter-comparison. [1]

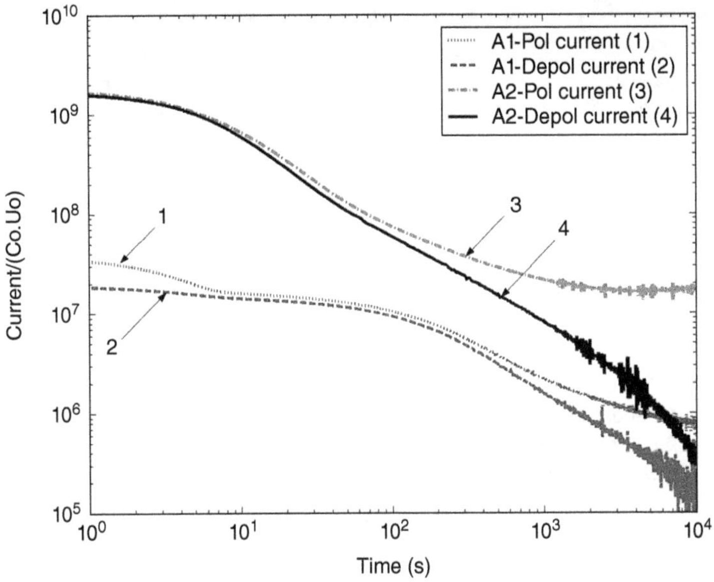

Figure 3A.68 Variation of polarization and depolarization currents for A1 and A2. 1 and 2, Pol/Depol current for A1; 3 and 4, Pol/Depol current for A2. *Source:* Ref. [1], Figure 9.

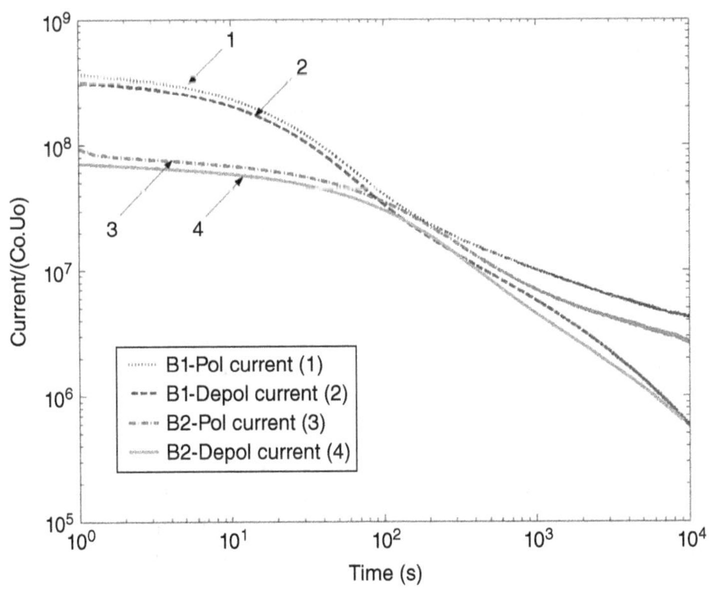

Figure 3A.69 Variation of polarization and depolarization currents for B1 and B2. 1 and 2, Pol/Depol current for B1; 3 and 4, Pol/Depol current for B2. *Source:* Ref. [1], Figure 10.

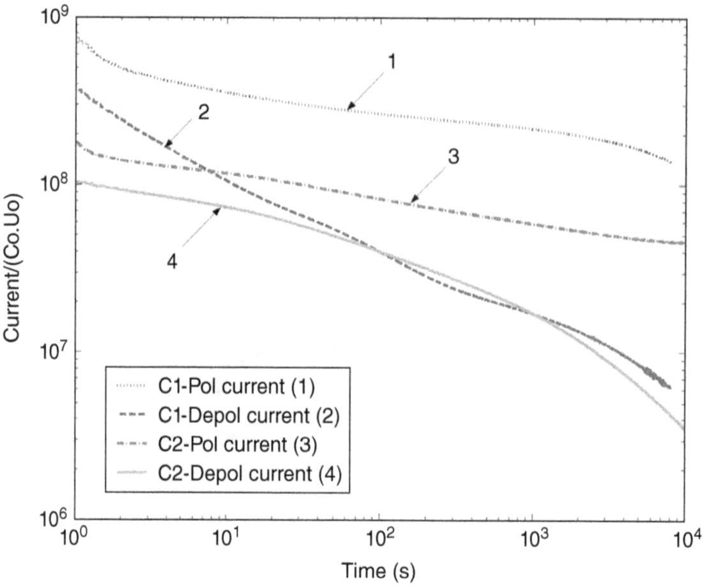

Figure 3A.70 Variation of polarization and depolarization currents for C1 and C2. 1 and 2, Pol/Depol current for C1; 3 and 4, Pol/Depol current for C2. *Source:* Ref. [1], Figure 11.

TRANSFORMERS A1 AND A2

It is obvious from [Figure 3A.68], that both polarization and depolarization currents of transformer A2 are higher than those of A1. Transformers with poor insulation condition normally have higher values of polarization/depolarization currents. Transformer A2 was suspected of having advanced insulation ageing due to the fact that one of the companion transformers of this group failed in service. A1 was relatively lightly loaded during its service life – it was expected that the condition of the insulation in A1 was better than A2.

The polarization and depolarization currents at longer times are lower for transformer A1 as compared to transformer A2. This is indicative of the fact that the paper insulation of A1 is less conductive than that of A2. The paper conductivity values for A1 and A2 are calculated and presented in [Table 3A.4]. The higher conductivity of paper in A2 than A1 indicates that the paper insulation in A2 is in more moist and degraded condition than A1. Oil conductivity values calculated from the polarization current during the initial period of time are also presented in [Table 3A.4]. The higher value of oil conductivity of A2 as compared to that of A1 indicates that the oil in A2 has higher moisture content and degradation by-products than A1. These results once again agree with the fact that A1 was lightly loaded during its service period, whereas A2 was taken out of service with suspected condition of advanced insulation degradation. Oil samples were collected for furan analysis by the HPLC technique. Results

Table 3A.4 Insulation condition of A1 and A2

Transformer		A1	A2
Conductivity (S/m)	Oil	3.0×10^{-13}	5.7×10^{-12}
	Paper	2.5×10^{-15}	3.0×10^{-13}
Capacitance (nF)		3.9	2.4
2-Furfuraldehyde (ppm)		0.02	1.03

Source: Ref. [1], Table 2.

of 2-furfuraldehyde are also shown in [Table 3A.4]. The furan content of A2 is more than 50 times higher than that of A1. A higher value of 2-furfuraldehyde in the oil analysis data also supports the inference of more degradation of the paper insulation in A2 than A1. [1]

TRANSFORMERS B1 AND B2

[Figure 3A.69] represents the polarization and depolarization current plots for B1 and B2, which essentially is the same transformer. Recently, oil reclamation was conducted for the transformer and we had the opportunity to test this transformer both before and after the oil reclamation process. According to the information available from the utility, the oil reclamation process basically consisted of recycling, filtering, and drying the oil through vacuum and at high temperature. The dry and filtered oil was pumped back to the tank and the whole process was continued in a closed cycle. In such a closed-cycle oil reclamation process, it is expected that some moisture and other degradation by-products will be driven off the oil. At the same time, when the dry oil is circulated through the transformer, some moisture will also migrate out of the paper towards the relatively dry oil. Thus, in such an oil reclamation process, the oil moisture content and other ageing by-products are expected to go down along with some improvement in the paper moisture condition as well.

As can be seen from [Figure 3A.69], both the polarization and depolarization currents for B1, which represents the transformer before oil reclamation, have higher values during the initial period than B2, which represents the transformer after oil reclamation. Oil conductivity values of B1 and B2 calculated from the initial values of polarization currents are shown in [Table 3A.5]. The oil moisture content values measured by Karl Fischer titration method are also presented in [Table 3A.5]. The initial lower magnitudes of polarization and depolarization currents and lower oil conductivity of B2 as compared to B1 indicates that after oil reclamation, the oil condition has improved as a whole. This observation is also confirmed by the lower oil moisture content value of B2 than B1, as shown in [Table 3A.5].

Table 3A.5 Insulation condition of B1 and B2

Transformer		B1	B2
Conductivity (S/m)	Oil	7.6×10^{-12}	1.9×10^{-12}
	Paper	3.5×10^{-14}	2.2×10^{-14}
Capacitance (nF)		3.3	3.2
2-Furfuraldehyde (ppm)		1.9	0.26
Moisture content	Oil (ppm)	28	10
	Paper (%)	5	3.5
Oil sampling temperature (°C)		23	25

Source: Ref. [1], Table 3.

As seen in [Figure 3A.69], the final values of polarization and depolarization currents are higher for B1 than B2. Paper conductivity values calculated from these current values are presented in [Table 3A.5]. It is found that the paper conductivity of B2 after oil reclamation is lower than the paper conductivity of B1 (before oil reclamation). [Table 3A.5] also contains the paper moisture content values determined using equilibrium charts from the measured oil moisture content values at the corresponding oil sampling temperatures. As seen in [Table 3A.5], the estimated paper moisture drops to a value of 3.5% after oil reclamation from the initial value of 5% before the oil reclamation. It however remains a fact that the accuracy of the equilibrium charts is questionable at the lower ranges of oil and paper moisture content values. Even then, it is not beyond expectation that during such an oil reclamation process, some moisture has migrated out of the paper insulation towards the relatively dry oil. [Table 3A.5] also contains the furan (2-furfuraldehyde) values measured in the oil samples using the HPLC technique. As expected, filtering and drying of the oil at high temperature under vacuum during the reclamation process have driven out some amount of furan and other ageing by-products from the oil.

In summary, it is observed that after the oil reclamation process, the conductivity values of both oil and paper are reduced. The reductions are, however, not very significant, only in the order of half. This suggests that oil reclamation did not significantly improve the ageing condition, though it drove some moisture and ageing by-products out of the oil and paper. One would expect that the reduction of conductivity should be in the order of 10 for a significant reduction in ageing and moisture condition. [1]

TRANSFORMERS C1 AND C2

Polarization and depolarization currents for transformers C1 and C2 are plotted in [Figure 3A.70]. The magnitudes of the polarization and

Table 3A.6 Insulation condition of C1 and C2

Transformer		C1	C2
Conductivity (S/m)	Oil	31.0×10^{-12}	3.6×10^{-12}
	Paper	2.3×10^{-12}	3.8×10^{-13}
Capacitance (nF)		9.2	10.2
2-Furfuraldehyde (ppm)		1.87	3.81
Moisture content	Oil (ppm)	36	26
	Paper (%)	4.2	3.2
Oil sampling temperature (°C)		40	40

Source: Ref. [1], Table 4.

depolarization currents of C1 are higher than those of C2. This indicates that the general condition of the insulation of C1 is worse than C2. This is not unexpected since transformer C1 has been in operation since 1936 as compared to C2, which was manufactured in 1959. As seen in [Figure 3A.70], there is quite a big difference between the polarization and depolarization currents of C1. This may be due to the extremely degraded and moist condition of insulation in C1. Another reason for this unusually large value of the polarization current may be due to the presence of some amount of ions in the oil and paper from previous excitation processes. However, the current values could not be reconfirmed by a second repeat measurement due to the limited period of actual testing time provided by the distribution company.

The polarization and depolarization currents during the initial period of time are larger in C1 than C2. The oil conductivities for C1 and C2 calculated are given in [Table 3A.6]. As expected, oil conductivity for C1, with more degraded insulation, is higher than that of C2. The values of the oil moisture contents measured by Karl Fischer titration method are also shown in [Table 3A.6]. The higher value of oil moisture content of C1 over C2 once again supports the observation that the oil condition of C1 is more degraded than that of C2.

[Figure 3A.70] also shows that the polarization and depolarization currents are high at larger values of time for C1 than C2. Paper conductivity values and paper moisture content values estimated from the measured oil moisture content values using the equilibrium charts are presented in [Table 3A.6]. As expected, both the paper conductivity and paper moisture content of C1 are higher than those of C2, indicating a more degraded condition of paper in C1 compared to C2. Furan measurements, however, showed an opposite trend for C1 and C2. Although both transformers have high 2-furfuraldehyde contents, this anomaly in the results could not be explained in any other way than attributing it to measurement errors and other associated inaccuracies. [1]

3A.6 PDC Measurement Equipment

A number of commercial and university/research organization in-house PDC measurement devices have been reported in the literature. For example, the authors of this book have developed several versions of PDC measurement devices. In their latest version, they have used a high-voltage power supply (up to 12 kV DC) to measure PDC currents, which can be used for transformers as well as for low-loss insulation materials (e.g., XLPE cables). Current measurement has traditionally been performed by an electrometer or similar device. Various commercial equipment is also available in the market, including Haefely Hipotronics' RVM 5462 [29], which has been used widely as an advanced automatic recovery voltage meter for diagnosis of oil–paper insulation. This device was primarily designed for recovery voltage measurement, but can also measure insulation resistance, polarization index, and polarization current. Alff Engineering has marketed the PDC-ANALYSER-1MOD [30] instrument for onsite measuring and analyzing the dielectric behavior of electrical insulation materials and insulation systems. Omicron has marketed the DIRANA-FDS-PDC dielectric response analyzer [31], which can determine the condition of power transformers, bushings, cables, and generators with dielectric response analysis.

3A.7 Summary of PDC Measurements

This chapter has provided a comprehensive understanding of polarization/depolarization current measurement methodologies. The basic theory behind PDC measurements is presented, along with the factors affecting the PDC measurements. PDC can be affected by many factors, including operating temperature, equilibrium of oil–paper moisture inside the transformer tank, polarization/depolarization measurement time, voltage magnitude, memory effect, etc. These factors have been discussed extensively, along with findings from major researchers around the world. Application of this technique for field transformers has also been explained, with several case studies. PDC techniques have been applied for current transformers, bushings, and electrical cables of both oil-impregnated paper and XLPE types, which are not discussed in this chapter. A number of attempts have been made to correlate time versus frequency domain dielectric measurements, as described in Part B of this chapter.

3A.8 Recovery Voltage Measurements (RVM)

3A.8.1 Introduction of Recovery Voltage Measurements [32]

When a direct voltage is applied to a dielectric for a long period of time and is then short-circuited for a short period, after opening the short circuit the charge bounded by the polarization will turn into free charges. This phenomenon is called the return or recovery voltage. The process of polarization and the equations to describe this process are given in Refs [3, 33].

When a dielectric material is charged with an electric field, the material becomes polarized. The total current density $j(t)$ is the summation of the displacement current density and the conduction current density, given by

$$j(t) = \sigma E(t) + \frac{dD}{dt} \tag{3A.24}$$

where σ is the DC conductivity and $D(t)$ is the electric displacement, given by

$$D(t) = \varepsilon E(t) + \Delta P(t) = \varepsilon_0 \varepsilon_r E(t) + \Delta P(t) \tag{3A.25}$$

where ε_0 is the vacuum permittivity and ε_r is the relative permittivity at power frequency (typically 4.5 for cellulose paper/pressboard and 2.2 for transformer hydrocarbon oil). The term ΔP(t) is related to the response function $f(t)$ by the convolution integral

$$\Delta P(t) = \varepsilon_0 \int_0^t f(t-\tau)E(\tau)d\tau \tag{3A.26}$$

If we expose the insulation to a step voltage at time t=0, the charging current density is given by

$$j_p = E(\sigma + \varepsilon_0 f(t)) \tag{3A.27}$$

If we consider the case where an insulation system with geometric capacitance C_0 is exposed to a step voltage U_a, the charging/polarization current can be given by

$$i_p = C_0 U_a \left\{ \frac{\sigma}{\varepsilon_0} + f(t) \right\} \tag{3A.28}$$

If the step voltage is now disconnected from the insulation, equation (3A.29) gives the discharging/depolarization currents:

$$i_d = -C_0 U_a \{ f(t) - f(t + t_p) \} \tag{3A.29}$$

The charging time (t_p) normally should be at least ten times longer than the time for which the response function is calculated, then the second term in equation (3A.29) can be neglected. So, the response function $f(t)$ becomes proportional to the depolarization current. Hence, the response function and conductivity can be calculated simultaneously by using polarization and depolarization currents. Very often the response function needs to be expressed in a parameterized form. The response function can be written in this general form:

$$f(t) = \frac{A}{\left(\frac{t}{t_0}\right)^n + \left(\frac{t}{t_0}\right)^m} \tag{3A.30}$$

The response function describes the fundamental memory property of any dielectric system and can provide significant information about the insulation material. After opening the short circuit, the charge bounded by the polarization will turn into free charges (i.e., a voltage will build up between the electrodes on the dielectric). This phenomenon is the return voltage or recovery voltage. The recovery voltage arises from the relaxation processes inside the dielectric material. The current density during the return voltage measurement is zero, and equation (3A.31) gives the expression of current density:

$$j(t) = \sigma E(t) + \frac{dD}{dt} = \sigma E(t) + \frac{d}{dt}(\varepsilon_0 \varepsilon_r E(t) + \Delta P(t))$$
$$j(t) = \sigma E(t) + \varepsilon_0 \varepsilon_r \frac{d}{dt} E(t) + \varepsilon_0 \frac{d}{dt}\left\{ \int_0^t f(t-\tau)E(\tau)d\tau \right\} \tag{3A.31}$$

where $E(t)$ is the electric field resulting from the recovery voltage build-up across the open-circuited dielectric. Equation (3A.31) shows that the recovery voltage depends on the conductivity σ, relative permittivity ε_r, and dielectric response function $f(t)$. These parameters are all affected by ageing and moisture in the insulation. The response function can be obtained from the polarization and depolarization currents. These currents depend on the geometric capacitance and on the applied step excitation. The response function and conductivity can be calculated from equations (3A.29) and (3A.30) if the geometric capacitance of the transformer composite insulation C_0 is known. If the proper geometry of the transformer oil–paper insulation is known, then by solving equation (3A.31), recovery voltage build-ups for a transformer can be estimated. The recovery voltage also depends on the applied electric field and, if the dielectric material is assumed to be linear, this problem is resolved easily for the interpretation of results [34]. A modeling tool can be very useful to investigate the impact of geometry on RV results.

3A.8.2 Determination of "Polarization Spectra" of Insulation System by RVM

Figure 3A.71 shows the RVM circuit for the determination of the polarization spectrum.

The measurements consist of three phases.

- First, switch S_1 is closed for time T_c so that the DC voltage source U_o applies a charge to the test object (transformer composite insulation). During the charging/polarization period, the molecules of the dielectric are polarized and aligned in the direction of the electric field.

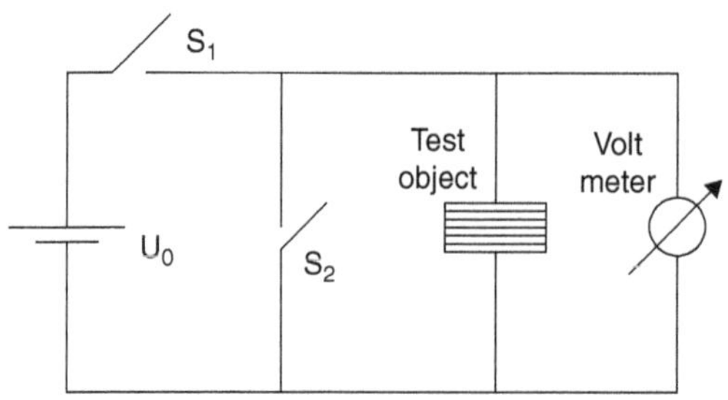

Figure 3A.71 Schematic diagram of RV measuring circuit.

- After the predefined charging time, switch S_1 is opened to remove the voltage source and switch S_2 is kept closed for time T_d (usually half of the charging time), so that the test object is short-circuited, causing the molecules to be partially depolarized.
- The third stage is the measurement of recovery voltage. After the discharging time T_d, the short circuit is removed and the remaining charge bounded in the insulation by the incomplete depolarization will turn into free charges (i.e., a voltage will build up). This residual polarization results in a voltage that is known as the recovery voltage or return voltage, which is normally measured by a very high input impedance voltmeter.

Figure 3A.72 shows the steps in one RV measurement cycle.

The most popularly used parameters of the recovery voltage profile are the maximum RV value V_{max} and the corresponding peak time T_{peak}, along with the initial slope dV/dt ($\tan \alpha$) of the RV curve, as shown in Figure 3A.73.

At the end of one RV measurement, the transformer is grounded again for a sufficiently long time to completely drain out any residual charge within the insulation system. Subsequently, the charging and discharging process is repeated using a sequence of increasing charging times, ranging from milliseconds to

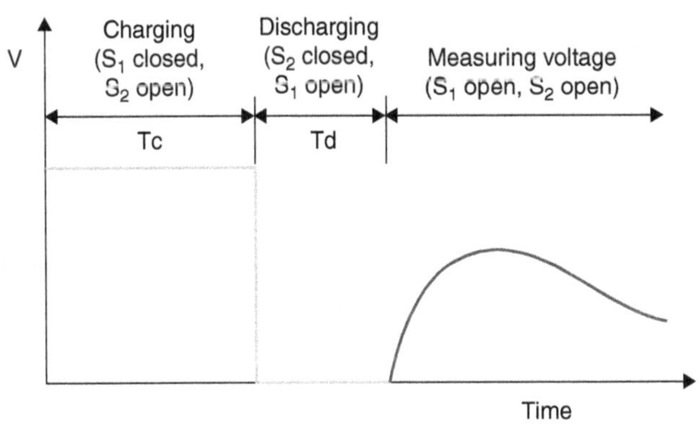

Figure 3A.72 Timing diagram of RVM procedure.

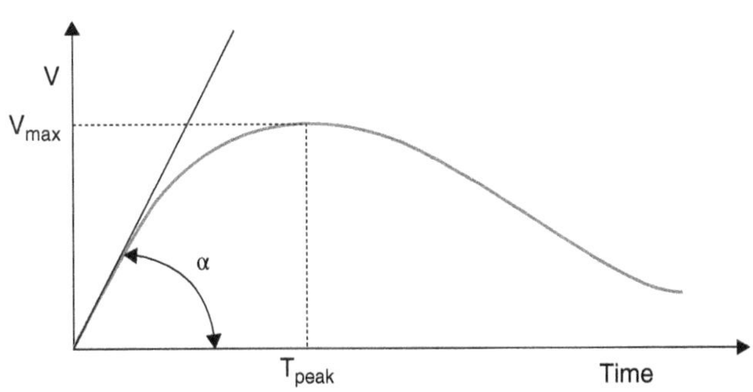

Figure 3A.73 RVM parameters.

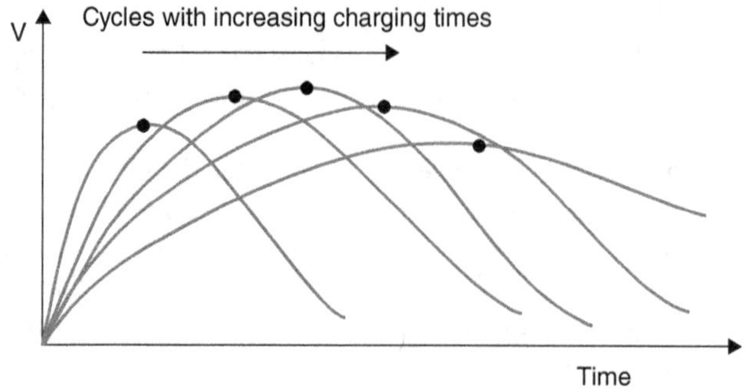

Figure 3A.74 RV profiles with increasing values of charging times (dots indicating maximum value of RV in each RVM cycle).

thousands of seconds. By doing so a series of RV curves are obtained, as shown in Figure 3A.74.

We take the maximum value V_{max} and the corresponding charging time T_c from each curve; then, the so-called "polarization spectrum" is obtained by plotting V_{max} versus T_c. For convenience, the x-axis is often plotted in terms of $\log(T_c)$. The time at which the peak occurs in the polarization spectrum is known as the dominant time constant (or central time constant) of the insulation system. Such a polarization spectrum is shown in Figure 3A.75.

A similar spectrum can be compiled using the initial slope values for each RVM and plotting those against the corresponding charging times in the x-axis. Such a spectrum derived from the individual RVM cycles, known as the initial slope spectrum, is shown in Figure 3A.76.

A third type of spectrum can be derived by plotting the T_{peak} values of individual RVM cycles against the corresponding charging time values. Such a spectrum is shown in Figure 3A.77.

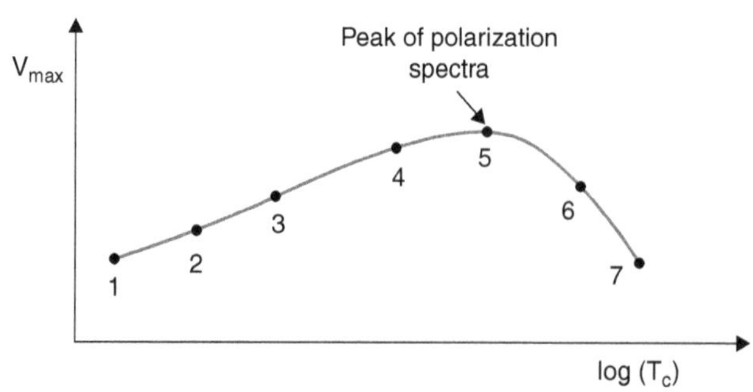

Figure 3A.75 Plot of polarization spectra: joining peaks of individual RVM cycles plotted against the corresponding charging times.

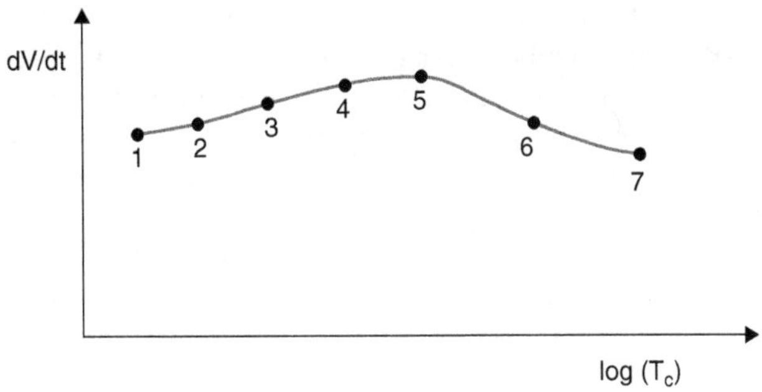

Figure 3A.76 Plot of initial slope spectrum.

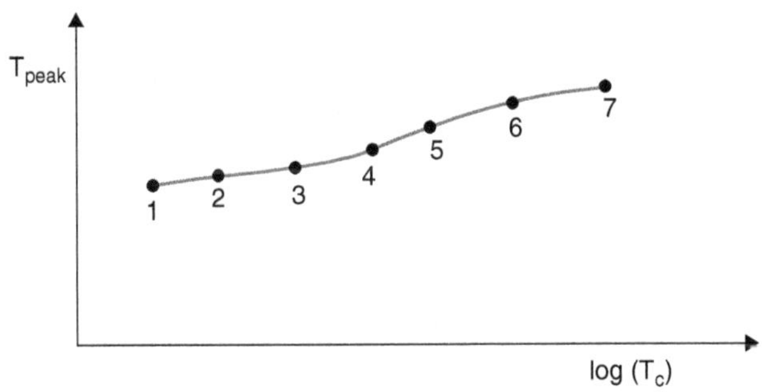

Figure 3A.77 Peak time spectrum.

3A.8.3 Interpretation of RVM Results

The dominant time constant is a significant characteristic of the polarization spectrum, influenced by several quantities, as reported in [36].

1) Increase in moisture content in insulation reduces the dominant time constant.
2) Ageing of solid insulation reduces the dominant time constant.
3) The higher the temperature, the lower is the dominant time constant.
4) The geometry of the insulation structure has an impact on the polarization spectrum.

3A.8.3.1 Effect of Moisture

3A.8.3.1.1 Moisture in Oil–Paper Insulation Systems

The solid insulation system stores moisture inside a transformer tank and the distribution between paper and oil is dependent on the operating temperature of the transformer. The insulation system of a transformer is quite complex, in which any water present is always in a dynamic state. The moisture dynamic is driven largely by variations in concentration, saturation levels, and temperature. It has been described before that high moisture content in transformers can cause accelerated ageing of the insulation system, reducing the dielectric strength of the oil and formation of bubbles. As a transformer is loaded, its operating temperature increases and moisture starts to migrate from the solid paper/pressboard insulation into the oil. When the transformer cools down again, moisture tends to return back to the paper/pressboard solid insulation. In most cases, it is possible for excessive moisture to remain in the oil because it migrates much more slowly back into the paper. This can cause saturation of the oil with moisture, and free water can form as droplets in the transformer. The moisture content of the transformer insulation increases significantly with the progression of ageing. In addition, moisture contained in the solid insulation is the primary cause of bubble evolution. High hotspot temperatures can greatly increase the risk of bubble formation, which may lead to breakdown in the solid insulation.

A large transformer (e.g., 240 MVA 275/132 kV) has a paper weight of 35 ton and an oil weight of 150 ton [35]. If this transformer is new and unaged, it is very common to have 0.5% moisture in the paper, which means it has 175 litre of water; with <5 ppm moisture in oil at 20°C it would have 0.75 litre of water. After 25 years in service, if the solid insulation has a moisture content of 2.8%, there would be 980 litre of water in the paper [35]. Moisture in oil has been measured by Karl Fischer titration for a long time, and then equilibrium curves are used to assess the moisture in paper. As mentioned before, a transformer in operation hardly reaches equilibrium due to a continuous change in load. Hence, an alternate non-intrusive method for moisture estimation of transformer solid insulation has been in demand for a long time. In the 1990s, the RV polarization spectrum was recommended by a number of researchers to assess the moisture content of the transformer solid insulation system.

3A.8.3.1.2 The Relationship between Moisture and Polarization Spectrum

High moisture content has been related to the peak maximum return voltage; having a shorter dominant time constant and a similarly low moisture content has been related to the peak maximum return voltage at a longer dominant time constant. A number of researchers have related the time taken to reach peak maximum return voltage with the moisture content of the insulation system.

3A.8.3.1.3 Laboratory Test Results

Figure 3A.78 [36] shows polarization spectra at constant temperature on oil-impregnated paper insulation models. The figure shows that the spectra shift to the left at higher moisture content on the insulation model and vice versa.

Csepes *et al.* [37] developed an experimental setup for controlling the moisture content of sample insulation and then performed RVM on the samples. Figure 3A.79 shows such spectra of recovery voltage with moisture content. The discharging time was half the charging time in these measurements. Again, it is evident that the dominant time constant has an inverse relationship with moisture content.

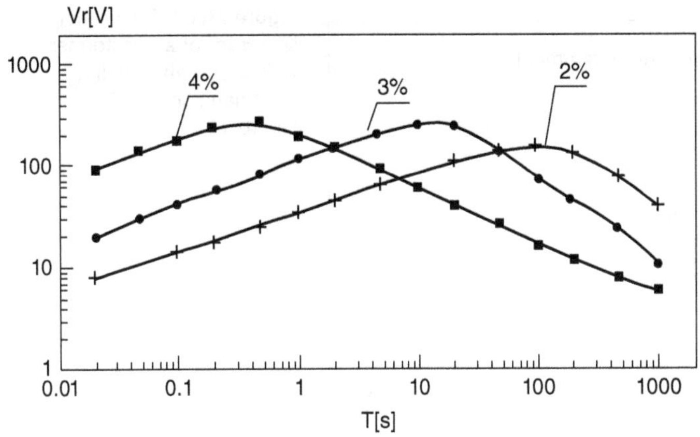

Figure 3A.78 Polarization spectra curves relative to paper moisture content at 25°C. *Source:* Ref. [37].

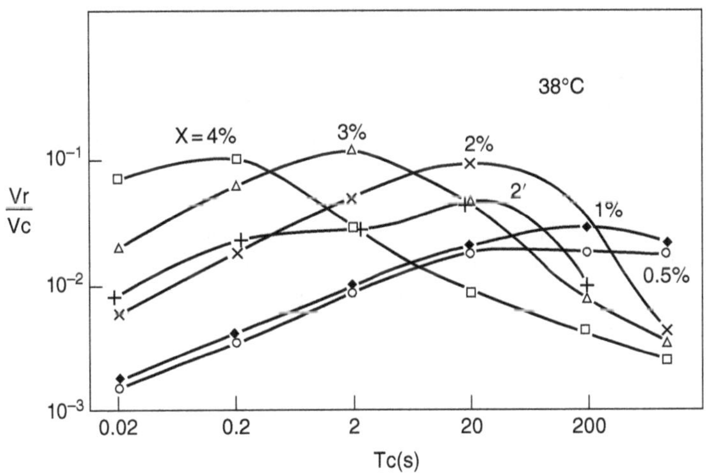

Figure 3A.79 Polarization spectra curves relative to paper moisture content. *Source:* Ref. [37], Figure 5.

3A.8.3.1.4 Field Test Results

The effects of moisture content of the insulation on polarization spectra were also reported in Ref. [37], where two transformers were tested before and after treatment and drying.

Figure 3A.80 shows the RV polarization spectra of a large power transformer before and after a successful drying treatment. It indicates that the RVM method is also useful for checking the effectiveness of a drying treatment.

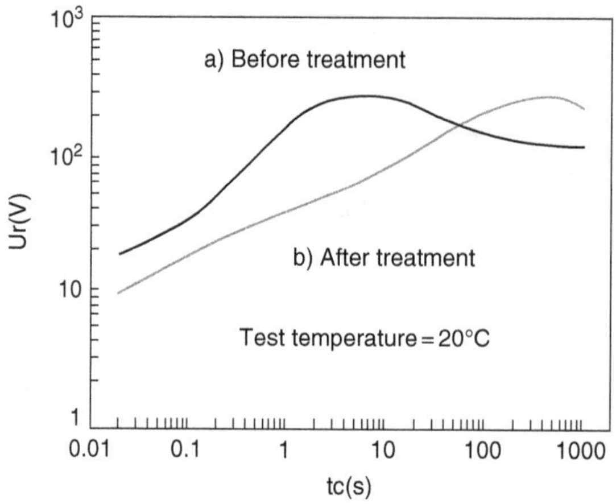

Figure 3A.80 Polarization spectrum of a transformer before and after drying treatment. *Source:* Ref. [37], Figure 13.

Before the treatment, the dominant time constant was a few seconds, while this became a few hundred seconds after the treatment. The reduction in moisture due to the treatment is reflected in Figure 3A.80 by a two order of magnitude increase in dominant time constant.

3A.8.3.2 Effect of Ageing

3A.8.3.2.1 Ageing of Oil–Paper Insulation Systems

Years of operation under elevated temperature causes the transformer insulation to age. Ageing is largely accelerated by elevated temperature, and particularly in the presence of high moisture. Wet insulation has a lower insulation resistance, causing higher leakage currents with increased losses. High moisture content also leads to resonance of the polarization with the mains frequency, causing an exponential increase in loss. Both these effects create heat, which leads to accelerated ageing, but most of the heating effect in a transformer is caused by loading. In addition, ageing by-products also contain moisture due to the breaking of cellulose chains, thereby enhancing the ageing process. Impurities, ageing products, and moisture increase both conduction and polarization at the oil–paper interface. The increased conductivity in turn enhances interfacial polarization in the oil–paper composite dielectric [38]. Overall, high moisture in the solid insulation leads to increased ageing, which is strongly affected by the operational temperature. Paper depolymerization due to high-temperature ageing can increase the moisture content in the oil–paper insulation system and produce higher concentrations of furans. The polarization process is influenced largely by moisture and insulation ageing by-products.

3A.8.3.2.2 The Relationship between Insulation Ageing and Polarization Spectrum

Ageing causes depolymerization of paper that generates water, acids, and gases. Depolymerization products in the paper can be polarized. The polarization process has been correlated directly with moisture content and indirectly with ageing by-products. A number of authors have attempted to relate the recovery voltage to a composite effect of moisture, ageing, and DP.

3A.8.3.2.3 Laboratory Test Results

Figure 3A.81 [39] shows RVM results from accelerated ageing over time. It can be observed that the RV spectrum shifts to the left (lower time constants) for higher-duration ageing. Figure 3A.82 [37] shows the impact of ageing on the RV polarization spectrum. The different degrees of ageing are identified by different values of DP. The insulation was aged at 120°C for up to 4000 h. After the ageing process, all the measurements were performed at 60°C. The most-aged paper showed a DP of 400, while the medium and least-aged samples showed DPs of 600 and 1200, respectively. These measurements show the impact of ageing on the dominant time constant, which shows similar behavior with increasing moisture content in insulation systems.

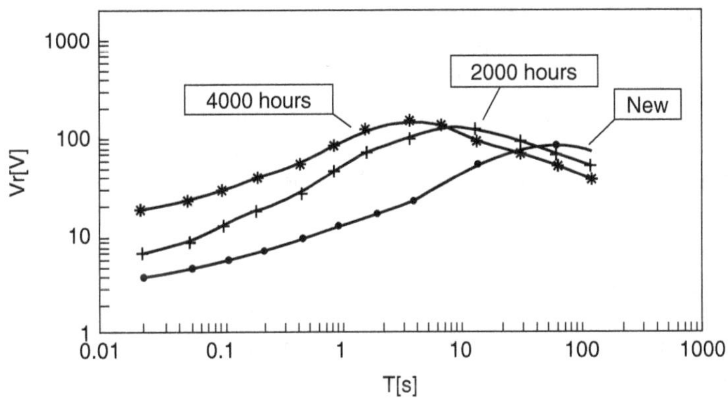

Figure 3A.81 Polarization spectra curves relative to ageing at 120°C. *Source:* Ref. [39], Figure 9.

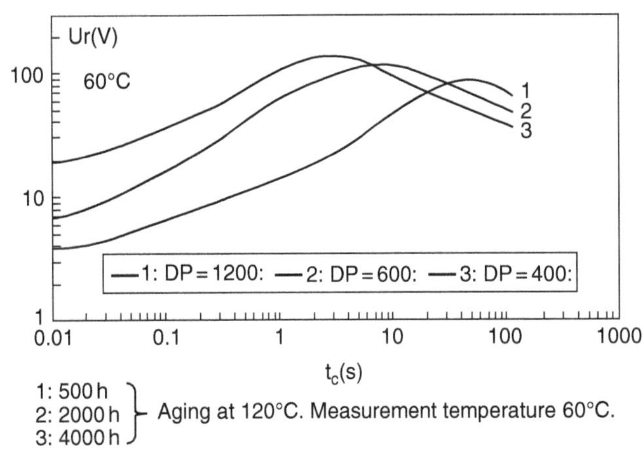

Figure 3A.82 Variation of polarization spectra with ageing by-product. *Source:* Ref. [37], Figure 7.

3A.8.3.2.4 Field Test Results

Some authors have presented cases where the comparison of polarization spectra from different transformers gave an immediate qualitative comment on the state of their insulation systems [4, 35, 39]. A transformer with very dry and unaged insulation will exhibit a dominant time constant of several hundreds or even thousands of seconds, while a transformer with a fairly high moisture content and aged condition of insulation will tend to have a dominant time constant in the order of seconds – as seen in Figure 3A.83 [35].

Saha and Yao [32] and Osvath and Zahn [39] presented RVM results on a number of power transformers under different degrees of ageing. The rating of one of the transformers was 30 MVA between primary and secondary, 10 MVA between primary and tertiary, and 10 MVA between secondary and tertiary. The voltage ratings were primary=132 kV, secondary=33 kV, and tertiary=11 kV. The year of manufacture was 1959. The transformer was in service for 25 years and then it was not in operation over the 10 years before RVM was performed. Two different RV measurements were carried out for this transformer. The first was carried out for the bulk insulation between the primary and secondary windings, and the second was carried out for the bulk insulation between the secondary and tertiary windings. The ambient temperature during RV measurements was 31°C. The polarization spectrum obtained is shown in Figure 3A.84 [40], and its values are given in Table 3A.7.

From the RV spectra in Figure 3A.84, the insulation paper has shown a certain degree of degradation. The paper has been contaminated by ageing products. The oil moisture content measured was, however, low (10 ppm). Since the dominant time constants are relatively low (compared with the usual few hundred seconds for new insulation), the general paper insulation condition in these windings displays evidence of some degradation. Comparing the RV parameters and spectra in Test 2 with those of Test 1, the dominant time constant of 30 s is smaller in Test 2. This indicates that the general oil–paper insulation condition between secondary

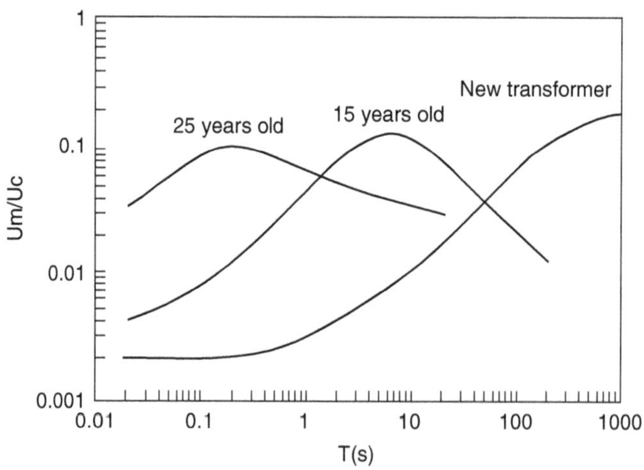

Figure 3A.83 Polarization spectra for new and aged transformers. *Source:* Ref. [35], Figure 4.

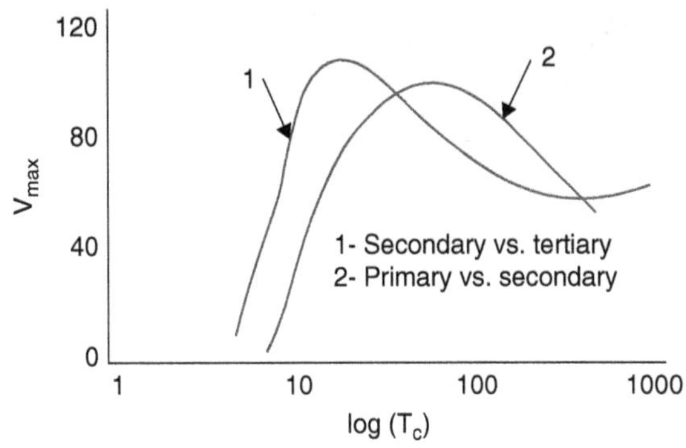

Figure 3A.84 Polarization spectra between two different windings of the same transformer. *Source:* Ref. [40].

Table 3A.7 Summary of polarization spectra results

Winding	V_{max} (V)	Dominant time constant (s)	Oil moisture (ppm)
Primary–secondary	99	58	10
Secondary–tertiary	100	30	10

Source: Ref. [40].

and tertiary shows more evidence of degradation than the condition of the insulation between primary and secondary windings.

The next series of transformers of different ages, as reported in Ref. [32], are detailed in Table 3A.8.

The spectra of maximum return voltage for transformers A–F are presented in Figure 3A.85. A summary of RVM results is presented in Table 3A.9.

It can be observed from Table 3A.9 that there was a considerable variation in peak maximum return voltage for different transformers. In this work, the authors have used a different definition of dominant time constant, which is the time to reach the peak maximum RV (not the charging time of the peak maximum RV).

Table 3A.8 Details of transformers under investigation

Identity	MVA capacity	HV (kV)	LV (kV)	Year of manufacture
A	1	10.5	0.415	1952
B	5	33	11	1960
C	5	33	11	1966
D	10	33	11	1983
E	3.15	11	3.3	1993
F	25	33	11	1997

Source: Ref. [32].

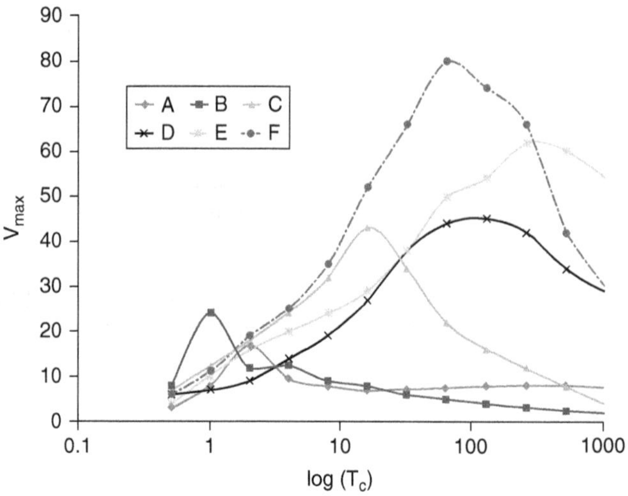

Figure 3A.85 Polarization spectra of different transformers under different operating conditions. *Source:* Ref. [32].

Table 3A.9 Summary of RVM results

Transformer	V_{max} (V)	Dominant time constant (s)	Oil moisture (ppm)
A	15.5	0.4	40
B	23	1.9	28
C	45	9.4	18
D	45.3	114.6	8
E	62.3	62	14
F	82	88	12

Source: Ref. [26].

They have also referred sometimes to the central time constant instead of the dominant time constant. Transformers A and B are the oldest and have the highest levels of oil moisture content. As expected, the dominant time constants of transformers A and B are the lowest. The middle-aged transformer C has an intermediate value of dominant time constant. The relatively new transformers D, E, and F have dominant time constant values in the higher range. From this table, a clear correlation between dominant time constant and ageing/moisture content of the different transformers can be observed.

3A.8.3.3 Effect of Temperature on RV Measurement

The RV method has also been reported to be sensitive to temperature, like the PDC measurement. A number of authors have reported this in the analysis of RV measurements with varying temperatures.

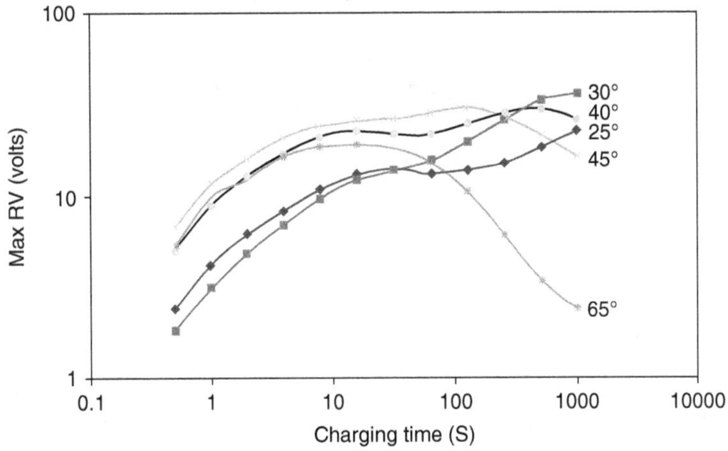

Figure 3A.86 RV spectra plotted against charging time at different temperatures. *Source:* Ref. [25], Figure 7.

3A.8.3.3.1 Laboratory Test Results

Saha and Purkait [25] performed experiments under controlled laboratory conditions to demonstrate the effects of temperature on RV measurements. The same sample oil–paper insulation system was tested at different temperatures. Polarization spectra obtained from those tests are shown in Figure 3A.86.

It is clear from this figure that for the same insulation, the dominant time constant decreases to lower values at higher temperatures. Similar findings were reported by other authors. Hongyan and Birlasekaran [40] measured RVs at different operating temperatures from 20°C to 60°C with an interval of 10°C. The peak was found to shift toward lower charging time when the temperature was increased. At 20°C there was no definite peak and the response was almost flat. Then, as the temperature was increased, it changed from 80 s at 30°C to 9 s at 60°C. They also found a shift to lower values of the charging time corresponding to the peak maximum RV with an increase in temperature. The peak maximum RV was not affected much by the change in temperature.

3A.8.3.3.2 Field Test Results

Similar findings were reported by researchers performing tests on transformers in the field. Figure 3A.87 [36] shows polarization spectra of a 10/0.4 kV transformer measured at various temperatures. The higher the measurement temperature, the more the dominant time constant shifts toward a lower value, which is consistent with laboratory measurements.

Kozlovskis and Rozenkrons [42] proposed a scheme to compare the RV spectra obtained at different temperatures by scaling the spectra up (or down) to a fixed reference temperature. Figure 3A.88 presents the RV spectra dependence on temperature for an average loss transformer. Two peaks are clearly observed for the measurement at 20°C, which is described by the authors as being due to separate oil and paper relaxation processes, respectively. Similar to other findings, the RV

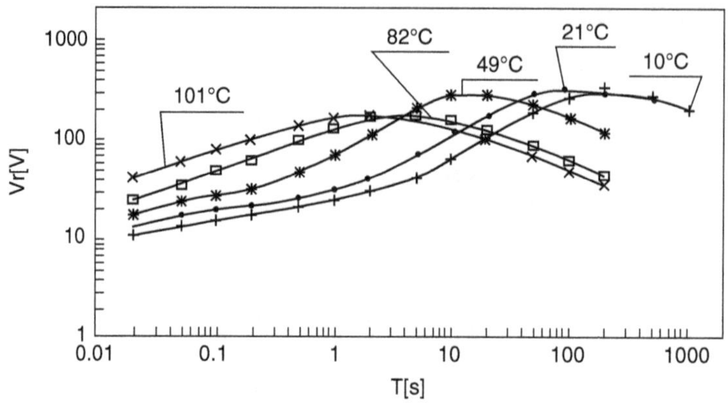

Figure 3A.87 Temperature dependence of polarization spectra curves. *Source:* Ref. [36], Figure 7.

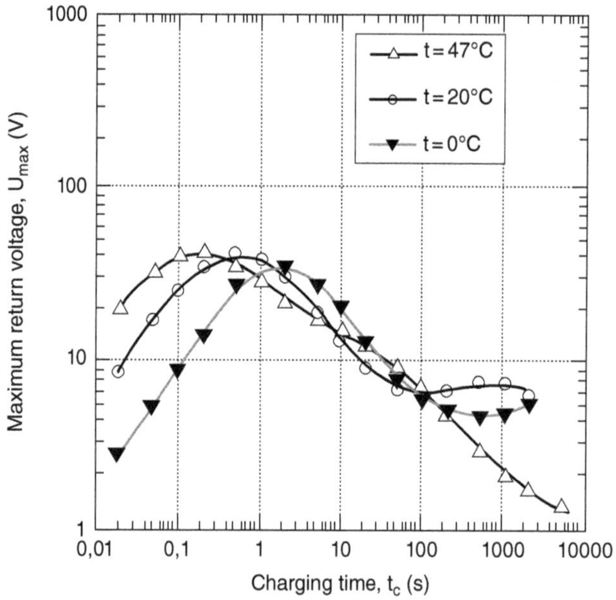

Figure 3A.88 Return voltage spectra dependence on temperature for an average-loss transformer. *Source:* Ref. [42], Figure 1.

spectrum shifts toward lower time constants for higher-temperature measurements. Two RV peaks in the measurement were reported; the second peak moved faster than the first peak and was spread over higher temperatures [42].

Kozlovskis and Rozenkrons [42] proposed an exponential dependence of the RV spectrum on temperature, which is described as a shift of RV spectrum with respect to charging time. The proposed model is

$$t_{c\,\text{at}\,T2} = t_{c\,\text{at}\,T1}.\exp(-a.\Delta T) \tag{3A.32}$$

where $t_{c\ at\ T1}$ is the charging time of a reference measurement at temperature T_1, $t_{c\ at\ T2}$ is a new shifted value of charging time for corresponding maximum return voltage measured at temperature T_2. ΔT is the temperature difference:

$$\Delta T = T_2 - T_1 \tag{3A.33}$$

The constant a is different for oil and paper, representing the difference in their polarization processes.

From the above findings it can be concluded that the RV spectrum depends on moisture, ageing by-products, and measurement temperature. In addition, oil and solid insulation have separate impacts on RV spectrum measurements. In laboratory experiments, it is easy to understand an individual process and hence easy to explain the results. However, the temperature dynamics of transformer insulation and corresponding moisture migration between oil and paper insulation make the interpretation of results more complicated for field transformers due to the convolution of the above dependencies.

3A.8.3.4 Effect of Insulation Geometry on RV Measurement

The effects of insulation geometry on RV measurement have been studied by many researchers, including the authors of this book [32]. The CIGRE Task Force [27] studied the effect of geometry on RV measurements comprehensively using the pancake model, as described before. The pancake model was designed to understand the influence of insulation geometry on dielectric responses, as the geometry of the insulation structure may not necessarily be known for old field transformers. The previously explained X–Y or X model of insulation system can provide some idea of the insulation structure. Figure 3A.89 [32] shows RVM spectra of maximum voltage and initial slope of the pancake model for two different geometrical arrangements. The CIGRE Task Force [27] concluded that:

> To summarize the results, one may say that dielectric response measurements on oil–paper insulation are sensitive not only to the state of the paper, but also to the oil conductivity and the geometry of the test object. As shown, a risk for errors in the interpretation of the RVM data exists when these parameters are disregarded, resulting in overestimates of the water content. This may be of importance, especially when making measurements on new or relatively new transformers in which the oil conductivity is unusually high. In a new transformer, there will not usually be a moisture balance between the oil and the pressboard. The oil is comparatively wetter and will with time, dry out due to moisture absorption in the cellulose. This results in a decrease in the conductivity of the oil. Furthermore, results from measurements on transformers with unconventional design (non-typical oil/paper ratios in the insulation) should also be treated with great care. [27]

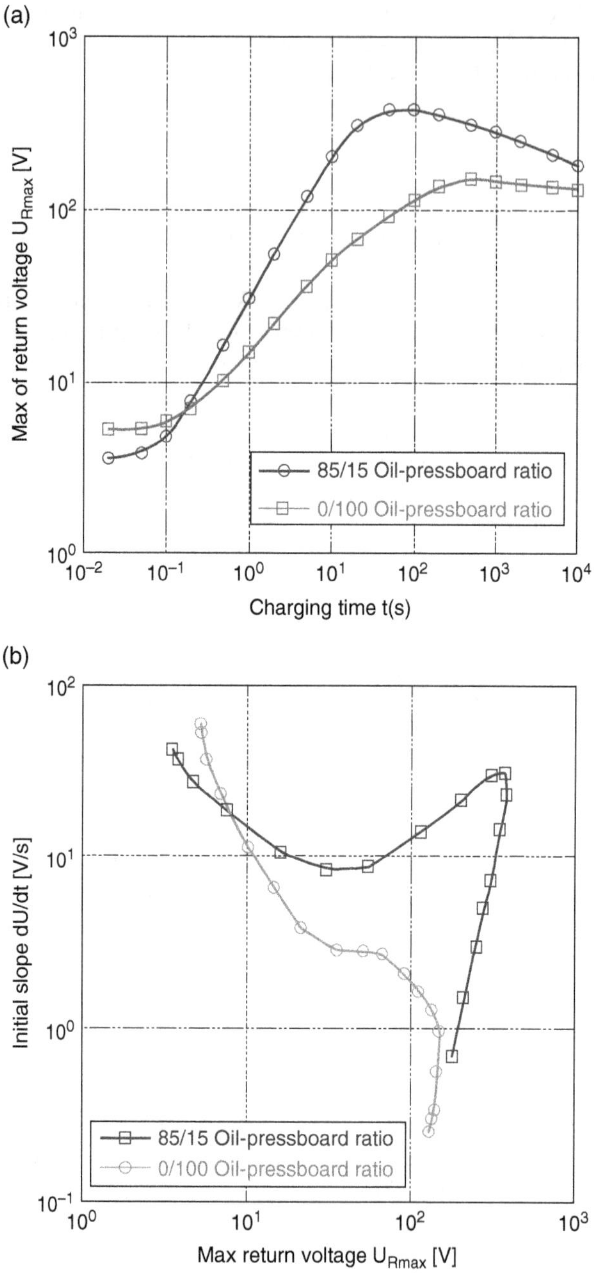

Figure 3A.89 RVM polarization spectra of maximum voltage and initial slope of the pancake model for two different geometrical arrangements, measured at 2000 V. *Source:* Ref. [32].

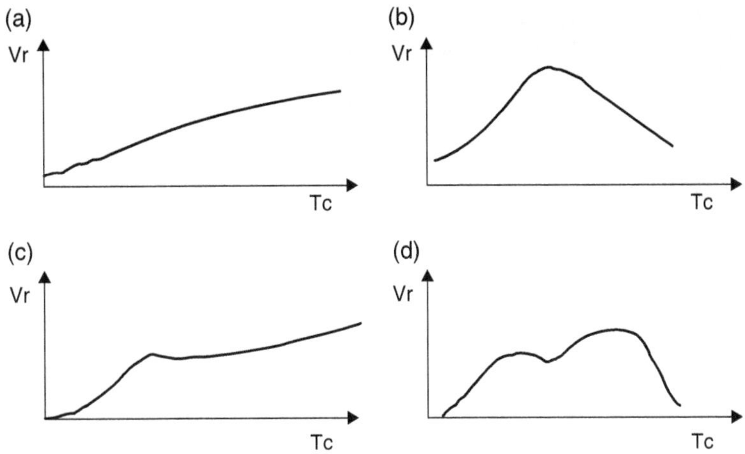

Figure 3A.90 Examples of RVM: (a) and (b) show "standard" spectra; (c) and (d) show "non-standard" spectra. *Source:* Ref. [43], Figure 1.

3A.8.4 Confounding Factors During Interpretation of RVM Results

During the last 25 years, numerous papers have been written to explain the RVM interpretation schemes. Most papers related RV spectra with moisture, and the researchers related to one peak of the RV spectra. Some authors explained all possible factors and their influence on the RV spectra, as well as with non-standard RV spectra. van Bolhuis *et al.* [43] divided the results of RVM tests into two basic categories: standard and non-standard spectra (see Figure 3A.90).

Standard RV spectra have been related as homogeneous, stable, and uniform in terms of temperature, moisture content, and ageing, showing some global maxima. In these cases, the dominant time constant can deliver a reasonable estimate of the actual condition of the insulation, provided the temperature of measurement, oil conductivity, and geometry information are considered. In contrast, RV spectra with multiple peaks, local maxima, flat curves, and measurement discontinuities fall into non-standard categories and their interpretation is more complicated. A number of non-standard RV spectra are given by van Bolhuis *et al.* [43], as shown in Figures 3A.91–3A.94, which are self-explanatory. Hence, while analyzing non-standard RV spectra, one needs to carefully consider impacts of oil quality, local low DP, and local moisture ingress conditions, along with standard factors affecting the RV spectra.

3A.8.5 Other Factors

For a proper assessment of the condition of a transformer by RV measurement, one should find and use the loading, operating, and manufacturing history of the transformer. Some authors have reported the influence of rain, humidity, and dew on RV measurements. This variation in RV measurement at high humidity could

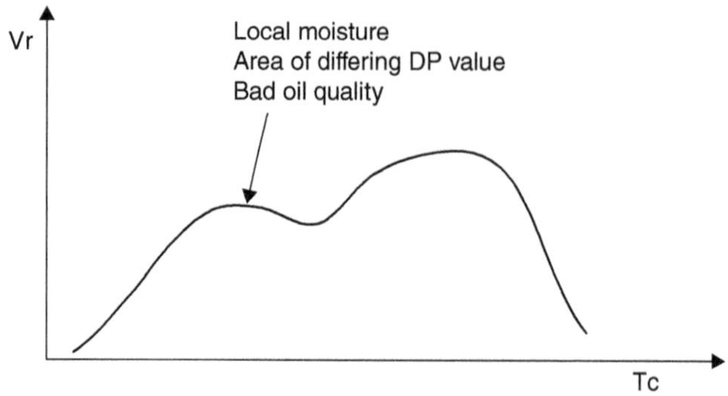

Figure 3A.91 Local moisture ingress, as area of insulation paper of different DP value and bad oil quality can cause a local maximum in the RVM spectrum. *Source:* Ref. [43], Figure 3.

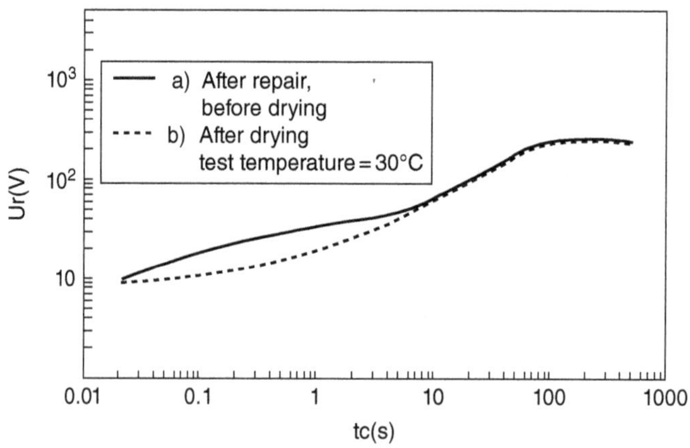

Figure 3A.92 Localized moisture ingress during onsite repairs on a transformer. The second curve shows the RVM spectrum after drying. *Source:* Ref. [43], Figure 6; Ref. [37], Figure 12.

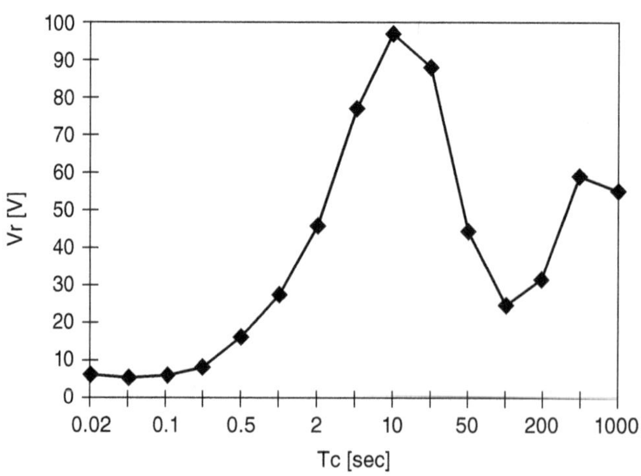

Figure 3A.93 Areas of differing DP value of the paper can cause multiple peaks. *Source:* Ref. [43], Figure 7.

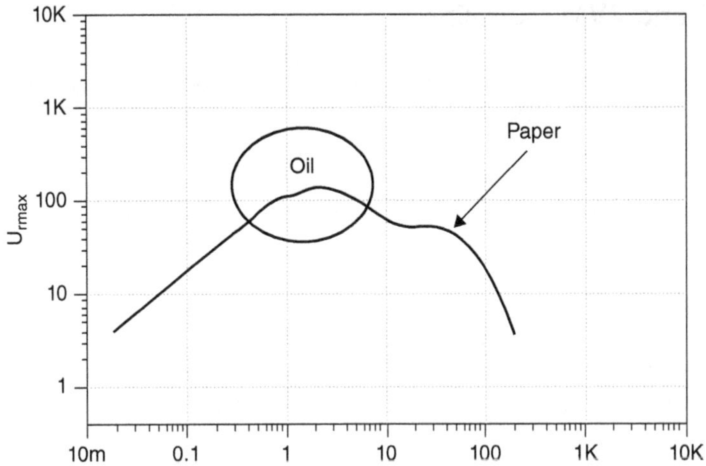

Figure 3A.94 RVM spectrum of the test sample with bad-quality oil. *Source:* Ref. [43], Figure 9.

be explained as due to any change in the leakage current across the bushing, which could be due to excessive moisture on the surface of the bushing, which may have generated local conduction. In a similar way, high atmospheric humidity can also be present during the early morning due to dew covering the external insulation of the transformer bushings. It was reported that an insufficient grounding period and residual static charge can affect RV measurements. Another serious influencing factor on RV measurement happens under field conditions when the test is conducted in an uneven electromagnetic environment. Among other outdoor-influenced factors of RVM, the effect of wind was reported by some authors. Any change in the shunt capacitance or insulation resistance during the RV measurement can make interpretation difficult.

3A.8.6 Conclusions

The main strength of the RVM test is in its simplicity and non-evasiveness. It does not require any intrusion into the transformer tank; neither has it required any physical samples to be taken out of the transformer. While the RVM test can provide useful information from an isolated test on a single transformer, its real value still lies in making comparisons of data between similar transformers of varying ages, or in periodic testing of the same transformer to detect an ageing rate. RVM can be a fairly lengthy test when dealing with dry insulation having long time constants. Too simplistic interpretation schemes should be avoided, and appropriate factors (e.g., loading and previous history, oil condition, geometry, external disturbance, and temperature) need to be considered for an accurate insulation interpretation. Modeling of RVM data and comparison with field measurements can also be an additional tool for assessing insulation condition. Additionally, a comparison between different diagnosis techniques – including with chemical-based techniques – can provide useful information.

3A.9 PDC and RVM Modeling

Two different modeling approaches are frequently used for modeling transformer insulation using PDC and or RV measurements and their interpretations. One of them is based on dielectric response equations, as described in Section 3A.3 for modeling PDC analysis and 3A.9 for RV analysis. Over the last two decades, several researchers have proposed a number of equivalent circuits for modeling the transformer oil–paper insulation system for a better understanding of the dielectric response.

Saha *et al.* [44] proposed a "black box" model to identify the individual effects of paper and oil condition on the model parameters without requiring any prior knowledge about their relative arrangements. The other models discussed in the literature are heavily dependent on correct knowledge of the design, geometry, and arrangement of insulation within the transformer. Such information may not be available easily from the utility companies – making the proposed modeling task difficult. Dielectric polarization processes can be modeled by a parallel arrangement of branches, each containing a series connection of resistor and capacitor, as shown in the circuit of Figure 3A.95. The conduction current in the insulation is due to the insulation resistance R_0, and C_0 represents the geometric capacitance of the insulation system.

As discussed in Ref. [44] for this model, most of the circuit parameters have been derived from measured polarization and depolarization currents using a curve-fitting technique. The capacitance is determined by conventional capacitance measurement techniques at power frequency. The insulation resistance is calculated from the difference between polarization and depolarization currents at larger values of time. The depolarization current is comprised of a summation of various relaxation mechanisms that appear at different locations within the transformer insulation. The relaxation currents can thus be modeled as the sum of exponentials of the various relaxation mechanisms. The individual elements with the corresponding time constants can then be determined by fitting

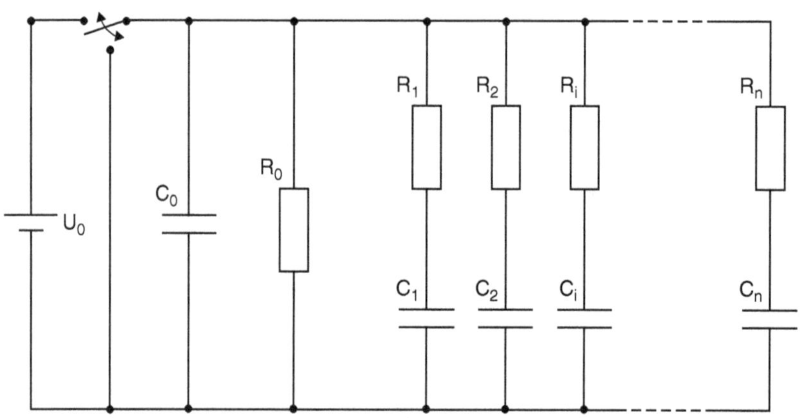

Figure 3A.95 Equivalent circuit to model a linear dielectric. *Source:* Ref. [44], Figure 1.

the depolarization current. The detailed procedure can be found in Ref. [44]. In this paper, the authors reported that the number of branches in most practical modeling purposes varied from six to ten, depending on the nature of the depolarization current. The ranges of values of the individual time constants, resistances, and capacitances are found to be dependent on the oil and paper conditions. The values of resistances and capacitances corresponding to the smaller time-constant branches are found to be related more to the condition of the oil, whereas the condition of the paper insulation is found to be responsible for affecting the higher time-constant branches.

Once the model parameters were estimated as described above, dielectric responses such as the polarization, depolarization currents, and return voltage were simulated [44]. Good agreement was reported between the measured PDC and RV spectra and the corresponding simulated curves. The approach presented in this work could be helpful for the interpretation and correlating the dielectric responses to the basic components of the insulation model. The variations of the PDC and RV spectra of the values of the equivalent model parameters have been investigated. In this way, the dielectric behavior of a complex insulation system has been explained from the properties of the basic model. The oil and paper conditions are found to affect the parameters of the equivalent model in a definite and distinct way, thereby enabling correct and transparent interpretation of the dielectric diagnosis results [44].

A case study is shown here from Ref. [44], where a transformer was undergoing an oil-reclamation process and tests were performed both before and after oil reclamation to study effect of the reclamation process on dielectric response. According to the information available from the utility, the oil-reclamation process basically consisted of recycling, filtering, and drying the oil through vacuum and at high temperature. The dry and filtered oil was pumped back to the tank and the whole process was continued in a closed cycle. In such a closed-cycle oil-reclamation process, it is expected that some moisture and other degradation by-products will be driven off the oil. At the same time, when dry oil is circulated through the transformer, some moisture will also migrate out of the paper toward the relatively dry oil. Thus, in such an oil-reclamation process, the oil moisture content and other ageing by-products are expected to go down, along with some improvement in the paper moisture condition as well. The oil moisture content values both before and after oil reclamation, measured by Karl Fischer titration, are presented in Table 3A.10. This table also contains the paper moisture content

Table 3A.10 Oil and paper moisture content before and after oil reclamation

Transformer		Before oil reclamation	After oil reclamation
Moisture content	Oil (ppm)	28	10
	Paper (%)	5	3.5
Oil sampling temperature (°C)		23	25

Source: Ref. [44].

values determined using the equilibrium chart from the measured oil moisture content values at the corresponding oil sampling temperatures [44].

Using the method in Ref. [44], polarization and depolarization currents and RV spectra were modeled and compared with the measured results, which showed good accuracy and are shown in Figures 3A.96 and 3A.97.

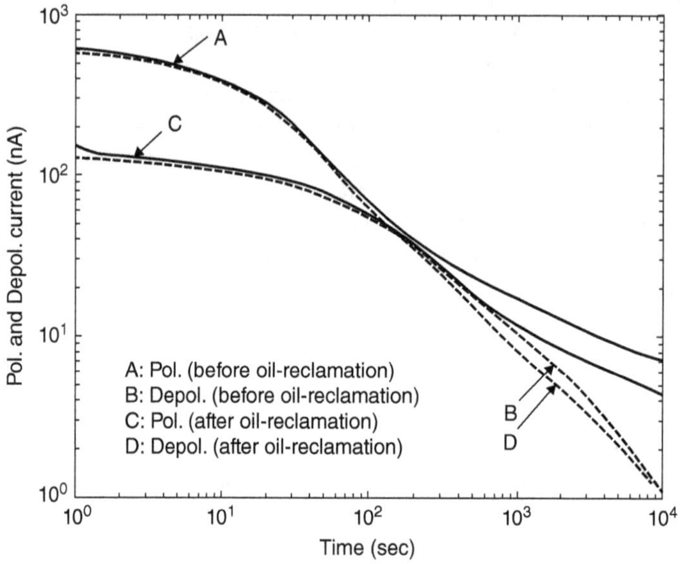

Figure 3A.96 Polarization and depolarization currents before and after oil reclamation. *Source:* Ref. [44], Figure 17.

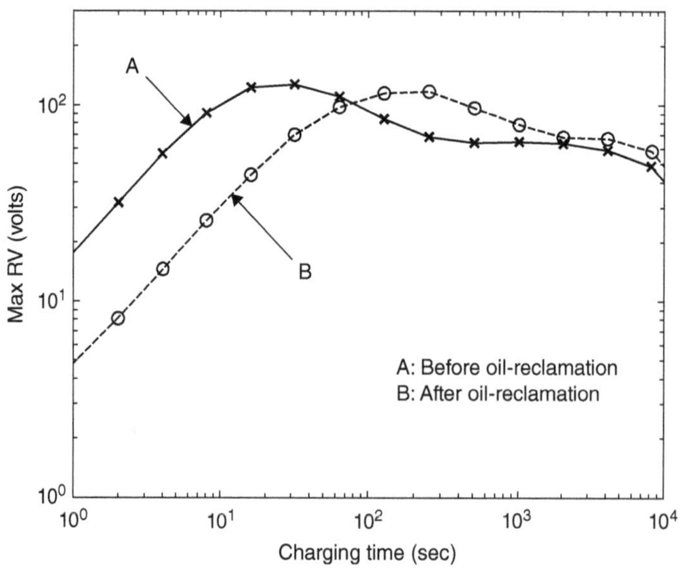

Figure 3A.97 RV spectra before and after oil reclamation. *Source:* Ref. [44], Figure 18.

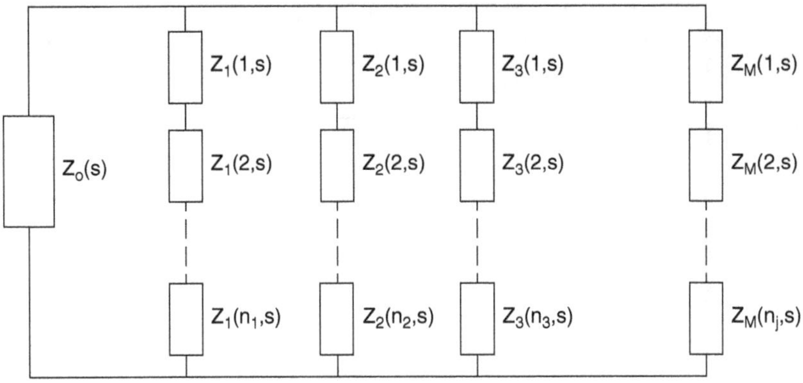

Figure 3A.98 Modified Debye model. *Source:* Ref. [45], Figure 4.

Baral and Chakravorti [45] proposed a modified Debye model (MDM), as shown in Figure 3A.98, which has been reported to take into account the effect of the non-uniform ageing in cellulosic insulation to better represent overall transformer insulation. They studied the MDM extensively for the insulation model, and verified that the proposed method was capable of predicting paper moisture content with acceptable accuracy compared with the conventional Debye model.

Several researchers have reported expert systems (ESs) for power system insulation diagnosis using different diagnostic techniques. Saha and Purkait [46] proposed an expert system based on the principle of transmission of human expert knowledge into the system, which is then used to evaluate diagnostic measurement data from commonly used offline diagnostic methods for diagnosis of transformer insulation. The proposed expert system used RVM and PDC techniques, which can help an inexperienced user to competently evaluate the state of transformer insulation system. Two knowledge bases are developed for the algorithm – the PDC knowledge base and the RVM database. Public domain information is retrieved to form these databases, along with the authors' own collection of hundreds of accelerated ageing results. The fundamental structure of the database is shown in Figure 3A.99. The proposed scheme has been tested with some field measurements, and has shown promising results. Further work has been in progress and is reported elsewhere by Saha *et al.*, in CIGRE and other publications, using machine learning techniques and developing a health index for transformer insulation assessments, which will be reported in Chapter 9.

Sarkar *et al.* [47] developed an ES, where they used dissipation factor, oil moisture content, central time constant (CTC), peak value of RV spectrum, and PDC data. This ES algorithm finds an optimized insulation model and specifically uses data from PDC and RV measurement data. The authors tested the ES for laboratory samples, as well as for a number of generator transformers. They reported good agreement between predicted and measured values for insulation diagnosis. They also discussed some major issues, which need to be considered for better performance of the ES.

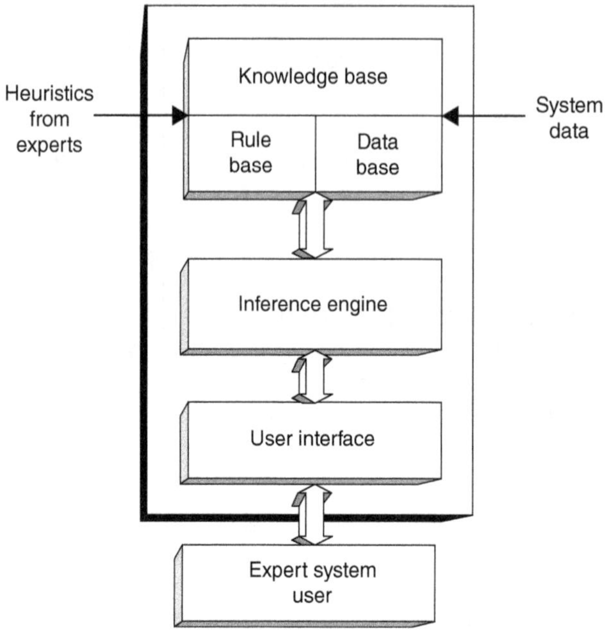

Figure 3A.99 Structure of ES. *Source:* Ref. [46], Figure 1.

3A.10 Final Comments on PDC and RV Measurements

One of the main criticisms of PDC and RV methods is the long time requirements for their measurements. Unless the transformer under test is going for a major overhaul, it is very difficult to disconnect a transformer for half a day or longer for PDC and/or RVM measurements. In recent years, PDC measurements have gained in popularity and acceptance more than RVM measurements. Usually, the polarization current is measured for 10,000 s, followed by a discharging process of 10,000 s for the depolarization current measurement. Saha *et al.* [48] investigated an alternative means to reduce the test time without losing significant accuracy of conductivity calculations compared with traditional PDC measurements, based on a method proposed by Küchler *et al.* [49]. It is reported that, for standard measurement conditions, if the test voltage and temperature are kept constant and the insulation is discharged completely before any testing is commenced, the difference between polarization and depolarization currents seems to be almost constant after about 1000 s. This means that if a PDC measurement is done with 2000 s charging and 2000 s discharging, then the difference in current may be constant, which could be used to estimate the paper conductivity. Similarly, the charge difference seems to be a straight line with a constant slope. Recently, Küchler *et al.* [49] reported a similar observation in the context of bushing insulation. Saha *et al.* [48] investigated the magnitude of error involved in paper conductivity calculations using short-duration PDC measurements from the concept of current difference and charge difference. A systematic experimental study was conducted on a model transformer and on a distribution transformer to select an optimal PDC test duration. The paper conductivities were calculated using all the methods, and it was found that a PDC test duration of 3000 s/3000 s can be used without losing much accuracy. Details about this methodology are presented in Ref. [48].

References

1 Saha TK, Purkait P. Investigation of polarization and depolarization current measurements for the assessment of oil–paper insulation of aged transformers. *IEEE Transactions on Dielectrics and Electrical Insulation* 2004; 11(1): 144–154.

2 Gäfvert U, Adeen L, Tapper M, Ghasemi P, Jonsson B. Dielectric spectroscopy in time and frequency domain applied to diagnostics of power transformers. 6th International Conference on Properties and Applications of Dielectric Materials, Xi'an, China, June 21–26, 2000.

3 Jonscher AK. *Dielectric Relaxation in Solids.* London: Chelsea Dielectric Press, 1983.

4 Saha TK. Review of time-domain polarisation measurements for assessing insulation condition in aged transformers. *IEEE Transactions on Power Delivery* 2003; 18(4): 1293–1301.

5 Ildstad E, Gäfvert U, Tharning P. Relation between return voltage and other methods for measurements of dielectric response. Conference Record of the IEEE International Symposium on Electrical Insulation, Pittsburgh, PA, June 5–8, 1994, pp. 25–28.

6 Leibfried T, Kachler AJ. Insulation diagnostics on power transformers using the polarisation and depolarisation current (PDC) analysis. Conference Record of the IEEE International Symposium on Electrical Insulation, Boston, MA, April 7–10, 2002, pp. 170–173.

7 Saha TK, Purkait P. 2003. Some precautions for the field users of PDC measurement for transformer insulation condition assessment. Proceedings of the IEEE Power Engineering Society General Meeting, Toronto, Ont., July 13–17, 2003.

8 Zaengl WS. Dielectric spectroscopy in time and frequency domain for HV power equipment. I. Theoretical considerations. *IEEE Electrical Insulation Magazine* 2003; 19(5): 5–19.

9 Homagk C, Leibfried T. Practical experience on transformer insulation: Condition assessment. Proceedings of 8th International Conference on Properties and Applications of Dielectric Materials, Bali, June 26–30, 2006, pp. 238–241.

10 Saha TK, Purkait P. Understanding the impacts of moisture and thermal ageing on transformers insulation by dielectric response and molecular weight measurements. *IEEE Transactions on Dielectrics and Electrical Insulation* 2008; 15(2): 568–582.

11 CIGRE Working Group 15.01, Task Force 09. Dielectric response methods for diagnostics of power transformers. *ELECTRA* 2002; 202: 24–35.

12 CIGRE Brochure 254, Report of the TF D1.01.09. *Dielectric Response Methods for Diagnostics of Power Transformers*. CIGRE, Paris, 2002.

13 Urbani GM. Untersuchung zum Einsatz der RVMMethode für die Bewertung von Oel-Papier Isolationssystemen. HAEFELY TEST AG Symposium, October 11–12, 2000.

14 Zaengl WS. Applications of dielectric spectroscopy in time and frequency domain for HV power equipment. *IEEE Electrical Insulation Magazine* 2003; 19(6): 9–22.

15 Gubanski SM *et al.* Reliable diagnostics of HV transformer insulation for safety assurance of power transmission system. *REDIATOOL – a European Research Project*. Paper No. D1-207, CIGRE, Paris, 2006.

16 Wei J-L, Zhang G-J, Xu H, Peng H-D, Wang S-Q, Dong M. Novel characteristic parameters for oil–paper insulation assessment from differential time-domain spectroscopy based on polarization and depolarization current measurement. *IEEE Transactions on Dielectrics and Electrical Insulation* 2011; 18(6): 1918–1928.

17 Koch M, Tenbohlen S. Diagnostics of oil–paper insulations using relaxation currents. Proceedings of the XIVth International Symposium on High Voltage Engineering, Beijing, August 25–29, 2005.

18 Fofana I, Hemmatjou H, Meghnefi F. Effect of thermal transient on the polarization and depolarization current measurements of oil–paper insulation. *IEEE Transactions on Dielectrics and Electrical Insulation* 2011; 18(2): 513–520.

19 Der Houhanessian V, Zaengl WS. Application of relaxation current measurements to on-site diagnosis of power transformers. IEEE Annual Report – Conference on Electrical Insulation and Dielectric Phenomena, 1997, pp. 45–51.

20 Frimpong G, Gäfvert U, Fuhr J. Measurement and modelling of dielectric response of composite oil–paper insulation. Proceedings of 5th International Conference on Properties and Applications of Dielectric Materials, Seoul, May 25–30, 1997.

21 Saha TK, Purkait P. *Investigating the impacts of ageing and moisture on dielectric response of oil/paper insulation systems*. International Symposium on High Voltage Engineering, Amsterdam, 2003.

22 Saha TK, Purkait R. Impact of the condition of oil on the polarisation based diagnostics for assessing the condition of transformers insulation. Proceedings of the IEEE Power Engineering Society General Meeting, San Francisco, CA, June 12–16, 2005.

23 Der Houhanessian V, Zaengl WS. Time domain measurements of dielectric response in oil–paper insulation systems. Conference Record of the IEEE International Symposium on Electrical Insulation, Montreal, Que., June 16–19, 1996, pp. 47–52.

24 Noirhommel B, David E, Garbi H, Lessard MC, Boissonneault R. Application of dielectric response techniques for the condition assessment of power transformers. IEEE Annual Report – Conference on Electrical Insulation and Dielectric Phenomena, 2005, pp. 273–276.

25 Saha TK, Purkait P. Investigations of temperature effects on the dielectric response measurements of transformer oil–paper insulation system. *IEEE Transactions on Power Delivery* 2008; 23(1): 252–260.

26 Der Houhanessian V, Zaengl WS. On-site diagnosis of power transformers by means of relaxation current measurements. Conference Record of the IEEE International Symposium on Electrical Insulation, Arlington, VA, June 7–10, 1998, pp. 28–34.

27 CIGRE Task Force. Dielectric response methods for diagnostics of power transformers. *IEEE Electrical Insulation Magazine* 2003; 19(3): 12–18.

28 Hassig M, Braunlich R, Gysi R, Alff J-J, Der Houhanessian V, Zaengl WS. On-site applications of advanced diagnosis methods for quality assessment of insulation of power transformers. IEEE Annual Report – Conference on Electrical Insulation and Dielectric Phenomena, 2001, pp. 441–447.

29 RVM 5462 product details. Available online at: www.haefely-hipotronics.com/products/product/recovery-voltage-meter-5462/.

30 ALFF Engineering PDC Analyser product details. Available online at: www.alff-engineering.ch/index.php?option=com_content&view=article&id=4&Itemid=112.

31 DIRANA-FDS-PDC Dielectric Response Analyser product details. Available online at: www.omicron.at/en/products/all/primary-testing-monitoring/dirana/.

32 Saha TK, Yao ZT. Experience with return voltage measurements for assessing insulation conditions in service-aged transformers. *IEEE Transactions on Power Delivery* 2003; 18(1): 128–135.

33 Gäfvert U *et al. Modelling of dielectric measurements on power transformers*. Paper No. 15-103, CIGRE, Paris, 1998.

34 Saha TK, Darveniza M, Hill DJT, Le TT. Electrical and chemical diagnostics of transformers insulation. Part A: Aged transformers samples. *IEEE Transactions on Power Delivery* 1997; 12(4): 1547–1554.

35 Webb M. Transformer condition assessment with the Tettex recovery voltage method. Digest No. 1996/186, IEE Colloquium on Monitors and Condition Assessment Equipment, 1996, pp. 1–5.

36 Osvath P, Csepes G, Webb M. Polarisation spectrum analysis for diagnosis of insulation systems. Sixth International Conference on Dielectric Materials, Measurements and Applications, 1992, pp. 270–273.

37 Csepes G, Hamos I, Brooks R, Karius V. Practical foundations of the RVM (recovery voltage method) for oil/paper insulation diagnosis. IEEE Annual Report – Conference on Electrical Insulation and Dielectric Phenomena, 1998, pp. 345–355.

38 van Bolhuis JP, Gulski E, Smit JJ. Monitoring and diagnostic of transformer solid insulation. *IEEE Transactions on Power Delivery* 2002; 17(2): 528–536.

39 Osvath P, Zahn H. Polarisation spectrum analysis for diagnosis of oil/paper insulation systems. Conference Record of the IEEE International Symposium on Electrical Insulation, Pittsburgh, PA, June 5–8, 1994, pp. 155–161.

40 Saha TK, Darveniza M, Yao ZT. Field experience with return voltage measurements for assessing insulation condition in transformers and its comparison with accelerated aged insulation samples. Eleventh International Symposium on High-Voltage Engineering, London, August 22–27, 1999.

41 Hongyan C, Birlasekaran S. Temperature dependent relaxation studies on oil-filled transformer. Conference Record of the IEEE International Symposium on Electrical Insulation, Boston, MA, April 7–10, 2002, pp. 174–178.

42 Kozlovskis A, Rozenkrons J. Temperature dependence of return voltage characteristics. *IEEE Transactions on Power Delivery* 1999; 14(3): 705–708.

43 van Bolhuis JP, Gulski E, Smit JJ, Urbani GM, Verhaart HFA. Development of knowledge rules for RVM for interpretation of the condition of transformer insulation. Conference Record of the IEEE International Symposium on Electrical Insulation, Anaheim, CA, April 2–5, 2000, pp. 267–270.

44 Saha TK, Purkait P, Müller F. Deriving an equivalent circuit of transformers insulation for understanding the dielectric response measurements. *IEEE Transactions on Power Delivery* 2005; 20(1): 149–157.

45 Baral A, Chakravorti S. Prediction of moisture present in cellulosic part of power transformer insulation using transfer function of modified Debye model. *IEEE Transactions on Dielectrics and Electrical Insulation* 2014; 21(3): 1368–1375.

46 Saha TK, Purkait P. Investigation of an expert system for the condition assessment of transformer insulation based on dielectric response measurements. *IEEE Transactions on Power Delivery* 2004; 19(3): 1127–1134.

47 Sarkar S, Sharma T, Baral A, Chatterjee B, Dey D, Chakravorti S. An expert system approach for transformer insulation diagnosis combining conventional diagnostic tests and PDC, RVM data. *IEEE Transactions on Dielectrics and Electrical Insulation* 2014; 21(2): 882–891.

48 Saha TK, Pradhan MK, Yew JH. Optimal time selection for the polarisation and depolarisation current measurement for power transformer insulation diagnosis. Proceedings of the IEEE Power Engineering Society General Meeting, Tampa, FL, June 24–28, 2007.

49 Küchler A, Hullmandel F, Bohm K, Neumann C, Koch N, Loppach K *et al.* *Condition assessment of aged transformer bushing insulations.* CIGRE, Paris, 2006.

Part B: Frequency Domain Dielectric Spectroscopy

Dielectric response measurement in the frequency domain is a well-established technique to study the polarization behavior of insulation systems, especially when they are formed by a complex mixture of insulation materials. In this technique, slow polarization mechanisms in the insulation system are studied by measuring the magnitude and phase of current due to a sinusoidal voltage excitation. Typically, in transformer applications, FDS measurements are conducted between 0.1 mHz and 1 kHz. This technique enables measurement of the frequency-dependent capacitance, conductivity, and loss tangent of an insulation system, which can be used to estimate the average moisture content and oil conductivity of the insulation through various modeling techniques. A major advantage of this technique is that measurements are typically performed at low voltage levels, as noise suppression is relatively easier in narrow band measurements.

3B.1 Theory of FDS [1]

Dielectric polarization of most of the insulation materials changes linearly with an applied electric field. Assuming that the insulation material is isotropic and uniform, the induced dielectric polarization P in an insulation system is related to the applied electric field E by the equation

$$P(t) = \varepsilon_0 \chi E(t) \tag{3B.1}$$

where ε_0 is the electric permittivity of free space and χ is the electric susceptibility of the dielectric material. Susceptibility of a dielectric material is a measure of polarizability in response to an electric field. This, in turn, determines the electric permittivity of the material and thus influences many other phenomena governed by the dielectric properties of the medium. Therefore, the susceptibility of a medium is related to its relative permittivity ε_r by

$$\chi = \varepsilon_r - 1 \tag{3B.2}$$

Owing to the induced dielectric polarization, the dielectric displacement is increased. The electric displacement D is related to the dielectric polarization P by

$$\begin{aligned} D(t) &= \varepsilon_0 E(t) + P(t) \\ &= \varepsilon_0 E(t) + \varepsilon_0 \chi E(t) \\ &= \varepsilon_0 (1 + \chi) E(t) \\ &= \varepsilon_0 \varepsilon_r E(t) \end{aligned} \tag{3B.3}$$

The current density $J(t)$ inside a dielectric material due to a step excitation from an external source is given by [2–4]

$$J(t) = \sigma E(t) + \partial D(t)/\partial t \tag{3B.4}$$

where σ is the DC conductivity of the dielectric material.

An analytical transition can be made from time to frequency domain (angular frequency ω) using a Fourier transform, assuming the dielectric to be linear, homogeneous, and isotropic, in addition to the electric field being purely sinusoidal:

$$\begin{aligned} J(\omega) &= \sigma E(\omega) + j\omega D(\omega) \\ &= \sigma E(\omega) + j\omega \varepsilon_0 \varepsilon_r E(\omega) \\ &= j\omega \varepsilon_0 \left(\varepsilon_r + \frac{\sigma}{j\omega \varepsilon_0} \right) E(\omega) \\ &= j\omega \varepsilon_0 \bar{\varepsilon}(\omega) E(\omega) \end{aligned} \tag{3B.5}$$

where $\bar{\varepsilon}(\omega)$ is the complex permittivity: $\bar{\varepsilon}(\omega) = \varepsilon'(\omega) - j\varepsilon''(\omega)$.

An expression for the current density can also be written:

$$
\begin{aligned}
J(\omega) &= j\omega\varepsilon_0\left(\varepsilon_r + \frac{\sigma}{j\omega\varepsilon_0}\right)E(\omega) \\
&= j\omega\varepsilon_0\left(1+\bar{\chi}-j\frac{\sigma}{\omega\varepsilon_0}\right)E(\omega)
\end{aligned}
\tag{3B.6}
$$

where $\bar{\chi}$ is the complex susceptibility: $\bar{\chi}=\chi'-j\chi''$.

Thus,

$$
\begin{aligned}
J(\omega) &= j\omega\varepsilon_0\left(1+\chi'-j\left(\frac{\sigma}{\omega\varepsilon_0}+\chi''\right)\right)E(\omega) \\
&= j\omega\varepsilon_0(\varepsilon'-j\varepsilon'')E(\omega) \\
J(\omega) &= j\omega\varepsilon_0\bar{\varepsilon}E(\omega)
\end{aligned}
\tag{3B.7}
$$

The complex permittivity $\bar{\varepsilon}$ can then be formulated having as real part

$$
\varepsilon'(\omega) = \mathrm{Re}(\bar{\varepsilon}(\omega)) = 1+\chi'(\omega) \tag{3B.8}
$$

and as imaginary part

$$
\varepsilon''(\omega) = \mathrm{Im}(\bar{\varepsilon}(\omega)) = \frac{\sigma}{\varepsilon_0\omega}+\chi''(\omega) \tag{3B.9}
$$

which contains the DC conduction as well.

Assuming a parallel-plate capacitor between the electrodes, the geometric capacitance of the dielectric material can be given by

$$
C_0 = \varepsilon_0 \times \frac{A}{d} \tag{3B.10}
$$

with ε_0 being the permittivity of free space, A the cross-sectional area of the dielectric, and d its thickness.

The current density expression given in equation (3B.1) can be rewritten as

$$
J(\omega) = j\omega C_0 \times \frac{d}{A}\left(1+\chi'-j\left(\frac{\sigma}{\omega\varepsilon_0}+\chi''\right)\right)E(\omega) \tag{3B.11}
$$

or

$$
J(\omega)\times A = j\omega C_0\left(1+\chi'-j\left(\frac{\sigma}{\omega\varepsilon_0}+\chi''\right)\right)E(\omega)\times d \tag{3B.12}
$$

or

$$
I(\omega) = j\omega C_0\left(1+\chi'-j\left(\frac{\sigma}{\omega\varepsilon_0}+\chi''\right)\right)U(\omega) \tag{3B.13}
$$

where I is the current through the dielectric due to the external voltage U across the insulation.

In the relation

$$I(\omega) = j\omega C_0\left(1 + \chi' - j\left(\frac{\sigma}{\omega\varepsilon_0} + \chi''\right)\right)U(\omega) \tag{3B.14}$$

or

$$I(\omega) = j\omega\bar{C}(\omega)U(\omega) \tag{3B.15}$$

we can define a complex capacitance $\bar{C}(\omega) = C'(\omega) - jC''(\omega)$, where $C'(\omega) = C_0(1 + \chi'(\omega))$ represents the real capacitance and $C''(\omega) = C_0\left(\frac{\sigma}{\omega\varepsilon_0} + \chi''\right)$ is the imaginary capacitance, which corresponds to energy loss in the insulation material. Energy loss is a combination of resistive losses due to the conduction effect and dielectric losses due to the inertia of bound and space charges when they are displaced by an electric field.

It can be seen in equation (3B.14) that the contribution of conduction effect to imaginary capacitance $C''(\omega)$ is inversely proportional to frequency. Therefore, the effect of conductivity on loss becomes more dominant at low frequencies. However, most of the solid insulation materials – such as paper/pressboard – possess low-frequency dispersion (LFD), which increases the dielectric loss at low frequencies. A common technique available to separate the effect of conduction from LFD is called the Krammers–Kronig (K–K) transformation, as given in the following equations:

$$\chi'(\omega) = \frac{2}{\pi}\lim_{a\to\infty}\int_0^a \frac{x\chi''(x)}{x^2 - \omega^2}dx \tag{3B.16}$$

$$\chi''(\omega) = \frac{2\omega}{\pi}\lim_{a\to\infty}\int_0^a \frac{\chi'(x)}{x^2 - \omega^2}dx \tag{3B.17}$$

K–K transformation interrelates the imaginary and real components of susceptibility. As the real part of the susceptibility is independent of conductivity, one can use it to calculate the imaginary part of susceptibility using K–K transformation. The difference between measured total loss and calculated imaginary part of susceptibility is the dielectric loss due to conduction.

As per equation (3B.14), the measured dielectric response depends on the geometrical capacitance of the insulation system. Consequently, the dielectric responses of transformers are highly influenced by the insulation geometry, which depends on various factors such as power level, voltage level, manufacturer, winding type, etc. Apparently, when geometrical information on the insulation system of a transformer is not readily available, as in most cases for old transformers, the absolute dielectric response cannot be compared without compensating for the geometrical capacitance. Therefore, in many field applications, the dielectric dissipation factor tan δ, which is independent of geometry, has been used widely.

The dielectric dissipation factor or loss factor is defined as the ratio of the imaginary part of the impedance to the real part of the impedance. In terms of complex capacitance and permittivity, it can be expressed as

$$\tan\delta(\omega) = \frac{\varepsilon''(\omega)}{\varepsilon'(\omega)} = \frac{C''(\omega)}{C'(\omega)} = \frac{\dfrac{\sigma}{\omega\varepsilon_0} + \chi''}{1+\chi'} \tag{3B.18}$$

Though DDF is an attractive parameter to compare different insulation systems with, it is worth mentioning that combining the two parameters could lose some critical information and therefore in certain FDS analysis techniques, permittivity and loss are utilized as two separate parameters.

3B.2 FDS of Oil–Paper Insulation

The majority of power transformer insulation, especially between the high and low-voltage windings, is formed by a combination of oil and oil-impregnated pressboard. The dielectric response of such a complex system depends on the response of individual materials, as well as the way they are combined together. Therefore, the FDS of insulation systems is useful to extract the properties of each material and the physical arrangement of an insulation system, through appropriate modeling.

Mineral oil is highly conductive compared with impregnated pressboard/paper insulation. Owing to conductivity differences when such a system is subjected to an electrical stress, charge accumulation typically occurs at the interface of the two materials. This phenomenon is known as interfacial polarization or the Maxwell–Wagner effect, which is pronounced at low-frequency excitations for oil–paper insulation systems. Therefore, low-frequency FDS is a useful tool for identifying the significance of interfacial polarization in complex insulation systems, such as transformer insulation [5].

Mineral oil is a nonpolar substance and therefore its dielectric dispersion can be neglected at frequencies below 1 MHz. Therefore, the FDS of oil is typically modeled by a constant permittivity value ($\varepsilon_r = 2.2$) and the DC conductivity σ_{dc} assuming a frequency-independent susceptibility at low frequencies. As oil conductivity is highly reliant on the presence of ionic constituents, it is used as a quality measure of a transformer insulation system. In contrast, pressboard is a polar material whose dielectric dispersion depends mainly on the content of polarizable substances, such as water, which resides in it. Therefore, the FDS of oil-impregnated pressboard is strongly dependent on the presence of moisture and other ageing products in the insulation.

Accordingly, the FDS of transformer insulation reflects the dielectric response of oil, paper, and the insulation geometry. However, these parameters dominate in the total response at different frequencies. Dominant influences of pressboard, oil, and geometry at different frequencies on the total response are illustrated in Figure 3B.1.

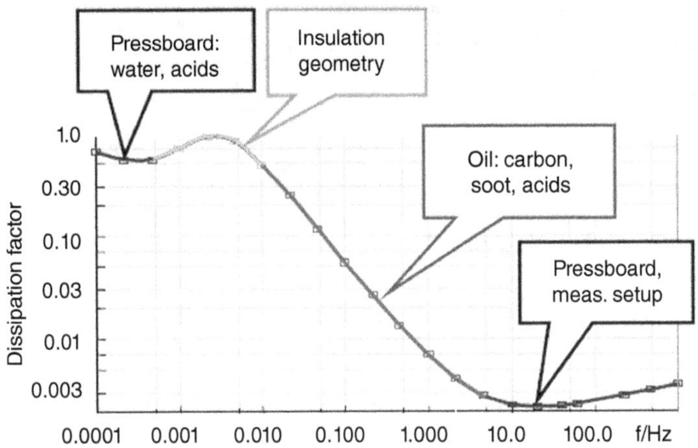

Figure 3B.1 Variation of loss factor of transformer insulation with frequency. *Source:* Ref. [6], Figure 10.

As seen in this figure, the frequency range above 10 Hz is dominated mostly by the condition of the solid insulation, with additional influences of inductances of the coils, measurement cables, and connection techniques also being present. The steep slope in the frequency region 0.01–1 Hz is caused mainly by oil conductivity. Dissolved conductive ageing by-products, soot, and high molecular weight acids increase the oil conductivity and thus influence this area [6]. The insulation geometry (ratio of oil to pressboard) determines the local maximum or "hump" visible at a lower frequency. The higher the ratio of oil to pressboard, the more dominating is this effect. The properties of cellulose appear once again at frequencies around 1 mHz, reflecting the moisture content, manufacturing process, presence of low molecular weight acids, and other conductive ageing by-products. The frequency ranges indicated, however, may vary in a wide range in other cases with varying degrees of moisture, oil conductivity, insulation geometry, temperature, and amount of conductive ageing by-products.

3B.3 FDS Measurement Procedure for Transformers

FDS measurement on transformer insulation is a relatively new technique to the power industry. Instrumentation for these measurements emerged in the market in the late 1990s. IDA 200, an instrument from General Electric Programma Inc., was the first instrumentation available in the market for FDS measurements on transformers. Various versions of IDA have been released during the last 15 years. DIRANA from Omicron is another device to measure dielectric response by combining the FDS and PDC responses of transformer insulation. These devices operate based on the same fundamental principal. However, they may each have their own merits as well as disadvantages, based on operational mechanisms and technical limitations. They are incorporated with modeling software to interpret the dielectric response measurements.

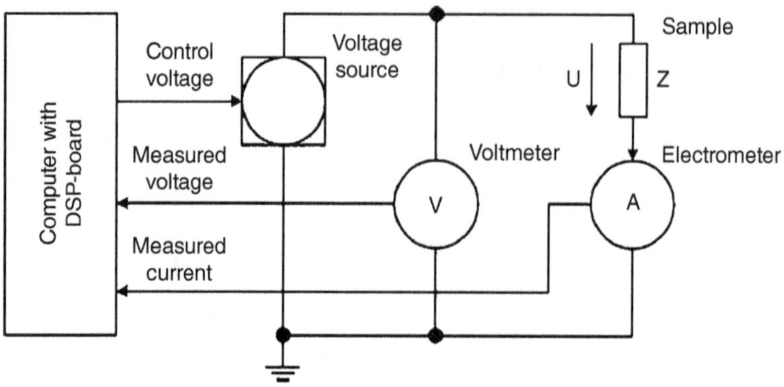

Figure 3B.2 Basic FDS measurement circuit using the IDA 200 system. *Source:* Ref. [7], Figure 2.

In a typical FDS measurement system, a relatively low voltage of up to 200 V with a supply frequency ranging between 0.1 mHz and 1 kHz is applied. The system measures the current through the insulation (magnitude and phase angle) at a given sinusoidal excitation. Measured current and voltage will be used to calculate the capacitance and dissipation factor of the insulation. FDS can be performed either as a frequency sweep or as a voltage sweep. In a frequency sweep, measurements are performed at set frequencies while the excitation voltage is kept at a constant peak level. Measurements always run from the highest frequency to the lowest, to minimize the polarization memory effect. This reduces the effect of polarization by the previous measurements on the subsequent measurement. The voltage sweep measurement is performed at a constant frequency, while the voltage peak varies from lowest to highest set level. In transformer applications, the frequency sweep is more common as it clearly elaborates the dielectric dispersion of the insulation. The voltage sweep is useful for checking nonlinear polarization phenomena in the insulation system. A schematic block diagram of a basic FDS measurement circuit system is illustrated in Figure 3B.2.

Three-electrode measurement, which enables diversion of the leakage current to ground through a guard electrode, is typically used in FDS measurements to improve the accuracy of measurements. A three-electrode FDS measurement configuration for a single-phase transformer is shown in Figure 3B.3. In the case of a three-phase transformer, all the terminals of the HV and LV windings are short-circuited separately and connected to the FDS measuring device, similar to the single-phase connection. The grounded tank is connected to the guard terminal, as it was for single-phase measurements.

The three-electrode configuration presented here is applicable for measurements between ungrounded terminals. In the case of transformers, the typical FDS measurement is performed between HV and LV (both are ungrounded) and the presented measurement configuration is utilized. Nevertheless, the latest FDS instrumentation provides support to measure the FDS between winding terminals and the tank (grounded), with minimum effect of leakage current to ground on the measurement accuracy.

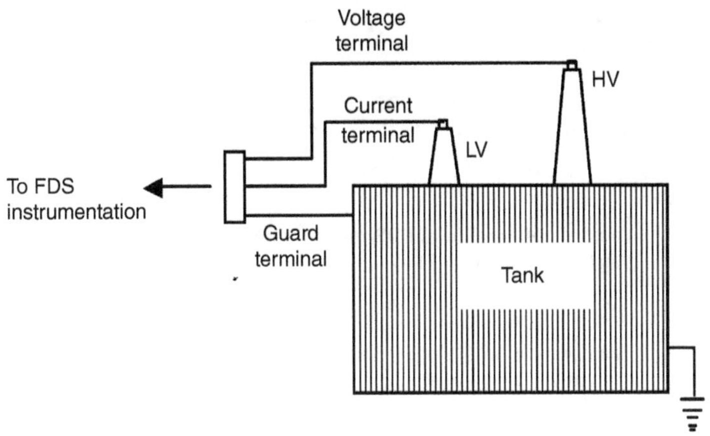

Figure 3B.3 FDS measurement configuration with guard electrode. *Source:* Ref. [8].

3B.4 Factors Affecting FDS Measurement on Transformers

Apart from the internal insulation condition of the transformer, there are various other internal and external factors that could affect the FDS response, potentially leading to a wrong interpretation of transformer condition. Therefore, it is essential to minimize the effect of those factors, or introduce proper calibration against them after carefully studying the overall measurement system.

3B.4.1 Temperature Variations

Dielectric response is highly temperature dependent. Therefore, the correct measurement temperature should be known for reliable interpretation. In the case of power transformers, the temperature is not uniform inside the transformer. On top of that, the transformer temperature changes with loading condition and ambient temperature. Typically, the transformer temperature is above ambient temperature when it is taken out of operation. However, due to the limited time available to keep the unit offline, as in most practical cases, one cannot wait until the temperature reaches a steady state before starting FDS measurements. When measurements are started at an elevated temperature, this will gradually decrease during the measurements. The rate of temperature decrease depends on the initial temperature, internal factors such as the heat capacity of the unit, and external factors such as the ambient temperature. In a tropical environment the temperature drop could be a couple of degrees, but could be as high as 15°C in cold temperate conditions. This effect has been studied in Ref. [9], and it is clearly shown that the temperature variation is dominant at frequencies below 1 Hz, as illustrated in Figure 3B.4. Based on the authors' experience of FDS data analysis, it is recommended to check the temperature variation during measurement – appropriate temperature corrections should be introduced in the FDS data analysis.

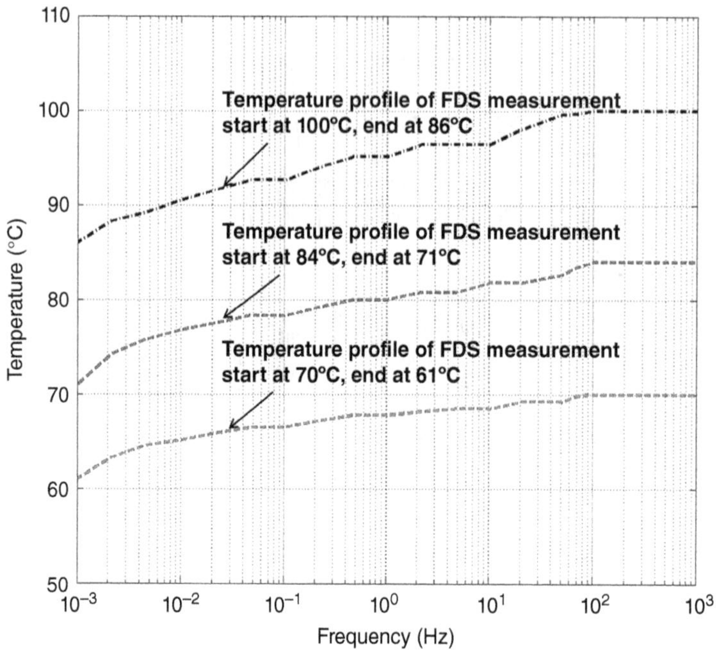

Figure 3B.4 Temperature variation profile during FDS measurements. *Source:* Ref. [9], Figure 13.

The typical temperature variation of a transformer from top to bottom could be high as 20–30°C. This could also introduce errors in the FDS analysis. However, one can easily use the average temperature of the winding in the analysis, since this technique is only capable of estimating the average moisture content in the solid insulation.

3B.4.2 Leakage Currents

In general, it is recommended to completely isolate the transformer from the system for FDS measurements. This enables the reduction of unnecessary leakage current paths and overloading of the measuring instrument source through stray capacitances. However, in certain circumstances a transformer cannot be completely isolated and it is disconnected from the system only by opening the isolators. Therefore, cables and isolators are still connected to the transformer terminals. These additional devices increase the stray capacitance and hence introduce additional leakage paths, which could affect the FDS measurements.

3B.4.2.1 Measurement between HV and LV

As shown in Figure 3B.5, the measurement electrode connected to the LV will be short-circuited to ground if the voltage transformer (VT) is still connected. Therefore, during FDS measurements, all VTs should be disconnected from the terminals. In this measurement configuration the guard electrode collects all the internal leakages and creep of the bushings. Hence, the FDS response between

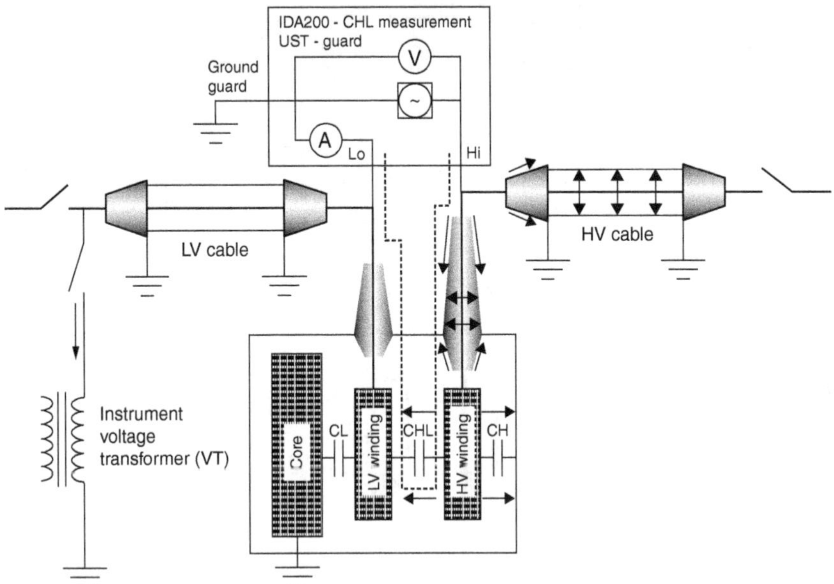

Figure 3B.5 FDS measurement between HV and LV. *Source:* Ref. [5].

HV and LV can be measured with good accuracy. Therefore, for FDS measurements between HV and LV it is not compulsory to completely remove the cable connections from the terminals. However, one has to consider the loading capability of the instrument source, as additional cabling will increase the stray capacitance and the source may not be able to supply the current required to charge the complete system.

3B.4.2.2 Measurement between HV/LV and Ground

Figure 3B.6 presents the measurement configuration to measure the dielectric response between HV and tank (ground). A similar configuration is used to measure the response between LV and tank, but the electrodes on LV and HV should be interchanged. As shown in the figure, all current passing to ground is measured as the insulation current. Therefore, in this configuration it is impossible to avoid measuring the external and internal leakage currents. Therefore, if one needs to measure the response of HV–ground or LV–ground insulation it is essential to isolate the transformer completely from the system and clean the bushing surfaces adequately.

3B.5 Interpretation of FDS Results

Plots often generated from measurements performed by FDS equipment are tan δ, C', C'', ε', and ε'' versus frequency. One such sample plot of dissipation factor versus frequency is shown in Figure 3B.1 [6]. As pointed out in the foregoing section, FDS results are influenced by several parameters including condition of oil and

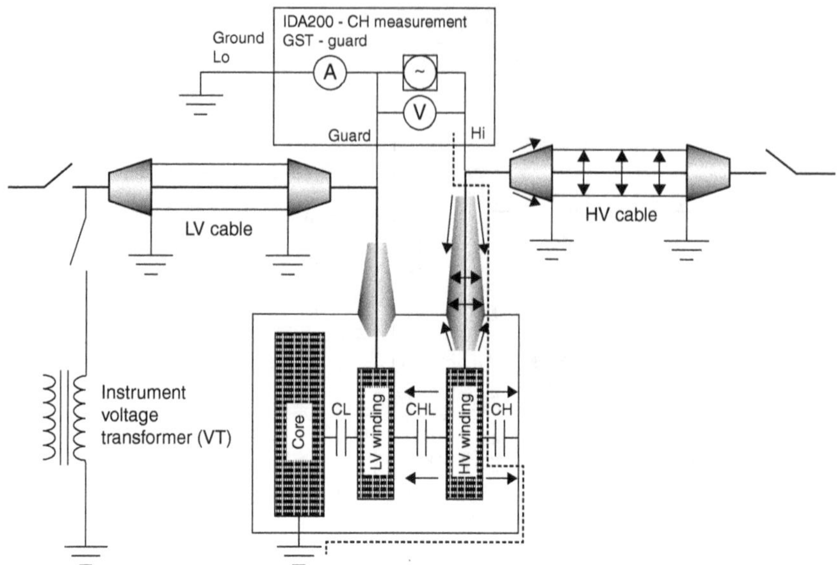

Figure 3B.6 FDS measurement between HV and ground. *Source:* Ref. [5].

paper/pressboard, presence of acids and other ageing-related impurities, moisture content, temperature, and insulation geometry. As will be discussed in more detail during the remaining sections of this chapter, different portions of the plot are influenced by different parameters of the insulation system, such as moisture content, ageing, acids, geometry, etc.

3B.5.1 Effect of Moisture

Moisture is a critical factor for paper ageing. Hence, to assess the condition of a complex oil–paper system such as that in power transformers, it is a requirement to identify the moisture content in the paper insulation. Taking a paper sample from the transformer for testing is generally impractical for a quick field test. Measurement of the absolute moisture content of a sample of the oil in ppm (e.g., by means of a Karl Fischer titration test) cannot generally provide any conclusive information, since the oil–paper moisture equilibrium is temperature dependent and can take a long time to stabilize. Therefore, a non-intrusive method allowing estimation of the moisture content of the transformer solid insulation is essential. Such a method will then permit a judgment to be made on the state of ageing of the insulation system, and can be used to determine a maximum safe operating temperature for the transformer in order to avoid accelerating the further ageing of the paper. Dielectric response measurements including FDS are extremely useful and are a scientifically proven measurement technique used to estimate the moisture content in transformer insulation through a simple onsite test, without the need to open the tank or take out any oil or paper samples. FDS can corroborate with time domain spectroscopy tests and further chemical analysis to reconfirm values of estimated moisture content in insulation.

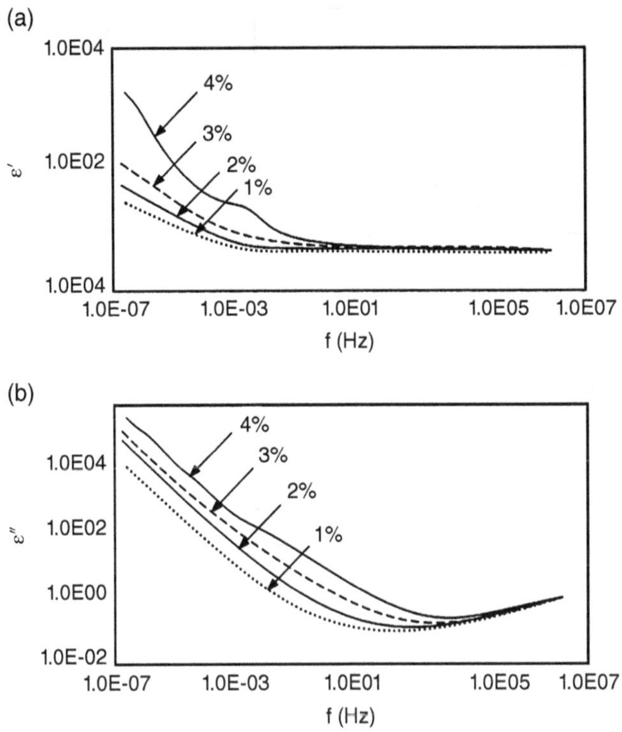

Figure 3B.7 Variation of real and imaginary part of complex permittivity at different moisture contents, at constant temperature of 20°C. *Source:* Ref. [10], Figure 3.

3B.5.1.1 Laboratory Test Results

Estimation of moisture content in transformer insulation through FDS is an indirect technique and needs to be scientifically proven through laboratory measurements. Gielniak *et al.* [10] have reported the FDS measurement results on 1.5 mm thick Weidmann pressboard samples with moisture contents of 1%, 2%, 3%, and 4% in a frequency span of 0.1 mHz to 1 MHz. As can be seen in Figure 3B.7 [10], the permittivity and losses (real and imaginary parts of complex permittivity, respectively) have significant variations with varying moisture levels. The results presented were obtained at a constant temperature of 20°C.

As shown in the figure, both permittivity and loss at low frequencies (below 1 Hz) exhibit significant variations with moisture content. A similar effect of moisture in paper on FDS response was reported in Ref. [11], and has been reproduced in Figure 3B.8.

In both studies, it is clearly indicated that moisture in paper causes both permittivity and the losses to increase, while the minimum of loss is shifted toward higher frequencies with increasing moisture content. A "bump" in the permittivity plot can also be observed at high moisture content, which may be attributed to a probable small relaxation peak such as quasi-DC conduction (q-dc) [12].

The effect of moisture content of paper on the frequency spectrum of the dissipation factor is also demonstrated in Figure 3B.9 [13]. Measurements were

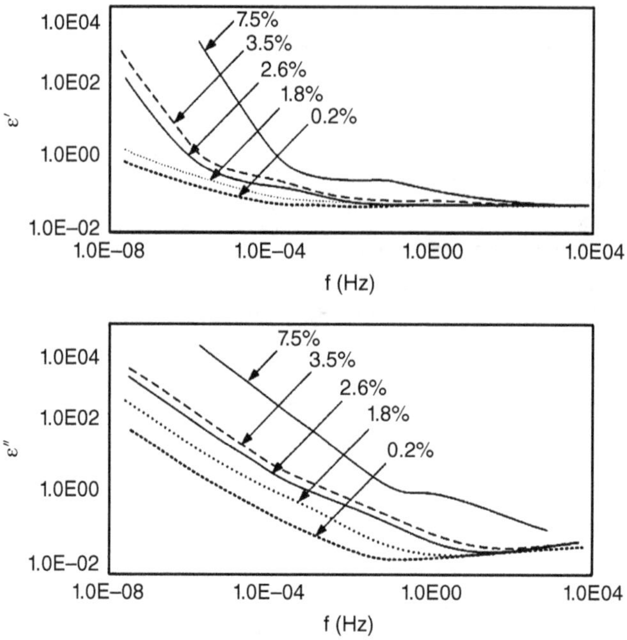

Figure 3B.8 FDS plots for paper with different moisture levels. *Source:* Ref. [11], Figure 2.

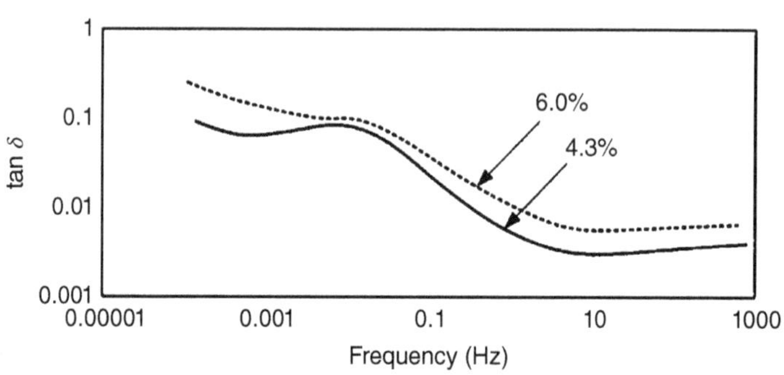

Figure 3B.9 Dissipation factor as a function of frequency at different moisture contents. *Source:* Ref. [13], Figure 4.

performed on two new oil–paper insulation samples, with 4.3% and 6% moisture content, respectively. Both curves have a local peak frequency and magnitude, which changes with moisture content. It can be seen from Figure 3B.9 that the dissipation factor increases with increasing moisture content. Another observation is that the dissipation factor values at two different moisture contents are very close in magnitude for the frequency band 0.001 to 0.1 Hz.

Similar observations were reported in [14], where the dissipation factor spectrum with respect to frequency was obtained for two different oil–paper samples

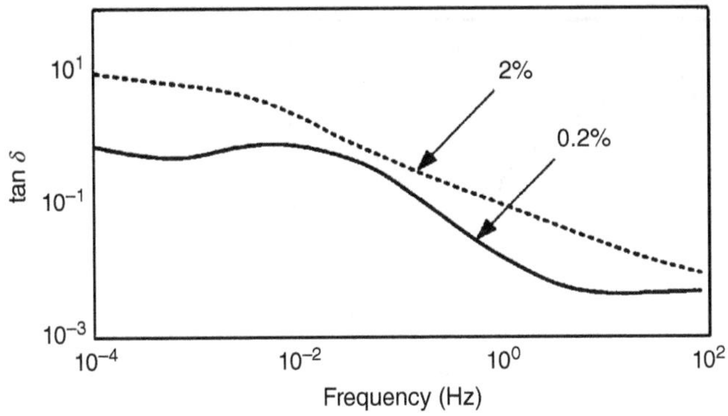

Figure 3B.10 Dissipation factor as a function of frequency at different moisture contents. *Source:* Ref. [14], Figure 9.

at 0.2% and 2% moisture content, respectively. The results, shown in Figure 3B.10, confirm that the FDS plot of the dissipation factor lies higher for the insulation with higher moisture content. The DDF values presented in Figure 3B.10 are given in percentages, and therefore when comparing with the values given in Figure 3B.9 should be converted to an absolute value by dividing by 100.

The above plots and similar findings reported by various researchers [2, 15, 16] have pointed out the fact that both permittivity and losses in the insulation greatly increase with increasing moisture content. FDS plots have a higher trajectory for insulation with higher moisture content. Such deviations in loss and permittivity values for insulation with different moisture contents are prominent in the low-frequency range (0.0001–1 Hz).

3B.5.1.2 Field Test Results

The effects on FDS due to the moisture content of power transformer insulation have been reported in various publications [6, 17–22], where the application of FDS for condition assessment of insulation under different conditions were presented.

In Ref. [18] it was reported about one transformer that was tested before and after an oil refurbishment process. The transformer was rated at 31.5 MVA, 132/33 kV and was manufactured in the year 1986. Figure 3B.11 shows the frequency-dependent loss factor of the transformer before and after its oil refurbishment. It is clearly seen from the figure that the FDS spectrum changed significantly and losses increased after oil refurbishment, showing a completely different dielectric relaxation process after refurbishment. One would expect that moisture contained in the insulation would be driven off during the oil refurbishment process, and therefore the losses after refurbishment would have been lower than those obtained before oil refurbishment. However, the figure depicts the situation as reversed in the lower-frequency range, where the expected "drier"

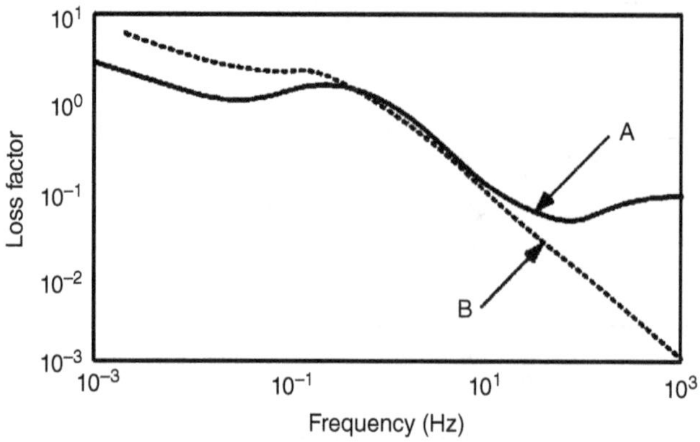

Figure 3B.11 Frequency-dependent loss factor of a transformer (A) before and (B) after oil refurbishment. *Source:* Ref. [18], Figure 7.

plot lies above the expected "moist" plot. One possible reason for this, as envisaged by the authors, could be the quasi-equilibrium conditions of the transformer oil–paper insulation system. It is supposed that during oil refurbishment, some of the moisture present inside the solid insulation moved out to the insulation surface, which was subsequently driven away by the oil and removed from the transformer insulation. In spite of that, even when the refurbishment process is over, a still considerable amount of moisture could be trapped on the insulation surface. Under such a quasi-equilibrium condition, the moisture present on the surface affects the interfacial polarization mechanism, which in turn influences subsequent FDS measurements.

A similar variation of response measurements was observed during measurements carried out by other researchers [6, 20–22], before and after oil refurbishment.

Saha *et al.* [23] presented a report on FDS measurements performed on four transformers installed in two substations. Short descriptions of the four transformers are presented in Table 3B.1.

The variation of tan δ with respect to frequency for the four transformers is reproduced in Figures 3B.12 and 3B.13.

Table 3B.1 Transformers under test

Transformer	Description	MVA	Voltage	Year of manufacture
A	Generator transformer	390	330/23	1991
B	Generator transformer	390	330/23	1992
C	Substation transformer	45	23/11.5	1983
D	Substation transformer	45	23/11.5	1983

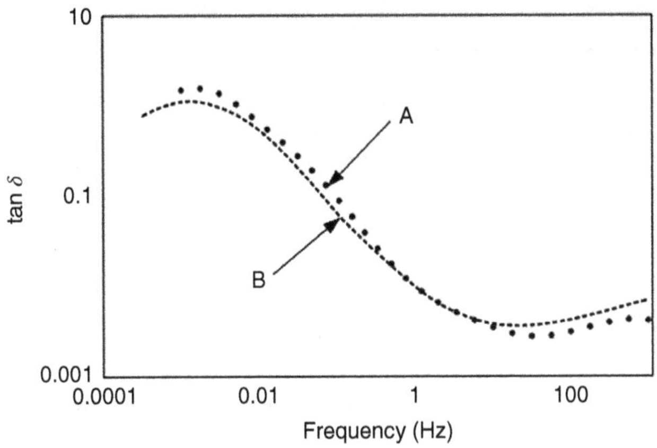

Figure 3B.12 Dielectric loss tan δ variations over frequency range 0.1 mHz to 1 kHz for transformers A and B. *Source:* Ref. [23], Figure 5.

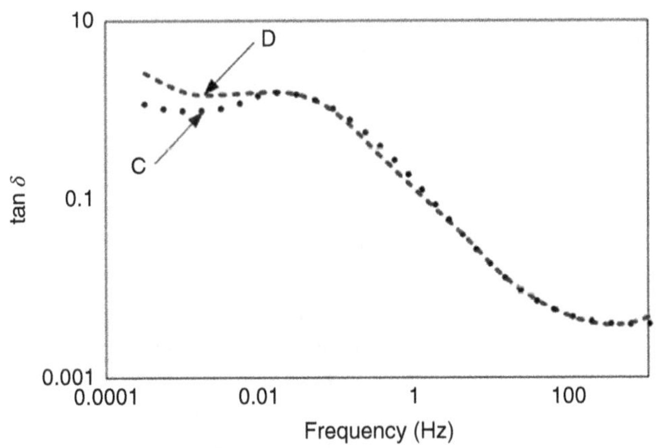

Figure 3B.13 Dielectric loss tan δ variations over frequency range 0.1 mHz to 1 kHz for transformers C and D. *Source:* Ref. [23], Figure 7.

The results presented in different studies [23–25] show that the loss tangent has a minimum, which tends to increase with moisture content. Table 3B.2 shows the values of loss tangent and corresponding frequency for the four transformers considered in Ref. [23]. As the DDF is independent of the geometry, such information is useful for a coarse estimation of the moisture content in the insulation based on FDS measurements.

It can be observed from Table 3B.2 that transformer A has the lowest value of minimum DDF compared with the other three transformers. This can be attributed to lower moisture content of transformer A with respect to the others. Transformer D has the highest minimum DDF among the group. It is thus expected that transformer D has higher moisture content than the others. However, to estimate the moisture content in these transformers through FDS, a

Table 3B.2 Summary of FDS results

Transformer	Minimum tan δ	Frequency at minimum tan δ (Hz)
A	0.0025	33
B	0.0035	10
C	0.0039	333.3
D	0.0041	333.3

proper modeling which accounts for all factors affecting the dielectric response is required.

The observations on the above field measurements lead to the conclusion that the moisture dynamics inside an oil–paper insulation system is a complex process, especially in field measurement conditions. The relations between moisture content and FDS spectra are not necessarily straightforward in field measurement cases compared with controlled laboratory tests. Therefore, it is essential to use appropriate modeling techniques, which account for all factors affecting the dielectric response of transformer insulation, to accurately estimate the moisture content in transformer solid insulation.

3B.5.2 Effect of Ageing and Ageing By-products

Another important factor characterizing the state of insulation is its ageing status. Insulation ageing may lead to a decrease in DP value of the solid insulation, generation of moisture and acids, increase in conductivity values, production of ageing by-products, etc. All these factors compound to influence the FDS measurement results. Various laboratory-based studies have confirmed that ageing of the transformer insulation and ageing by-products – such as low molecular acids have an impact on the dielectric responses [2, 17, 26]. Furthermore, the results of those laboratory studies have been incorporated in moisture evaluation software to estimate the moisture content in field-installed power transformers.

3B.5.2.1 Laboratory Test Results

FDS measurements performed on artificially aged oil–paper insulation samples in laboratories have been reported by several researchers. Linhjell *et al.* [2, 26] and Blennow *et al.* [17] discussed FDS results on artificially aged oil-impregnated paper samples with different initial moisture contents. Figures 3B.14 and 3B.15 show the effect of ageing on oil-impregnated Kraft paper with different levels of initial moisture content.

It can be observed from these figures that the FDS curve initially decreases from the unaged reference when the ageing time is less. At longer ageing times, however, the FDS plot lies well above the unaged reference curve. This phenomenon is predominant when the initial moisture content of the insulation is higher. During the initial period of ageing, moisture is consumed by the hydrolysis reaction and

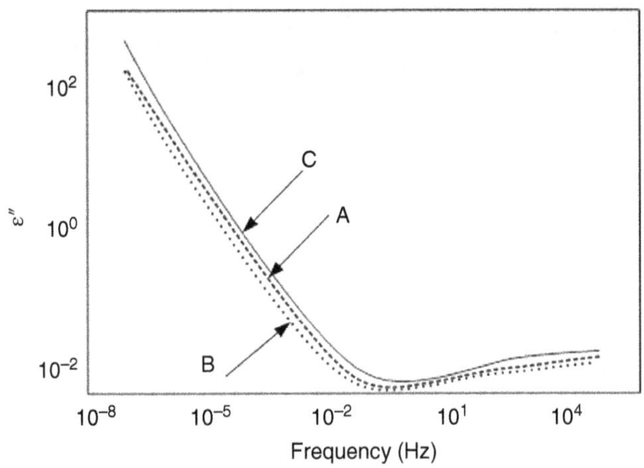

Figure 3B.14 Effect of ageing on Kraft paper with 0.2% initial moisture content: (A) unaged; (B) 94 hours aged DP 840; (C) 1029 hours aged DP 360. *Source:* Ref. [2], Figure 17.

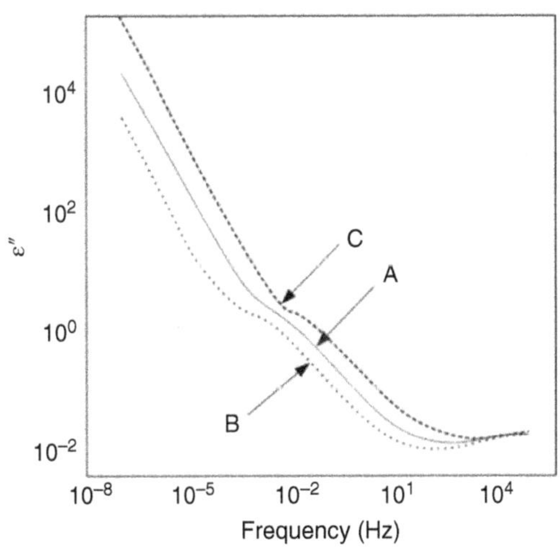

Figure 3B.15 Effect of ageing on Kraft paper with 4% initial moisture content: (A) unaged; (B) 72 hours aged DP 1000; (C) 2005 hours aged DP 280. *Source:* Ref. [2], Figure 19.

the full ageing process has not yet been completed. Also, during the initial stage of ageing the amount of acids produced is not significant. Therefore, one could expect a small drop in losses, although this does not imply that the quality of the insulation has improved. When the ageing progresses, both moisture and low molecular acids are formed, leading to an increase in dielectric losses of the insulation. The vertical or horizontal shifts of FDS plots at different ageing levels are, however, still not well understood. Such shifts are attributed to compounding effects of DP level, acidity, and moisture level present.

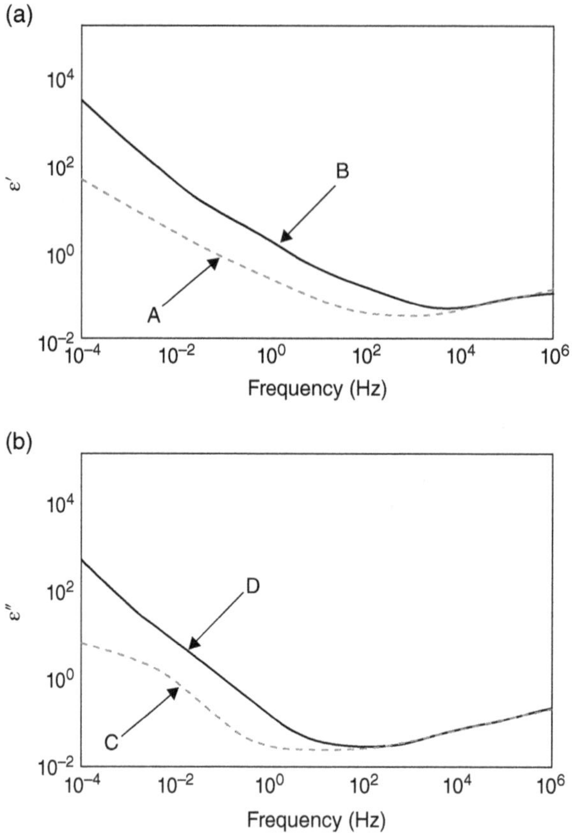

Figure 3B.16 Frequency response of aged and unaged impregnated pressboard samples: (a) initial moisture content 4%; (b) initial moisture content 0.6%. A and C, unaged; B and D, aged. *Source:* Ref. [16], Figure 16.

The foregoing discussions arrive at the observation that the ageing of a paper/pressboard sample with initial moisture content results in noticeable deviations in shape and absolute value of the FDS plot (complex permittivity, complex capacitance, and dielectric loss) from the unaged reference plot. In contrast, when a dry sample is aged, such deviations are less significant. These situations can be explained using Figure 3B.16. Two pressboard samples were used, one very dry with moisture content less than 1% and the other relatively wet with initial moisture content of 4%. Both samples were aged for 800 hours at 130°C.

As in earlier cases, deviation of the FDS plots between aged and unaged samples could be attributed to reduced DP value, increased conductivity value, or generation of moisture and acidic by-products during the ageing process.

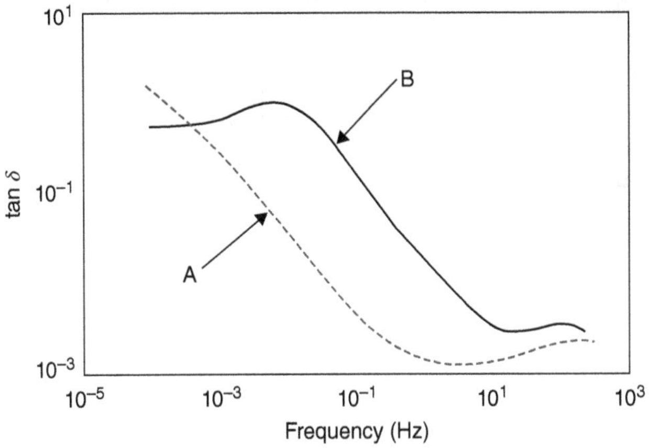

Figure 3B.17 Dissipation factor spectra for two differently aged transformers: (A) less aged; (B) more aged. *Source:* Ref. [27], Figure 3.

3B.5.2.2 Field Test Results

Transformers installed in the field age naturally with years of service. Figure 3B.17 shows the dissipation factor frequency spectra for two transformers, one with a breather in operation since 1973 and the other with a rubber membrane in operation since 1977. The first transformer, in addition to being in operation for a longer period, has a higher chance of ageing since it did not have a closed breathing system as in the second case.

It is observed from this figure that the spectrum of transformer B is shifted horizontally approximately two decades in frequency with respect to the spectrum of transformer A. The main reason for this difference is a change in conductivity due to generation of moisture and other acidic by-products of ageing.

FDS of the complex capacitance of two differently aged transformers was presented in Ref. [20]. Transformer T1 was lying spare in a substation, while transformer T2 was loaded during its operation and thus was expected to be aged. The variation of real and imaginary parts of the complex capacitance with respect to frequency for the two transformers (A and B) are shown in Figure 3B.18.

As seen in this figure, the dielectric losses (represented by the imaginary part C'' of the complex capacitance) for the aged transformer T2 are higher than those for the spare transformer T1 over the entire measured frequency range. However, the dispersive character of the real part C' of the complex capacitance is higher for T2 than T1 in the low-frequency region only, which may be due to limited instrument accuracy. The instrument accuracy is reasonable down to a certain limit defined by the manufacturer. In the case of IDA 200, if the measured capacitance is lower than 1 pF, the accuracy of the measured results is not guaranteed.

Ekanayake *et al.* [18] presented FDS results on three service-aged three-phase autotransformers. Details of these transformers are given in Table 3B.3.

Figure 3B.19 shows the variation of tan δ with respect to frequency for the three transformers A1, A2, and A3. It was reported that none of the transformers were

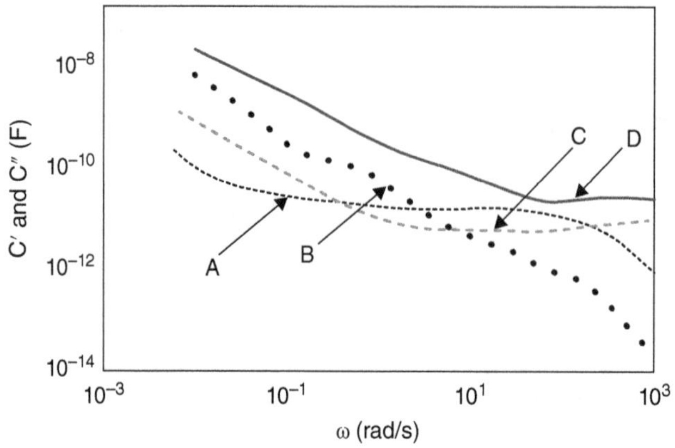

Figure 3B.18 FDS results: (A) C' for transformer T1; (B) C' for transformer T2; (C) C'' for transformer T1; (D) C'' for transformer T2. *Source:* Ref. [20], Figure 7.

Table 3B.3 Details of transformers under test

Transformer	MVA	Voltage (kV)	Year of manufacture	Year of measurement
A1	31.5	132/33	1986	2003
A2	31.5	132/33	1986	2003
A3	31.5	132/33	1999	2008

Source: Ref. [18], Table I.

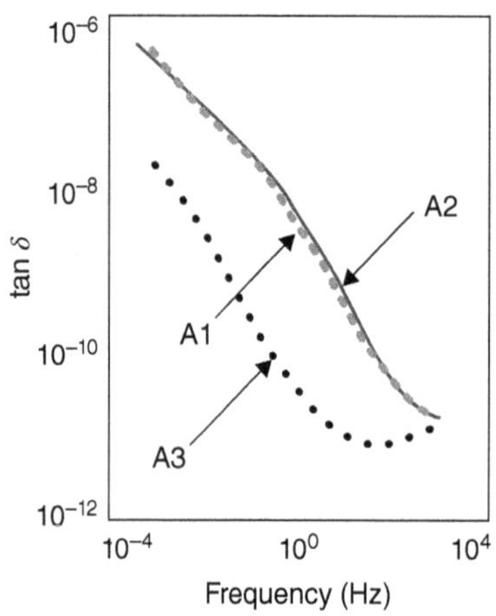

Figure 3B.19 Variation of loss with frequency for transformers A1, A2, and A3. *Source:* Ref. [19], Figure 5.

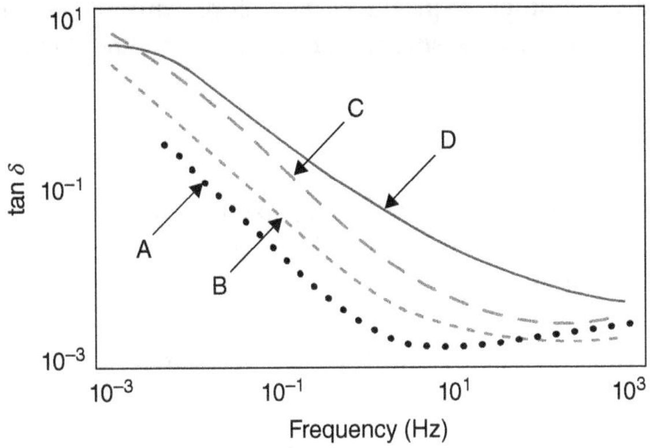

Figure 3B.20 Frequency-dependent DDF for transformers of different service times. *Source:* Ref. [15], Figure 4.

overloaded. The only cause of ageing may be due to temperature cycles undergone by the transformers during normal routine operation over the years.

One can clearly see that the dielectric loss of A3 is significantly lower than that of A1 and A2. Hence, it can be concluded that the insulation of both A1 and A2 is aged significantly compared with that of A3.

FDS test results on a group of differently aged transformers are shown in Figure 3B.20 [14]. The transformers under investigation were in service for different periods, as detailed in Table 3B.4.

Based on the responses, transformer A, which is practically new, has the lowest DDF at low frequencies. Higher the age, higher the DDF typically is, indicating higher dielectric loss. The minimum DDF value of the individual plots is found to shift to higher frequencies with increasing transformer service time. This phenomenon indicates that the value and location of the minimum DDF can potentially be used to characterize the state of the transformer insulation.

The foregoing discussions lead to the observation that in general, the dielectric loss spectrum and the complex capacitance spectrum will have higher absolute magnitudes when the transformer is aged. The minimum value of DDF for a given FDS plot will have its minimum at lower values and also at lower frequency ranges when the insulation is relatively less aged or dry. However, insulation ageing is a complex phenomenon and has complicated influences of depolymerization,

Table 3B.4 Service details of transformers tested

Transformer	Years in service
A	New
B	13
C	19
D	34

generation of moisture, acids, and other by-products. Therefore, without proper modeling, it is impossible to relate FDS measurement results to the state of insulation ageing.

3B.5.3 Effect of Acids

During operation of a transformer, deterioration of both paper and oil is caused by a number of factors including oxidation accelerated by high temperatures, air, and moisture [28]. Oxidation of oil and cellulose leads to formation of acids, peroxides, and additional water.

There are primarily two kinds of acid found in aged transformer insulation [29, 30]:

- hydrophilic acids such as formic and acetic acids formed by ageing of cellulose;
- hydrophobic acids such as naphthenic and stearic acids produced due to the oxidation of insulating oils.

The low molecular weight "light" acids (hydrophilic acids) consisting of formic and acetic acids are water-soluble and prefer to stay in the paper [2, 31, 32] and contribute to ageing. In contrast, high molecular weight "heavy" acids (hydrophobic acids) consisting of naphthenic and stearic acids are fat-soluble and prefer to stay in the oil. Recent publications [33] suggest that low molecular weight acids are more reactive than high molecular weight acids [34]. As acids penetrate into paper, this supports the degradation reaction and thus results in a loss of tensile strength and degree of polymerization. On top of that, other insulation materials such as polymer coatings also become brittle.

In real transformers, long-chain acids arise predominantly from degradation of the oil, while short-chain acids are expected to be generated by degradation of paper [26, 31]. The effect of acids on the frequency spectrum of the permittivity of paper is demonstrated in Figure 3B.21.

As can be seen in the figure, low molecular weight acids cause the FDS plot to shift toward higher frequencies. The minimum value of permittivity is also found to have a markedly higher value with acids than without acids. In contrast, the FDS curves of paper samples with high molecular weight acids almost overlap the curves without acids. This is clearly in line with the observation in Refs [20, 29] that lighter acids prefer to stick to the paper, whereas heavier acids stay mostly with the oil.

In Figure 3B.21 [2, 12, 26, 31] it is also demonstrated that light acids cause an increase in ε'' value and a shift of the minimum ε'' point toward higher frequencies. This effect is similar to the effect of moisture on the dielectric response of paper. Therefore, one could assume that the moisture causes the corresponding shift and could overestimate the moisture content in the insulation.

3B.5.4 Effect of Temperature

Temperature strongly affects the dielectric response of an insulation system. An increase in temperature increases the electrical conductivity. The conductivity can increase in two ways [20], either due to an increase in the number of charge carriers or due to an increase in the mobility of charge carriers. Increased

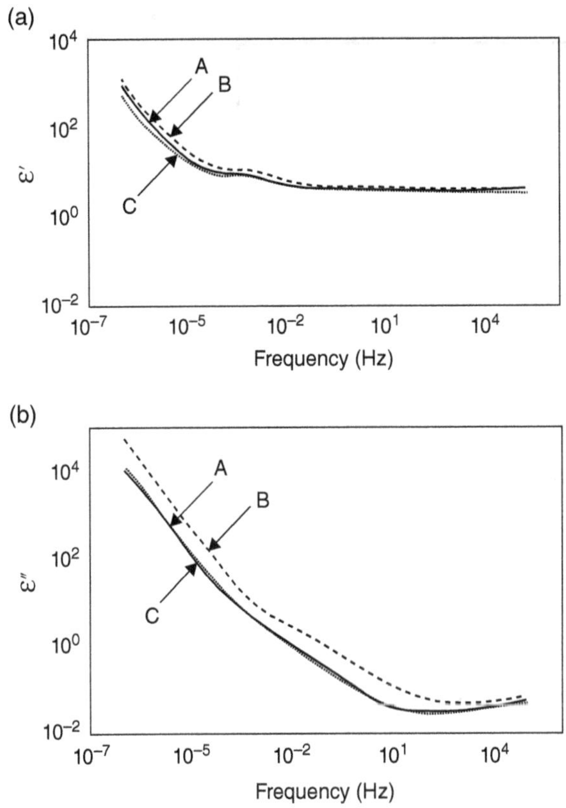

Figure 3B.21 Frequency response of permittivity of paper with acids: (A) without acid; (B) with light acid; (C) with heavy acid. *Source:* Ref. [2], Figure 12.

temperature can lead to thermal ionization, giving rise to an increased number of charge carriers. In addition to this, increased temperature disturbs the distribution of moisture in oil and paper. All of these contribute to influencing the temperature on FDS response.

The general shape of dissipation factor spectra of oil-impregnated pressboard at any temperature depends on two factors: oil conductivity and insulation resistance (conductivity and polarization) of the pressboard.

The oil conductivity depends on the mobility and density of ions in oil, which are formed by dissociation of electrolytic impurities present in the oil. At higher temperatures, fluidity (inverse of viscosity) and dissociation of electrolytic impurities to ions will increase, leading to an increase in conductivity with temperature.

In contrast, paper conductivity depends mainly on the amount of moisture in the pressboard. At higher temperatures, the polarization and conduction will increase due to two superimposed effects: an increase in hopping charge carrier density and an increase in the mobility of charge carriers residing in pressboard insulation [35]. Owing to an increase in thermal vibration of mobile charge carriers in the system with increasing temperature, one could expect higher conduction and polarization levels of pressboard at elevated temperatures.

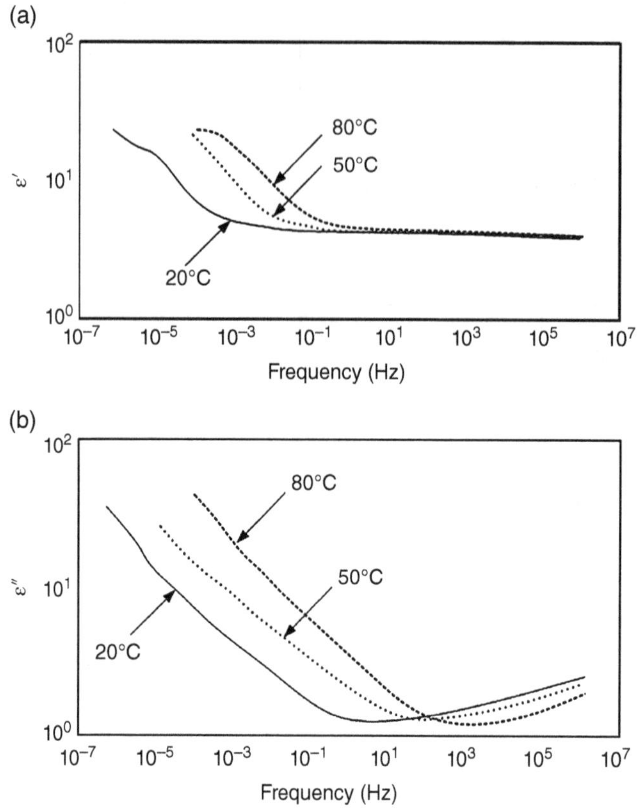

Figure 3B.22 Measured dielectric response at different temperatures. *Source:* Ref. [16], Figure 15.

Figure 3B.22 presents measured responses of permittivity with respect to frequency at three different temperatures: 20°C, 50°C, and 80°C.

Figure 3B.22(a) shows that ε' values are higher at higher temperatures. There is a tendency for the responses to have a dielectric relaxation peak at low frequencies and corresponding frequency shifts toward high frequency with increasing temperature. Similarly, Figure 3B.22(b) shows that ε'' values are also higher at higher temperatures, with the frequency corresponding to minimum ε'' also being higher at higher temperature levels.

FDS results as the variation of DDF with frequency obtained on oil–paper insulation samples at different temperatures in another study are shown in Figure 3B.23. Variations in dielectric loss confirm the observation that the temperature has a great effect on the FDS. Furthermore, one could notice that with increasing temperature, responses shift horizontally along the frequency axis toward higher frequencies. This behavior has been well documented by various researchers, and the frequency shift has been used to calculate a generalized response curve called the master curve [5, 8]. The Arrhenius theory has been used to calculate the activation energy corresponding to the frequency shift of the

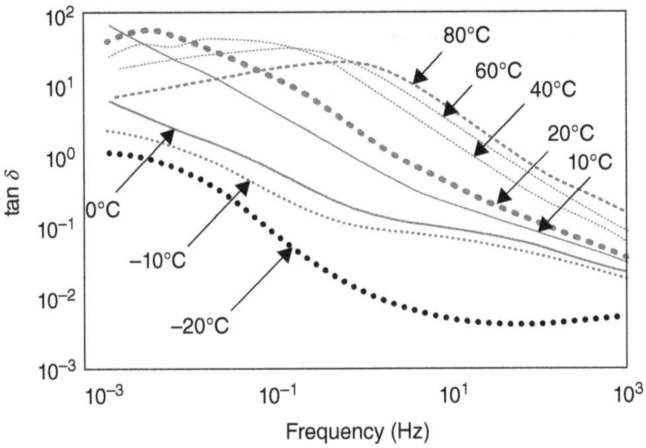

Figure 3B.23 Tan δ plots at different temperature. *Source:* Ref. [15], Figure 6.

response due to temperature variation [36]. This allows estimation of the FDS of the insulation at any given temperature. Therefore, the master curve concept has been used widely in FDS-based diagnostic software proposed in various research studies, as well as in products available in the market.

The dielectric responses of modeled transformer insulation with combined oil channels and paper layers in series were reported in Ref. [37]. Figure 3B.24 shows the variation of the real and imaginary parts of complex capacitance against frequency at various temperatures. As can be seen, both real and imaginary parts of the complex capacitance are found to increase with increasing temperature.

Figure 3B.25 shows the variation of DDF (tan δ) at five different temperature levels obtained from Ref. [37]. Similar to dielectric loss, the DDF of the oil–paper insulation system also increases with temperature and with increasing temperature, responses shift horizontally along the frequency axis toward higher frequencies. Maximum and minimum values of the dissipation factor with respect to temperature within the measured frequency range are given in Table 3B.5. From the data presented in this table, it can be observed that both maximum and minimum values of dissipation factor within the measured frequency range increase with increasing temperature. Similar test results and conclusions were arrived at in other studies [7, 14, 29]. However, it is important to note that the measured minimum DDF is not the absolute/local minimum of the curve, except for the lowest measured temperature (25°C).

Finally, the FDS plot of dielectric loss for a 110 kV power transformer at different temperatures is reproduced here from Ref. [15] and presented in Figure 3B.26. As per the authors of Ref. [15], this transformer had been kept out of service for more than a month before the measurements were performed at two different ambient conditions. One could assume that moisture in the paper would be similar on both occasions, as the transformer was at ambient conditions for a reasonable time and the amount of moisture migration between paper and oil is insignificant compared with the amount of moisture trapped inside the paper. Therefore, the observed variation in response is caused by the temperature

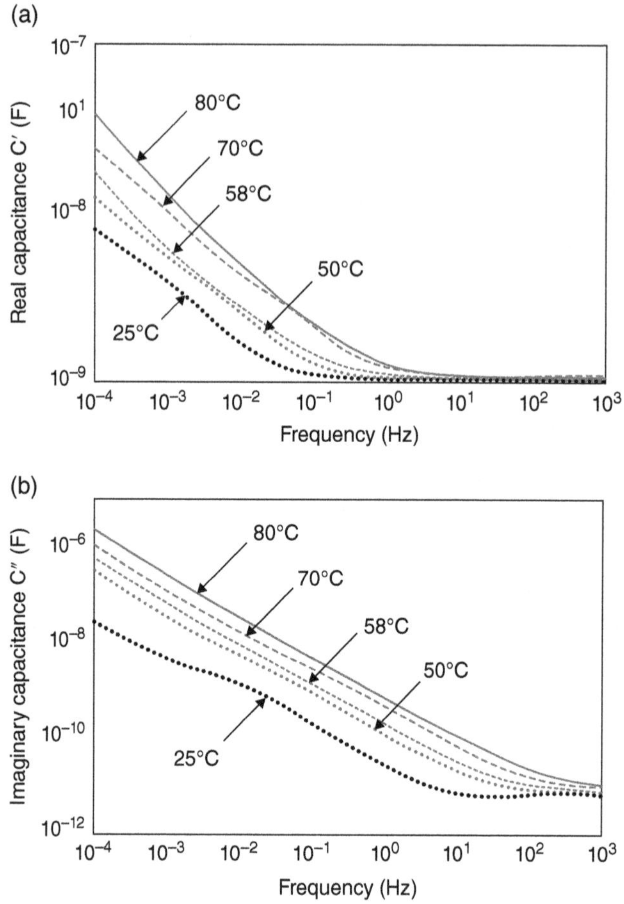

Figure 3B.24 (a) Real capacitance at different temperatures. (b) Imaginary capacitance at different temperatures. *Source:* Ref. [37], Figures 5, 6.

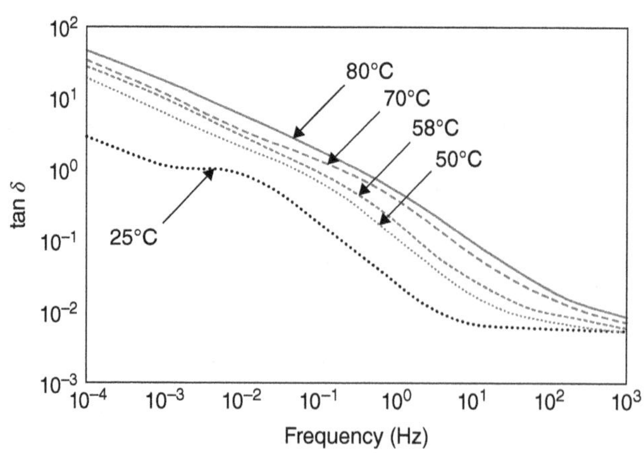

Figure 3B.25 Dissipation factor tan δ at different temperatures. *Source:* Ref. [37], Figure 4.

Table 3B.5 Maximum and minimum dissipation factors with respect to varying temperature

Temperature (°C)	Minimum tan δ	Maximum tan δ
25	0.006709	2.8866
50	0.005799	21.114
58	0.006400	29.339
70	0.007489	35.605
80	0.008199	48.181

Source: Ref. [37], part of Table II.

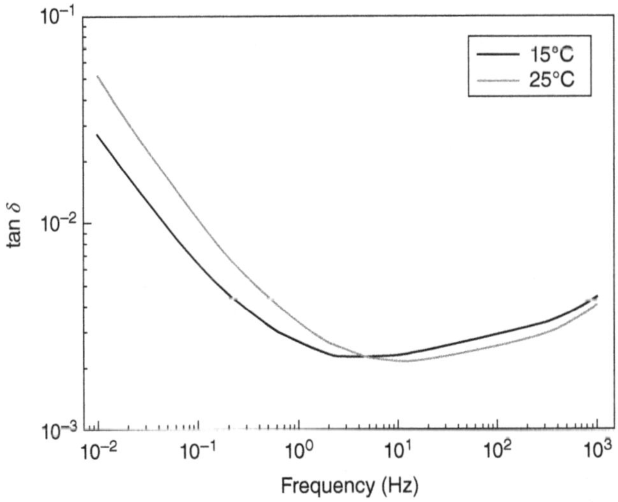

Figure 3B.26 Dissipation factor tan δ at different temperatures of a field-installed unit.
Source: Ref. [15], Figure 7.

difference. As one would anticipate, the curve corresponding to 25°C shifted to the right with respect to the curve for 15°C, while the shape remained unchanged.

This once again indicates that temperature is a crucial piece of information for the assessment of insulation condition and should be recorded accurately. Before drawing any conclusive inference about the state of insulation, it is thus of paramount importance to make the necessary temperature corrections.

3B.5.5 Effect of Insulation Geometry

Power transformer main insulation (between HV and LV) is formed by pressboard and oil. Pressboard is used as the barriers and spacers, whereas oil channels in the

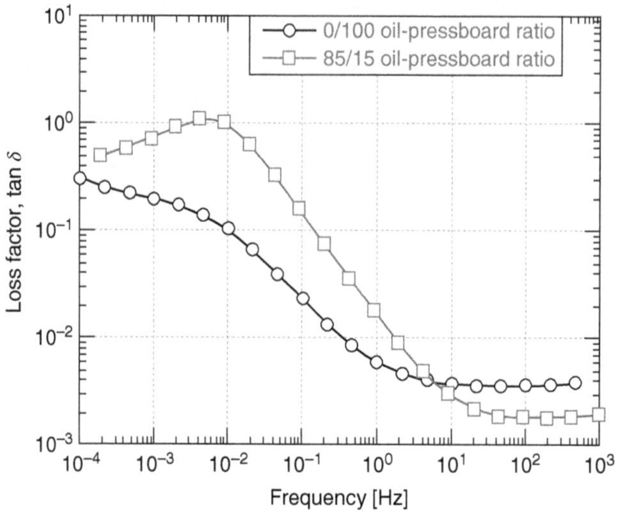

Figure 3B.27 Tan δ plots at different geometrical structure of insulation for oil/pressboard ratios 0:100 and 85:15. *Source:* Ref. [38], Figure 5(b).

ducts are used as the cooling medium. The amount of pressboard and number of oil channels used in each transformer design depend on various factors – such as rated voltage, power level, cooling type, manufacturer's design choices, etc. Therefore, the proportion of oil, paper, and pressboard in the insulation assembly is more often dissimilar in different transformers. Careful studies are hence necessary to understand the effect of insulation geometry on FDS measurements.

Experimental results of dielectric loss with respect to frequency at different geometrical arrangements of oil and pressboard in a sample transformer winding have been reported in Ref. [38]. Figure 3B.27 shows FDS results of DDF with an oil/pressboard ratio of 0:100 and 85:15. It is seen that in the response without an oil duct, the relaxation peak seen in the DDF is less pronounced compared with insulation using oil channels.

To investigate further the effect of geometry on dielectric response of the insulation, an experimental setup was developed at the University of Queensland, Australia that emulates the winding and insulation structure of a real transformer. The model was based on the insulation arrangement mostly available in a power transformer, as shown in Figure 3B.28. This insulation model is a scaled-up version of a transformer insulation duct with simplified geometry. Geometrical arrangements and relative amounts of oil duct, pressboard barrier, and spacers can be varied in a controlled manner to study their effects on FDS measurement results.

The geometry presented in this figure is a simplified model of a real transformer insulation. This cylindrical insulation geometry can be represented by a lumped parallel-plate capacitor model, called the X–Y model [39, 40]. In the X–Y model all barriers, spacers, and oil channels are individually combined together based on the relative amount of spacers and barriers in the duct. As shown in

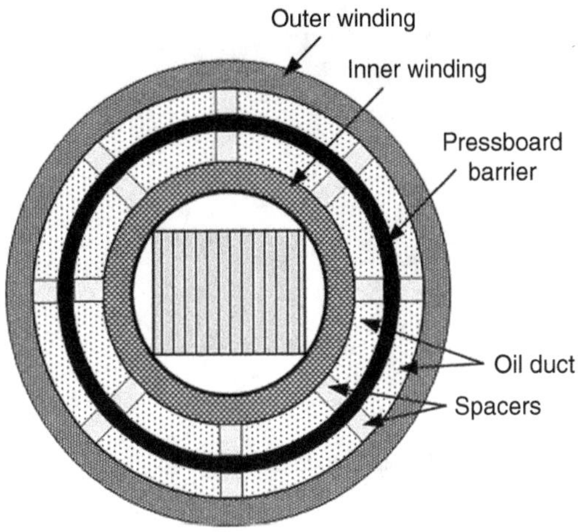

Figure 3B.28 Cross-section of transformer major insulation structure.

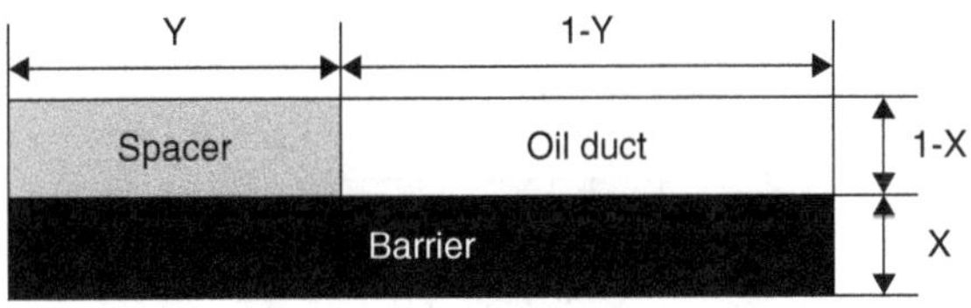

Figure 3B.29 *X–Y* model of transformer insulation geometry.

Figure 3B.29, parameter *X* is defined as the ratio of the sum of all the thicknesses of all barriers in the duct lumped together and divided by the duct width. In addition, the spacer coverage *Y* is defined as the total width of all the spacers divided by the total length of the periphery of the duct. Typical values of *X* in a power transformer are in the range 0.2–0.5 [39].

The effect of geometry on the FDS has been studied in Ref. [29], where the dielectric response of the insulation model presented in Figure 3B.29 has been measured while parameter *X* has been varied from 0.63 to 0.15. The outcome of the measured responses is presented in Figure 3B.30. A closer look at the spectra shown in the figure reveals that the critical point (occurring in the frequency band of 10^{-2} to 1 Hz, as labeled in the figure) shifts toward higher frequency with increasing values of *X*. Furthermore, the spectra have significant differences at the high-frequency end (1 kHz), which have been reported to be linked to the geometry of the insulation [29].

The corresponding spectra for the real and imaginary parts of the complex capacitance are presented in Figures 3B.31 and 3B.32, respectively.

As can be seen from Figure 3B.32, the imaginary capacitance at 0.1 mHz increases as the value of *X* decreases. Another important observation is that both DDF and loss curves exhibit a minimum point at high frequency (close to 1 kHz) with decreasing *X*. This can be considered as a possible indication of the value of *X*

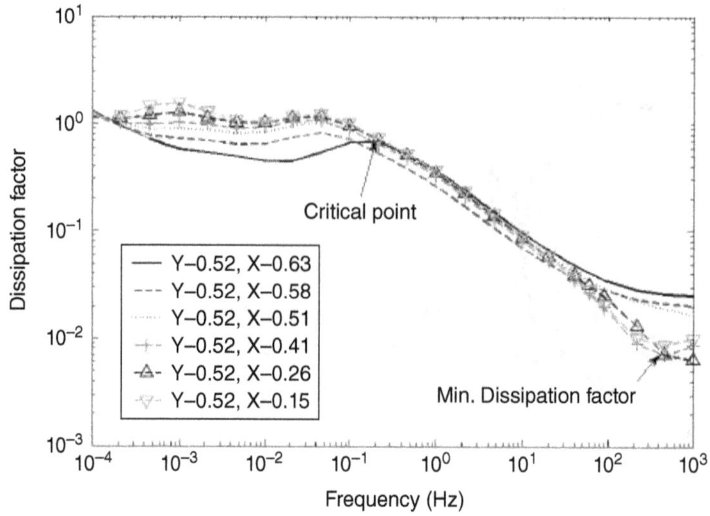

Figure 3B.30 Effects of barriers on dissipation factor when $Y = 0.52$. *Source:* Ref. [29], Figure 11.

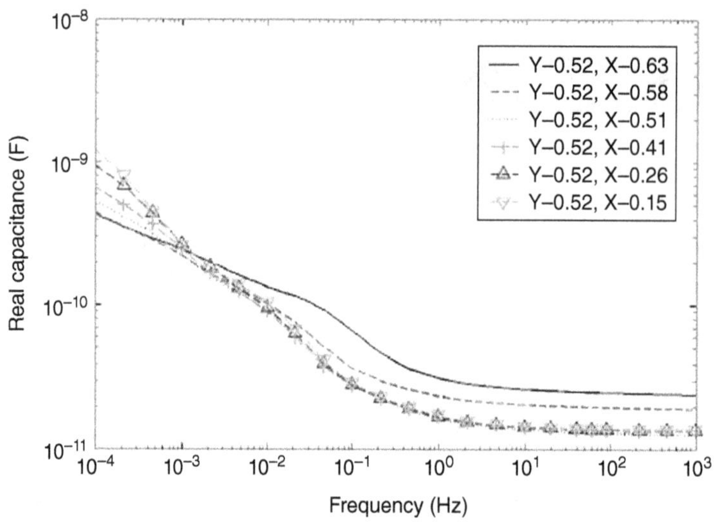

Figure 3B.31 Effects of barriers on real capacitance when $Y = 0.52$. *Source:* Ref. [29], Figure 12.

(i.e., relative thickness of barrier). Similarly, the effects of spacer dimension were also studied by changing the numbers of pressboard strips, which results in different values of Y.

The dissipation factor curves obtained from different amounts of spacers are plotted in Figure 3B.33. It can be seen that as the amount of spacers reduces, the critical point in the dissipation factor curve shifts toward higher frequencies.

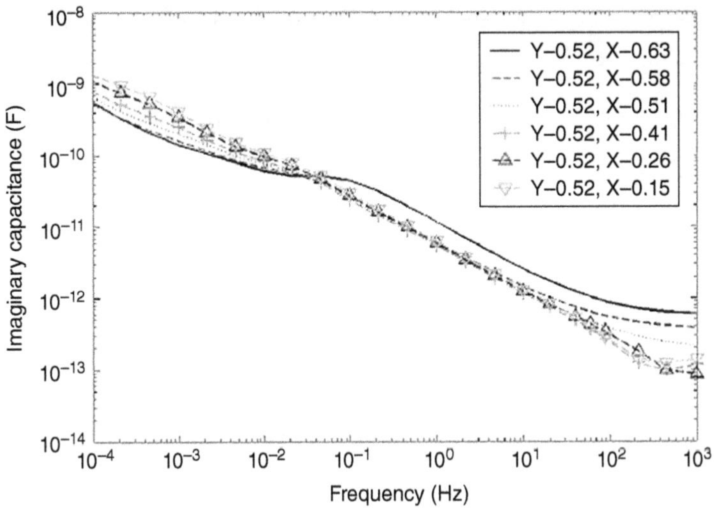

Figure 3B.32 Effects of barriers on imaginary capacitance when $Y = 0.52$. *Source:* Ref. [29], Figure 13.

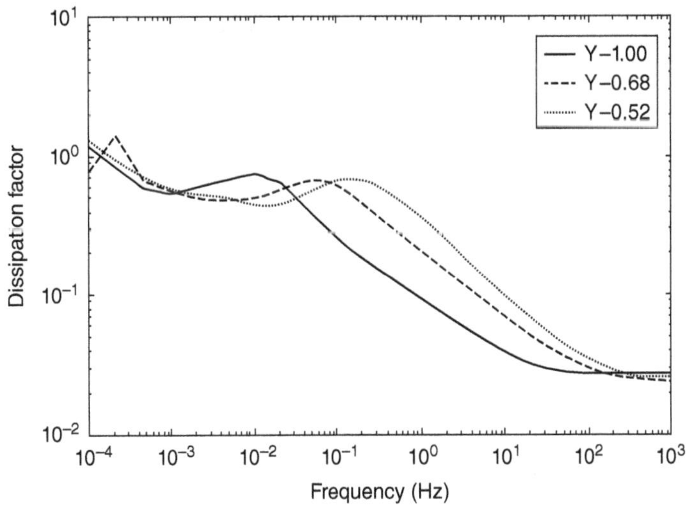

Figure 3B.33 Effects of spacers on dissipation factor when $X = 1.00$. *Source:* Ref. [29], Figure 14.

Real and imaginary capacitance curves are presented in Figures 3B.34 and 3B.35, respectively. From Figure 3B.34, the real capacitance curves can be seen to increase as the amount of spacer decreases. This phenomenon is similar to the effects of barriers as observed in Figure 3B.30.

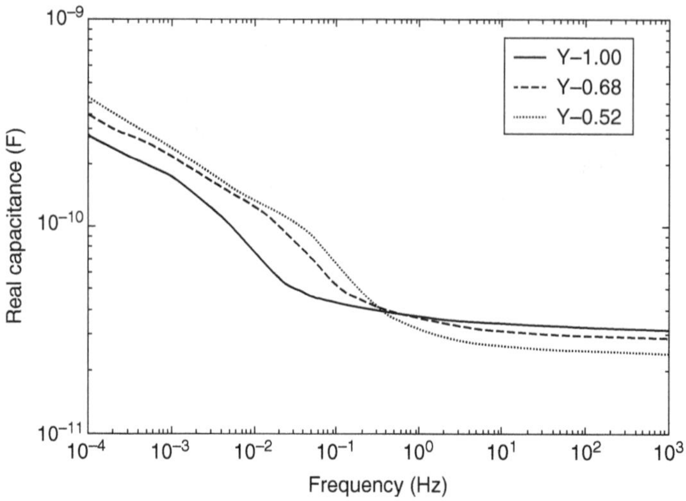

Figure 3B.34 Effects of spacers on real capacitance when $X = 1.00$. *Source:* Ref. [29], Figure 15.

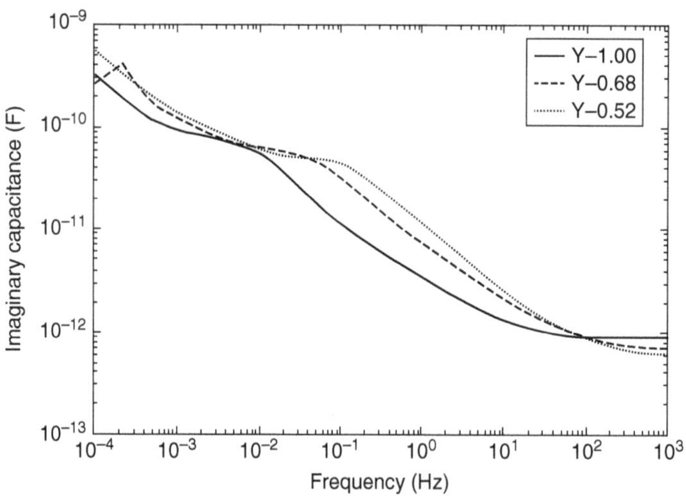

Figure 3B.35 Effects of spacers on imaginary capacitance when $X = 1.00$. *Source:* Ref. [29], Figure 16.

The foregoing discussions indicate that the insulation geometry has a certain influence on the nature of FDS plots to be used for the assessment of insulation condition. Thus, before drawing any conclusive inference about the state of insulation, it is thus recommended to obtain some idea about the insulation geometry of the test object.

3B.6 Modeling FDS of Transformer Insulation

Various interpretation techniques have been developed to estimate the insulation condition through FDS response measurements on high-voltage insulation [39–41]. Out of these techniques, the functional approach where an analytical expression is fitted to the frequency-dependent dielectric response by selecting suitable coefficients for the expression is widely applicable for diagnosing transformer insulation. The so-called *X*–*Y* modeling is a functional approach, which is developed further to consider important factors such as geometrical arrangement and temperature in the model. In this approach, the simplified *X*–*Y* insulation model shown in Figure 3B.29 is used to interpret the insulation geometry between HV and LV of a core-type power transformer.

Based on the definition of parameters *X* and *Y*:

$$X = \frac{\text{thickness of total barriers}}{\text{width of the duct}}$$

$$Y = \frac{\text{total width of the spacer}}{\text{periphery of the duct}}$$

The expected FDS response across the *X*–*Y* geometry can be calculated when the individual dielectric responses of oil, spacers, and barriers are known, using the simple series and parallel combination of impedances. The resultant response can be calculated using

$$\hat{\varepsilon}(\omega, T)_{\text{duct}} = \frac{Y}{\dfrac{1-X}{\hat{\varepsilon}_{\text{spacer}}} + \dfrac{X}{\hat{\varepsilon}_{\text{barrier}}}} + \frac{1-Y}{\dfrac{1-X}{\hat{\varepsilon}_{\text{oil}}} + \dfrac{X}{\hat{\varepsilon}_{\text{barrier}}}} \tag{3B.19}$$

where

$$\hat{\varepsilon}_{\text{spacer}} - \hat{\varepsilon}_{\text{barrier}} = \hat{\varepsilon}_{\text{pressboard}}(\omega, T)$$
$$\hat{\varepsilon}_{\text{oil}} = \varepsilon_{\text{r,oil}} - j\frac{\sigma(T)}{\varepsilon_0 \omega} \tag{3B.20}$$

As shown in equation (3B.19), the response of the nonpolar oil is modeled by a constant value of DC conductivity σ and permittivity $\varepsilon_r = 2.2$. The barriers and spacers are typically formed by the same pressboard and therefore the dielectric responses of both materials are equal to the response of the pressboard. The dielectric response of impregnated pressboard has significant dispersion – depending on moisture, temperature, and ageing – in the frequency range of interest. Therefore, the response of the pressboard is hard to represent accurately using an analytical expression such as for oil. Consequently, the response of the pressboard is described by its measured FDS under given conditions. In this modeling technique, FDS measurements on well-defined pressboard samples are used to form a database of FDS responses at different temperatures under different moisture levels.

Once the dielectric response data is available and if *X* and *Y* are known, equation (3B.18) can be used to derive the FDS response of a transformer for a given moisture level at a given temperature. Since the database does not cover the response of the pressboard for all temperatures and moisture levels, appropriate interpolation or extrapolation techniques are used to calculate the FDS in arbitrary conditions.

To evaluate the status of the insulation, the derived response from the model is compared with the response from the FDS measurement on the tested transformer. An appropriate algorithm is used to minimize the error between the derived and measured responses by changing the amount of moisture content in the pressboard and the conductivity of the oil. The temperature corresponding to the modeled response is fixed at the measurement temperature. Therefore, it is always important to record the temperature during FDS measurements.

As presented above, the X–Y modeling technique is useful for predicting the moisture content in pressboard and the oil conductivity of the investigated transformer. However, in most cases X, Y and the transformer geometrical capacitance are unknown. In such situations, these parameters are also estimated from the software by considering their effect on the response. Nevertheless, it increases the number of unknowns in the optimization algorithm, leading to reduced accuracy of the estimated quantities.

We have also conducted many experiments in the laboratory to understand the ageing behavior of oil–paper insulation systems [42]. Our study suggested that FDS measurements are primarily affected by moisture, and shows non-uniform variations of ageing due to the moisture dynamics during the ageing process. FDS measurements are influenced by ageing by-products dissolved in the oil, and have some ability to distinguish ageing conditions from just moisture effects. However, from laboratory measurements it was found that the impact of ageing products and ageing conditions on FDS measurements is comparatively smaller than that of moisture, and hence in the case of a wet pressboard sample it was difficult to separate the impact of ageing from that of moisture [42].

3B.7 Frequency Domain to Time Dielectric Response Measurement

It has been reported by many authors that if the dielectric material is linear, homogeneous, and isotropic, then information obtained from the time domain or frequency domain measurements is the same [23, 38]. In fact, the transformation from one domain to the other is possible with high accuracy. Several methods are available in the literature to perform this conversion from one domain to the other. We have also attempted a method of converting this from one domain to the other using the extended Debye model, as shown in Figure 3B.36 [23].

The impedance of the circuit in this figure can be obtained from the following:

$$Z = \frac{1}{\frac{1}{Z_0} + \frac{1}{Z_1} + \frac{1}{Z_2} + \cdots + \frac{1}{Z_n}} \tag{3B.21}$$

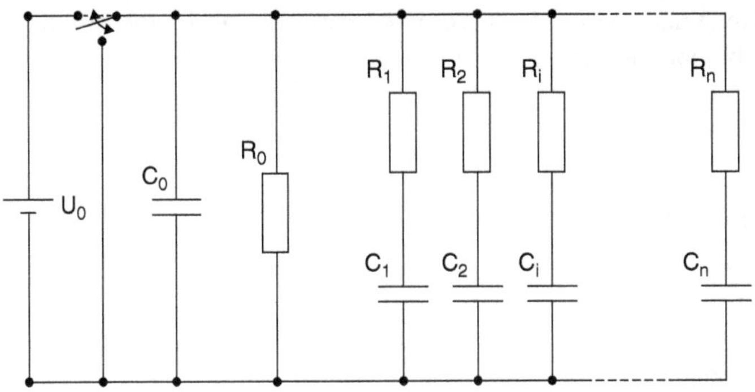

Figure 3B.36 Equivalent circuit to model a linear dielectric material. *Source:* Ref. [23], Figure 2.

where

$$Z_0 = \frac{1}{\frac{1}{R_0} + j\omega C_0}$$

and

$$Z_n = R_n + \frac{1}{j\omega C_n}$$

Here, n is the number of parallel branches of the series RC network. If the applied voltage is an alternating signal at frequency ω, then the measured capacitance is a complex quantity whose real and imaginary parts correspond to the real and imaginary components of the complex permittivity

$$C(\omega) = C'(\omega) - jC''(\omega) = (A/w)\cdot(\varepsilon'(\omega) - j\varepsilon''(\omega)) \tag{3B.22}$$

where A is the plate area of the capacitance, ε is the permittivity, and w is the distance between two plates. The DDF or loss factor can be defined as in equation (3B.18). For the FDS measurement, U_0 in Figure 3B.36 would be replaced by $U(\omega)$, where $U(\omega)$ is the input signal to the circuit as a function of frequency.

The current is then given by

$$I(\omega) = j\omega\cdot C(\omega)\cdot U(\omega) \tag{3B.23}$$

The current can also be given by

$$I(\omega) = \frac{U(\omega)}{Z(\omega)} = Y(\omega)\cdot U(\omega) \tag{3B.24}$$

Using equations (3B.23) and (3B.24), we obtain

$$C(\omega) = Y(\omega)/j\omega \tag{3B.25}$$

The reciprocal of the impedance $Z(\omega)$ is the admittance $Y(\omega)$, given by

$$Y(\omega) = \frac{1}{Z(\omega)} \tag{3B.26}$$

The real and imaginary parts of the complex capacitance for the circuit in Figure 3B.36 can be obtained as

$$C'(\omega) = C_0 + \sum_{i=1}^{n} \frac{C_i}{1 + (\omega R_i C_i)^2}$$
$$C''(\omega) = \frac{1}{\omega R_0} + \sum_{i=1}^{n} \frac{\omega R_i C_i^2}{1 + (\omega R_i C_i)^2} \quad (3B.27)$$

Now, $Z(\omega)$ can be obtained from the following equation, where C' and C'' are given in equation (3B.27):

$$Z(\omega) = \frac{1}{C(\omega)j\omega} = \frac{1}{j\omega[C'(\omega) - jC''(\omega)]} \quad (3B.28)$$

The problem is to take the measured real and imaginary parts of the complex capacitance versus frequency data from the FDS measurements in the form shown as follows, where ω is a set of test frequencies (rad/sec) and Z_n is the measured impedance at ω_n:

$$\omega = \begin{bmatrix} \omega_1 \\ \omega_2 \\ \cdots \\ \omega_n \end{bmatrix}, Z = \begin{bmatrix} Z_1 \\ Z_2 \\ \cdots \\ Z_n \end{bmatrix} \quad (3B.29)$$

There are a number of different model forms available in the literature to solve this by transfer function-based, pole-zero, and state-space techniques. We have used the gain, zero-pole form with MatLab "zpk," given in detail in Ref. [23]. Once a model is estimated, conversion of the RC model form and prediction of the RVM/PDC model estimation can be carried out as presented in Ref. [23], with a number of case studies. Once the mathematical relationship between different dielectric diagnostics can be properly established, the analysis of insulation condition could be better understood.

3B.8 Summary

FDS measurement theory and techniques have been explained in detail. Then, the factors affecting FDS measurement data have been highlighted, together with results from laboratory-based controlled experiments as well as from field transformer measurements. All the factors are analyzed critically, including the effect of connection arrangement, moisture, ageing, acidity, temperature, and geometry of the oil–paper insulation system. Finally, the importance of modeling for correct interpretation of FDS measurements on field transformers has been highlighted. A methodology for time domain and frequency domain conversion has been discussed, along with some case studies.

References

1 Zaengl WS. Dielectric spectroscopy in time and frequency domain for HV power equipment. Part I: Theoretical considerations. *IEEE Electrical Insulation Magazine* 2003; 19(5): 5–19.

2 Linhjell D, Lundgaard L, Gäfvert U. Dielectric response of mineral oil impregnated cellulose and the impact of aging. *IEEE Transactions on Dielectrics and Electrical Insulation* 2007; 14(1): 156–169.

3 CIGRE Brochure No. 254, SC TF D1.01.09. *Dielectric Response Methods for Diagnostics of Power Transformers*. CIGRE, Paris, 2004.

4 Jonscher AK. *Dielectric Relaxation in Solids*. London: Chelsea Dielectrics Press, 1983.

5 Ekanayake C. *Diagnosis of moisture in transformer insulation*. Ph.D. dissertation, Chalmers University of Technology, 2005.

6 Jaya M, Leibfried T, Koch M. Information within the dielectric response of power transformers for wide frequency ranges. IEEE International Symposium on Electrical Insulation (ISEI), 2010, pp. 1–5.

7 Wang S, Wei J, Yang S, Dong M, Zhang G. Temperature and thermal aging effects on the frequency domain spectroscopy measurement of oil–paper insulation. Proceedings of the 9th IEEE International Conference on Properties and Applications of Dielectric Materials, Harbin, China, July 19–23, 2009, pp. 329–332.

8 Ekanayake C. *Application of dielectric spectroscopy for estimating moisture content in power transformers*. Technical Licentiate thesis, Chalmers University of Technology, 2003.

9 Ekanayake C, Saha T, Ma H, Martin D. Condition assessment of model transformers under accelerated ageing experiments. CIGRE Study Committee B3 & Study Committee D1 Colloquium, Brisbane, Australia, September 9–11, 2013.

10 Gielniak J, Ekanayake C, Walczak K, Graczkowski A, Gubanski S. Dielectric responses of new and aged transformer pressboard in dry and wet states. Proceedings of International Symposium on Electrical Insulating Materials, Kitakyushu, Japan, June 5–9, 2005, pp. 386–389.

11 Linhjell D, Hestad OL, Gäfvert U, Lundgaard LE. Dielectric response of oil–impregnated cellulose from 0.1 mHz to 3 MHz. IEEE International Conference on Dielectric Liquids (ICDL), 2005, pp. 277–280.

12 Bandara K, Ekanayake C, Saha TK. Modelling the dielectric response measurements of transformer oil. *IEEE Transactions on Dielectrics and Electrical Insulation* 2015; 22(2): 1283–1291.

13 Hassanl O, Shyegani AA, Borsi H, Gockenbach E, Abu-Elzahab EM, Gilany MI. Detection of oil–pressboard insulation aging with dielectric spectroscopy in time and frequency domain measurements. International Conference on Solid Dielectrics, Toulouse, France, July 5–9, 2004.

14 Yew JH, Pradhan MK, Saha TK. Effects of moisture and temperature on the frequency domain spectroscopy analysis of power transformer insulation. IEEE Power and Energy Society General Meeting – Conversion and Delivery of Electrical Energy in the 21st Century, July 20–24, 2008.

15 Yang S, Dong M, Zhang G-j, Zhang Z. Condition evaluation of oil–paper insulation based on dielectric spectroscopy. IEEE Annual Report – Conference on Electrical Insulation and Dielectric Phenomena, Ann Arbor, MI, October 18–21, 2009, pp. 77–80.

16 Ekanayake C, Gubanski SM, Graczkowski A, Walczak K. Frequency response of oil impregnated pressboard and paper samples for estimating moisture in transformer insulation. *IEEE Transactions on Power Delivery* 2006; 21(3): 1309–1317.

17 Blennow J, Ekanayake C, Walczak K, Garcia B, Gubanski SM. Field experiences with measurements of dielectric response in frequency domain for power transformer diagnostics. *IEEE Transactions on Power Delivery* 2006; 21(2): 681–688.

18 Ekanayake C, Saha TK, Ma H, Allan D. Application of polarization based measurement techniques for diagnosis of field transformers. IEEE Power and Energy Society General Meeting, Detroit, MI, July 24–29, 2010, pp. 1–8.

19 Jadav RB. Investigation of improved interpretation schemes for transformer insulation diagnosis. Ph.D. thesis, School of Information Technology and Electrical Engineering, University of Queensland, 2013.

20 Ekanayake C, Gubanski SM, Mularachchi KD, Fernando MARM. Diagnostic of power transformers in Sri Lanka: Application of dielectric spectroscopy in frequency domain. Proceedings of IEEE Electrical Insulation Conference and Electrical Manufacturing & Coil Winding Conference, 2001, pp. 593–596.

21 Yang S, Dong M, Zhang G, Zhang Z. Application of dielectric response for oil-immersed transformer. Proceedings of the 9th IEEE International Conference on Properties and Applications of Dielectric Materials, Harbin, China, July 19–23, 2009, pp. 370–373.

22 Koch M, Krueger M, Tenbohlen S. On-site methods for reliable moisture determination in power transformers. IEEE PES Transmission and Distribution Conference and Exposition, 2010, pp. 1–6.

23 Saha TK, Middleton R, Thomas A. Understanding frequency & time domain polarisation methods for the insulation condition assessment of power transformers. IEEE Power & Energy Society General Meeting 2009, PES '09, pp. 1–8.

24 Neimanis R, Saha TK, Eriksson R. Determination of moisture content in mass impregnated cable insulation using low frequency dielectric spectroscopy. Proceedings of the 2000 IEEE Power Engineering Society Summer Meeting, Seattle, WA, July 16–20, 2000, pp. 463–468.

25 Zaengl WS. Applications of dielectric spectroscopy in time and frequency domain for HV power equipment. *IEEE Electrical Insulation Magazine* 2003; 19(6): 9–22.

26 Linhjell D, Gäfvert U, Lundgaard LE. Dielectric response of oil-impregnated paper insulation: Variation with humidity and ageing level. IEEE Annual Report – Conference on Electrical Insulation and Dielectric Phenomena, 2004, pp. 262–266.

27 Gäfvert U, Adeen L, Tapper M, Ghasemi P, Jonsson B. Dielectric spectroscopy in time and frequency domain applied to diagnostics of power transformers. Proceedings of the 6th IEEE International Conference on Properties and Applications of Dielectric Materials, Xi'an, China, June 21–26, 2000, pp. 825–830.

28 Lundgaard LE, Hansen W, Linhjell D, Painter TJ. Aging of oil-impregnated paper in power transformers. *IEEE Transactions on Power Delivery* 2004; 19(1): 230–239.

29 Pradhan MK, Yew JH, Saha TK. Influence of the geometrical parameters of power transformer insulation on the frequency domain spectroscopy measurement. IEEE Power and Energy Society General Meeting – Conversion and Delivery of Electrical Energy in the 21st Century, July 20–24, 2008.

30 Lundgaard LE, Hansen W, Linhjell D, Painter TJ. Aging of oil-impregnated paper in power transformers. *IEEE Transactions on Power Delivery* 2004; 19(1): 230–239.

31 Ingebrigtsen S, Dahlund M, Hansen W, Linhjell D, Lundgaard LE. Solubility of carboxylic acids in paper (Kraft)–oil insulation systems. Annual Report – Conference on Electrical Insulation and Dielectric Phenomena 2004, CEIDP '04, pp. 253–257.

32 IEC 60814. *Insulating liquids: Oil-impregnated paper and pressboard: Determination of water by automatic coulometric Karl Fischer titration*, 2nd edn, Geneva, 1989.

33 Homagk C, Mössner K, Leibfried T. Investigation on degradation of power transformer solid insulation material. Annual Report – Conference on Electrical Insulation and Dielectric Phenomena 2008, CEIDP '08, pp. 75–78.

34 Lundgaard LE, Hansen W, Ingebrigtsen S. Ageing of mineral oil impregnated cellulose by acid catalysis. *IEEE Transactions on Dielectrics and Electrical Insulation* 2008; 15(2): 540–546.

35 Saha TK, Purkait P. Investigations of temperature effects on the dielectric response measurements of transformer oil–paper insulation system. *IEEE Transactions on Power Delivery* 2008; 23: 252–260.

36 Caruso H, Dasgupta A. A fundamental overview of accelerated testing analytic models. Proceedings of Annual Reliability and Maintainability Symposium, January 19–22, 1998, pp. 389–393.

37 Yew JH, Saha TK, Thomas AJ. Impact of temperature on the frequency domain dielectric spectroscopy for the diagnosis of power transformer insulation. IEEE Power Engineering Society General Meeting, Montreal, 2006.

38 Gubanski SM, Boss P, Csépes G, Der Houhanessian V, Filippini J, Guuinic P *et al.* Dielectric response methods for diagnostics of power transformers. *IEEE Electrical Insulation Magazine* 2003; 19(3): 12–18.

39 Gäfvert U, Frimpong G, Fuhr J. Modelling of dielectric measurements on power transformers. Proceedings of 37th Session – Large High Voltage Electric Systems, 1998.

40 Gäfvert U. Condition assessment of insulation systems. Nordic Insulation Symposium, Bergen, June 10–12, 1996, pp. 1–20.

41 Helgeson A. Analysis of dielectric response measurement methods and dielectric properties of resin-rich insulation during processing. Ph.D. thesis, Kungle Tekniska Högskolan, Stockholm, 2000.

42 Jadav RB, Ekanayake C, Saha TK. Understanding the impact of moisture and ageing of transformer insulation on frequency domain spectroscopy. *IEEE Transactions on Dielectrics and Electrical Insulation* 2014; 21(1): 369–379.

4

Dissolved Gas Analysis Interpretation and Intelligent Machine Learning Techniques

4.1 Introduction

A power transformer is one of the most critical pieces of equipment in an electricity grid, and its fault-free operation has a great influence on the reliability of power delivery. However, a power transformer's insulation can eventually deteriorate, since the transformer is continuously subjected to electrical, mechanical, and thermal stresses. Such insulation deterioration may lead to disruptive failure of a transformer [1–3]. Over the last several decades, a variety of chemical and electrical techniques have been developed for transformer insulation diagnosis, including: (1) oil characteristics tests such as breakdown voltage (BDV), acidity, dielectric dissipation factor (DDF), resistivity, 2-furfuraldehyde, water content, and dissolved gas analysis (DGA); (2) dielectric response measurement by frequency dielectric spectroscopy (FDS) and polarization and depolarization current (PDC) measurement; (3) frequency response analysis (FRA); and (4) partial discharge (PD) measurement [4–8].

Among the above techniques, transformer oil tests have been adopted widely by utilities to reveal various chemical and physical properties of insulating oil. Through the measurement of 2-furfuraldehyde and carbon oxides, oil tests can also provide some indication of the condition of the solid (pressboard and paper) insulation of a transformer. In the past 20 years, various interpretation methods have been developed for analyzing oil test results (oil characteristics) and detecting discharge and overheating faults occurring in the transformer insulation system [9]. However, there still exist some limitations of these conventional schemes. For example, the diagnosis results might be inconsistent while using different industrial standards. Sometimes the conventional schemes cannot produce the diagnosis results for every possible combination of dissolved gas ratios [10].

To overcome the disadvantages of the conventional DGA interpretation schemes, various intelligent diagnostic techniques – such as artificial neural networks (ANN) [11], fuzzy logic method [12, 13], neural–fuzzy system [14], and wavelet transforms [15] have been investigated extensively. These techniques have achieved some extent of success in transformer insulation diagnosis using oil characteristics. In contrast to several conventional interpretation schemes, these pattern recognition algorithms utilize not only the oil test data obtained from the transformer of interest, but also the historical oil test datasets collected from other transformers. By using historical datasets, a pattern recognition

Transformer Ageing: Monitoring and Estimation Techniques, First Edition.
Edited by Tapan Kumar Saha and Prithwiraj Purkait.

algorithm learns the underlying relationship between oil characteristics and the transformer's insulation, and it then applies such knowledge to make a diagnosis on the health condition of the transformer of interest.

This chapter provides a comprehensive study on intelligent DGA interpretation techniques and their applications in power transformer condition assessment. The chapter is organized as follows. Some major work and contributions from CIGRE in dissolved gas measurement and its interpretation are summarized in Section 4.2. The experience of using propene and propane as a new tool for DGA interpretation is discussed in Section 4.3. The drawbacks and uncertainties of the conventional DGA interpretations are identified in Section 4.4.

Section 4.5 proposes a general framework for applying intelligent pattern recognition techniques to transformer insulation diagnosis. A comprehensive literature review on various pattern recognition algorithms for transformer insulation diagnosis is provided in Section 4.6.

To overcome the drawbacks of the conventional pattern recognition techniques for health assessment of power transformers, a hybrid SMOTEBoost algorithm is introduced in Sections 4.7–4.10. This hybrid algorithm can facilitate pattern recognition algorithms, consistently achieving desirable diagnosis accuracy and generalization capability. A case study is presented in Sections 4.12–4.15 to illustrate the procedure of applying pattern recognition techniques and the SMOTEBoost algorithm to practical transformer insulation diagnosis. The chapter concludes in Section 4.16.

4.2 CIGRE Works Related to DGA

CIGRE has produced important technical documents on DGA in the past two decades. In Task Force (TF) 15.01.07 [5, 16], a new laboratory technique of gas extraction was developed during the late 1990s and subsequently introduced into IEC Standard 60567 in 2005 [17].

The typical gas concentrations and increase rates of dissolved gases collected from a large number of in-service transformers were investigated by Joint Task Force (JTF) D1.01/A2.11 (TF11) [18], and included in IEC Standard 60599 in 2007 [19]. TF11 also made extensive surveys concerning pre-failure concentrations in service, stray gassing of oils at relatively low temperatures, gas formation from paper, and gases trapped in paper insulation.

In 2008, TF D1-01-15 (recently renamed WG.D1.15 then WG D1.32) prepared a CIGRE Brochure [20] on gas monitoring devices, which are widely used for online detection of gas and fault formation in transformers. The findings of DGA for non-mineral oils and load tap changers, as well as modified DGA diagnosis schemes, are presented in Ref. [21].

4.3 Advancement of New Gases for Fault Diagnosis

Traditional DGA interpretation methods take into account only C1 and C2 hydrocarbons. It has been recognized that the concentrations of C3 hydrocarbons, including propene (C_3H_6) and propane (C_3H_8), can also provide some

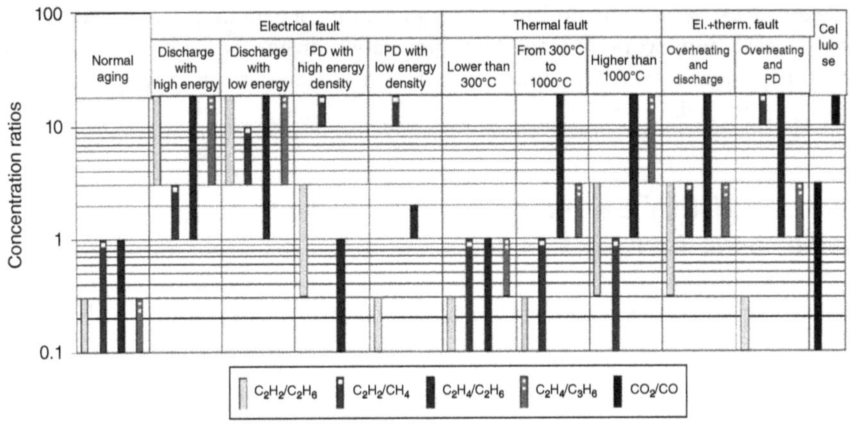

Figure 4.1 DGA interpretation scheme including C3 hydrocarbons. *Source:* Ref. [25], Figure 10.9.

complementary information and make the DGA diagnosis more precise [19, 22]. The DGA interpretation scheme from CIGRE which uses the gas ratio of C_2H_4/C_3H_6 to determine the incipient fault is shown in Figure 4.1. The interpretation scheme of C3 hydrocarbons is quite similar to the Rogers method. It has been reported that during the thermal decomposition of hydrocarbon, CH_4, C_2H_4, and C_3H_6 are evolved more than the other hydrocarbon gases [23] at a relatively higher temperature. The presence of C_3H_6 can be used as a potential indicator to detect thermal faults in transformers [24].

4.4 Uncertainty in Dissolved Gas Analysis

Although the DGA method has been used widely to identify incipient faults and assess the health condition of power transformers, some key issues still need to be investigated further for utilities to perform a reliable diagnosis on transformers. First, not all the combinations of gas ratios from measurements can be classified into certain fault types in different DGA interpretation schemes (e.g., Doernenburg ratio, Rogers ratio, IEC ratio). Inconsistent diagnostic results may be produced using different interpretation methods. Second, some DGA methods (e.g., Rogers ratio) may not be able to make diagnosis on certain faults (e.g., borderline cases), while other interpretation schemes can diagnose the potential faults in transformers. Thirdly, existing DGA interpretation schemes originate mainly from statistical analysis on historical DGA data and experience of experts and engineers. The relationship between measured gas concentrations and the different fault types may not be explored fully and presented in the conventional interpretation schemes. Therefore, some intelligent interpretation techniques – such as expert systems [26], ANNs [11], and fuzzy logic [13, 14] – are needed to determine the correlation between dissolved gases and transformers' insulation conditions.

Instead of using the above conventional schemes, this chapter investigates intelligent DGA diagnostic techniques and their application for transformer insulation diagnosis using oil characteristics. After presenting a generic pattern recognition application framework for transformer insulation diagnosis, a comprehensive literature review on various pattern recognition techniques reported in the literature is provided.

4.5 Statistical Learning-Based Intelligent Diagnostic Techniques

4.5.1 Pattern Recognition Framework for Transformer Diagnosis

Figure 4.2 illustrates the general application framework of pattern recognition algorithms. It consists of three steps: (1) data pre-processing; (2) feature selection and extraction; (3) classification or regression.

4.5.2 Data Pre-processing and Training Database Construction

Raw data for constructing the training database of a pattern recognition algorithm can often be inconsistent and noise-corrupted. Hence, it is necessary to apply various techniques to process the raw data before implementing any pattern recognition algorithm. Data pre-processing can be performed in the following ways:

- data cleaning for dealing with noise and removing redundant data;
- data transformation for converting the raw data into appropriate forms;
- data reduction for eliminating unnecessary attributes;
- data discretization for reducing the number of levels of an attribute of data.

4.5.3 Feature Extraction and Selection

Feature extraction aims to find the characteristic attributes (features) from the original data. This will enable pattern recognition algorithms to focus on those

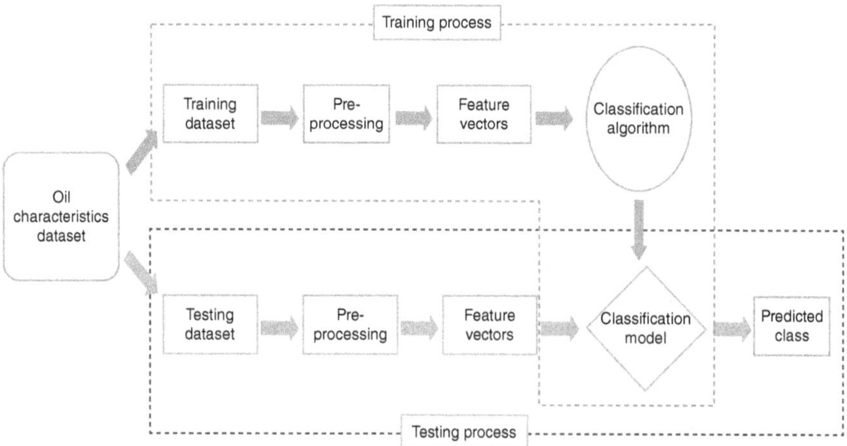

Figure 4.2 Framework of pattern recognition. *Source:* Ref. [27], Figure 1.

most relevant features for fault classification. In transformer insulation diagnosis using oil characteristics, the features may include the concentrations of dissolved gases and oil quality test results such as acidity, BDV, 2-furfuraldehyde, resistivity, DDF, and moisture concentration.

4.5.4 Classification and Regression

In the classification step, the pattern recognition algorithm makes use of a historic oil tests dataset to construct a mathematical model that approximates the relationship between the oil characteristics (e.g., features) and the categories of transformer insulation condition or the types of incipient faults in a transformer. Then, the above model is used to make classification on the insulation condition or the incipient fault of the transformer of interest into one of the categories previously determined from the historic dataset.

4.6 Review on Pattern Recognition Techniques for Transformer Insulation Diagnosis

This section reviews various pattern recognition techniques for transformer insulation diagnosis. The techniques include ANN, fuzzy logic, expert systems, decision-making algorithms, support vector machines (SVMs) and their optimization by population-based algorithms, and hybrid algorithms.

4.6.1 Artificial Neural Networks

ANNs can learn the potential relationship between input features and fault types of transformers through a training process. This is a significant advantage of ANNs, since the training dataset is usually constructed by a large number of actual samples from the real world and events, rather than being generated from human thoughts, which are usually involved in fuzzy logic or expert systems. However, the training process of ANNs can also be a major problem, as the construction of a training dataset requires the adequate representation of the domain knowledge of interest. Otherwise, the diagnosis accuracy of the network can be reduced significantly.

Over the last several decades, ANNs have been used extensively for transformer insulation diagnosis [11, 13, 14, 17–31]. Two steps are required for ANNs to complete the diagnosis task, including training and the classification process. During the training process, an ANN can construct a model to describe the dependency between the extracted features and the fault types through adjusting the weights between neurons and the thresholds of activation function of each neuron. Then, the well-established model is used to identify any new samples in one of the transformer insulation conditions predefined in the training process.

In Ref. [28], a two-stage ANN for transformer insulation condition assessment using dissolved gases was proposed. By integrating the gas concentration of CO_2, this network can identify the potential faults caused by the degradation of paper insulation. Tenfold cross-validation was employed in each of the two-stage networks to determine the optimal structure of ANN. In Ref. [29], Vanegas *et al.* proposed a network with two groups of input features: the first group based on the IEC gas ratios and the second group consisting of the concentrations of five dissolved gases. They demonstrated that the proposed network could attain better classification accuracy by using five gas concentrations as input features. A comparison between ANNs' efficiency for transformer insulation diagnosis using different DGA criteria was conducted by Guardado *et al.* [30]. The diagnosis criteria for training a neural network included five conventional interpretation schemes: Doernenburg ratio, modified Rogers ratio, Rogers ratio, IEC ratio, and DGA interpretation guidelines from California State University Sacramento (CSUS; this diagnosis criterion uses the individual gas concentrations). The authors reported that a network with three hidden layers could be more suitable for transformer insulation incipient fault detection. In Ref. [31], a hybrid of multinomial logistic regression model integrated with a back-propagation (BP) ANN algorithm was proposed to detect fault conditions of power transformers. The multinomial logistic regression model was used to determine the most relevant dissolved gases corresponding to certain types of incipient fault. The concentrations of these gases were used as input features for training. By using the hybrid approach, the diagnosis accuracy can be improved. Other types of ANN and their variants for transformer insulation diagnosis were also reported in the literature, including a self-organizing polynomial network (SOPN), which formulated the problem heuristically as a hierarchical architecture having multiple layers of low-order polynomials [32] and a reduced multivariate polynomial (RMP) network, in which the parameters are determined only by a predefined RMP's order without an iterative procedure [33].

In order to enhance the adaptation capability and accelerate the retraining process when new samples (data) become available, a probabilistic neural network (PNN) was adopted in Ref. [34]. In Ref. [35], a fuzzy learning vector quantization (FLVQ) network was also introduced. The FLVQ network uses a fuzzy classifier to segment historical DGA data into different categories of gas ratios. Then, for each category of gas ratio, a learning vector quantization (LVQ) network was trained to classify the types of incipient fault.

During the training process of normal ANNs, it is difficult to determine the network architecture and hidden layer neurons number. To solve this problem, various evolutionary algorithms were integrated into ANNs [36, 37]. Such hybrid algorithms can make better use of the excellent global search capability to find optimal interconnection weights, as well as bias terms of the networks simultaneously. Another popular extension theory was also introduced into the clustering algorithm for detecting transformer incipient faults [38, 39]. Instead of using the Euclidean distance, the extension distance was used in this method for measuring the similarity between different data points. One of the significant advantages in this method is that no artificial parameters are required to be tuned, and thus no learning process is needed.

To improve the data visualization capability, a self-organizing map (SOM)-based network was proposed in Ref. [40]. The SOM can also provide visualization on the evolution of an incipient fault by plotting the DGA trajectories using historical data. Similar to SOM, a number of auto-associative neural networks were implemented to provide visualization and clustering in transformer insulation diagnosis [41]. The basic k-nearest neighbor (kNN) algorithm and adaptive kNN algorithm were also applied for transformer insulation fault diagnosis [42, 43]. Moreover, a Bayesian network was adopted by some researchers for transformer fault diagnosis using DGA measurements data [44].

Owing to the advantages of a simpler network structure and better approximation ability, a radial basis function (RBF) neural network was investigated extensively and applied to transformer fault diagnosis. In Ref. [45], an integrated self-adaptive training-based RBF for transformer diagnosis was developed. This RBF network was constructed by making use of fuzzy C-means (FCM) clustering and quantum-inspired particle swarm optimization (QPSO). In recent years, a wavelet transform has also been incorporated into ANNs for analyzing DGA data. A three-layer ANN with evolving wavelet networks (EWNs) [46] and a genetic algorithm-tuned wavelet network (GAWN) [47] had been reported previously. In these two wavelet-integrated networks, the network parameters and weighting values were tuned automatically through an evolutionary algorithm-based optimization process.

Though they have been applied widely to transformer insulation diagnosis using oil characteristics, ANNs still suffer some inherited drawbacks. For example, they lack explanation capability on the established model and require a relatively large size of historic dataset for training the networks.

4.6.2 Fuzzy Logic System and Expert System

An expert system (ES) can mimic human expert behaviors to learn specific knowledge from an exclusive domain with a machine implementable form. An ES adopts this knowledge to facilitate decision making and justify its reasoning. The ES first extracts the inference symptoms from the domain knowledge and then presents the knowledge in different forms and structures, including production rules, frames, combinations of frames and rules, semantic nets, and objects [48]. As a transformer usually has a complex insulation structure and the degradation of the transformer involves different mechanisms under multiple stresses, potential faults of a transformer are normally diagnosed by engineers with solid knowledge and professional experience.

An ES tool or engine performs the decision making and inferences by formulating the original problem into a set of "if–then–else" rules [49], which requires constructing a set of consistent and complete rules. The procedures of rule generation are dominated by two specialists (i.e., a knowledge engineer and a domain expert). Some difficulties may exist in the rule generation from these two specialists, and the main reasons can be summarized as follows.

- First, the domain expert may lack the willingness to cooperate with the knowledge engineer. This is due to the fact that they are being asked to help develop a computer system that can replace them under certain circumstances.
- Another difficulty involving rule generation is that a human's knowledge is often expressed as natural language, which contains significant imprecise or vague information. This will result in more complex rule generation procedures to cover all important aspects of the knowledge.

Fuzzy logic (FL) can provide a simple way to extract convinced conclusions from information with vagueness, ambiguousness, or impreciseness. This is due to the advantages of FL theory in softening decision boundaries of different faults for transformers.

A number of researchers have applied FL techniques to diagnose potential faults of power transformers [50–53]. In Ref. [26], the authors developed an FL system for diagnosing transformer insulation and providing recommendations on maintenance actions. In this system, the fuzzy set concept was adopted to manage the uncertainties in key gas analysis, threshold, and gas ratio boundaries. In Ref. [54], the authors introduced a framework for conducting transformer diagnostics by applying fuzzy information theory. The fuzzy rules were integrated into a decision tree to give the diagnosis on transformer incipient faults. In Ref. [13], Su *et al.* proposed a fuzzy logic system to deal with the problem in which multiple faults occur simultaneously in a transformer insulation system. The proposed system can also indicate the severity of each fault. In Ref. [55], the authors adopted the acceptable/unacceptable norms of both the key gas concentrations and gas ratios in the implementation of the FL system. Their method can interpret the boundary cases, in which a transformer has nearly equal probability of having two different types of fault. In Ref. [56], Duraisamy *et al.* applied triangular, trapezoidal, and Gaussian membership functions to an FL system, which was then integrated with a BP ANN to diagnose transformer faults. The conventional IEC interpretation and gas concentrations were also used as references to establish the FL system.

In Ref. [57], Flores *et al.* combined an expert system with a Type-2 FL system to evaluate the transformer insulation using oil characteristics. This hybrid system can be used to determine whether paper insulation was a factor in any insulation faults, and it also allowed other features to be input into the pattern recognition algorithm. In Ref. [58], Abu-Siada *et al.* incorporated several DGA interpretation schemes into a single expert model to overcome individual experts' limitations. It was reported that such an approach can improve the diagnosis performance and also pave the way for standardizing DGA interpretation schemes. The major limitations of fuzzy logic and expert systems for transformer insulation diagnosis are due to the fact that the performance of these methods is highly dependent on the completeness of the predefined knowledge base. Neither fuzzy logic systems nor expert systems can automatically adjust the system parameters when new knowledge is incorporated.

4.6.3 Decision-Making Algorithms

Transformer insulation diagnosis is based on a variety of oil characteristics and is formulated as a multiple-attribute decision-making (MADM) problem. Two commonly adopted methods for solving this problem are evidence reasoning and gray theory.

Tang *et al.* [59] adopted an evidence reasoning algorithm to combine evidence and deal with uncertainties in transformer condition assessment. The algorithm provided an overall evaluation of transformer condition and ranked several transformers based on their necessities of maintenance. In Ref. [60], a fuzzy set

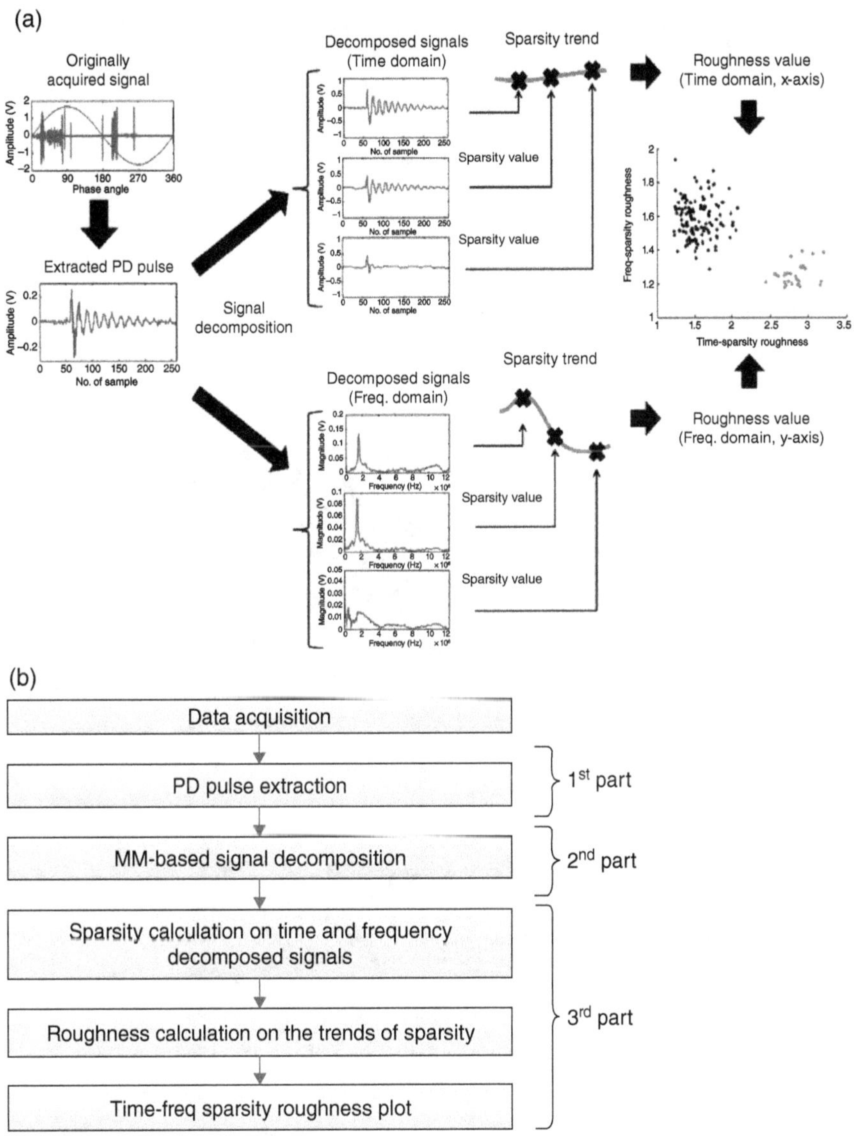

Figure 5.8 TF sparsity map for multiple PD source separation: (a) conceptual diagram and (b) flowchart.

Transformer Ageing: Monitoring and Estimation Techniques, First Edition.
Edited by Tapan Kumar Saha and Prithwiraj Purkait.

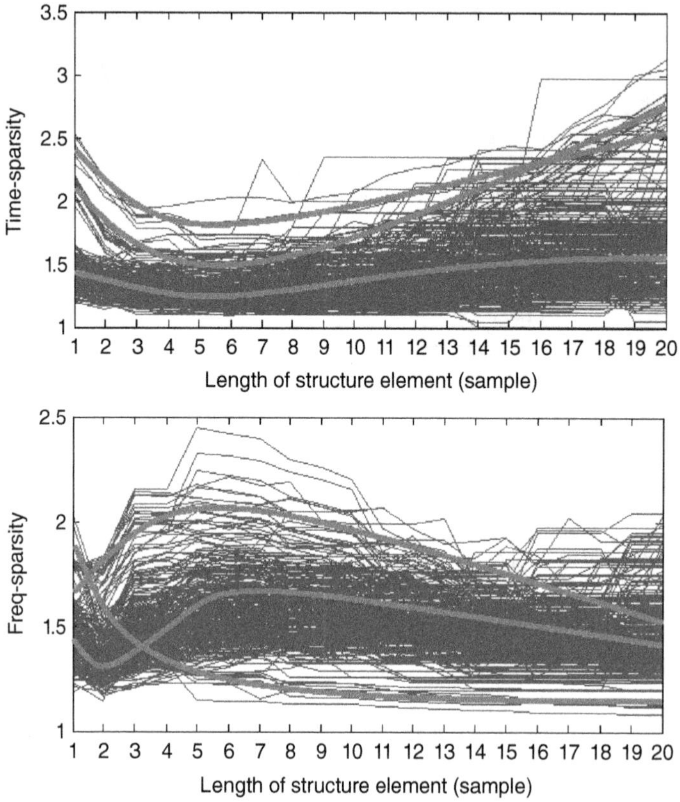

Figure 5.9 Sparsity trends for time (top) and frequency (bottom) domains of decomposed signals. Each blue line represents a sparsity trend of each PD pulse. Red lines represent the average of three different tendencies of the trends.

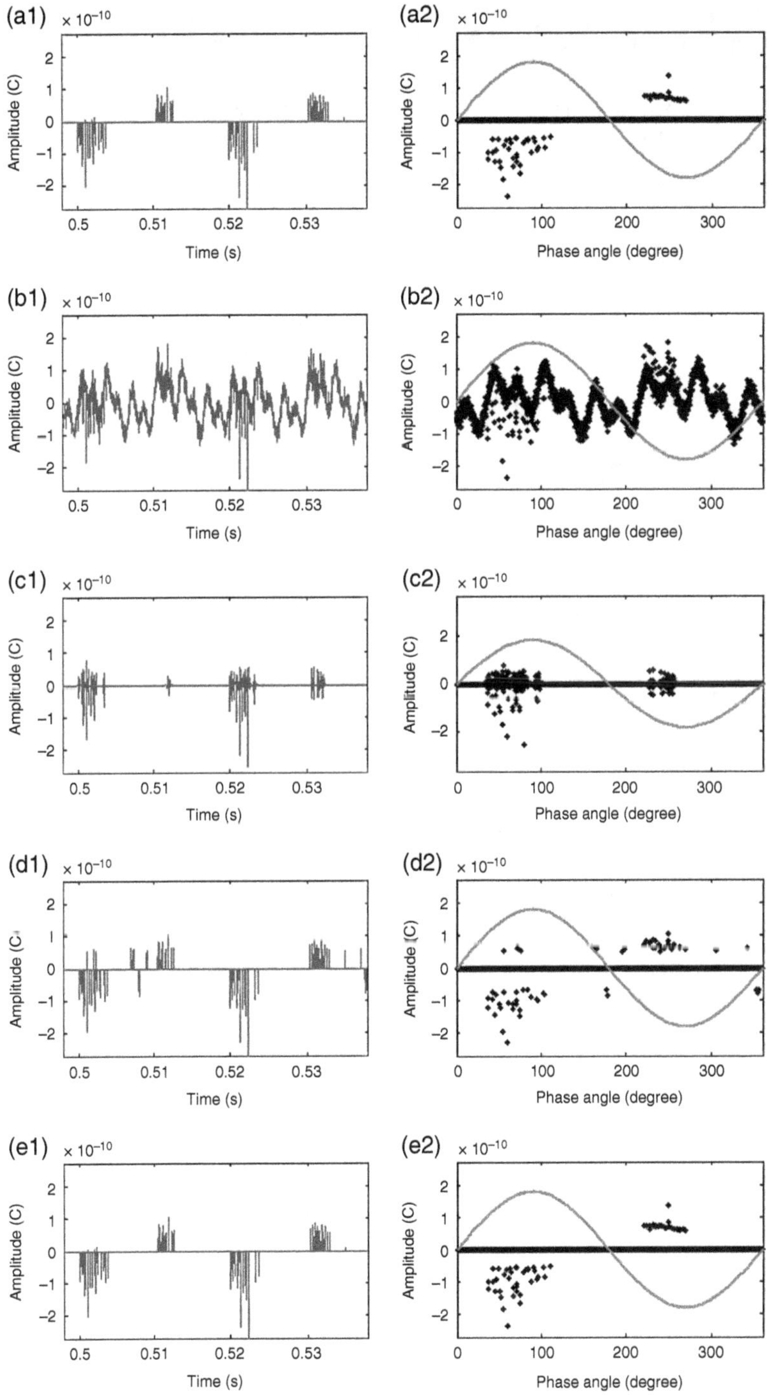

Figure 5.16 De-noising results of internal discharge. (a1)–(e1) Original signal, noise-corrupted signal (SNR = −5 dB), and de-noised signals of DWT (db5), EMD + AMT, and EEMD-based method. (a2)–(e2) Corresponding PRPD diagrams.

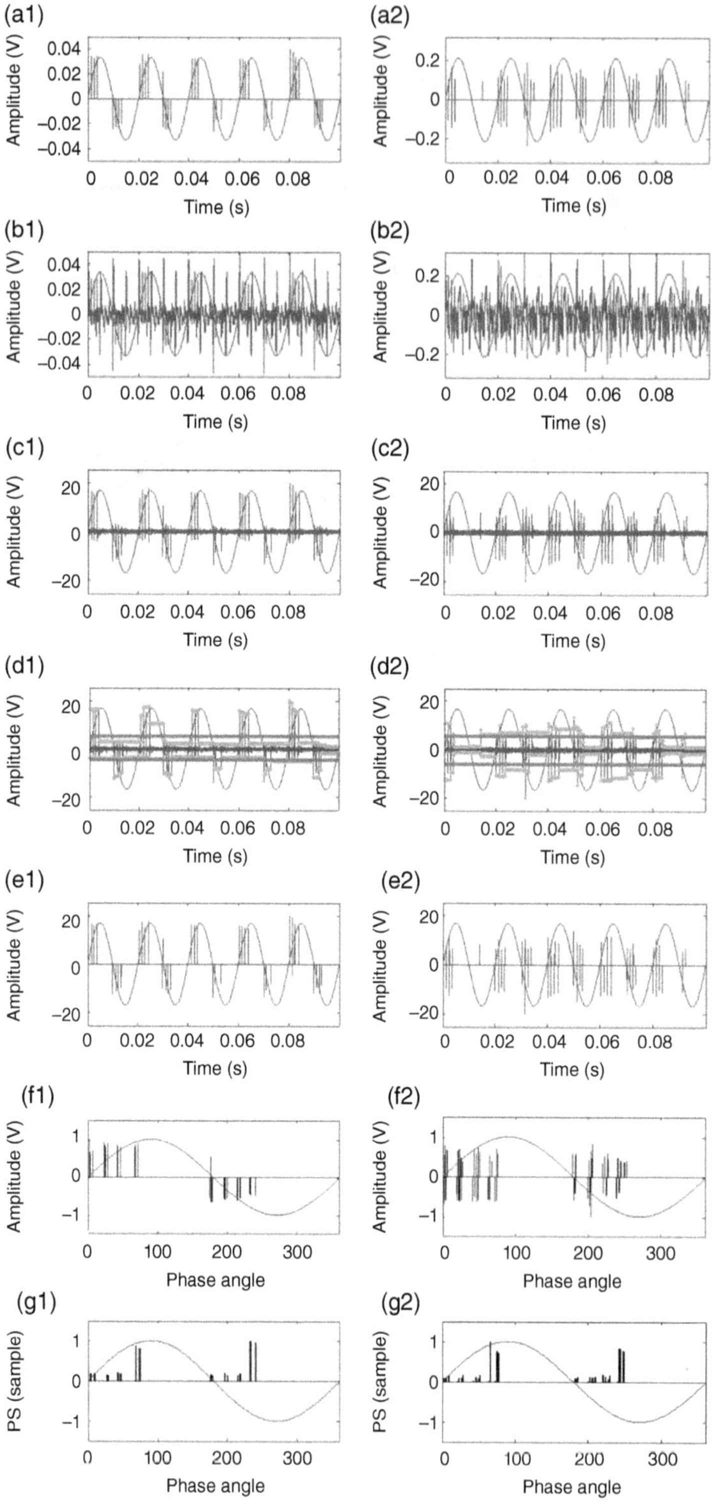

Figure 5.18 Results of discharge in transformer oil using (a1)–(g1) CT1 and (a2)–(g2) CT2. (a) Original signals, (b) noise-corrupted signals, (c) optimal equalized signals, (d) results of AMT on equalized signals, (e) de-noised signals, (f) PRPD diagram of original signals, and (g) PRPS diagram of de-noised signals.

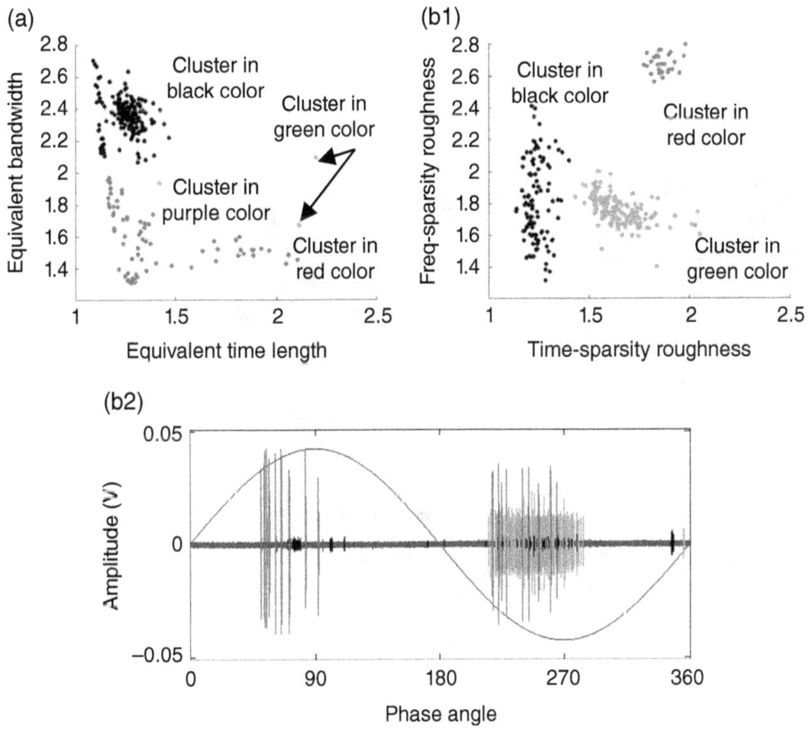

Figure 5.28 Results of multiple PD sources combining corona, surface discharge, and internal discharge. (a) Result of conventional TF map, (b1) result of TF sparsity map, and (b2) PRPD diagram based on the result of TF sparsity map.

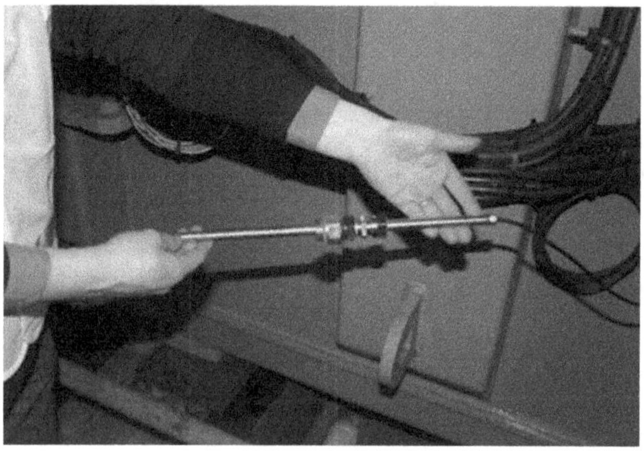

Figure 7.6 Example water activity probe. *Source:* Ref. [13], Figure 1.

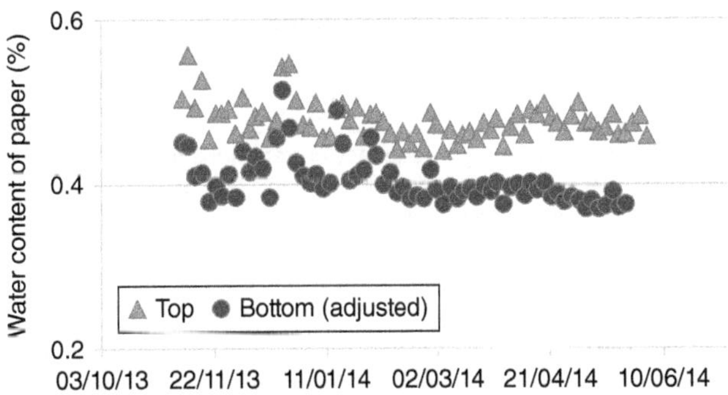

Figure 7.7 Comparison of measuring WCP using probes at different locations in transformer, corrected for temperature difference. *Source:* Ref. [13], Figure 10.

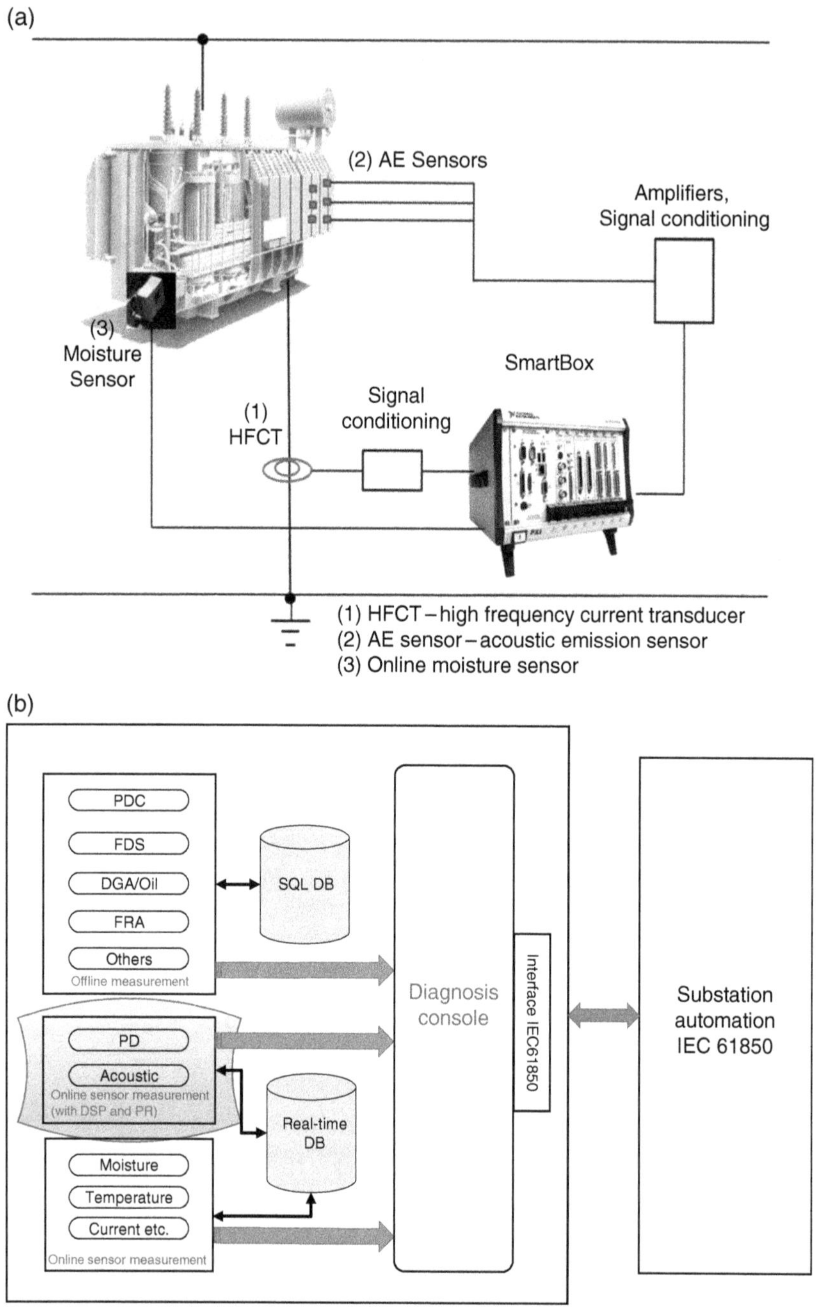

Figure 9.11 SmartBox – configurable hardware and software platform: (a) hardware modules; (b) system software architecture.

theory-based algorithm was first employed, which provided the diagnosis results as a set of possible types of fault with the probability of each type of fault. These diagnoses were then aggregated using an evidential reasoning algorithm. Based on an information fusion strategy, a multi-level and multi-aspect expert system was developed [61].

The gray model can perform pattern recognition using a relatively small size dataset and without involving formal statistical process and inference. A number of approaches combining gray theory and extension theory were developed to predict the trend of gas concentrations in transformer oil [62, 63].

4.6.4 Support Vector Machine and its Optimization by Population-Based Algorithms

There are also a number of other types of pattern recognition algorithm which have been applied for interpreting oil characteristics data. For example, SVM algorithms and population-based approaches such as genetic algorithms (GAs) and particle swarm optimization (PSO)-based programming.

A Parzen windows-based classifier was proposed in Ref. [64]. Parzen windows is a technique for nonparametric density estimation, which can be used for classification. This classifier is integrated with a PSO to search for the optimal parameters for a Parzen windows-based classifier. Over the last two decades, SVMs have been used for various classification problems. An SVM converts the data from the original space to a higher-dimensional feature space. In Ref. [65], the authors implemented a multi-layer SVM classifier and demonstrated its applicability to fault identification of transformers. In Ref. [66], a clonal selection algorithm was adopted to select the optimal input features and appropriate parameters for an SVM algorithm. In Ref. [67], an SVM algorithm was integrated with a GA to make the prediction on gas concentrations in transformer oil. This hybrid algorithm can prevent numeric instabilities such as overfitting or underfitting with the proper selection of SVM parameters using a GA. PSO can also be integrated into an SVM, with the PSO searching for optimal parameters for the SVM [68–70]. Some modifications have been made by adopting time-varying acceleration coefficients to improve the PSO convergence when searching for optimal parameters in an SVM [71]. Moreover, an artificial immune network classification (AINC) algorithm was proposed for transformer diagnosis [72].

4.6.5 Other Hybrid Algorithms

Attempts have been made to combine ANNs and fuzzy logic or expert systems in transformer insulation diagnosis. Such hybrid algorithms take advantage of the learning capability of ANNs, the knowledge formation of expert systems, and the uncertainty representation of fuzzy logic. Xu *et al.* [73] proposed a consultative mechanism to combine fuzzy logic and ANN. A combined expert system and ANN was also proposed in Ref. [74]. In the above two hybrid algorithms, the knowledge base was derived from IEEE/IEC interpretation schemes and also included experts' experiences. Miranda and Castro [75] proposed combining a neural network into a fuzzy system for the extraction of rules. Some other hybrid approaches proposed in the literature include: integration of conventional DGA

Table 4.1 Advantages and limitations of different pattern recognition algorithms

Methodology	Advantages	Limitations
Artificial neural network	1) Knowledge acquired through training. 2) Ability to learn comprehensive input–output correlations. 3) Excellent interpolation and extrapolation capability.	1) Accuracy determined by quality of training database. 2) Lack of rules in determining parameters and structures of the network. 3) Time-consuming training process.
Fuzzy logic	1) Makes decisions through imprecise evidence. 2) Softens decision boundaries to solve problems with uncertainties.	1) Requires human involvement in determining membership functions. 2) Lack of capability to learn directly from samples.
Expert system	1) Performs inference based on domain knowledge and specialized experience. 2) Imitates human behavior to make decisions.	1) Accuracy determined by knowledge from engineers and domain experts. 2) Lack of capability to discover new knowledge through autonomous learning.
Decision-making algorithm	1) Formulates diagnosis as a multiple-attribute decision-making problem. 2) Attains final decisions by aggregating multiple evidences.	1) Difficulty in assigning weights to various attributes. 2) Difficulty in aggregating overall assessment.
Support vector machine	1) Reduces dimensional complexity by kernel method. 2) Able to attain good accuracy with small size database.	1) Possibility of overfitting. 2) Difficulty in determining parameters of the network.
Hybrid algorithm	1) Combines advantages and computational intelligence of different methods.	1) Complex algorithm development.

interpretation schemes with ANNs and fuzzy logic [76]; integrated neural fuzzy approach [14]; association rule mining [77]; hybrid of fuzzy approach and evidential reasoning-based decision-making approach [78]. Other types of hybrid approach have also been proposed in the literature. For example, in Ref. [79] Dempster–Shafer evidential theory was combined with BP neural networks (NNs) and fuzzy logic. In Ref. [80], a neuro-fuzzy scheme and PNN were integrated.

Table 4.1 summarizes the advantages and limitations of different pattern recognition algorithms. From this table it can be seen that ANNs can learn potential relationships between the input features and fault types of transformers. However, the knowledge learned is still hidden in the network. FL has the advantage of utilizing the knowledge discovered from experts to perform fault diagnosis with uncertainties. However, the explicit membership functions in FL need to be defined by experts and they are not extracted directly from dissolved gas samples. The diagnosis accuracy of ES relies heavily on the quantity and quality of the obtained rules. An SVM can be utilized to identify incipient faults when DGA data is not sufficient.

Table 4.2 Dataset configurations and results of some representative pattern recognition algorithms in the literature

Model	Ref.	Samples	Training	Testing	# Features	# Classes	Accuracy (%)
ANN	[28]	40	N/A	N/A	5	4	86–95
	[29]	26	N/A	N/A	5/8/3	3	73–96
	[30]	150	117	33	3/5	8	87–100
	[33]	167	156	11	6	N/A	100
	[37]	820	N/A	N/A	3/5	5	90–93
	[41]	352	N/A	N/A	3	5	100
	[36]	630	N/A	N/A	3/5	5	91–95
	[34]	503	497	4	4	8	100
FL	[81]	711	N/A	N/A	3	8	92
	[82]	561	N/A	280	N/A	5	94
	[55]	20	N/A	N/A	9	9	100
ES	[74]	210	150	60	24	6	93–96
	[75]	318	N/A	88	3	5	99
	[26]	101	N/A	N/A	9	3	93
SVM	[65]	75	50	25	5	4	100
	[83]	142	N/A	N/A	5	4	94
DM	[84]	21	N/A	N/A	8	9	88
	[62]	46	N/A	N/A	10	10	97
	[39]	22	N/A	N/A	5	9	96–100
Hybrid	[64]	168	N/A	N/A	3/3/8	4	80
	[32]	711	N/A	N/A	3/5	8	87–97
	[35]	711	N/A	N/A	3	8	97

Difficulties still exist in determining the parameters of the SVM model and avoiding an overfitting problem. Hybrid artificial intelligence (AI) systems, which integrate the advantages from different computational intelligences, are developed as state-of-the-art intelligent tools for diagnosing transformer faults.

To provide a clear picture of the setup and performance of some state-of-the-art pattern recognition algorithms for transformer insulation diagnosis, we have summarized the data configuration presented in the literature, as shown in Table 4.2. In this table, the total sample numbers, training dataset size, testing dataset size, number of input features, and types of transformer insulation fault in terms of fault classes as well as the classification accuracies of different pattern recognition algorithms are compared in detail. It can be seen that the sample numbers of some datasets are very limited, and they contain only 20 samples. Moreover, the general guidelines for oil characteristics database preparation, training/testing datasets formation, cross-validation, testing, and evaluation in applying pattern recognition algorithms for diagnosing transformer conditions are also absent.

4.7 Hybrid Algorithm for Improving Power Transformer Insulation Diagnosis

The fault rate of power transformers in developed countries' utilities is reasonably low, and few transformers' insulation systems are in a critical condition. This implies that the records (samples) of some types of incipient fault or insulation condition (e.g., transformer insulation with critical condition) in these countries could be insufficient. In contrast, recording of transformer faults and DGA data in most developing countries is still in its initial stages. Thus, the samples in the training datasets discussed in Section 4.3 are unevenly scattered over the data space, where samples in some classes (minority classes) are significantly outnumbered by those in other classes (majority classes). Trained by such an unbalanced dataset, a pattern recognition algorithm could be in favor of the majority classes, which could even lead to misclassification of the minority classes.

To overcome the above-mentioned problem, Section 4.8 introduces a hybrid SMOTEBoost algorithm to facilitate a pattern recognition algorithm consistently achieving desirable diagnosis accuracy and generalization capability. In SMOTE-Boost, the synthetic over-sampling technique (SMOTE) is implemented to deal with the class imbalance problem, while the boosting approach is adopted to adaptively reweight and group data points in the training dataset. Moreover, the bootstrap method is also utilized to statistically equalize the sample numbers after executing SMOTE.

4.8 Synthetic Minority Oversampling Technique

Figure 4.3 depicts a schematic diagram of the SMOTE algorithm. For a particular sample in the minority class, SMOTE generates synthetic samples along the sections of the line which join the k nearest neighbors of this sample [85].

SMOTE is implemented in the following two steps [85].

1) Calculating the Euclidean distance between $\mathbf{X}$ and its k closest neighboring samples in the minority class. If k is set to a large value, it may increase the possibility of including outliers in the training dataset and lead to a large error rate. By contrast, if k is set to a very small value, the number of generated synthetic samples may not be enough to ensure the diversity of the minority class. By using a trial and error method, k is chosen to be five here.
2) Determining the number of new samples to be synthesized (over-sampling rate) based on the imbalance ratio of the original dataset. Suppose $\mathbf{x}_2$ is one of the closest neighbors of sample $\mathbf{X}$ and the Euclidean distance between $\mathbf{X}$ and $\mathbf{x}_2$ is $d(\mathbf{X}, \mathbf{x}_2)$, then the synthetic samples are

$$\mathbf{m}_i = \mathbf{X} + \text{rand}(0, 1) \times d(\mathbf{X}, \mathbf{x}_2) \tag{4.1}$$

where rand(0,1) is a random number in the range [0, 1].

However, the above standard SMOTE still poses some limitations. Since new samples are synthesized by random interpolation, it cannot guarantee that all

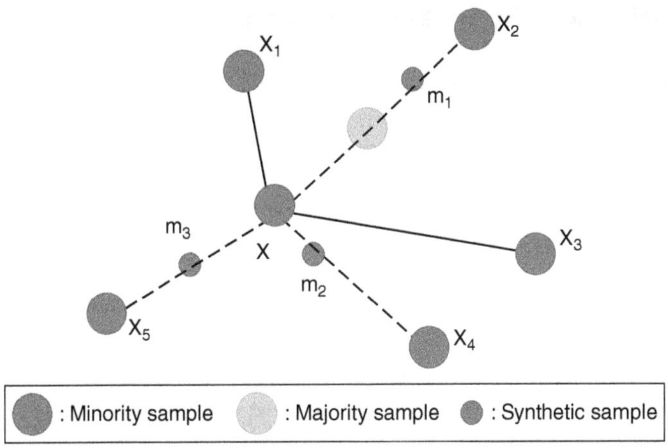

Figure 4.3 Schematic diagram of SMOTE algorithm, X denotes a sample and $X_1, \ldots, X_5$ are its five nearest neighbors. SMOTE generates new data m_i along the line between **X** and one of five neighbors. *Source:* Ref. [86], Figure 1.

these synthesized samples are distributed within the center area of the minority class. Some samples may be located in the center area of the minority class while others may become outliers, which are located remotely from the center of the minority class. Such a sample distribution may still introduce difficulties for a pattern recognition algorithm to make classification on fault types (or insulation conditions). To handle such limitations, a boosting procedure is proposed in this chapter, which is integrated with SMOTE [86].

4.9 Integrated SMOTEBoost Algorithm

Boosting is a meta-learning technique that "boosts" the performance of pattern recognition algorithms. Its basic premise is that multiple underperformance algorithms (termed "weak learners") can be combined to construct a high-performance ensemble (termed a "strong learner") [87]. During the initial training of the boosting, all samples are assigned an equal weight. Then, the incorrectly classified samples are given a higher weight to redirect the subsequent training to focus on these samples. The training iterations will continue until the ensemble of weak learners attains the desirable classification accuracy.

In SMOTEBoost, SMOTE is executed in each iteration of boosting to enable each classifier (pattern recognition algorithm) to attain high classification accuracy on the minority class. It should be noted that SMOTE is only invoked for processing the samples in the minority class. This implicitly increases these samples' weights. The synthesized samples will later be discarded after completing the training at iteration t (i.e., they are not added to the training dataset at iteration $t + 1$). Such an arrangement has the significant benefit of increasing data diversity for classifiers, as different sets of synthesized samples are provided to the classifiers at different iterations.

4.10 Hybrid of SMOTEBoost and Bootstrap

Figure 4.4 depicts the architecture of the SMOTEBoost and bootstrap hybrid algorithm. In this hybrid algorithm, SMOTE is first used to synthesize new samples for the minority classes in the training dataset. SMOTE may change the original sample distribution by synthesizing higher-cost samples. The number of synthetic samples is determined by SMOTE, until the appearance of different samples is proportional to their cost [88]. After SMOTE, sample numbers in minority and majority classes may still not be equal. To equalize the sample numbers statistically amongst majority and minority classes, bootstrap [89, 90] is adopted to redraw the samples in the majority class in the training dataset.

Bootstrap is a resampling technique, which can be used to approximately equalize the sample numbers in each class of a dataset. Suppose, after the SMOTE process, that n samples are distributed in m majority classes and $n_1, \ldots, n_t, \ldots, n_m$ are the sample numbers in each majority class. Let $n_{tr}, t = 1, \ldots, m$ be the desired sample numbers of each majority class. Bootstrap is performed for each class according to the ratio $n_{tr}/n_t, t = 1, \ldots, m$ to redraw new samples from the original majority class. Then, these samples are used to form the decreased size majority classes. Finally, these majority samples are combined with the minority samples generated by SMOTE to construct a class balance training dataset, where different classes have equal numbers of samples. This training dataset will be used for algorithm

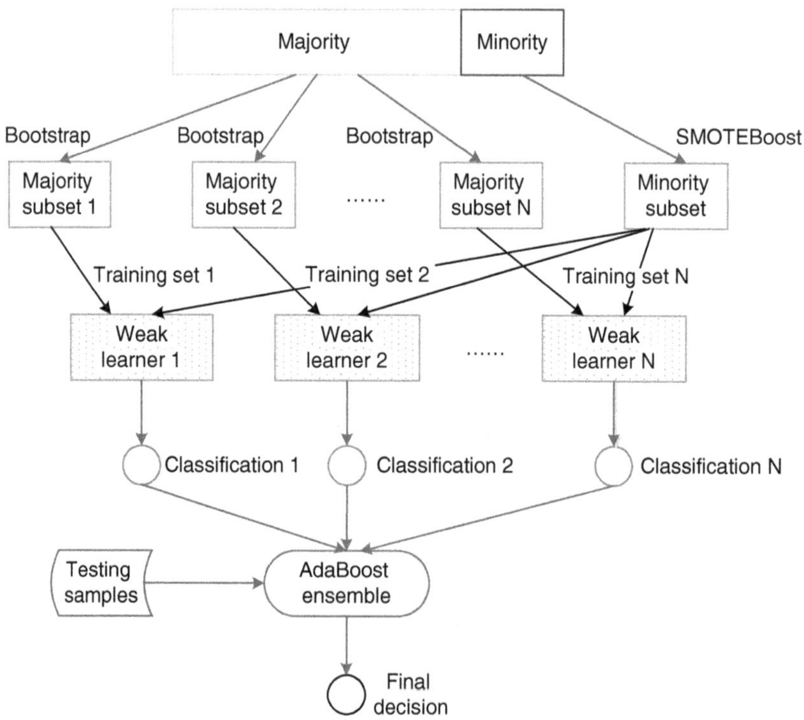

Figure 4.4 Architecture of hybrid SMOTEBoost and bootstrap algorithm. *Source:* Ref. [86], Figure 2.

training at the current iteration. It needs to be mentioned that the samples (in the majority classes) discarded by the bootstrap are put back into the dataset after completing the training at the current iteration, and can be used at the next iteration.

After being processed by SMOTE and bootstrap, both majority and minority classes now have statistically equalized samples and the original dataset becomes class balanced. This will help improve the classification performance of a pattern recognition algorithm, which integrates SMOTE and bootstrap as a preprocessing step. After such an integration, a boosting approach will be invoked to enable the above pattern recognition algorithm to make a classification on the fault types or insulation condition of the transformer of interest.

In the boosting approach, during each iteration a number of training subsets will be established by combining the samples from the majority and minority classes. By using these training subsets, a number of weak learners (e.g., any classifiers that only attain low classification accuracy during the training process) can be obtained at the current iteration. Then, a strong learner can be obtained by combining these weak leaners after certain iterations (i.e., a classifiers ensemble that attains high classification accuracy) [17]. Subsequently, this strong learner is applied to make classification on the fault types or conditions of the transformer insulation in the testing dataset. These testing samples were previously unknown for the strong learner. The above process (iterations) will be continued until a satisfactory performance is attained by the strong learner. Here, SMOTE, boosting, and bootstrap are integrated with a number of representative pattern recognition algorithms including C4.5, *k*NN, RBF network, and SVM for transformer incipient insulation condition fault classification. For the sake of integrity, the next section will provide a brief review of these four pattern recognition algorithms.

4.11 Collection of Pattern Recognition Algorithms Used for DGA

4.11.1 Support Vector Machine

To separate the samples in the original data space, SVM maps these samples into a high-dimensional hyperspace. In that hyperspace, SVM finds a hyperplane which can keep the separation of samples of different classes maximized, while minimizing the classification error [91]. To obtain the above hyperplane, the following optimization problem needs to be solved.

Minimize

$$\sum_{k=1}^{N} \alpha_k - \frac{1}{2}\sum_{k=1}^{N}\sum_{j=1}^{N} \alpha_k \alpha_j y_k y_j K\left(x_k, x_j\right) \tag{4.2}$$

subject to

$$\begin{gathered} \sum_{k=1}^{N} \alpha_k y_k = 0 \\ 0 \leq \alpha_k \leq \rho_k C,\ k = 1, \ldots, N \end{gathered} \tag{4.3}$$

where α_k is the Lagrange multiplier, C is the regularization parameter to make the balance between the margin maximization and classification error, and $K(x_k, x_j)$ is the kernel function formulated as $K(x_k,x_j) = \varphi(x_k)^T \varphi(x_j)$ [18]. After obtaining the above hyperplane, SVM determines the label of the class to which a new sample x^* belongs by using the equation $y = \text{sgn}\left[\sum_{k=1}^{N} \alpha_k y_k K(x_k,x^*) + b\right]$ (where N is the support vectors number and b is the bias).

4.11.2 *k*-Nearest Neighbor Algorithm

The *k*NN algorithm computes the Euclidean distances between a certain sample $\mathbf{x}^*$ and each sample in the training dataset. Suppose S_k is a sub-dataset which includes the K closest data points and K_i is the occurrence frequency of the ith class in S_k. Then the class of $\mathbf{x}^*$ is determined with the highest frequency in S_k [43, 92].

4.11.3 Decision Tree

A decision tree is a tree-like graph which consists of a number of decisions and the corresponding outcomes. In the graph, the internal nodes of the tree represent tests on input data and the leaf nodes represent the classes (fault types or insulation condition). The C4.5 decision tree adopts the information gain I(Y|X) to measure the reduction in uncertainty about the class label Y of a sample X given the value of this sample [93]. The uncertainty that attributes to the class label Y is calculated with its entropy, H(Y), and the uncertainty about Y given the value of X is denoted by the conditional entropy H(Y|X). C4.5 also adopts the gain ratio, which is (Y|X) = I(Y|X)/H(X) to measure the information gain with respect to the raw information contained in the X distribution.

4.11.4 Radial Basis Function Network

The architecture of the RBF network is presented in Figure 4.5 and it belongs to a family of ANNs. Generally, the RBF network contains input, hidden, and output layers.

In an RBF network, the interpolation function is in the form $f(\|x - c_i\|)$, where $\|.\|$ is a distance measure and c_i is the center of the class to which the sample x belongs. The Gaussian function $f(x) = \exp\left(-\frac{\|x - c_i\|^2}{2\sigma^2}\right)$ is adopted in many applications. For large enough neurons in the hidden layer, this can prove that it can be sufficiently approximated by the summation of RBFs, where each is located on a different point in the space [92]

$$y(x) = w_0 + \sum_{i=1}^{k} w_i \exp\left(-\frac{(x - c_i)^{\mathrm{T}}(x - c_i)}{2\sigma^2}\right) \tag{4.4}$$

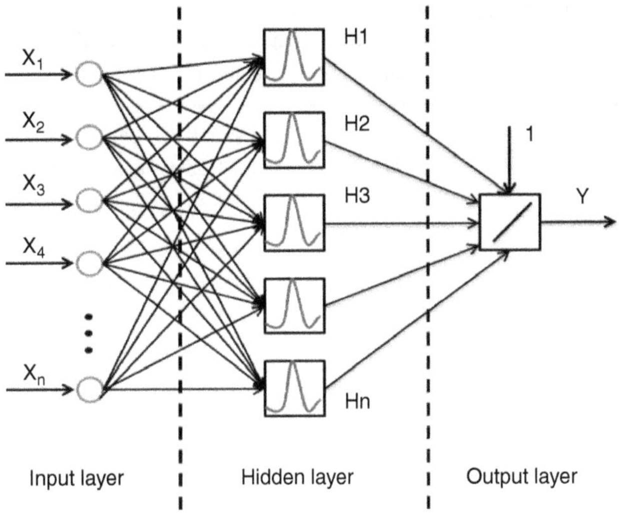

Figure 4.5 Schematic of radial basis function network. *Source:* Ref. [86], Figure 3.

4.12 Case Studies and Results

4.12.1 Training Dataset Construction

Here, eight oil characteristic datasets are used to verify the performance of the above four pattern recognition algorithms with integration of SMOTEBoost and bootstrap. Table 4.3 summarizes the configurations of these eight datasets. Among these datasets, datasets 1, 2, 3, 5, and 6 were collected from different utilities and dataset 4 was prepared from the literature [9]. Dataset 7 is the combination of datasets 1, 2, 3, and 4 with a few incompatible records removed. In dataset 8, about two thirds of the samples are from dataset 5 and one-third of the records are from dataset 6. The purpose of combining datasets collected from different sources is to evaluate the generalization capability of the above pattern recognition algorithms. The algorithm which is trained on one dataset but can achieve consistent classification accuracy on another dataset is regarded as having a higher generalization capability.

In Table 4.3, the features (input data of pattern recognition algorithms) of datasets 1, 2, 3, 4, and 7 are the concentrations of dissolved gases (i.e., H_2, CH_4, C_2H_6, C_2H_4, C_2H_2) and the transformer conditions – including normal deterioration, discharge, partial discharge, and thermal faults. The features in datasets 5, 6, and 8 include the concentrations of dissolved gases (i.e., C_2H_4, C_2H_2, C_2H_6, H_2, CH_4, CO, CO_2) and oil test data of water content, acidity, dielectric dissipation factor, 2-furfuraldehyde, resistivity, and breakdown voltage. In these three datasets, transformer insulation conditions are categorized into four levels of "excellent," "good," "fair," and "poor" condition [94]. The reason for selecting the gas concentration rather than the gas ratio as an input feature is that in some cases it may not be possible to calculate ratios if the concentration of a certain gas equals zero.

Table 4.4 describes the sample distributions of the original eight datasets. Tables 4.5 to 4.7 give the 95% confidence interval length of mean value (μ) and

Table 4.3 Configuration of eight datasets

Dataset	Source	Samples	Features
1	Utility	390	Dissolved gas concentrations
2	Utility	992	Dissolved gas concentrations
3	Utility	181	Dissolved gas concentrations
4	Publications	206	Dissolved gas concentrations
5	Utility	181	Dissolved gases + oil tests
6	Utility	882	Dissolved gases + oil tests
7	Utility and publications	1542	Dissolved gas concentrations
8	Utility	875	Dissolved gases + oil tests

Table 4.4 Sample distribution of original dataset

Dataset	Normal deterioration	Discharge	Thermal	Partial discharge
1	201	12	145	32
2	169	227	295	301
3	21	30	50	80
4	50	83	56	17
7	248	440	523	331
Dataset	**Excellent**	**Good**	**Fair**	**Poor**
5	80	50	21	30
6	165	188	266	263
8	80	316	284	195

Table 4.5 Mean value and standard deviation of five DGA datasets

	H_2		CH_4		C_2H_4		C_2H_6		C_2H_2	
Features	μ	σ	μ	σ	μ	σ	μ	σ	μ	σ
Dataset 1	8652	9382	1073	1546	42	160	227	318	265	1439
Dataset 2	39	104	75	209	51	415	53	68	2	14
Dataset 3	2931	7606	545	1368	443	1148	160	407	527	1392
Dataset 4	20533	23901	2944	4727	122	483	414	685	475	1938
Dataset 7	7795.41	8522.75	800.12	1535	65	439	141	266	110	808

Note: μ denotes the mean value and σ denotes the standard deviation.

Table 4.6 Mean value and standard deviation of three oil characteristics datasets

	H_2		CH_4		C_2H_4		C_2H_6		C_2H_2	
Features	μ	σ	μ	σ	μ	σ	μ	σ	μ	σ
Dataset 5	2932	7606	546	1369	443	1149	161	408	528	1392
Dataset 6	26	43	63	65	14	50	54	56	1	0.01
Dataset 8	6548	3911	570	553	65	471	105	163	55	414

Table 4.7 Mean value and standard deviation of three oil characteristics datasets

	H_2O		Acidity		BDV		Furfural		DDF	
Features	μ	σ	μ	σ	μ	σ	μ	σ	μ	σ
Dataset 5	17	11	0.1	0.1	55	19	0.8	1.2	0.1	0.1
Dataset 6	2	2	0.1	0.1	43	30	0.1	0.1	2	5
Dataset 8	5	7	0.1	0.2	28	34	0.2	0.7	3	6

standard deviation (σ) of each feature in the original five DGA datasets and three oil characteristics datasets. From Tables 4.5 to 4.7, it can be seen that there is a large distribution in both mean value and standard deviation for all input samples in each class. Therefore, it can be assumed that it would be sufficiently difficult to conduct a classification task based on the available DGA and oil characteristics datasets.

To demonstrate the effectiveness of SMOTE and bootstrap, Figure 4.6 provides the visualizations of sample distributions of dataset 1 before and after applying SMOTE. It can be seen that dataset 1 is class imbalanced, in which certain classes significantly outnumber other classes. After performing SMOTE, the sample numbers in different classes tend to be equal (normal deterioration, 252; discharge, 253; thermal fault, 250; partial discharge, 258). It can be observed from Figure 4.6(b) that SMOTE does not replicate the samples in the majority class but synthesizes new samples for the minority class to achieve approximately equal distribution of samples throughout the whole dataset. Table 4.8 presents the sample distributions of eight datasets after applying SMOTE and bootstrap on the original datasets.

4.13 Numeric Experiment Setup

To evaluate the performance of four pattern recognition algorithms (C4.5, *k*NN, RBF, and SVM) with and without integrating SMOTEBoost and bootstrap, the following training and testing procedure is followed. Firstly, each dataset of Table 4.3 is divided into a training sub-dataset (70% samples) and a testing

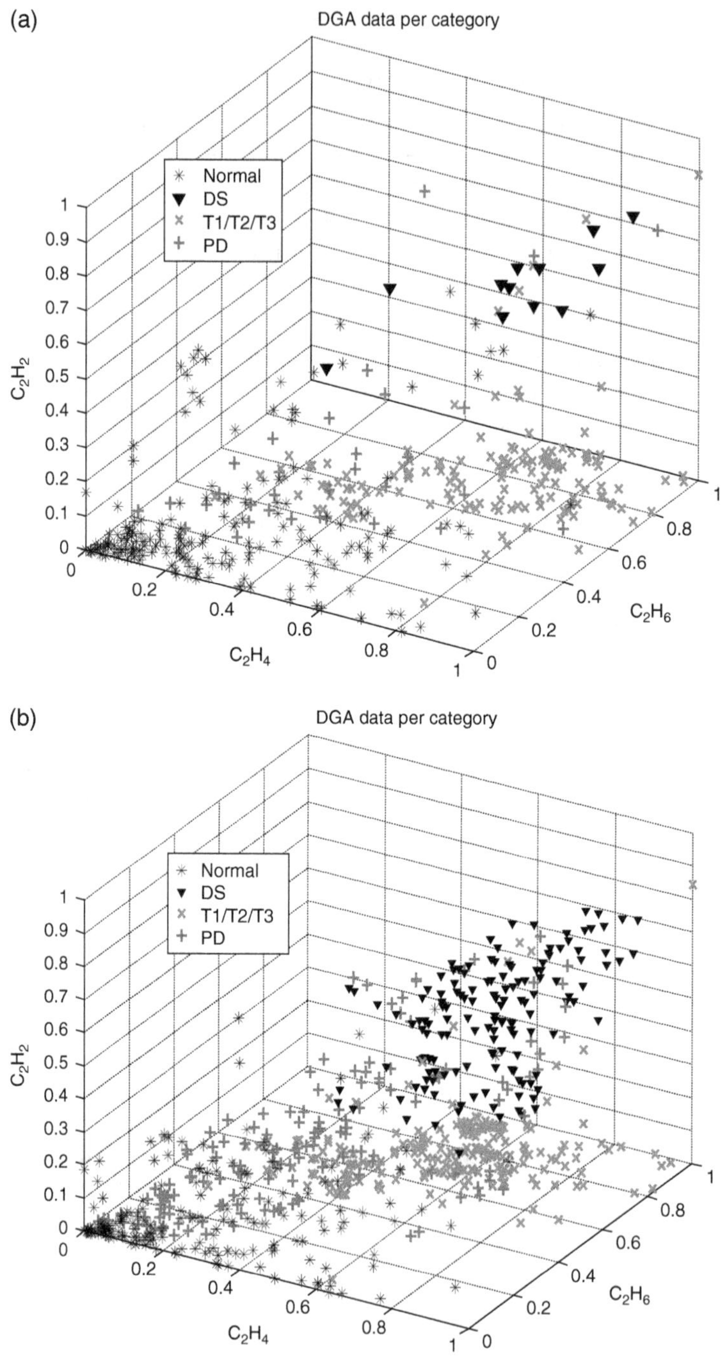

Figure 4.6 Sample distribution of dataset 1 in Table 4.3: (a) original dataset; (b) after processing by SMOTE and bootstrap. DS = discharge faults, OT = thermal faults, PD = partial discharge, normal deterioration. *Source:* Ref. [86], Figure 4.

Table 4.8 Sample distribution after SMOTE and bootstrapping

Dataset	Normal deterioration	Discharge	Thermal	Partial discharge
1	252	253	250	258
2	411	410	412	411
3	116	118	117	115
4	99	100	98	99
7	729	728	728	730
Dataset	**Excellent**	**Good**	**Fair**	**Poor**
5	107	108	107	108
6	378	379	377	378
8	436	435	437	435

sub-dataset (30% samples). The samples in both datasets are normalized within [0, 1]. Then, tenfold cross-validation is performed on the training dataset to determine the optimal values of the above four algorithms, including the hidden neurons number in RBF, the neighbors number for *k*NN, the parameters C and γ in SVM. For k-fold cross-validation, the original dataset D is randomly split into k mutually exclusive subsets $D_1, D_2, \ldots, D_k$ with approximately equal size. The algorithm is then trained on $D_1, \ldots, D_{t-1}$ and validated on D_t, $t \in \{1, 2, \ldots, k\}$. A total of k trained models can be obtained. The parameters of the trained model which has the highest classification accuracy are chosen to build the final model for performing classification on the testing dataset.

After finding the optimal values of the above parameters, pattern recognition algorithms are trained on the whole training sub-dataset. Subsequently, the well-trained algorithms are used to label each sample of the testing sub-dataset into particular transformer insulation conditions (excellent, good, fair, poor) or incipient fault types (discharge, partial discharge, thermal fault). The above procedure is repeated 50 times (runs). In each run, the classification accuracy is computed according to $N'_{\text{total}}/N_{\text{total}}$ (where N'_{total} denotes the number of samples which are correctly classified and N_{total} denotes the total number of samples of the whole dataset). The overall accuracy is the average over 50 runs. The software routines in [95, 96] are tailored for the algorithmic implementation in this section [86].

4.14 Discussion on Classification Accuracy

Table 4.9 presents the classification performance of the SVM algorithm with and without integrating SMOTEBoost and bootstrap on the eight datasets. In this table, the overall classification accuracy and the classification accuracy on each individual class (fault types or insulation conditions) are provided. This will help to verify the capability of SMOTEBoost in facilitating the pattern recognition

Table 4.9 Comparison of classification accuracy of SVM over eight datasets (%)

	Without integrating SMOTEBoost and bootstrap					Integrating SMOTEBoost and bootstrap				
Accuracy	**Class 1**	**Class 2**	**Class 3**	**Class 4**	**Overall**	**Class 1**	**Class 2**	**Class 3**	**Class 4**	**Overall**
Dataset 1	85	67	87	70	82	100	100	100	100	100
Dataset 2	60	63	97	89	88	100	100	100	93	97
Dataset 3	56	50	83	61	58	100	100	100	100	100
Dataset 4	83	88	88	60	84	100	100	100	100	100
Dataset 5	83	70	58	78	76	100	100	100	100	100
Dataset 6	43	81	98	84	74	100	96	100	97	98
Dataset 7	54	75	71	85	71	100	86	90	91	91
Dataset 8	85	96	92	96	94	100	98	100	94	98

Note: For datasets 1, 2, 3, 4, and 7, class 1 to class 4 denote "Normal condition," "Discharge fault," "Thermal fault," and "Partial discharge fault," respectively. For datasets 5, 6, and 8, class 1 to class 4 denote transformer insulation conditions of "Excellent," "Good," "Fair," and "Poor."

algorithm to achieve consistent performance on both majority and minority classes.

From Table 4.9 it can be seen that the original SVM algorithm (e.g., without integration) can attain a reasonable overall classification accuracy over most datasets. To make a comparison of the diagnosis accuracy between pattern recognition algorithms and conventional diagnostic methods, the diagnosis accuracy using the Duval triangle method is also calculated on dataset 5, where the overall accuracy is 52%. This demonstrates the effectiveness of pattern recognition algorithms in improving insulation condition diagnosis of power transformers.

From Table 4.9 it can be seen that the classification accuracy of SVM on the samples belonging to the minority class is rather low. For datasets 3, 6, and 7, the original SVM attains less than or slightly above 50% accuracy on class 1. This is because these three datasets are class imbalance datasets, in which class 1 has fewer samples compared with other classes. In contrast, the SVM algorithm integrating SMOTEBoost and bootstrap algorithms consistently attains higher overall classification accuracy and accuracy on each class as well. For datasets 3, 6, and 7, this hybrid algorithm can achieve close to 100% accuracy for class 1. This demonstrates the advantages of SMOTEBoost and bootstrap in dealing with class imbalance problems and improving SVM classification accuracy. Figure 4.7 illustrates the classification error rate of the SVM with the integration of SMOTEBoost and bootstrap on dataset 6 over 30 iterations.

From Figure 4.7 it can be seen that after five iterations the overall error rate decreases to 5% (0.05, top graph), while the error rate on the minority class has changed to 15% (0.15, bottom graph). At the 30th iteration, the overall error rate and the error rate of each individual class are reduced to around 2% (0.02). In this figure, the oscillation in error rate is due to overfitting at some iterations in the

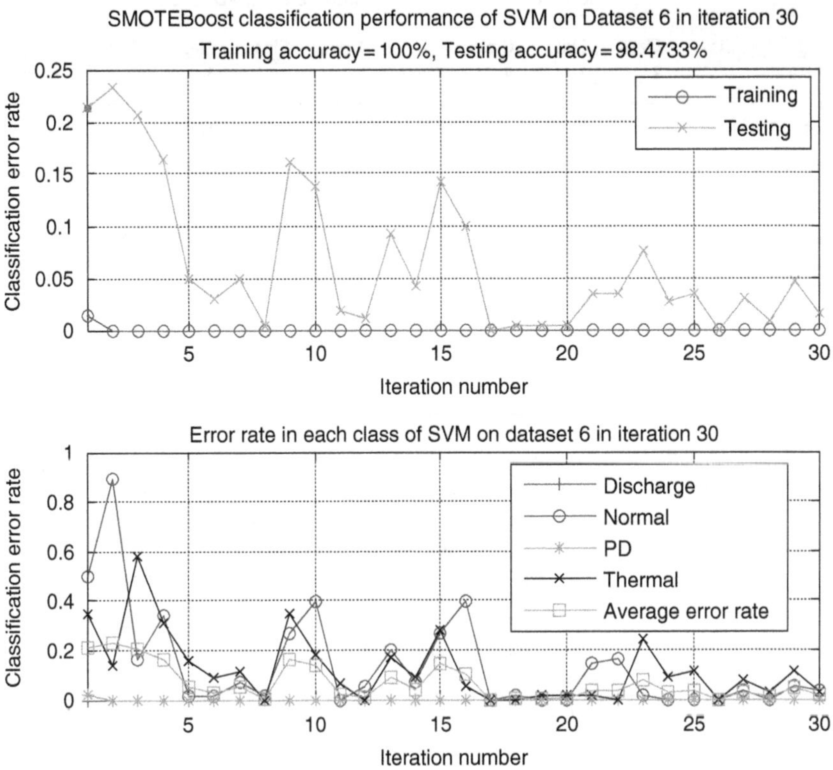

Figure 4.7 Classification error rate of the SVM with integration of SMOTEBoost using oil characteristics dataset 6. *Source:* Ref. [86], Figure 5.

training stage. Therefore, properly selecting the iteration number is very important. In this work, the iteration number is set to 30 based on the trade-off between computational cost and error rate.

The classification results of the other three algorithms (i.e., *k*NN, RBF, and C4.5) with and without integration of SMOTEBoost and bootstrap over the eight datasets are shown in Tables 4.10, 4.11, and 4.12, respectively. It can be observed that these three algorithms with integration of SMOTEBoost and bootstrap can achieve higher classification accuracies compared with those without the integration of SMOTEBoost and bootstrap. The only exception is the *k*NN algorithm with integration of SMOTEBoost and bootstrap on dataset 8 (Table 4.10). The error rate of this hybrid algorithm on dataset 8 over 30 iterations is presented in Figure 4.8. It can be observed from this figure that *k*NN after integrating SMOTEBoost and bootstrap experiences a more serious overfitting problem for the minority class ("thermal" in the bottom graph) at the 30th training iteration. The error rate of the minority class is increased to 30% (0.3) at the 30th iteration from less than 5% (0.05) at the 29th iteration.

This may be due to the fact that the performance of *k*NN is normally determined by the nearest neighbors of the samples rather than the samples themselves. At each iteration, SMOTE generates new samples for the minority class

Table 4.10 Comparison of classification accuracy of *k*NN over eight datasets (%)

Accuracy	Without integrating SMOTEBoost and bootstrap					Integrating SMOTEBoost and bootstrap				
	Class 1	Class 2	Class 3	Class 4	Overall	Class 1	Class 2	Class 3	Class 4	Overall
Dataset 1	86	97	91	78	87	100	100	100	89	99
Dataset 2	70	56	98	89	87	100	88	100	94	96
Dataset 3	71	43	75	78	65	100	93	100	89	96
Dataset 4	77	94	75	80	83	100	92	94	80	93
Dataset 5	83	93	67	100	87	100	100	100	89	98
Dataset 6	80	53	97	94	82	100	93	98	88	95
Dataset 7	54	84	87	84	79	100	84	86	91	89
Dataset 8	92	92	97	99	95	100	95	69	98	88

Table 4.11 Comparison of classification accuracy of RBF over eight datasets (%)

Accuracy	Without integrating SMOTEBoost and bootstrap					Integrating SMOTEBoost and bootstrap				
	Class 1	Class 2	Class 3	Class 4	Overall	Class 1	Class 2	Class 3	Class 4	Overall
Dataset 1	87	67	95	84	89	100	100	100	100	100
Dataset 2	93	31	41	65	74	100	100	100	84	95
Dataset 3	81	33	83	67	66	100	87	100	100	96
Dataset 4	80	85	28	90	69	100	100	88	100	97
Dataset 5	83	87	75	61	80	100	93	100	100	98
Dataset 6	83	49	99	93	82	100	100	100	96	99
Dataset 7	71	85	88	86	84	97	89	83	92	89
Dataset 8	90	94	96	95	95	100	97	94	95	96

Table 4.12 Comparison of classification accuracy of C4.5 decision tree over eight datasets (%)

Accuracy	Without integrating SMOTEBoost and bootstrap					Integrating SMOTEBoost and bootstrap				
	Class 1	Class 2	Class 3	Class 4	Overall	Class 1	Class 2	Class 3	Class 4	Overall
Dataset 1	90	60	93	76	89	100	100	100	67	97
Dataset 2	62	32	99	88	82	100	46	98	100	87
Dataset 3	85	50	58	83	72	100	93	100	100	98
Dataset 4	80	90	50	30	72	100	96	100	100	98
Dataset 5	94	90	83	94	92	100	73	100	100	93
Dataset 6	7	64	99	97	63	100	98	94	100	98
Dataset 7	72	86	88	84	84	100	89	83	92	91
Dataset 8	83	96	97	94	95	100	97	94	95	96

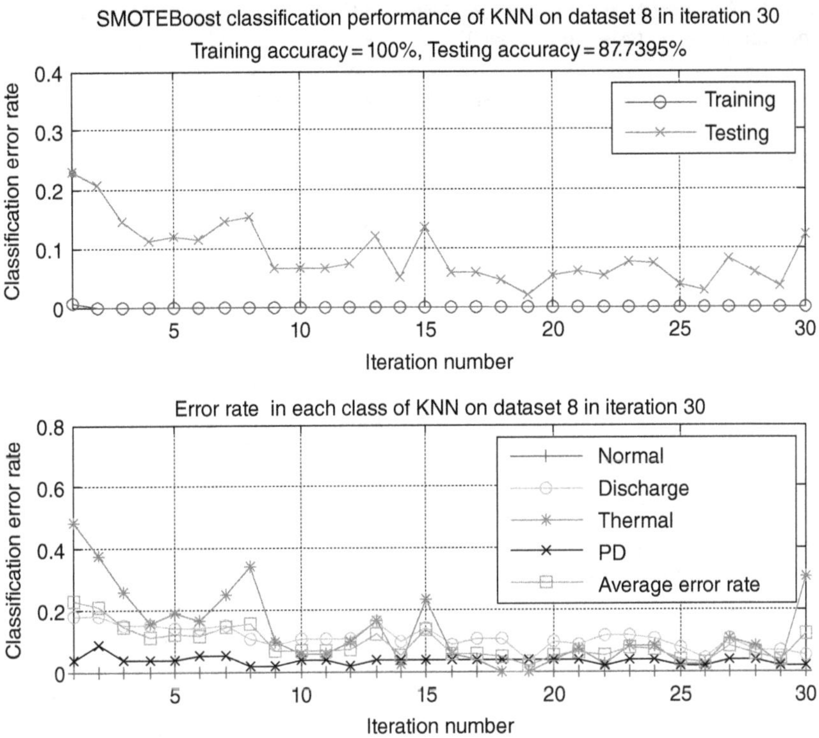

Figure 4.8 Classification error rate of the *k*NN classifier with integration of SMOTEBoost and bootstrap. *Source:* Ref. [86], Figure 6.

and these new samples can be influenced by their neighbors, which might be embedded with noise. This can compromise the classification accuracy of *k*NN on the minority class and in turn the overall classification accuracy.

4.15 Generalization Capability Validation

To evaluate the generalization ability of the above four algorithms with and without the integration of SMOTEBoost and bootstrap, these algorithms are trained on one dataset and tested on another dataset. The two datasets are collected from two different sources (e.g., from utility companies).

Tables 4.13 and 4.14 present the results of such cross-training and testing. In Table 4.13 the four algorithms are trained on dataset 4 but tested on dataset 1 (Table 4.3). In Table 4.14, the four algorithms are trained on dataset 6 but tested on dataset 5.

From Tables 4.13 and 4.14 it can be observed that the original algorithms (i.e., not integrating SMOTEBoost and bootstrap) fail to make classification on the testing dataset. This indicates that the generalization capability of the four original algorithms is limited. This is because inconsistencies exist in the training datasets collected from different utilities, and the four algorithms are unable to deal with

Table 4.13 Comparison of generalization ability of different pattern recognition algorithms (trained on dataset 4, tested on dataset 1)

Accuracy (%)		Normal	DS	OT*	PD	Overall
Original	C4.5	3	58	6	84	13
	*k*NN	33	100	62	82	50
	RBF	31	92	68	74	50
	SVM	35	100	68	79	52
SMOTEBoost	C4.5	100	100	99	97	99
	*k*NN	100	100	92	97	97
	RBF	75	92	97	88	85
	SVM	100	100	100	100	100

*OT = thermal fault in transformer insulation.

Table 4.14 Comparison of generalization ability of different pattern recognition algorithms (trained on dataset 6, tested on dataset 5)

Accuracy		Excellent	Good	Fair	Poor	Overall
Original	C4.5	33	0	0	0	14
	*k*NN	29	31	95	0	33
	RBF	29	11	29	2	20
	SVM	3	4	100	0	14
SMOTEBoost	C4.5	100	100	100	97	98
	*k*NN	100	84	95	35	84
	RBF	100	58	100	74	84
	SVM	100	100	100	90	98

such inconsistencies. However, by integrating SMOTEBoost and bootstrap, these algorithms can attain much higher classification accuracies on the testing dataset. The comparative results indicate that SMOTEBoost and bootstrap can facilitate the pattern recognition algorithm by improving its generalization capability for power transformer diagnosis using oil test data.

4.16 Summary

This chapter has described intelligent diagnostic techniques for power transformer insulation diagnosis using DGA data by a number of machine learning algorithms. The conventional DGA interpretation schemes were reviewed in Chapter 2, and the uncertainties and drawbacks of the existing DGA interpretations are outlined briefly in this chapter. A general pattern recognition application framework is proposed, involving the process of oil characteristics database preparation, training/testing datasets formation, cross-validation, testing, and evaluation in applying a pattern recognition algorithm for diagnosing transformer insulation. A comprehensive literature review on state-of-the-art pattern recognition techniques for transformer insulation diagnosis is also provided. To overcome the drawbacks of the conventional pattern recognition techniques for assessing the insulation condition of power transformers due to the unbalanced training dataset, a novel data preprocessing approach, a hybrid of SMOTEBoost and bootstrap, is proposed to enhance pattern recognition algorithms to consistently achieve desirable classification accuracy in power transformer diagnosis using oil test data. The SMOTEBoost and bootstrap algorithms were integrated with a number of pattern recognition algorithms to classify the types of incipient fault or the conditions of the transformer insulation system on eight oil test datasets collected from different utilities. The results demonstrated that the proposed approach can improve the generalization capability of pattern recognition algorithms when these algorithms are applied to transformer diagnosis.

References

1 Wang M, Vandermaar AJ, Srivastava KD. Review of condition assessment of power transformers in service. *IEEE Electrical Insulation Magazine* 2002; 18: 12–25.

2 Naderian A, Piercy R, Cress S, Service J, Fan W. An approach to power transformer asset management using health index. *IEEE Electrical Insulation Magazine* 2009; 25: 20–34.

3 Saha TK. Review of modern diagnostic techniques for assessing insulation condition in aged transformers. *IEEE Transactions on Dielectrics and Electrical Insulation* 2003; 10: 903–917.

4 Saha TK, Purkait P. Investigation of polarization and depolarization current measurements for the assessment of oil–paper insulation of aged transformers. *IEEE Transactions on Dielectrics and Electrical Insulation* 2004; 11: 144–154.

5 Duval M. New techniques for dissolved gas-in-oil analysis. *IEEE Electrical Insulation Magazine* 2003; 19: 6–15.

6 Wang Z, Li J, Sofian DM. Interpretation of transformer FRA responses. Part I: Influence of winding structure. *IEEE Transactions on Power Delivery* 2009; 24: 703–710.

7 Ekanayake C, Gubanski SM, Graczkowski A, Walczak K. Frequency response of oil impregnated pressboard and paper samples for estimating moisture in transformer insulation. *IEEE Transactions on Power Delivery* 2006; 21: 1309–1317.

8 Strachan SM, Rudd S, McArthur SDJ, Judd MD, Meijer S, Gulski E. Knowledge-based diagnosis of partial discharges in power transformers. *IEEE Transactions on Dielectrics and Electrical Insulation* 2008; 15: 259–268.

9 Duval M, DePablo A. Interpretation of gas-in-oil analysis using new IEC publication 60599 and IEC TC 10 databases. *IEEE Electrical Insulation Magazine* 2001; 17: 31–41.

10 Ma H, Ekanayake C, Saha TK. Power transformer fault diagnosis under measurement originated uncertainties. *IEEE Transactions on Dielectrics and Electrical Insulation* 2012; 19: 1982–1990.
11 Da Silva IN, Imamura MM, de Souza AN. Application of neural networks to the analysis of dissolved gases in insulating oil used in transformers. Proceedings of IEEE International Conference on Systems, Man and Cybernetics, Nashville, TN, October 8–11, 2000, pp. 2643–2648.
12 Xu Z. Multivalued logic and fuzzy logic – their relationship, minimization, and application to fault diagnosis. *IEEE Transactions on Computing* 1984; C-33: 679–681.
13 Su Q, Mi C, Lai LL, Austin P. Fuzzy dissolved gas analysis method for the diagnosis of multiple incipient faults in a transformer. *IEEE Transactions on Power Systems* 2000; 15: 593–598.
14 Naresh R, Sharma V, Vashisth M. An integrated neural fuzzy approach for fault diagnosis of transformers. *IEEE Transactions on Power Delivery* 2008; 23: 2017–2024.
15 Chen W, Pan C, Yun Y, Liu Y. Wavelet networks in power transformers diagnosis using dissolved gas analysis. *IEEE Transactions on Power Delivery* 2009; 24: 187–194.
16 CIGRE TF D1.01.07. New techniques for dissolved gas-in-oil analysis. *Electra* 2001; 198: 20–27.
17 IEC 60567. Guide for the Sampling of Gases and of Oil from Oil-Filled Electrical Equipment and for the Analysis of Free and Dissolved Gases, 2005.
18 CIGRE Joint TF D1.01./A2.11. Recent Developments in DGA Interpretation, 2006.
19 IEC 60599 (Amendment I to Ed. 2.0) Mineral Oil-Impregnated Electrical Equipment in Service – Guide to the Interpretation of Dissolved and Free Gases Analysis, 2007.
20 Duval M. Report on Gas Monitors for Oil-Filled Electrical Equipment. CIGRE Brochure, 2009.
21 CIGRE TF D1.32. DGA in Non-Mineral Oils and Load Tap Changers and Improved DGA Diagnosis Criteria, 2010.
22 Łosiewicz I, Wasiukowa L, Słowikowska H, Słowikowski J. Contribution to the diagnosis of life management of medium voltage oil transformers over 1 MVA. Proceedings of the 17th International Conference on Electricity Distribution, 2003, pp. 1–6.
23 Tong-in A. New partial discharge diagnostic methods for transformer. Proceedings of CIGRE Paris Session, 2014.
24 Christopoulos C, Wright A. *Electrical Power System Protection*. Berlin: Springer, 1999.
25 López-Fernández XM, Ertan HB, Turowski J. *Transformers: Analysis, Design, and Measurement*. Boca Raton, FL: CRC Press, 2012, p. 254.
26 Lin CE, Ling JM, Huang CL. An expert system for transformer fault-diagnosis using dissolved-gas analysis. *IEEE Transactions on Power Delivery* 1993; 8: 231–238.
27 Cui Y, Ma H, Saha TK. Pattern recognition techniques for power transformer insulation diagnosis – a comparative study. Part 1: Framework, literature, and illustration. *International Transactions on Electrical Energy Systems* 2014; 25(10): 2247–2259.

28 Zhang Y, Ding X, Liu Y, Griffin PJ. An artificial neural network approach to transformer fault diagnosis. *IEEE Transactions on Power Delivery* 1996; 11: 1836–1841.

29 Vanegas O, Mizuno Y, Naito K, Kamiya T. Diagnosis of oil-insulated power apparatus by using neural network simulation. *IEEE Transactions on Dielectrics and Electrical Insulation* 1997; 4: 290–299.

30 Guardado JL, Naredo JL, Moreno P, Fuerte CR. A comparative study of neural network efficiency in power transformers diagnosis using dissolved gas analysis. *IEEE Transactions on Power Delivery* 2001; 16: 643–647.

31 Yang MT, Hu LS. Intelligent fault types diagnostic system for dissolved gas analysis of oil-immersed power transformer. *IEEE Transactions on Dielectrics and Electrical Insulation* 2013; 20: 2317–2324.

32 Yang HT, Huang YC. Intelligent decision support for diagnosis of incipient transformer faults using self-organizing polynomial networks. *IEEE Transactions on Power Systems* 1998; 13: 946–952.

33 Wu H, Li X, Wu D. RMP neural network based dissolved gas analyser for fault diagnostic of oil-filled electrical equipment. *IEEE Transactions on Dielectrics and Electrical Insulation* 2011; 18: 495–498.

34 Lin WM, Lin CH, Tasy MX. Transformer-fault diagnosis by integrating field data and standard codes with training enhancible adaptive probabilistic network. *IEE Proceedings C: Generation, Transmission and Distribution* 2005; 152: 335–341.

35 Yang H, Liao C, Chou J. Fuzzy learning vector quantization networks for power transformer condition assessment. *IEEE Transactions on Dielectrics and Electrical Insulation* 2001; 8: 143–149.

36 Huang YC. Condition assessment of power transformers using genetic-based neural networks. *IEE Proceedings C: Generation, Transmission and Distribution* 2003; 150: 19–24.

37 Huang YC. Evolving neural nets for fault diagnosis of power transformers. *IEEE Transactions on Power Delivery* 2003; 18: 843–848.

38 Wang MH. Extension neural network for power transformer incipient fault diagnosis. *IEE Proceedings C: Generation, Transmission and Distribution* 2003; 150: 679–685.

39 Wang MH. A novel extension method for transformer fault diagnosis. *IEEE Transactions on Power Delivery* 2003; 18: 164–169.

40 Thang KF, Aggarwal RK, McGrail AJ, Esp DG. Analysis of power transformer dissolved gas data using the self-organizing map. *IEEE Transactions on Power Delivery* 2003; 18: 1241–1248.

41 Miranda V, Castro A, Lima S. Diagnosing faults in power transformers with auto associative neural networks and mean shift. *IEEE Transactions on Power Delivery* 2012; 27: 1350–1357.

42 Sun S, Huang R. An adaptive k-nearest neighbour algorithm. Proceedings of 7th International Conference on Fuzzy Systems and Knowledge Discovery (FSKD), 2010, pp. 91–94.

43 Cover T, Hart P. Nearest neighbour pattern classification. *IEEE Transactions on Information Theory* 1967; 13: 21–27.

44 Carita A, Leite LC, Junior A, Godoy RB, Sauer L. Bayesian networks applied to failure diagnosis in power transformer. *IEEE Latin American Transactions* 2013; 11: 1075–1082.

45 Meng K, Dong ZY, Wang DH, Wong KP. A self-adaptive RBF neural network classifier for transformer fault analysis. *IEEE Transactions on Power Systems* 2010; 25: 1350–1360.

46 Huang YC, Huang CM. Evolving wavelet networks for power transformer condition monitoring. *IEEE Transactions on Power Delivery* 2002; 17: 412–416.

47 Huang YC. A new data mining approach to dissolved gas analysis of oil-insulated power apparatus. *IEEE Transactions on Power Delivery* 2003; 18: 1257–1261.

48 Huang Y, Huang C, Sun H. Data mining for oil-insulated power transformers: An advanced literature survey. *Wiley Interdisciplinary Reviews: Data Mining and Knowledge Discovery* 2012; 2: 138–148.

49 Kirtley JL, Hagman WH, Lesieutre BC, Boyd MJ, Warren EP, Chou HP *et al.* Monitoring the health of power transformers. *IEEE Computer Applications in Power* 1996; 9: 18–23.

50 Huang YC, Sun HC. Dissolved gas analysis of mineral oil for power transformer fault diagnosis using fuzzy logic. *IEEE Transactions on Dielectrics and Electrical Insulation* 2013; 20: 974–981.

51 Dhote NK, Helonde JB. Fuzzy algorithm for power transformer diagnostics. *Advances in Fuzzy Systems* 2013; 421621–421627.

52 David IG, Rajaram M, Gnanasaravanan A. Power transformer faults identification using fuzzy based dissolved gas analysis method. *Journal of Electrical Engineering* 2013; 13: 5.

53 Dhote NK, Helonde JB. Improvement in transformer diagnosis by DGA using fuzzy logic. *Journal of Electrical Engineering and Technology* 2014; 9: 615–621.

54 Tomsovic K, Tapper M, Ingvarsson T. A fuzzy information approach to integrating different transformer diagnostic methods. *IEEE Transactions on Power Delivery* 1993; 8: 1638–1646.

55 Islam SM, Wu T, Ledwich G. Novel fuzzy logic approach to transformer fault diagnosis. *IEEE Transactions on Dielectrics and Electrical Insulation* 2000; 7: 177–186.

56 Duraisamy V, Devarajan N, Somasundareswari D, Vasanth A, Sivanandam SN. Neuro fuzzy schemes for fault detection in power transformer. *Applied Soft Computing* 2007; 7: 534–539.

57 Flores WC, Mombello EE, Jardini JA, Rattá G, Corvo AM. Expert system for the assessment of power transformer insulation condition based on type-2 fuzzy logic systems. *Expert Systems with Applications* 2011; 38: 8119–8127.

58 Abu-Siada A, Hmood S, Islam S. A new fuzzy logic approach for consistent interpretation of dissolved gas-in-oil analysis. *IEEE Transactions on Dielectrics and Electrical Insulation* 2013; 20: 2343–2349.

59 Tang WH, Spurgeon K, Wu QH, Richardson ZJ. An evidential reasoning approach to transformer condition assessments. *IEEE Transactions on Power Delivery* 2004; 19: 1696–1703.

60 Spurgeon K, Tang WH, Wu QH, Richardson ZJ, Moss G. Dissolved gas analysis using evidential reasoning. *IEE Proceedings C: Science, Measurement and Technology* 2005; 152: 110–117.

61 Wang XL, Li QM, Li CR, Yang R, Su Q. Reliability assessment of the fault diagnosis methodologies for transformers and a new diagnostic scheme based on fault info

integration. *IEEE Transactions on Dielectrics and Electrical Insulation* 2013; 20: 2292–2298.

62 Wang MH. Grey-extension method for incipient fault forecasting of oil-immersed power transformer. *Electrical Power Composite Systems* 2004; 32: 959–975.

63 Wang MH, Hung CP. Novel grey model for the prediction of trend of dissolved gases in oil-filled power apparatus. *Electric Power System Research* 2003; 67: 53–58.

64 Tang WH, Goulermas JY, Wu QH, Richardson ZJ, Fitch J. A probabilistic classifier for transformer dissolved gas analysis with a particle swarm optimizer. *IEEE Transactions on Power Delivery* 2008; 23: 751–759.

65 Lv G, Cheng H, Zhai H, Dong L. Fault diagnosis of power transformer based on multi-layer SVM classifier. *Electric Power System Research* 2005; 75: 9–15.

66 Cho M, Lee T, Gau S, Shih C. Power transformer fault diagnosis using support vector machines and artificial neural networks with clonal selection algorithms optimization. In Gabrys B, Howlett RJ, Jain LC (eds), *Knowledge-Based Intelligent Information and Engineering Systems*, Vol. 4251. Berlin: Springer, 2006, pp. 179–186.

67 Fei S, Sun Y. Forecasting dissolved gases content in power transformer oil based on support vector machine with genetic algorithm. *Electric Power System Research* 2008; 78: 507–514.

68 Liao R, Zheng H, Grzybowski S, Yang L, Tang C, Zhang Y. Fuzzy information granulated particle swarm optimisation-support vector machine regression for the trend forecasting of dissolved gases in oil-filled transformers. *IET Electric Power Applications* 2011; 5: 230–237.

69 Lee T, Cho M, Fang F. Features selection of SVM and ANN using particle swarm optimization for power transformers incipient fault symptom diagnosis. *International Journal of Computational Intelligence Research* 2007; 3: 60–65.

70 Lee T, Cho M, Shieh C, Lee H, Fang F. Particle swarm optimization-based SVM for incipient fault classification of power transformers. In Esposito F, Ras ZW, Malerba D, Semeraro G (eds), *Foundations of Intelligence Systems*, Vol. 4203. Berlin: Springer, 2006, pp. 84–90.

71 Liao RJ, Zheng HB, Grzybowski S, Yang LJ. A multiclass SVM-based classifier for transformer fault diagnosis using a particle swarm optimizer with time-varying acceleration coefficients. *International Transactions on Electrical Energy Systems* 2013; 23: 181–190.

72 Xiong H, Sun C. Artificial immune network classification algorithm for fault diagnosis of power transformer. *IEEE Transactions on Power Delivery* 2007; 22: 930–935.

73 Xu W, Wang D, Zhou Z, Chen H. Fault diagnosis of power transformers: Application of fuzzy set theory, expert systems and artificial neural networks. *IEE Proceedings C: Science, Measurement and Technology* 1997; 144: 39–44.

74 Wang Z, Liu Y, Griffin P. A combined ANN and expert system tool for transformer fault diagnosis. *IEEE Transactions on Power Delivery* 1998; 13: 1224–1229.

75 Miranda V, Castro A. Improving the IEC table for transformer failure diagnosis with knowledge extraction from neural networks. *IEEE Transactions on Power Delivery* 2005; 20: 2509–2516.

76 Morais D, Rolim J. A hybrid tool for detection of incipient faults in transformers based on the dissolved gas analysis of insulating oil. *IEEE Transactions on Power Delivery* 2006; 21: 673–680.

77 Yang Z, Tang W, Shintemirov A, Wu Q. Association rule mining-based dissolved gas analysis for fault diagnosis of power transformers. *IEEE Transactions on Systems, Man and Cybernetics Part C – Applications and Reviews* 2009; 39: 597–610.

78 Liao R, Zheng H, Grzybowski S, Yang L, Zhang Y, Liao Y. An integrated decision-making model for condition assessment of power transformers using fuzzy approach and evidential reasoning. *IEEE Transactions on Power Delivery* 2011; 26: 1111–1118.

79 Bhalla D, Bansal RK, Gupta HO. Integrating AI based DGA fault diagnosis using Dempster–Shafer theory. *International Journal of Electric Power & Energy Systems* 2013; 48: 31–38.

80 Malik H, Yadav AK, Mishra S, Mehto T. Application of neuro-fuzzy scheme to investigate the winding insulation paper deterioration in oil-immersed power transformer. *International Journal of Electric Power & Energy Systems* 2013; 53: 256–271.

81 Huang YC, Yang HT, Huang CL. Developing a new transformer fault diagnosis system through evolutionary fuzzy logic. *IEEE Transactions on Power Delivery* 1997; 12: 761–767.

82 Yang HT, Liao CC. Adaptive fuzzy diagnosis system for dissolved gas analysis of power transformers. *IEEE Transactions on Power Delivery* 1999; 14: 1342–1350.

83 Fei S, Zhang X. Fault diagnosis of power transformer based on support vector machine with genetic algorithm. *Expert System Applications* 2009; 36: 11352–11357.

84 Wang M, Tseng Y, Chen H, Chao K. A novel clustering algorithm based on the extension theory and genetic algorithm. *Expert System Applications* 2009; 36: 8269–8276.

85 Chawla NV, Bowyer FW, Hall LO, Kegelmeyer WP. SMOTE: Synthetic minority over-sampling technique. *Journal of Artificial Intelligence Research* 2002; 16: 321–357.

86 Cui Y, Ma H, Saha T. Improvement of power transformer insulation diagnosis using oil characteristics data pre-processed by SMOTEBoost technique. *IEEE Transactions on Dielectrics and Electrical Insulation* 2014; 21: 2363–2373.

87 Ratsch G, Onoda T, Muller KR. Soft margins for AdaBoost. *Machine Learning* 2001; 42: 287–320.

88 Kai MT. An instance-weighting method to induce cost-sensitive trees. *IEEE Transactions on Knowledge Data Engineering* 2002; 14: 659–665.

89 Zoubir AM, Boashash B. Bootstrap and its application in signal processing. *IEEE Signal Processing Magazine* 1998; 15: 56–76.

90 Efron B, Tibshirani R. *An Introduction to the Bootstrap*. London: Chapman & Hall, 1993.

91 Cristianini N, Shawe-Taylor J. *An Introduction to Support Vector Machines and Other Kernel-Based Learning Methods*. Cambridge: Cambridge University Press, 2001.

92 Bishop CM. *Pattern Recognition and Machine Learning.* New York: Springer, 2006.

93 Navada A, Ansari AN, Patil S, Sonkamble BA. Overview of use of decision tree algorithms in machine learning. Proceedings of IEEE Control and System Graduate Research Colloquium (ICSGRC), Shah Alam, Malaysia, June 27–28, 2011, pp. 37–42.

94 Ashkezari AD, Ma H, Saha TK, Ekanayake C. Application of fuzzy support vector machine for determining the health index of the insulation system of in-service power transformers. *IEEE Transactions on Dielectrics and Electrical Insulation* 2013; 20: 965–973.

95 Nabney IT. *Netlab Algorithms for Pattern Recognition.* New York: Springer, 1995.

96 Chang CC, Lin CJ. LIBSVM: A library for support vector machines. Available at: www.csie.ntu.edu.tw/~cjlin/libsvm.

5

Advanced Signal Processing Techniques for Partial Discharge Measurement

5.1 Partial Discharge in Power Transformer

A large number of power transformers in many countries have approached or even exceeded their designed lives. For instance, National Grid, UK operates with approximately 780 transformers over 100 MVA and about half of these transformers were aged 40 years or more in 2010 [1]. In Hydro-Québec, Canada, the average age of the whole transformer fleet was over 30 years in 2012 [2]. In New South Wales, Australia, Endeavour Energy reported that almost half of its power transformers were aged 36 years or more in 2007 [3]. Any failure of a power transformer can be disastrous. Therefore, the condition of transformers needs to be monitored and evaluated continuously.

A transformer's condition is largely dependent on its insulation system. Many surveys have reported that insulation problems are one of the key sources of transformer failure [4–7]. Therefore, various diagnostic methods have been deployed in utilities for assessing transformer insulation conditions over the past decades. Among these methods, Partial discharge (PD) measurement is widely adopted due to its capability for providing continuous online condition monitoring of a transformer without disturbing its normal operation. Through the PD measurement, many types of insulation defect can be revealed. A summary of the insulation defects that can induce PD in a power transformer is given below [8, 9]:

- contamination of liquid and solid insulation due to moisture;
- air trapping in insulation due to poor vacuum of oil;
- gas bubble formation due to overheating and discharges;
- floating metal particles in oil;
- protrusion in HV structures;
- overstressed connections;
- poor earthing.

PD is a rather complicated phenomenon and stochastic in nature. Techniques for properly performing PD measurements of a transformer, effectively analyzing the measured PD signals, and subsequently making an informed condition assessment on a transformer's insulation system are much needed. These techniques are covered in detail in this chapter.

Transformer Ageing: Monitoring and Estimation Techniques, First Edition.
Edited by Tapan Kumar Saha and Prithwiraj Purkait.

5.2 Overview of PD Analysis

The ultimate goal of PD analysis is to evaluate the characteristics (extensities, types, and locations) of PD events inside a transformer. This can be helpful in evaluating the health condition of a transformer insulation system. The major steps of PD signal analysis are shown in Figure 5.1.

After acquiring a signal from a field transformer by a PD measurement system such as an inductive measurement system using a high-frequency current transformer (HFCT) and a capacitive measurement system (i.e., IEC 60270 [10]), performing signal processing on the acquired signal is essential. This is mainly due to the contamination of an acquired signal by extensive noise. In PD measurements, common types of noise include discrete spectral interference (DSI), stochastic noise, and white noise, as listed in Table 5.1 [11–14].

Among these three types of noise, DSI is a narrow-band noise and the others are wide-band noise [14]. Depending on the source of noise, DSI consists of frequency components up to around 26 MHz for AM radio, 300 MHz for FM radio, and 1880 MHz for mobile communications. In general, this type of noise can have a higher amplitude compared to PD signals [15]. Since DSI is periodic with fixed frequencies, it may be removed by digital filtering, such as a notch filter [16]. However, it is a non-adaptive method since two parameters, forgetting factor and bandwidth, need to be predetermined [16, 17].

Stochastic noise is considerably difficult to remove, due to its similarity with PD signals. One approach to remove stochastic noise is to apply artificial neural networks (ANNs) to recognize different types of PD signal and noise [18]. However, this approach requires constructing a database that consists of fingerprints of different types of PD signal for training algorithms. It may be limited by the availability of fingerprints for proper training the algorithm.

White noise always exists during online PD measurement of transformers. It possesses a flat power spectrum with equal power at all frequencies. This type of noise can be suppressed by gating methods when they have lower amplitude compared to the PD signals.

The noise distorts PD signals, which can significantly influence the performance of the later steps of PD signal analysis. Therefore, the first step of PD signal analysis is de-noising, which aims at removing the noise and extracting the PD signals. De-noising can be achieved by using hardware, but software-based methods provide an adaptive means of removing various types of noise. Some typical software-based methods are signal decomposition, blind processing, and graphics-based techniques. These techniques will be detailed in this chapter.

In practice, signals acquired by PD measurement systems may consist of signals induced by a number of PD sources, which can co-exist in a transformer and discharge simultaneously. PD source separation is executed to separate the PD sources and reveal individual PD patterns related to each source. This can be performed by utilizing the time and frequency characteristics of PD signals to form distinctive clusters in a time/frequency (TF) map, with each cluster corresponding to a single PD source. This technique will be detailed in this chapter.

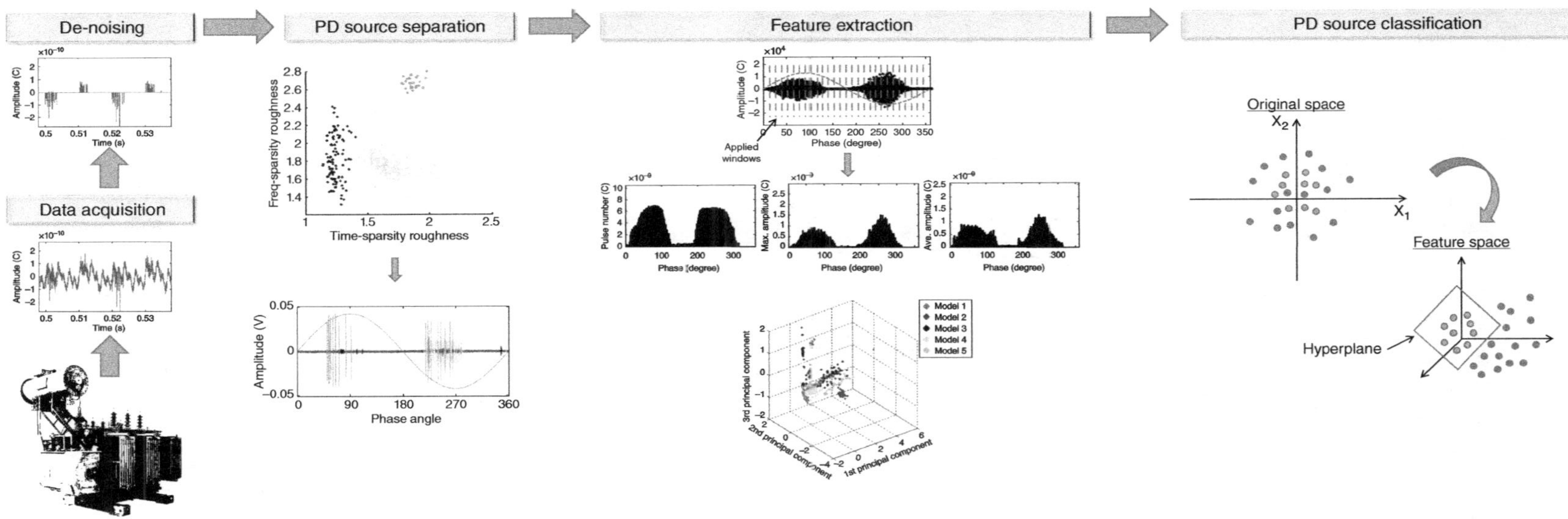

Figure 5.1 Major steps of PD signal analysis.

Table 5.1 Types of noise in PD measurements

Types of noise	Sources of noise
DSI	Communication systems Radio transmissions (e.g., AM and FM)
Stochastic noise	Switching operations (e.g., thyristor) Sparking Automobile spark plugs Electrical noise from detection circuit Arcing between metallic contacts Lightning
White noise	Amplifier Ambience

After PD source separation, individual PD sources can be described by a set of unique features (characteristic attributes), which are normally obtained from phase-resolved PD (PRPD) diagrams. A PRPD diagram consists of the distribution of PD pulse numbers, maximum PD pulse amplitude, and average PD pulse amplitude with reference to the phase angle of AC voltage. Features such as the first four moments (mean, standard deviation, skewness, and kurtosis) can be extracted from these distributions to act as fingerprints to a particular PD source. The features can also be extracted from decomposed signals (using different signal decomposition techniques) in different frequency scales to provide a more accurate representation of PD patterns. The effectiveness of different feature extraction techniques for assisting PD source classification will be examined in this chapter.

5.3 PD Measurement Methods

The occurrence of PD events is often accompanied by electromagnetic transients, chemical reactions, emission of sound, pressure waves, light, and heat [19]. These phenomena are utilized to detect the presence of PD events and acquire the PD signals for analysis.

Conventional PD measurement methods are described in IEC 60270 [10]. Some non-conventional methods are associated with electromagnetic and acoustic PD detection [19]. Table 5.2 summarizes various PD measurement methods.

The circuitry diagram of the conventional PD measurement system is depicted in Figure 5.2. In this figure, a HV power supply is required which should exhibit a low background noise level. An analog filter F can also be introduced to further reduce the noise. In the figure, C_a represents a test object. C_k is a coupling capacitor, which acts as a high impedance for power frequency of the voltage supply and a low impedance for high frequencies of PD signals generated from the test object [20]. The measuring impedance Z is used to convert PD current to voltage for signal acquisition by a PD detector [21]. The measuring impedance is either an RC or RLC circuit, in which the measured voltage pulses appear to be unidirectional or attenuated oscillatory of waveforms, respectively. This conventional PD

Table 5.2 PD measurement methods

Conventional methods	Non-conventional methods
Narrow-band detection	Electromagnetic transients detection
Wide-band detection	High frequency (HF)/very HF (VHF) (3 ~ 300 MHz)
	Ultra HF (UHF) (300 MHz ~ 3 GHz)
	AE detection (10 ~ 300 kHz)
	Optical occurrences detection
	Chemical compounds detection

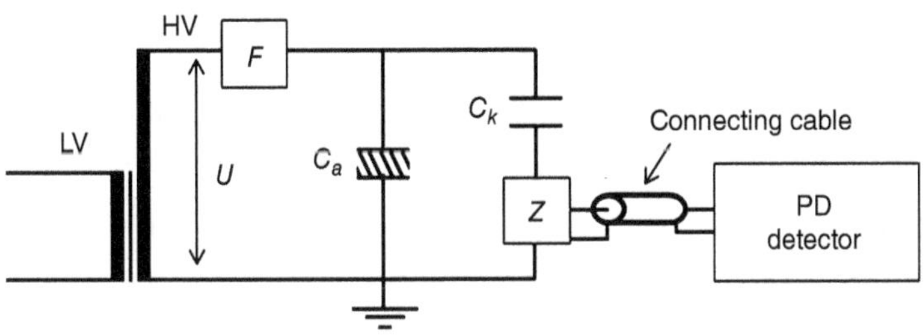

Figure 5.2 Conventional PD measurement method (defined in IEC 60270).

measurement method is also called "direct coupling method" or "capacitive method."

During PD measurements, the acquired PD signals may be contaminated by numerous noise sources. Therefore, PD measurements should be carried out in a frequency range where the noise level is as low as possible. According to IEC 60270 [10], the PD measurements can be performed in a narrow band or wide band. In a narrow-band system, the bandwidth Δf is recommended to satisfy

$$9\,\text{kHz} \le \Delta f \le 30\,\text{kHz} \tag{5.1}$$

In a wide-band PD system, transient PD pulses can be recorded with higher time resolution [11]. Based on IEC 60270, the bandwidth Δf of a wide-band system is recommended to satisfy

$$100\,\text{kHz} \le \Delta f \le 400\,\text{kHz} \tag{5.2}$$

Figure 5.3 shows a PRPD diagram acquired using a commercial product (Omicron MPD 600), which complies with IEC 60270.

Though the above capacitive system is capable of acquiring PD signals with relatively low noise, in practice the non-conventional PD measurements can provide a more flexible means for online PD measurements. In non-conventional PD measurements, a UHF antenna and inductive sensors such as a Rogowski coil and HFCT are adopted to acquire PD signals [22–25]. The acquired PD signals have various frequency components, depending on the types of PD source and insulation material [26]. A UHF PD measurement system is normally applied for the frequency range 300 MHz ~ 3 GHz (refer to Section 5.7 for a brief review of UHF).

Inductive sensors (HFCT, Rogowski coil) are clamped on a grounding wire or neutral of a transformer to measure high-frequency currents induced from PD.

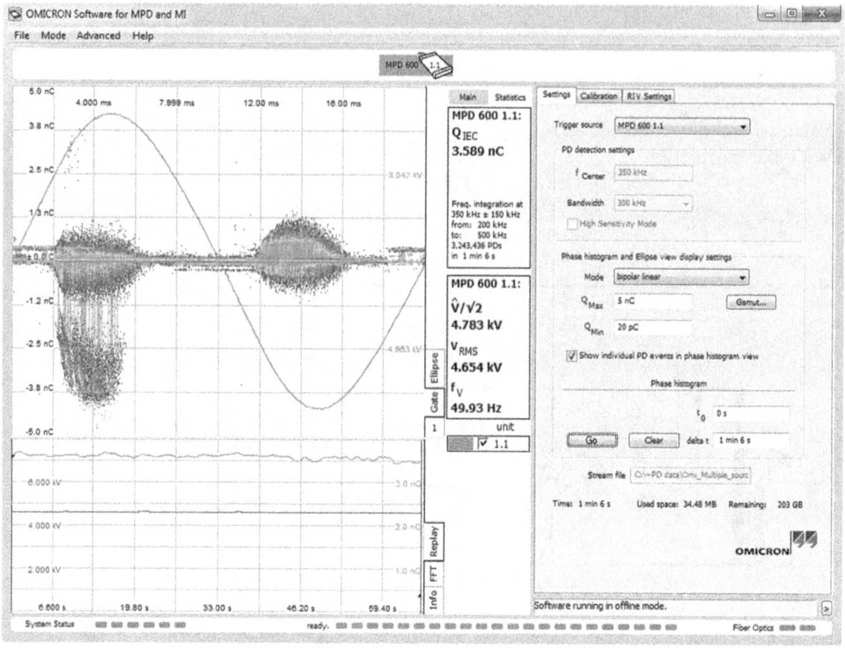

Figure 5.3 An example of an acquired PD signal using Omicron MPD 600, which complies with IEC 60270.

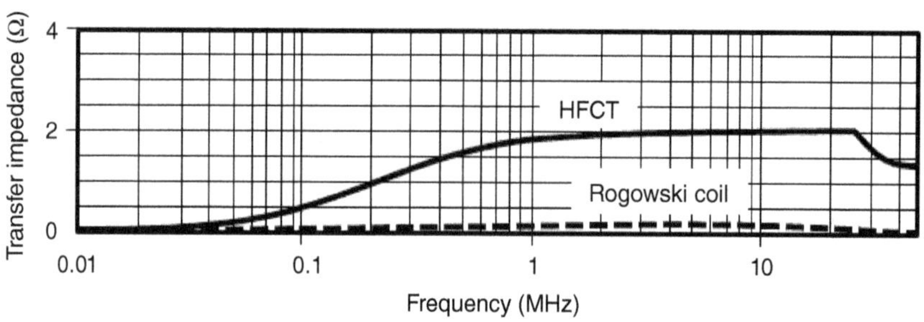

Figure 5.4 Frequency responses of Rogowski coil and HFCT (model: 140/100HC from HVPD).

They can also be clamped on HV leads, but their installations require a temporary disconnection of transformers from the power grid [27]. Rogowski coils are constructed with inductive coils wound on a non-magnetic core, such as plastic for flexibility on clamping. The required number of turns is decided largely by the expected inductance. Since the amplitudes of voltage outputs for PD signals are inductance dependent, changing the number of turns can adjust the amplitudes [22–24]. However, compared to HFCTs, Rogowski coils are considerably less sensitive and have a lower bandwidth. This is because HFCTs use a ferrite core [27].

Figure 5.4 shows a comparison of frequency responses (output voltages to input currents) between a Rogowski coil and a HFCT. From the figure, it can be seen that the sensitivity of the HFCT is much higher than that of the Rogowski coil. Though HFCTs have higher sensitivity, the use of a ferrite core implies that they are nonlinear, and this may introduce some difficulties for calibration [28]. Nevertheless, using a split-core type of HFCT can ease installation in practice.

An acoustic method is often used for PD source localization. This is achieved by attaching a number of acoustic sensors on a transformer tank (refer to Section 5.7 for PD source localization by acoustic sensors) [29].

Except for the conventional PD measurement method (IEC 60270), there is a lack of standardized guidelines for the non-conventional methods. Therefore, a number of working groups from different organizations are now working on standardizing the PD measurements and the related interpretation techniques. Those working groups include CIGRE WG D1.37, IEC WG14, and CIGRE WG D1. 29. For instance, the remit of CIGRE WG D1.29 includes a survey of PD measurement methods on their practical applications, suggestions of benchmarks for evaluating different PD measurement methods, collation of signatures from different PD sources, and determination of severity of PD sources [30].

With consideration of the above issues, this chapter is aimed at providing comprehensive PD signal analysis techniques for utilities to conduct PD measurements of transformers.

5.4 Advanced Signal Processing Techniques

This section presents a number of recently developed techniques for PD analysis, summarized as:

- signal decomposition;
- blind processing;
- TF map;
- PD pattern representation;
- other signal processing techniques.

5.4.1 Signal Decomposition

Since a PD signal is a type of transient signal, conventional time or frequency domain processing techniques may not represent the PD signals accurately. Consequently, these techniques cannot always effectively extract PD signals from extensive noise. Therefore, signal decomposition in joint time and frequency domains has been developed for PD signal analysis [31, 32]. In the following sections, wavelet transform (WT), empirical mode decomposition (EMD), ensemble EMD (EEMD), and MM are introduced.

5.4.1.1 Wavelet Transform (WT)

WT decomposes a signal into time and frequency domains with multi-resolutions. In WT, a wavelet is a waveform having zero-mean amplitude and short time duration [33]. This wave-like signal is named a wavelet function (also called a "mother wavelet"). It generally satisfies $\int_{-\infty}^{\infty} \Psi(t)dt = 0$ and $\int_{-\infty}^{\infty} |\Psi(t)|^2 dt < \infty$, where Ψ is the wavelet function [34]. Among various wavelet functions,

Daubechies (db), biorthogonal (bior), and Coiflets (coif) are adopted widely in PD signal processing [11–14, 35–41].

After selecting a proper wavelet function, WT applies shifting and scaling operations on the wavelet function, which in turn decompose the original signal into a series of wavelet coefficients [34]:

$$\Psi_{a,b}(t) = \frac{1}{\sqrt{|a|}}\Psi\left(\frac{t-b}{a}\right) \quad \text{s.t. } a,b \in \mathbb{R} \tag{5.3}$$

where a is the scale factor, adopted to compress and stretch the wavelet function; b is the translation factor, adopted to shift the wavelet function along the time axis. Through the scale and translation factors, WT can achieve both time and frequency representations of the signal.

In PD analysis, Discrete WT (DWT) is normally used in the form [34]

$$W_{j,k} = \sum_{n \in Z} X(n) 2^{(-j/2)} \Psi\left(2^{-j} n - k\right) \tag{5.4}$$

where $X(n)$ is the discrete function of a signal, j and k are integers. For signal reconstruction, inverse DWT is performed as [34]

$$X(n) = \sum_{j \in Z} \sum_{k \in Z} W_{j,k} \Psi_{j,k}(n) \tag{5.5}$$

5.4.1.2 Empirical Mode Decomposition (EMD) and Ensemble EMD (EEMD)

EMD is an adaptive decomposition technique. Without specifying any wavelet function as WT does, EMD transforms the original signal into a number of mono-component signals (termed intrinsic mode functions, IMFs) with different frequency scales, using the original signal's characteristic time scale [42]. The decomposed signals (i.e., IMFs) can later be used to rebuild the original signal precisely. To remove riding waves and smooth uneven amplitudes that might be embedded in IMFs, a shifting process is used in EMD. Therefore, an IMF can be treated as a mono-component signal [43].

It is assumed that $x(t)$ is a signal being processed in the sifting process. First, local extrema of $x(t)$ are defined. Then, cubic spline interpolations are performed on both maxima and minima to calculate the upper and lower envelopes [$x_U(t)$ and $x_L(t)$]. The mean of the above two envelopes is denoted $m(t) = (x_U(t) + x_L(t))/2$. The iteration for deciding IMF is $I_{ni}(t) = x(t) - m_{ni}(t)$, where $I_{ni}(t)$ is the nth IMF at the ith iteration. $I_{ni}(t)$ is the first IMF if it complies with the mono-component condition. Otherwise, the iteration will be continued on $I_{ni}(t)$. After extracting the first IMF, which embraces the original signal's highest-frequency component, the iteration will be continued on the residual $r_n(t) = x(t) - I_{ni}(t)$, replacing $x(t)$ by $r_n(t)$ and repeating the above process until the last IMF is obtained. A precise reconstruction $x_R(t)$ can be obtained by a linear combination:

$$x_R(t) = \sum_{(n=1)}^{(N-1)} I_{ni}(t) + r_N(t) \tag{5.6}$$

where N is the total number of IMFs and r is the last IMF or residual.

Ideally, an IMF is a signal with mono-component. However, due to the problem of mode mixing in the above sifting process, an IMF often embraces signal components with dramatically disparate frequency scales. The signal components with similar frequency scales may also reside in different IMFs [44]. Therefore, EEMD was proposed.

Through adding white noise to the original signal, EEMD forces the sifting process to make different frequency scales collate in the proper IMFs dictated by a dyadic filter bank [44]. After adding the white noise, EEMD extracts IMFs as EMD does. Once the extraction of the first ensemble is complete, a set of IMFs with noise is generated. Then, a new white noise with the same standard deviation of the amplitude as the previously added noise is injected into the original signal again and the EMD process is continued on this new noisy signal. The resulting IMFs are the average value of the ensembles on each IMF. Since white noise is a zero-mean random signal, the average value can cancel the errors produced by the noise.

5.4.1.3 Mathematical Morphology (MM)

MM applies mathematical operators between the original signal and a structure element, which is a predefined geometric shape with finite length as shown in Figure 5.5. Morphological features, which are the result of MM operation on the original signals, are then extracted.

Two basic operators in MM are erosion and dilation, which can be expressed by Minkowski subtraction and addition, respectively. By combining these two operators, another two operators (opening and closing) are created. Assuming x_n is a signal with a discrete function in the domain $X = (0,1,2,\ldots,N-1)$ and s_n is a structure element with a discrete function over the domain $S = (0,1,2,\ldots,M-1)$ [45], the four operators are [46]

$$(X \ominus S)_n = \min\left(X_{(n+m)} - S_m\right), \quad m \in 0,1,2,\ldots,M-1 \tag{5.7}$$

$$(X \oplus S)_n = \max\left(X_{(n-m)} + S_m\right), \quad m \in 0,1,2,\ldots,M-1 \tag{5.8}$$

$$(X \circ S)_n = (X \ominus S \oplus S)_n \tag{5.9}$$

$$(X \cdot S)_n = (X \oplus S \ominus S)_n \tag{5.10}$$

where $\ominus$, $\oplus$, $\circ$, and $\cdot$ denote erosion, dilation, opening, and closing operators, respectively. Among the above four operators, opening and closing operators can preserve negative and positive peaks of signals, respectively. This means that the amplitudes of negative (positive) peaks of the original signals will be unchanged before and after execution of opening (closing) operation. Therefore, they are chosen for signal decomposition to maintain the integrity of decomposed signals. Also, a flat shape structure element is used due to easy implementation. Previous research proved that the shapes of structure elements do not affect signal

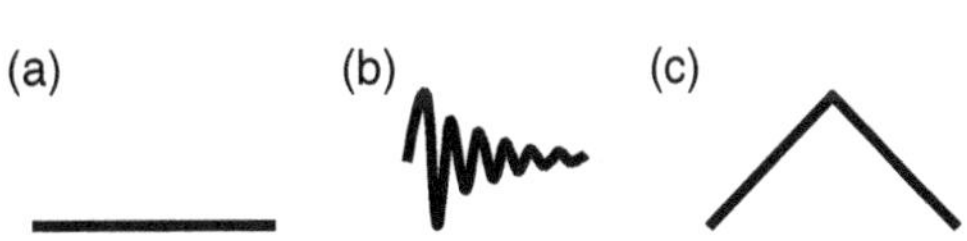

Figure 5.5 Types of structure element: (a) flat; (b) sinusoidal; and (c) triangular structure elements.

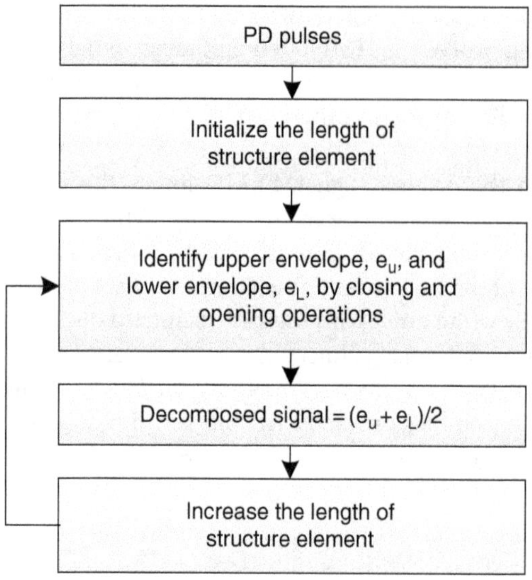

Figure 5.6 The MM-based signal decomposition method.

analysis much [47]. The flowchart of MM-based signal decomposition for individual PD pulses is shown in Figure 5.6.

As shown in this figure, the MM-based signal decomposition starts with an initialization of the structure element length, which will be used to construct upper and lower envelopes with closing and opening operators, respectively. The decomposed signal at the current length of the structure element is obtained by averaging the upper and lower envelopes. Then, the length of the structure element is increased by one and the above process is repeated. Since more pulses (including both shorter and longer intervals) are covered by the envelopes if the lengths of structure elements become longer, changing the lengths could alter the frequencies of envelopes. In this decomposition method, the lengths were set from 1 to 20 samples to provide 20 different frequency bands of PD pulses.

5.4.2 Blind Processing

Blind processing has been applied to diverse applications, but there are a few applications in PD analysis. In this section, blind equalization (BE), which is one of the blind processing techniques, is introduced.

5.4.2.1 Blind Equalization (BE) System

BE can recover a source signal without requiring a training process to identify the signal. Figure 5.7 depicts the schematic model of a single-input single-output (SISO) BE system (dotted line) [48]. BE only requires the received signal and some statistical parameters of the source signal. From the SISO BE system, the equalized signal is the output and it can reveal the source signal.

In this figure, the source signal is an independent and identically distributed (i.i.d.) non-Gaussian random signal having zero mean, second-order cumulant $\sigma^2 = E[s^2(k)]$, third-order cumulant $c_3 = E[s^3(k)]$, and fourth-order cumulant

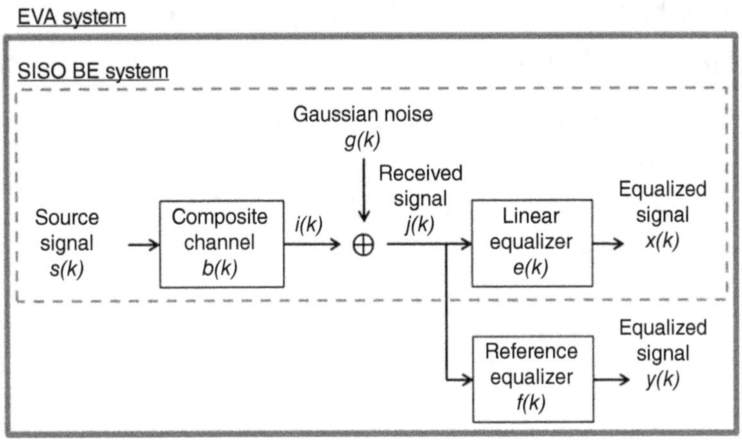

Figure 5.7 Schematic models of SISO BE and EVA systems.

$c_4 = E[s^4(k)] - 3\{E[s^2(k)]\}^2$, where $E[\cdot]$ denotes expectation. It assumes the fourth-order cumulant is nonzero ($c_4 \neq 0$) [49].

The source signal in Figure 5.7 is influenced by an unknown composite channel, which is considered as a time-invariant system with finite impulse response (FIR) $b(k) = [b(0), \ldots, b(n)]$, where n is the filter order. After the signal is linearly distorted, the distorted signal is mixed with zero-mean Gaussian noise, which is statistically independent of source signal, to produce a received signal

$$j(k) = b(k)^* s(k) + g(k) \tag{5.11}$$

where * denotes the convolution operator. For signal recovery, a linear equalizer with FIR $e(k) = [e(0), \ldots, e(n)]$ is used to reconstruct the equalized signal

$$x(k) = e(k)^* j(k) \tag{5.12}$$

To obtain accurate signal recovery, a combined impulse response of equalization should only contain one nonzero unit magnitude component [50]

$$u(k) = b(k)^* e(k) = \delta(k - k_0) \tag{5.13}$$

where $u(k)$ is the combined system, k_0 is the time delay, and $\delta(k)$ is the Dirac delta function. After equalization, the obtained equalized signal is closed to the delayed source signal in terms of mean square error (MSE), defined as [49]

$$\text{MSE}(e, k_0) = E\{[x(k) - s(k - k_0)]^2\} = \min \tag{5.14}$$

5.4.2.2 Eigenvector Algorithm (EVA) for Blind Equalization (BE)

The structure of EVA [49] is similar to that of BE (Figure 5.7), but a reference equalizer is added. This reference equalizer can generate implicit signals as a reference for an iteration process to find an optimal equalized signal. EVA seeks the maximum fourth-order cross-cumulant in the equalization:

$$\max|c_4^{xy}(0,0,0)| \quad \text{s.t. } r_{xx}(0) = \sigma_s^2 \tag{5.15}$$

Replacing $x(k)$ by $j(k)$ [i.e., $x(k) = j(k)^* e(k)$], equation (5.15) becomes

$$\max\left|e^* C_4^{yj} e\right| \quad \text{s.t. } e^* R_{jj} e = \sigma_s^2 \tag{5.16}$$

where * denotes the conjugate transpose function, R_{jj} is the autocorrelation matrix of the received signal, and C_4^{yj} is an $(n+1) \times (n+1)$ Hermitian cross-cumulant matrix:

$$\begin{aligned} C_4^{yj} = {} & E\{[y(k)]^2 jj^*\} - E\{[y(k)]^2\}E\{jj^*\} - E[y(k)j]E[y^*(k)j^*] \\ & - E[y^*(k)j]E[y(k)j^*] \end{aligned} \tag{5.17}$$

The above approach makes use of a statistical property (i.e., the fourth-order cumulant) of both the output of the reference equalizer and the received signal. By optimizing equation (5.16), this leads to a generalized eigenvector problem

$$C_4^{yj} e_{\text{EVA}} = \lambda R_{jj} e_{\text{EVA}} \tag{5.18}$$

The coefficient vector $e_{\text{EVA}} = [e_{\text{EVA}}(0), \ldots, e_{\text{EVA}}(n)]^T$ can be obtained by selecting the eigenvectors of $R_{jj}{}^{-1} C_4^{yj}$ with the largest eigenvalue λ.

EVA is implemented by an iteration process as follows [49]:

Step 1. *Initialization*. Set reference equalizer $f^{(0)}(k) = \delta(k - \lfloor n/2 \rfloor)$ and iteration number $i = 0$, where $\lfloor\ \rfloor$ is the floor function and n is the equalizer length.
Step 2. Estimate autocorrelation matrix R_{jj}.
Step 3. Find $y(k) = j(k)^* f^{(i)}(k)$ and estimate cross-cumulant matrix C_4^{yj}.
Step 4. Find coefficient $e^{(i)}{}_{\text{EVA}}$ by choosing eigenvectors of $R_{jj}{}^{-1} C_4^{yj}$ with the largest eigenvalue λ in $C_4^{yj} e^{(i)}{}_{\text{EVA}}(k) = \lambda R_{jj} e^{(i)}{}_{\text{EVA}}(k)$.
Step 5. Let the reference equalizer equal e_{EVA} [i.e., $f^{(i)}(k) = e^{(i)}{}_{\text{EVA}}(k)$], increase $i = i + 1$, and go to Step 3 until i reaches its maximum value (e.g., $i = I$).

The above signal decomposition and blind processing techniques are used mainly for PD signal de-noising. In the next section, a TF map is detailed for PD source separation.

5.4.3 Time/Frequency (TF) Map

Conventionally, a TF map is constructed based on equivalent time length and bandwidth of individual PD pulses for multiple PD source separation [51–54]. However, the TF map may not always capture accurately the characteristics embedded in the shape of a PD pulse. Therefore, in this chapter, a TF sparsity map is proposed [32].

5.4.3.1 Equivalent Time and Bandwidth

A TF map by means of equivalent time length and bandwidth of a PD pulse is a widely adopted approach for separating multiple PD sources [51–54]. For a PD pulse p, with N samples, a time reference is defined as

$$t_0 = \frac{\sum_{i=0}^{N} t_i p_i(t_i)^2}{\sum_{i=0}^{N} p_i(t_i)^2} \tag{5.19}$$

where t_i denotes the time instant of acquiring the ith sample. The equivalent time length of a PD pulse is defined as

$$T^2 = \frac{\sum_{i=0}^{N} (t_i - t_0)^2 p_i(t_i)^2}{\sum_{i=0}^{N} p_i(t_i)^2} \tag{5.20}$$

If P is the frequency value obtained from Fourier transform, the equivalent bandwidth of a PD pulse is defined as

$$F^2 = \frac{\sum_{i=0}^{N/2} f_i^2 |P_i(f_i)|^2}{\sum_{i=0}^{N/2} |P_i(f_i)|^2} \tag{5.21}$$

where f_i denotes the frequency at the ith sample.

A TF map is constructed by projecting T^2 and F^2 of each PD pulse onto a two-dimensional (2D) plane. On this plane, the projected data points, which belong to PD pulses having similar shape, will merge together to form a distinct group. Thus, a number of groups can be formed and each group corresponds to a unique PD source. This method provides a mechanism for describing an individual PD pulse in terms of time and frequency representations. However, as shown in equations (5.20) and (5.21), pulse shapes cannot always be represented precisely, since only the overall amplitude/magnitude in the time/frequency domain of a PD pulse is considered. As a consequence, sometimes the PD pulses generated by the same PD source, which are supposed to have similar values of T^2 and F^2 and locate closely on the TF plane, can be scattered in different groups. In the next section, an improved version of the TF map, the TF sparsity map, is described.

5.4.3.2 Time/Frequency Sparsity Map

Figure 5.8 presents a conceptual diagram [Figure 5.8(a)] and a flowchart of the TF sparsity map [Figure 5.8(b)].

The construction of a TF sparsity map has three stages. The first stage is extracting PD pulses from measured signals at one acquisition. The second stage is the MM-based signal decomposition for decomposing the extracted PD pulses in time and frequency domains. The third stage is the calculation of sparsity values, sparsity trends, and roughness values for constructing a TF sparsity map.

Sparsity is defined as

$$S(p) = \frac{\sqrt{1/N \sum_{i=1}^{N} p^2(i)}}{1/N \sum_{i=1}^{N} |p(i)|} = \sqrt{N}\frac{\sqrt{\sum_{i=1}^{N} p^2(i)}}{\sum_{i=1}^{N} |p(i)|} \tag{5.22}$$

where p is the PD pulse with N samples. $\sqrt{\sum_{i=1}^{N} p^2(i)}$ and $\sum_{i=1}^{N} |p(i)|$ denote norms.

Figure 5.9 shows sparsity values calculated from time and frequency domains of the decomposed signals. In this figure, the signals were generated by three

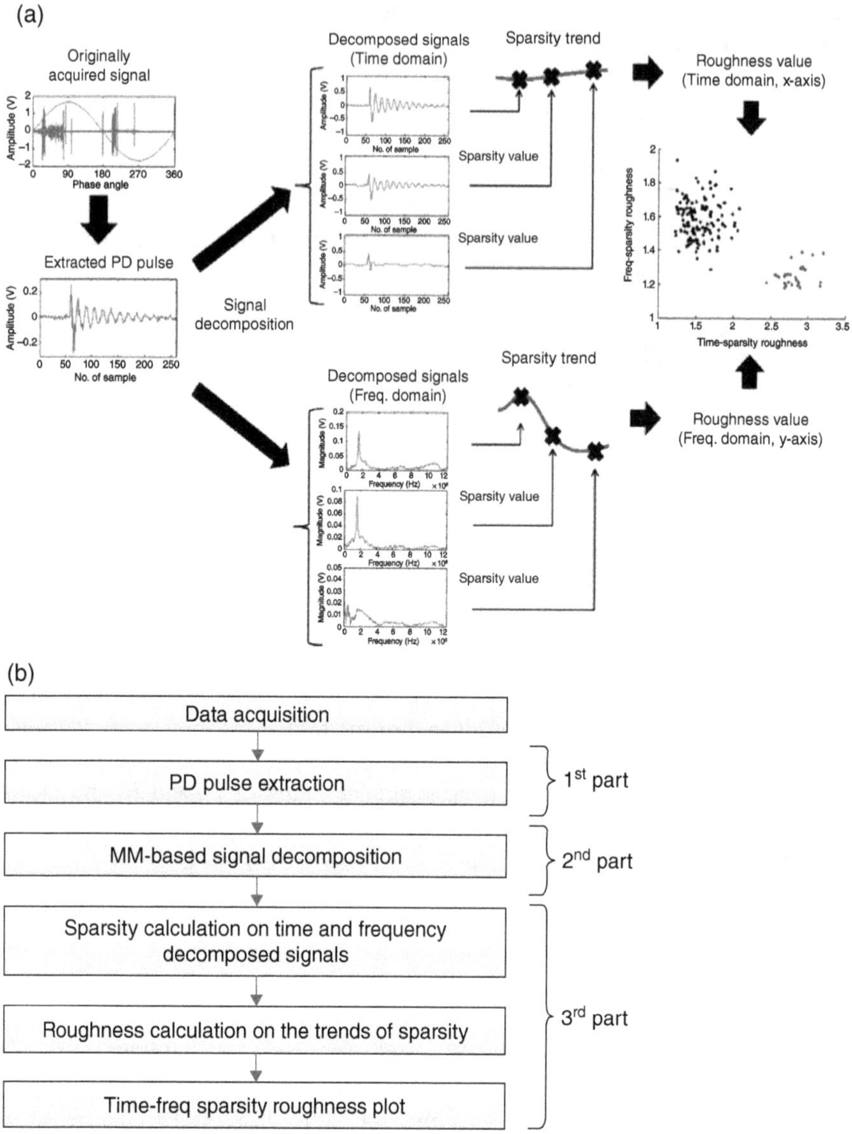

Figure 5.8 TF sparsity map for multiple PD source separation: (a) conceptual diagram and (b) flowchart. (*See insert for color representation of the figure.*)

different PD sources including discharge in transformer oil, corona, and surface discharge. After considering sparsity values for all decomposed signals, each PD source exhibits a particular trend of sparsity values (solid gray lines). This trend is named the sparsity trend. In the figure, solid gray lines represent the average of three different tendencies of the trends, with each trend related to one type of PD source.

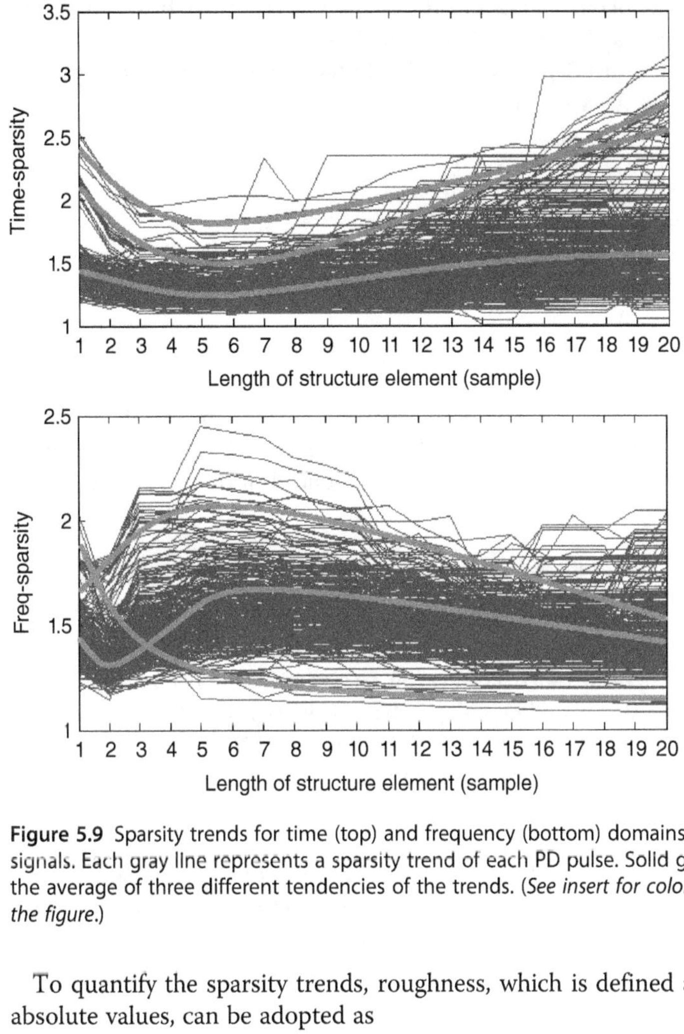

Figure 5.9 Sparsity trends for time (top) and frequency (bottom) domains of decomposed signals. Each gray line represents a sparsity trend of each PD pulse. Solid gray lines represent the average of three different tendencies of the trends. (*See insert for color representation of the figure.*)

To quantify the sparsity trends, roughness, which is defined as an average of absolute values, can be adopted as

$$R = \frac{1}{L}\sum_{i=1}^{L}|t(i)| \tag{5.23}$$

where $t(i)$ is the sparsity trend for the ith length of the structure element and L is the number of lengths of structure elements.

The values of roughness are used to describe variation of sparsity trends in time and frequency domains to represent PD pulses. Therefore, each PD pulse is represented by two roughness values (one in the time domain and another in the frequency domain). After projecting roughness values of each PD pulse onto a 2D map, the TF sparsity map is obtained. After PD pulses are represented by the roughness values and the values are projected onto a TF sparsity map, each group of PD pulses on the TF sparsity map represents a single PD source. Then, a density-based spatial clustering of application with noise (DBSCAN) is used for clustering PD pulses on the TF sparsity map [55, 56]. DBSCAN performs clustering based on the density of data points in a volume, which is decided by the minimum number of data points $N_{\min}$ and neighborhood radius ε. $N_{\min} = 2$ is set to ensure even a small amount of PD pulses belonging to the same PD source can be clustered. For a data set $X(m \times 2)$, its neighborhood radius is defined as [57]

$$\varepsilon = \sqrt{\frac{prod\left(\max\limits_{1\le i\le m}(X) - \min\limits_{1\le i\le m}(X)\right)N_{\min}\Gamma(2)}{m\pi}} \tag{5.24}$$

where $prod$ is the product of the array and Γ is the gamma function, $\Gamma(x) = \int_0^{\infty} t^{x-1}e^{-t}dt$.

After PD source separation, each individual PD source can have its particular pattern (mixture PD sources also have their own patterns). This pattern can be represented by various methods, which are introduced in the next section.

5.4.4 PD Pattern Representation

5.4.4.1 Phase-Resolved Pulse Sequence

In PD measurements, PD activity is represented as a sequence of pulses with their intensity and time of occurrence. For PD source classification, PD patterns should exhibit a clear consistency and be independent of the types of PD sensor used in the measurements. Conventional PRPD diagrams can be suitable for PD source classification, given that test arrangements are standardized and the relation between test voltages and discharge activities is known a priori. However, PRPD diagrams can be influenced by a number of factors.

PD signals are wide-band signals and different types of PD sensor can have different measuring bandwidths. This can influence PD pulse resolutions. Noise resistivity and sensitivity also depend strongly on the measuring bandwidths. Therefore, discrepancy may exist among PRPD diagrams obtained from different types of PD sensor for the same PD source.

Therefore, a phase-resolved pulse sequence (PRPS) diagram has been investigated as a complement to the PRPD diagram for PD source classification [48]. Figure 5.10(a) shows parameters that are used for pulse sequence (PS) analysis in a single power cycle. There are a number of methods to analyze PS. In Ref. [58], PS was analyzed based on voltage differences Δh between sequential PD pulses. This indicates the amount of voltage required to ignite a PD pulse. In Ref. [59], a PRPS diagram was developed based on the slope $m = \Delta h/\Delta w$ to examine voltage gradients for initiating a PD pulse. Although these methods can improve the consistency of PD pattern analysis, the amplitudes of applied voltages are required.

Figure 5.10(b) illustrates a recently developed PRPS diagram [48], which uses the phase angle between two consecutive PD pulses as a quantitative measure ("amplitude"). It does not require the amplitude values of applied voltages. Each power cycle is divided into 360 windows and the PD pulse amplitudes are averaged in each window. Then, in each cycle, the differences in pulse phase angles are

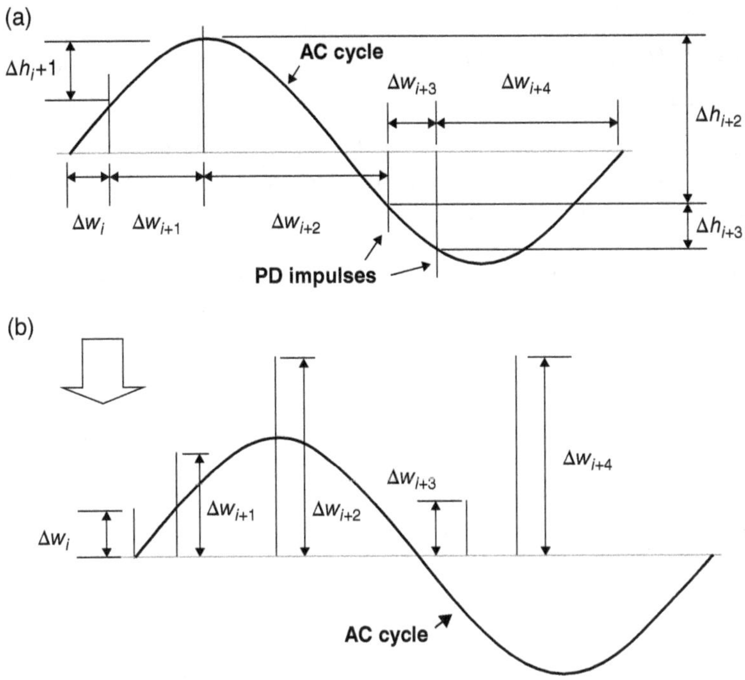

Figure 5.10 (a) Parameters of PS analysis and (b) the recently developed PRPS diagram.

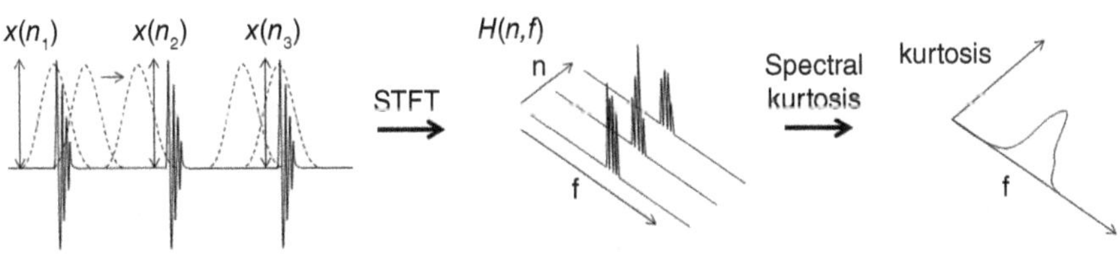

Figure 5.11 Interpretation of spectral kurtosis [62].

calculated and converted into amplitudes in a PRPS diagram. Finally, these amplitudes are normalized to the $[-1, 1]$ interval.

5.4.4.2 Kurtogram Construction Using Spectral Kurtosis

A kurtogram, which is constructed by spectral kurtosis, is another PD pattern representation method. Spectral kurtosis describes the impulsiveness of a signal in the frequency domain [60]. To compute spectral kurtosis, a short-time Fourier transform (STFT) with a window shifting along the time domain of a PD signal is used to obtain a complex envelope function $H(n,f)$ (Figure 5.11) [61]. The complex envelope function can be defined as the decomposition of a non-stationary random signal $x(n)$, where $n \in \mathbb{Z}$, which matches the characteristic of a PD signal by the Wold–Cramer representation [61]

$$x(n) = \int_{-1/2}^{+1/2} H(n,f)e^{j2\pi fn} dZ_x(f) \tag{5.25}$$

where $dZ_x(f)$ is a spectral increment. Taking the square of the complex envelope function represents the power spectrum at each time instant (i.e., n) of a PD signal. The power spectrum of the whole time period of the signal is obtained by taking the average of all the power spectra [62].

A kurtosis value for each frequency scale (right graph of Figure 5.11) is obtained by calculating the fourth moment of $H(n,f)$ [as shown in the numerator of equation (5.26)] at each time position and then taking the average of these values. The final kurtosis value is a normalization of the average value by the square of the mean square value [62]. In equation (5.26), the minus 2 makes the value of kurtosis zero for a Gaussian signal. The above process is performed based on a window shifting along a PD signal to find the spectral kurtosis for one level. By repeating the process for different sizes of window, different levels of spectral kurtosis are formed and a kurtogram is constructed.

$$K(f) = \frac{\langle |H(n,f)|^4 \rangle}{\langle |H(n,f)|^2 \rangle^2} - 2 \tag{5.26}$$

In addition to the aforementioned signal processing techniques, there are several more recently developed techniques for PD signal analysis. These techniques are presented in the next section and their applications are shown in later sections of this chapter.

5.4.5 Other Signal Processing Techniques

5.4.5.1 Fractal Dimension and Entropy

Fractal dimension and entropy are used to remove stochastic noise embedded in the measured PD signal. The fractal dimension is referred to as the changes of a pattern's details with respect to the scales used for measuring this pattern. A number of methods including box-counting, variance, and spectral methods have been used to calculate fractal dimension [63]. In this chapter, box counting is introduced for its simplicity and efficiency [64, 65]. The fractal dimension (FD) of a signal in Euclidean space is defined as [66]

$$\mathrm{FD} = \lim_{\varepsilon \to 0} \frac{\log N(\varepsilon)}{\log(1/\varepsilon)} \tag{5.27}$$

where $N(\varepsilon)$ is the least number of boxes with side length ε to cover the signal.

A PD source caused by a particular insulation defect can exhibit a unique PD pattern. Owing to the fractal dimension's ability to represent various patterns and describe complex shapes using different sizes of box (i.e., side lengths of boxes), it has been applied to feature extraction and subsequent PD source recognition [66, 67]. In Ref. [65], the fractal dimension was used to quantify the severity of stochastic noise when the noise and PD signals were located in different power cycles. After the quantification, the noise and PD signals can have various values of fractal dimension. This enables the separation of the noise and PD signals and subsequent removal of the noise.

When the noise and PD signals appear in the same power cycles, fractal dimension alone cannot extract PD signals. To deal with this situation, entropy is introduced [65]. Entropy measures the disorder in a random variable. A larger value of entropy relates to more chaotic data [68]. For a signal $X = [x_1, x_2, \ldots, x_n]$, its entropy $H(X)$ is defined as

$$H(X) = -\sum_{i=1}^{n} p(x_i)\log_2 p(x_i) \tag{5.28}$$

where $p(x_i)$ is the probability mass function of x_i.

5.4.5.2 Pre-whitening

The pre-whitening process is used to enhance the blind processing technique in PD signal de-noising. It is performed based on an autoregressive (AR) model, which provides a linear prediction of the noise in the acquired signals in PD measurements. The predicted value (or amplitude) of the noise $\hat{y}(n)$ at the current time step is calculated based on the weighted sum of p values (or amplitudes) of an acquired signal $x(n)$ at previous time steps [62]:

$$\hat{y}(n) = -\sum_{k=1}^{p} a(k)x(n-k) \tag{5.29}$$

where k denotes previous time steps from 1 to p, and the $a(k)$ are coefficients which can be calculated from the solution of Yule–Walker equations. The actual value of the acquired signal $x(n)$ at the current time step is the summation of the predicted value of the noise $\hat{y}(n)$ and a residual part $e(n)$ [62]:

$$x(n) = \hat{y}(n) + e(n) \tag{5.30}$$

The aforementioned process is based on a linear prediction of the noise. After subtracting the predicted value of noise $\hat{y}(n)$ from the acquired signal $x(n)$, the term $e(n)$ in equation (5.30) contains the remaining stationary white noise and transient pulses, which are related to PD pulses. Thus, it can be considered that the impulsiveness of a signal (i.e., transient PD pulses) is enhanced [69], since a large portion of noise can be removed by the above pre-whitening process. In equation (5.30), since $e(n)$ is considered as a pre-whitened signal, this process is named a pre-whitening process.

5.4.5.3 Mathematical Morphology for Thresholding in PD Signal Extraction

MM is used not only for signal decomposition, but also for defining thresholds in PD signal extraction. Based on the theory of MM, AMT is developed to automatically define the lengths of flat structure elements for creating thresholds, as shown in Figure 5.12 [31].

A series of lengths $L_1, L_2, \ldots, L_N$, is selected for flat structure elements to form upper and lower envelopes by using closing and opening operators, respectively. The lengths are multiples of the sample number in one AC power cycle divided by

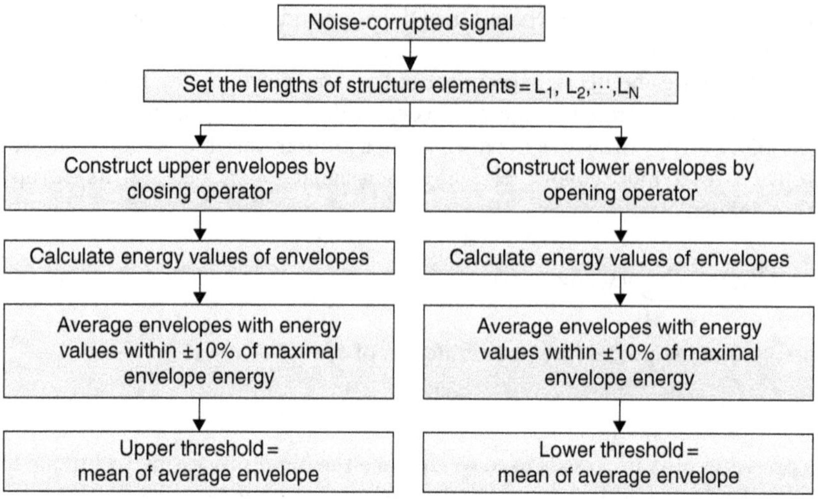

Figure 5.12 AMT method for thresholding.

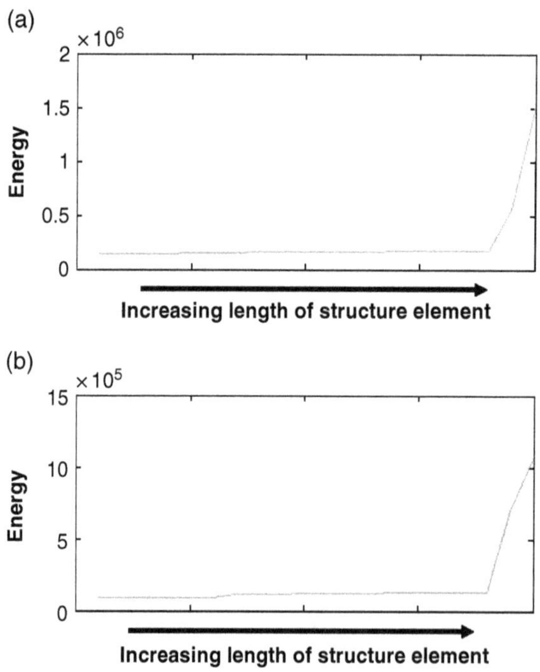

Figure 5.13 Energy values of (a) upper and (b) lower envelopes.

32 to maintain a reasonable computational time. After generating the envelopes, energy values – defined as the sum of amplitudes squared for each sample of envelopes – are calculated. An example of the energy values for both upper and lower envelopes using flat structure elements with different lengths is shown in Figure 5.13.

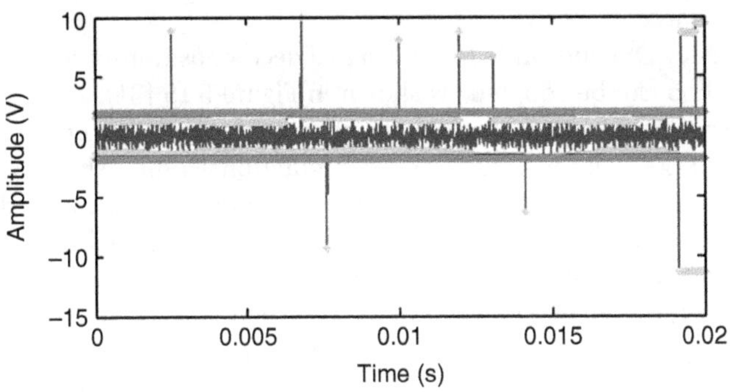

Figure 5.14 Results of AMT for random pulses [average upper and lower envelopes (gray dots) and optimal thresholds (dark gray lines)].

It can be seen from this figure that the energy values increase slowly with increasing lengths of structure elements until the lengths attain relatively large values. The relatively small increase of energy values at the initial and middle stages in Figure 5.13 are due to the envelopes covering pulses with either small intervals or large intervals but small amplitudes. However, when the lengths become longer, the envelopes start to cover pulses with both large intervals and amplitudes. This leads to a significant increase in energy values at the later stage. After calculating the energy values of envelopes, the envelopes whose energy values are equivalent to ±10% of maximal envelope energy are averaged to form an overall envelope. Thresholds are obtained by taking the mean value of each average envelope and used to remove white noise, as shown in Figure 5.14.

After describing the advanced signal processing techniques, in the next section their applications in PD signal analysis are provided with several case studies. This is to verify the effectiveness of these techniques in practice.

5.5 Application of Advanced Signal Processing Techniques for PD Analysis

As shown in Figure 5.1, PD signal analysis includes (1) PD signal de-noising, (2) PD source separation, and (3) feature extraction for PD source classification. In this section, the application of the signal processing techniques on these three steps is demonstrated.

5.5.1 PD Signal De-noising

In this section, the de-noising techniques are focused on the three major types of noise: white noise, DSI, and stochastic noise.

5.5.1.1 De-noising of Low-Frequency DSI and White Noise

DSI can have a wide range of frequency bands generated from communication systems and radio transmissions. In the following sections, a PD signal de-noising method is presented to remove low-frequency DSI and white noise.

5.5.1.1.1 De-noising Method

To remove low-frequency DSI and white noise, a signal decomposition-based PD signal de-noising method can be adopted, as shown in Figure 5.15 [31].

Firstly, EEMD is executed to decompose a noise-corrupted signal. The ensemble number was set to 300 without sacrificing computational time, while the decomposition performance can be maintained; the amplitude of the injected white noise was set to 0.2 standard deviations of the signal, as suggested in Ref. [44] to maintain the decomposition performance. According to Ref. [70], an IMF generally satisfies the Cauchy condition after five iterations in the sifting process. More iterations may not change the IMFs significantly. Thus, 10 iterations were used to guarantee the stability and convergence of IMFs. After decomposition by EEMD, kurtosis was applied to select the IMFs with PD pulses due to the sensitivity of transient signals. In other words, a larger kurtosis value refers to a signal consisting of abruptly changed pulses, while a smaller kurtosis value refers to a slowly fluctuating signal or a signal consisting of evenly distributed amplitudes (e.g., white noise and periodic signal). The kurtosis k is defined as

$$k = \frac{\sum_{i=1}^{L} (I_i - \bar{I})^4}{(L-1)\sigma^4} \tag{5.31}$$

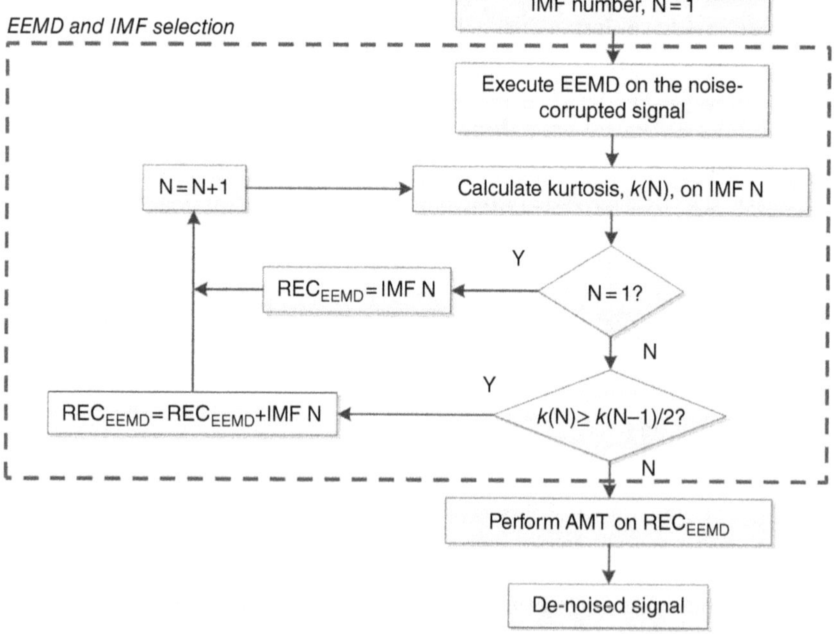

Figure 5.15 The signal decomposition-based (EEMD-based) PD signal de-noising method.

where $\bar{I}$, L, and σ are the mean, length, and standard deviation of an IMF, respectively.

An IMF selection process starts from the first IMF, which comprises the highest-frequency component of a noise-corrupted signal. If the kurtosis of an IMF is suddenly dropped to half that of the previous IMF, this implies that the IMF with the lower kurtosis value consists of fluctuations in low-frequency noise and/or white noise. Therefore, it is considered as pure noise without containing any PD pulses and discarded. Otherwise, it is added to the previous IMF for reconstruction and the selection will be continued on the remaining IMFs. Such kurtosis-based IMF selection can eliminate the low-frequency IMFs generated from the end effect of EMD.

After the above selection process, signal reconstruction can be performed. Since some noise with the same frequency scales of the selected IMFs may be included in the reconstructed signal, the AMT can be employed to create upper and lower thresholds of the signal. A de-noised signal can then be produced by applying hard thresholding.

5.5.1.1.2 De-noising Results

To evaluate the performance of the signal decomposition-based PD signal extraction, three measures including MSE M, correlation coefficient γ, and pulse number error PE were adopted and defined [31]:

$$M = \frac{1}{L}\sum_{i=1}^{L}(O_i - D_i)^2 \tag{5.32}$$

$$\gamma = \frac{\sum_{i=1}^{L}(O_i - \bar{O})(D_i - \bar{D})}{\sqrt{\sum_{i=1}^{L}(O_i - \bar{O})^2 \sum_{i=1}^{L}(D_i - \bar{D})^2}} \tag{5.33}$$

$$\text{PE} = |P_o - P_d| \tag{5.34}$$

where L is the length of signal, O denotes the original signal, D denotes the de-noised signal, $\bar{O}$ and $\bar{D}$ are the mean values of O and D, respectively, P_o is the number of pulses in the original signal, and P_d is the number of pulses in the de-noised signal. The measures M and γ are used to indicate the similarity of PD patterns, whereas PE is for examining the integrity of PD pulses.

The de-noising methods introduced in the above sections are used to remove both DSI and white noise in the following case studies. In the case studies, both DSI and white noise are injected into the measured PD signal. The DSI consists of three sinusoidal signals with randomly selected amplitudes and frequencies. The frequencies of DSI are up to 1 kHz to simulate signal fluctuations that could appear in online PD measurements. White noise is simulated with different SNRs. The de-noising results are compared using three different types of signal decomposition technique: EMD, EEMD, and DWT. The results and evaluations of the three methods are shown in Figure 5.16 and Table 5.3.

The DWT-based method applied a commonly used threshold to wavelet coefficients with five decomposition levels [11]. The EMD-based method used the same approach as the EEMD-based method (Figure 5.15), but the decomposition

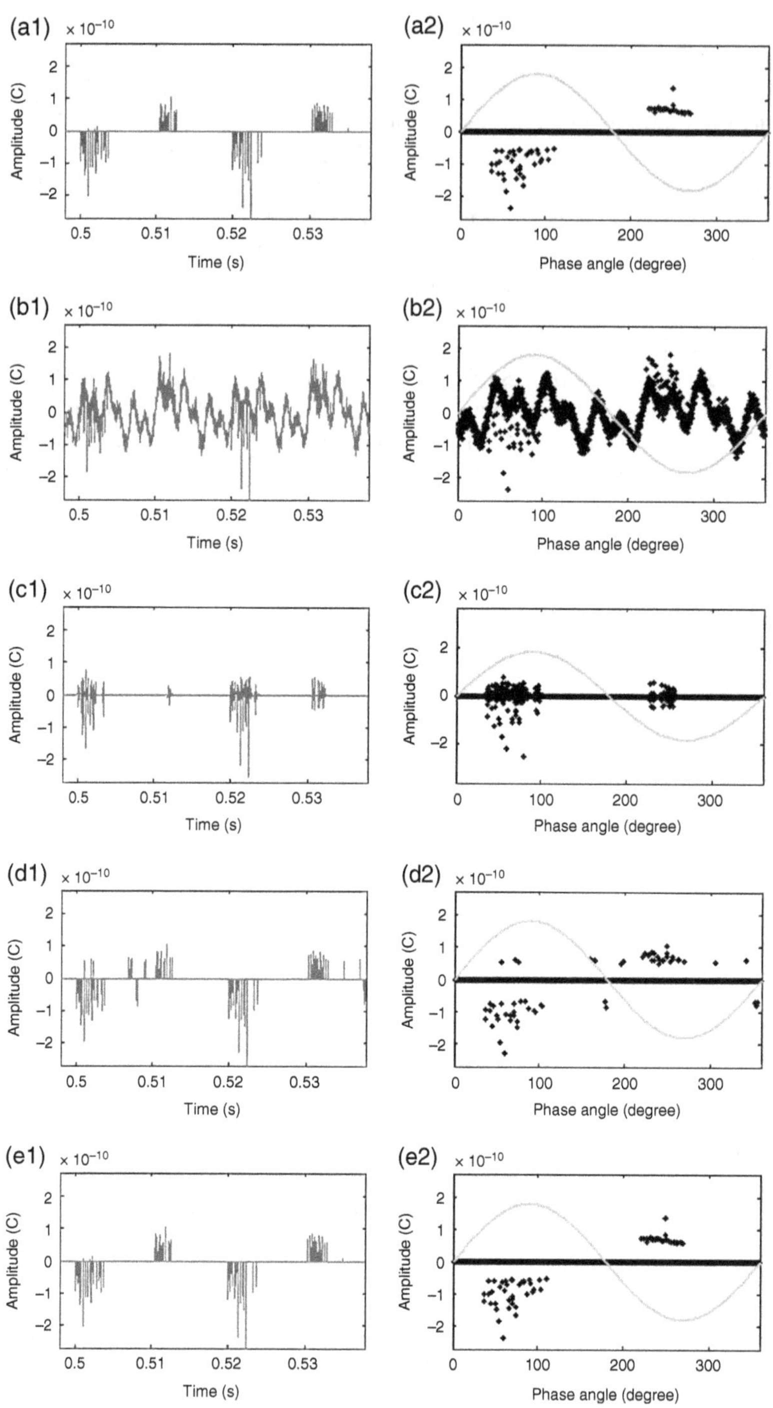

Figure 5.16 De-noising results of internal discharge. (a1)–(e1) Original signal, noise-corrupted signal (SNR = −5 dB), and de-noised signals of DWT (db5), EMD + AMT, and EEMD-based method. (a2)–(e2) Corresponding PRPD diagrams. (*See insert for color representation of the figure.*)

Table 5.3 Comparisons of measures on PD signals acquired from PD source models

	Comparison with DWT			Comparison with EMD + AMT		
Model no.	↓ M (%)	↑ γ (%)	↓PE (%)	↓ M (%)	↑ γ (%)	↓PE (%)
1	74	32	99	44	6	0
2	81	14	99	87	24	92
3	81	29	96	81	27	40
4	91	23	96	87	12	14

Note: Model 1 – internal discharge, Model 2 – discharge due to floating particles, Model 3 – discharge in transformer oil, Model 4 – corona.

of EEMD was replaced by EMD. This is termed EMD + AMT. Figure 5.16 presents both time and PRPD diagrams of original signal, noise-corrupted signal, and de-noised signals in two power cycles for internal discharge. Table 5.3 compares the de-noising performances of DWT, EMD + AMT, and the EEMD-based method on signals (SNR = −5 dB) obtained from different PD source models for 100 power cycles. The signals were acquired from the capacitive measurement system.

From Figure 5.16(c1), it can be seen that the polarities of most pulses after de-noising by DWT are blurred. A whole cluster of PD pulses at the right-hand side shift from positive to the middle position; and a large number of pulses disappear. The results can also be observed in the corresponding PRPD diagram [Figure 5.16 (c2)]. It can also be seen that some pulses are missing and noise still remains in the de-noised signal obtained by EMD + AMT [Figure 5.16(d1)]. The results reveal the superiority of the EEMD-based method. It can effectively extract PD signals from noise and preserve the amplitudes and locations of PD pulses.

Table 5.3 also shows that the EEMD-based method still outperforms the others in all four PD source models. A higher reduction of pulse number error can be noticed in the DWT compared to EMD + AMT. The results imply that the DWT-based technique may not be suitable for PD signal de-noising, since additional pulses can be induced in the de-noised signals.

From the above results, the signal decomposition-based de-noising method is proven to be able to remove low-frequency DSI and white noise. In the next section, two methods based on the blind processing technique are verified to remove high-frequency DSI and white noise. PRPS and kurtogram are also examined for accurate PD pattern representation.

5.5.1.2 De-noising of High-Frequency DSI and White Noise with PD Pattern Representation

5.5.1.2.1 De-noising and PD Pattern Representation Methods – First Method

To remove high-frequency DSI and white noise as well as to provide an accurate PD pattern representation after de-noising, two BE-based PD signal de-noising and pattern representation methods can be used [48]. The flowchart of the first method is shown in Figure 5.17 [48]. It uses an advanced EVA for PD signal de-noising on DSI and AMT for removing white noise. A PRPS diagram is then constructed from the de-noised signals and used for PD pattern representation.

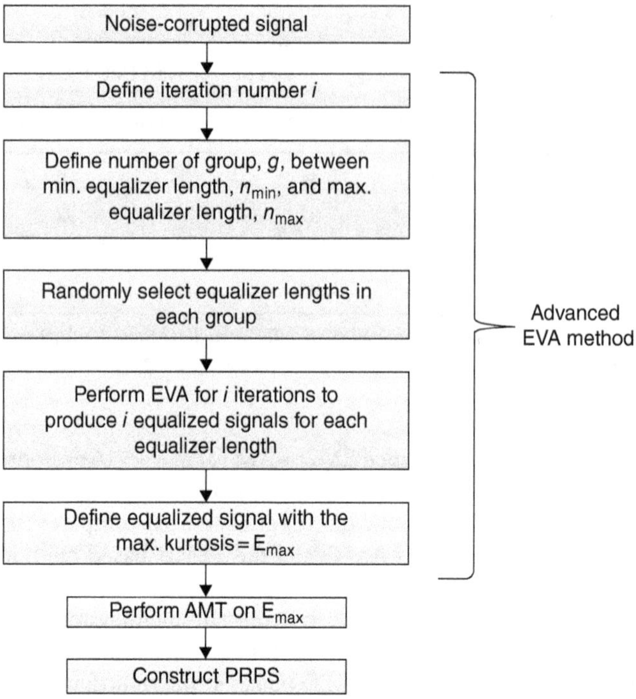

Figure 5.17 BE-based PD signal de-noising and pattern representation (first) method.

Equalized signals (recovered signals) in EVA are obtained by an iteration process. In this process, the selection of equalizer length is crucial. Improper selection of the length can incur the generation of unwanted equalized signals [71, 72]. In Ref. [71], all possible equalizer lengths were employed to seek optimal results on signal recovery. Although the results can reveal proper equalizer lengths, it is time-consuming and may not be suitable for online PD measurements of field transformers.

The BE-based method includes a novel approach for providing an adaptive selection of equalizer lengths. In this approach, a noise-corrupted signal is processed by EVA. The autocorrelation matrix R_{jj} and cross-cumulant matrix C_4^{yj} in EVA are defined as

$$R_{jj} = \frac{\sum_{k=n}^{N-1} jj^T}{N-n} \tag{5.35}$$

$$C_4^{yj} = \frac{\sum_{k=n}^{N-1} y^2(k) jj^T}{N-n} - \frac{\left(\sum_{k=n}^{N-1} y^2(k)\right) R_{jj}}{N-n} - \left(\frac{\left(\sum_{k=n}^{N-1} y(k) j\right)\left(\sum_{k=n}^{N-1} y^T(k) j^T\right)}{(N-n)^2} + \frac{\left(\sum_{k=n}^{N-1} y^T(k) j\right)\left(\sum_{k=n}^{N-1} y(k) j^T\right)}{(N-n)^2} \right) \tag{5.36}$$

where $j^T = [j(k), j(k-1), \ldots, j(k-n)]$, T is the transpose function, N is the sample number, and n is the equalizer length. Equation (5.36) is derived based on Ref. [49]. The iteration number i for EVA is set to 10 to produce a reasonable number of equalized signals without sacrificing computational time. Then, a number of groups g are defined and evenly distributed between the minimum equalizer length $n_{\min}$ and the maximum equalizer length $n_{\max}$ to allocate the equalizer lengths in different ranges. Here, $n_{\min}$ and $n_{\max}$ were set to 10 and 100, respectively, and g was set to four. Three equalizer lengths were randomly selected in each group and the selection was subject to uniform distribution [72]. Such an arrangement is to ensure that the selected equalizer lengths cover the optimal equalizer length, which can be located in different groups.

Once the equalizer lengths are determined, EVA is performed to produce possible equalized signals aimed at removing the noise. After that, the equalized signal with the maximum kurtosis, $E_{\max}$, is chosen as an optimal equalized signal to extract PD pulses from the noise-corrupted signal. Since EVA seeks the maximum fourth-order cross-cumulant which refers to peakedness of signal distributions, PD pulses can be extracted from the equalized signals. After the EVA process, the noise can be removed even if its amplitude is larger than that of PD pulses. EVA can also remove white noise by reselecting the equalized signals. However, the de-noising performance of EVA on white noise is not consistent. Therefore, in the BE-based method, AMT is applied to eliminate the remaining white noise. Finally, a PRPS diagram is constructed from the de-noised signal.

5.5.1.2.2 De-noising and PD Pattern Representation Results – First method

According to Ref. [12], a variety of high-frequency DSI up to about 2 GHz could exist during field PD measurements at substations. Therefore, in case studies (using data collected from laboratory measurements), the noise was generated from a linear combination of 10 random frequencies up to 2 GHz with different amplitudes. The white noise was generated from the measurement environment. Here, it is worth mentioning that no artificial noise was added to the case studies of field measurements.

Figure 5.18 shows results of each processing step of the first proposed method on the PD signals acquired from the PD model of discharges in transformer oil using two HFCTs. CT1 has a frequency range from 350 kHz to 35 MHz and CT2 has a frequency range from 500 kHz to 50 MHz. After the noise-corrupted signals are processed by EVA with selected equalizer lengths, a series of equalized signals are generated for extracting PD pulses from the signals. Among all the equalized signals, some of them cannot reveal PD pulses. However, when the selection criterion is based on the kurtosis, an optimal equalized signal with the highest kurtosis value is chosen and can effectively extract PD pulses as shown in Figure 5.18(c). It can be seen that the phase angles (or time intervals) of the extracted pulses are the same as the original PD signals [Figure 5.18(a)].

Since white noise is still present in the optimal equalized signals, AMT is then applied for adaptively selecting upper and lower thresholds [Figure 5.18(d)]. The results of de-noised signals after AMT are shown in Figure 5.18(e), in which the phase angles of PD pulses are clearly identified. Figures 5.18(f) and (g) are PRPD

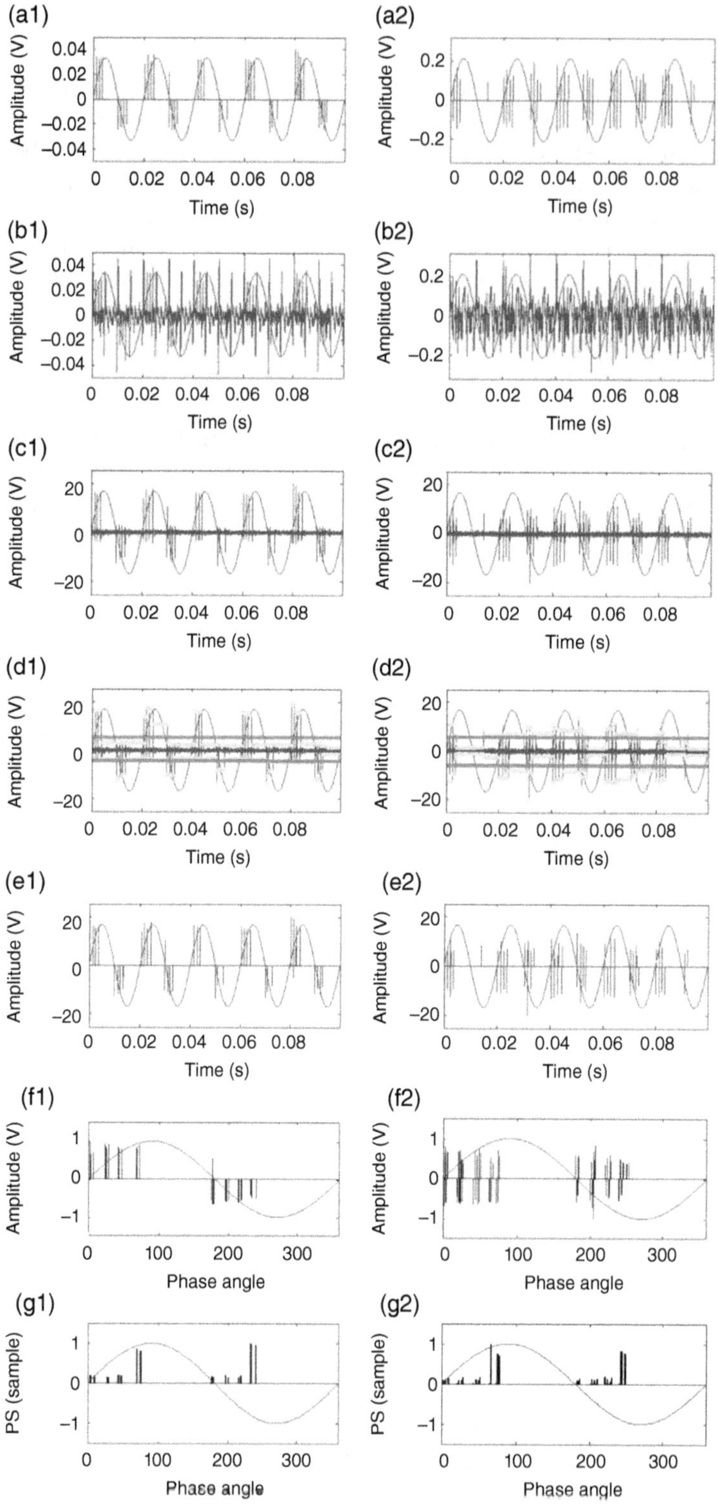

Figure 5.18 Results of discharge in transformer oil using (a1)–(g1) CT1 and (a2)–(g2) CT2. (a) Original signals, (b) noise-corrupted signals, (c) optimal equalized signals, (d) results of AMT on equalized signals, (e) de-noised signals, (f) PRPD diagram of original signals, and (g) PRPS diagram of de-noised signals. (*See insert for color representation of the figure.*)

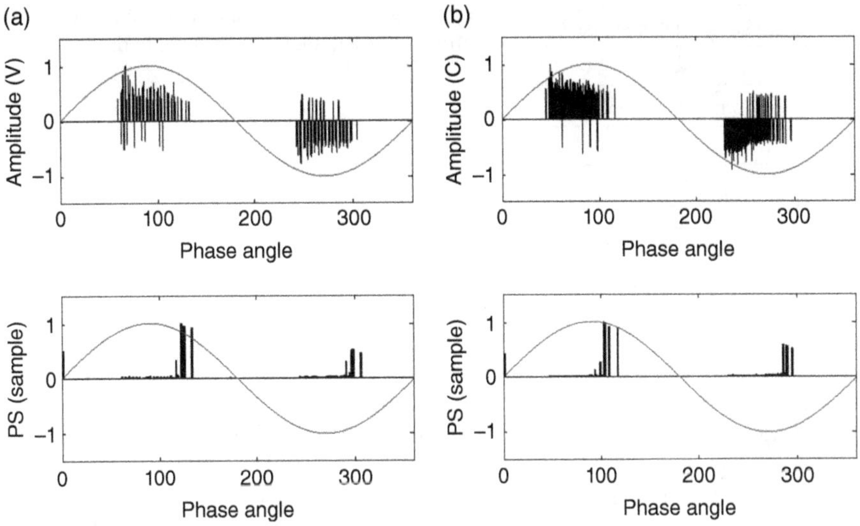

Figure 5.19 Results of PRPD and PRPS diagrams for a distribution transformer using (a) CT1 and (b) capacitive measurement.

diagrams of the original signals and PRPS diagrams of the de-noised signals, respectively. It can be seen that the PRPD diagrams fluctuate between the signals of the two CTs, while the PRPS diagrams are similar.

Field measurement and validation. To further validate the BE-based method, PD signals were acquired from a 10 kVA single-phase distribution transformer using the two HFCTs and capacitive measurement system. The results are shown in Figure 5.19 [48]. It can be seen that the PRPD patterns are different, while the PRPS patterns are similar between the PD signals of CT1 and capacitive measurement.

5.5.1.2.3 De-noising and PD Pattern Representation Methods – Second Method

In the second method for PD signal de-noising, a pre-whitening technique is applied before executing the advanced EVA method. This is to enhance the de-noising ability of EVA.

The flowchart of the second BE-based PD signal de-noising and pattern representation method is shown in Figure 5.20 [73].

As shown in Figure 5.20, the BE-based method starts with pre-whitening on a noise-corrupted PD signal. This is due to pre-whitening being able to enhance the impulsiveness of PD signals and thus facilitate the subsequent EVA process. Though the pre-whitening process can remove most of DSI, it may encounter difficulties if the noise amplitude is too large. This is due to the AR model used in the pre-whitening process not being able to provide an accurate prediction of the noise. In practical situations, the noise level is not always predictable. If small-amplitude PD signals mix with heavy noise, the pre-whitening process cannot always fulfil the task of removing the noise. Therefore, EVA is used to extract PD pulses from the remaining noise. After pre-whitening and EVA processes, a

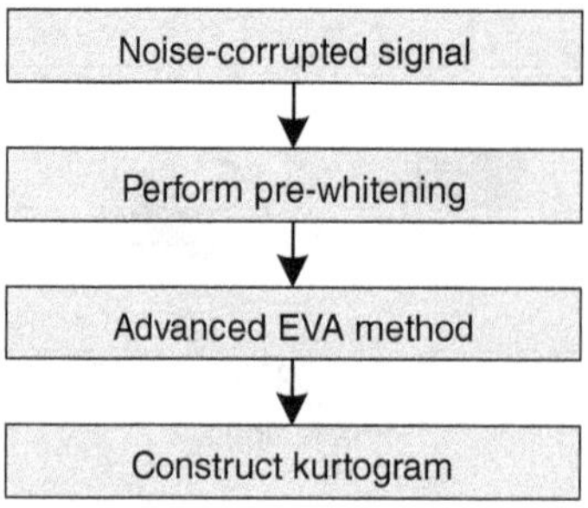

Figure 5.20 BE-based PD signal de-noising and pattern representation (second) method.

de-noised signal is plotted using a spectral kurtosis-based kurtogram for accurate PD pattern representation.

PD signals generated by different PD sources exhibit different patterns and are normally represented by PRPD diagrams [51, 74]. However, the patterns in PRPD diagrams can be overlapped if more than one PD source co-exist and occur simultaneously in a transformer [75, 76]. The patterns can also be influenced by different PD sensors and measurement systems. As such, in a PRPD diagram, PD patterns can be distorted and consequently PD sources causing discharges in HV apparatus cannot be identified and separated. Therefore, some extent of ambiguity can be induced in assessing HV apparatus conditions [75].

The kurtogram can be adopted to represent patterns of PD signals which have been de-noised by pre-whitening and EVA. Such a representation can be used to identify different PD sources. Though DSI can be removed by pre-whitening and EVA, white noise, which is normally found in field PD measurements, can still be present in the de-noised signals. However, a kurtogram is capable of representing PD patterns accurately even in the presence of white noise.

5.5.1.2.4 De-noising and PD Pattern Representation Results – Second Method

The noise investigated in the second method was the same as in the first method except that 10 more noise components of DSI with random frequencies and amplitudes were generated. The white noise was generated from the measurement environment. Figure 5.21 shows results of the proposed method for a signal acquired from the PD model of corona. The original and noise-corrupted signals are shown in Figures 5.21(a) and (b), respectively. As can be seen from the two figures, the amplitude of noise imposed on the original signal is very high. This demonstrates that the proposed method can extract PD signals and represent PD patterns when heavy noise is present. Figure 5.21(c) presents the signal obtained after a pre-whitening process on the noise-corrupted signal. It can be seen that after pre-whitening, the noise amplitude is much reduced. However, noise still suppresses the PD signal across the whole time span. Though the pre-whitening process cannot remove the entire noise, it does reveal PD pulses to some extent, which makes the subsequent EVA process easier. Figure 5.21 (d) shows the de-noised signal (i.e., equalized signal) after EVA. It can be seen that the PD pattern of the corona is extracted, although the remaining noise amplitude is higher than that of the original signal shown in Figure 5.21(a).

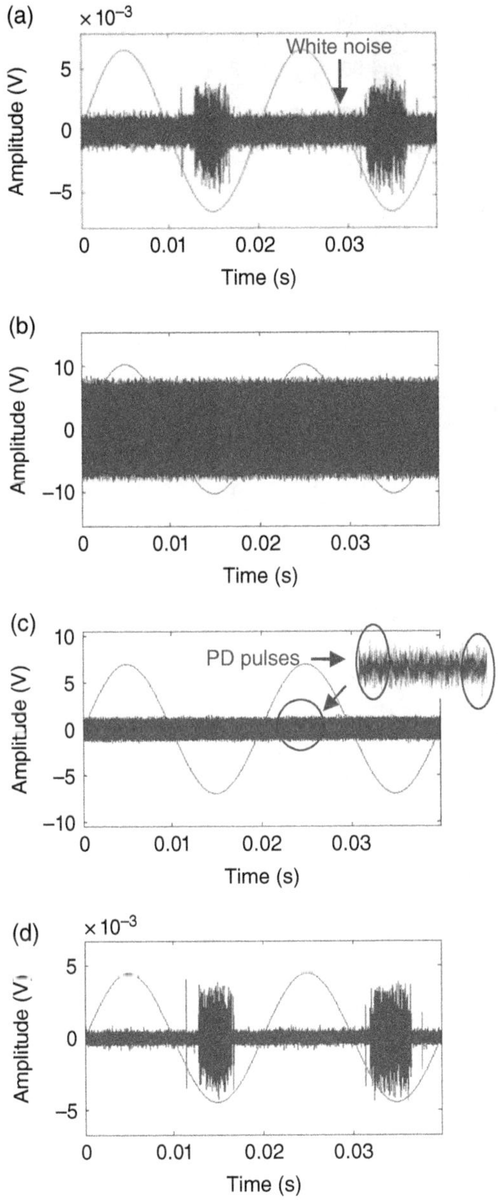

Figure 5.21 Results of corona. (a) Original signal, (b) noise-corrupted signal, (c) pre-whitened signal with magnified portion, (d) de-noised (equalized) signal, (e) kurtogram of original signal, and (f) kurtogram of de-noised signal.

Figure 5.21(e) shows a kurtogram constructed from the original signal [Figure 5.21 (a)], while Figure 5.21(f) shows a kurtogram of a de-noised signal [Figure 5.21(d)]. Comparing these two kurtograms, there is not much difference, although the noise amplitude in the de-noised signal is higher than that in the original signal. This demonstrates the consistency of kurtograms in representing PD patterns.

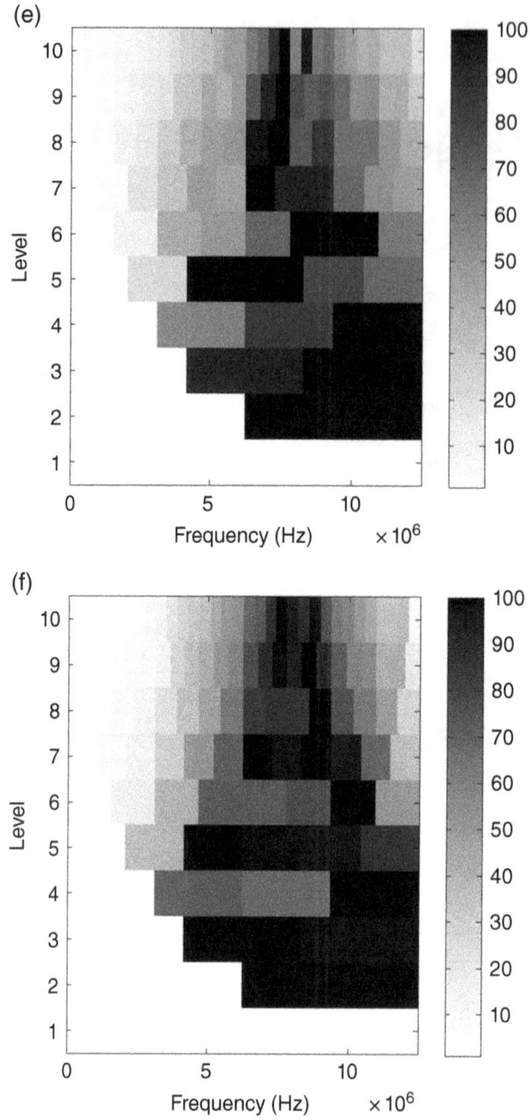

Figure 5.21 *(Continued)*

Field measurement and validation. The BE-based method has also been applied to signals acquired from a field PD measurement on a 10 MVA transformer (33 kV/11 kV) at the substation of a local distribution authority. As shown in Figure 5.22, the originally acquired signal is severely corrupted by a relatively low-frequency noise [Figure 5.22(a)]. This low-frequency noise was caused by AC harmonics. After applying pre-whitening, two pulse-shaped signals were uncovered in the positive power cycles [Figure 5.22(b)]. By applying EVA to the pre-whitened signal, two more pulses were revealed in each negative power cycle [Figure 5.22(c)]. The periodic signal may indicate that a single PD source exists

Figure 5.22 Results of a 10 MVA transformer. (a) Original signal, (b) pre-whitened signal, (c) de-noised (equalized) signal, and (d) kurtogram of de-noised signal.

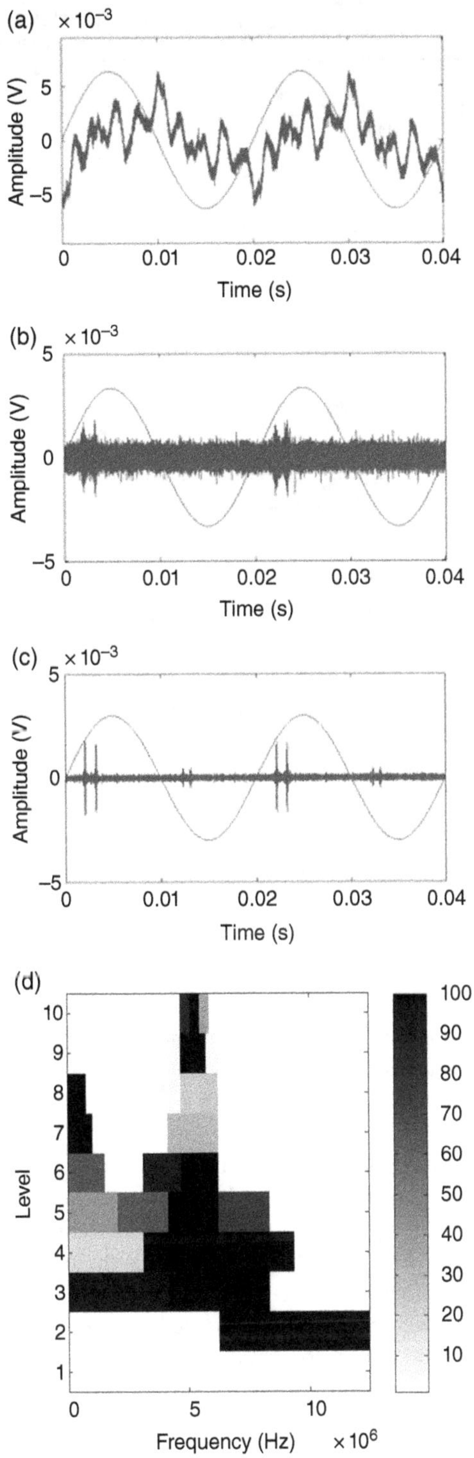

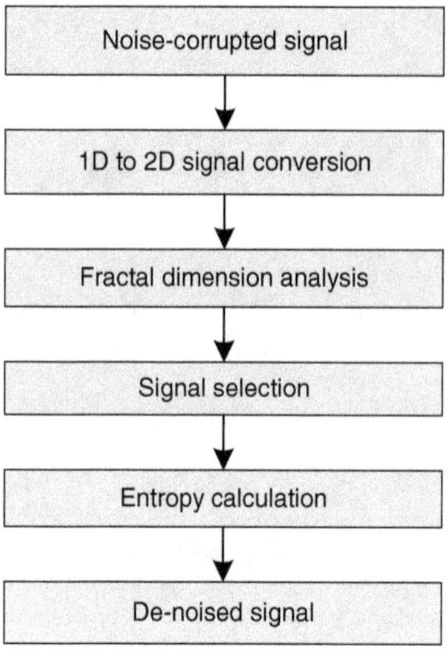

Figure 5.23 Graphics-based PD signal de-noising method.

inside the transformer. To further verify whether the pulses are related to PD events, oil samples were collected from this transformer and DGA was conducted. The DGA analysis using IEC 60599 indicates that there was a possibility of PD occurrence. Moreover, the result from Doernenburg ratios provided an indication of arcing. Thus, the pulses in Figure 5.22(c) have a high possibility of being generated by a PD event. After de-noising, a kurtogram was plotted in Figure 5.22(d). It can be seen that the pattern is different from those in Figures 5.21(e) and (f). This means that the PD event may not be related to the corona. To verify this pattern, a comparison with other kurtogram patterns from different PD sources should be performed.

5.5.1.3 De-noising of Stochastic Noise

To remove stochastic noise, a graphics-based PD signal de-noising method can be adopted. This method is a hybrid of fractal dimension and entropy, as shown in Figure 5.23 [65]. Firstly, an acquired one-dimensional (1D) noise-corrupted signal is transformed into a 2D signal. In this transformation, the original signal in a power cycle is converted into 512 × 512 pixels Then, the fractal dimension is calculated on the transformed 2D signal in each power cycle with different side lengths of boxes, 2^x, where $x = [1,2,\ldots,n]$. If severe noise is present in a particular power cycle, then the value of fractal dimension in the power cycle is different from that of others. Thus, it is able to indicate the presence of noise. Subsequently, the noise can be removed.

Since noise may be located in every or most of the power cycles with similar fractal dimension values, using fractal dimension alone may not be able to completely distinguish and separate PD signals from noise. Moreover, even in situations where fractal dimension can distinguish noise appearing at most of the power cycles, it is still not appropriate to discard all signals located in those power cycles. Therefore, entropy is calculated in each box within each power cycle.

For a signal generated by stochastic noise, if it locates in a particular box in a power cycle, it is unlikely that it will be present in the same box in other power cycles when compared with PD signals, due to the random nature of stochastic noise. Based on this observation, the entropy value of the noise in each box is not as high as that of PD signals. After adding entropy values of the same box for every power cycle together and using colors to represent the sum (the highest to lowest values are represented by a color map ranging from red, orange, yellow, cyan, to blue), PD signals (i.e., de-noised signals) can be identified and subsequently extracted.

5.5.1.3.1 De-noising Results

Figure 5.24 presents de-noising results for a PD measurement performed on the PD source model of internal discharge. From Figure 5.24(a) (original PD signal), it can be observed that the PD signal locates periodically in most of the power cycles. Severe stochastic noise occurs at the first and second power cycles, while relatively insignificant noise can be found at the third power cycle. The stochastic noise was generated by occasional discharges from the HV transformer supplying voltage to the PD measurement and the coupling capacitor in the measurement circuit. This is because both the HV transformer and the coupling capacitor have been used for more than 20 years and cannot be regarded as discharge-free.

Figure 5.24(b) presents normalized results of fractal dimension in each power cycle with different side lengths of boxes. The numbers in the figure represent the power cycle numbers. It can be seen that fractal dimensions from the first to third cycles are separated from those of the other cycles in most of the side lengths. The fractal dimension of the third cycle containing minor noise is close to those of the fourth to tenth cycles. By contrast, the fractal dimensions of the first and second cycles containing severe noise are relatively distant from those of the fourth to tenth cycles. These results can be used to distinguish and separate PD signals from noise.

By rejecting the signals in the first three power cycles and combining the signals in the remaining cycles into one cycle, a PRPD pattern is obtained [Figure 5.24(d)]. Figure 5.24(c) is the original PRPD diagram without applying the proposed de-noising method. It can be seen that the PD pattern in the original PRPD diagram is contaminated by stochastic noise. This demonstrates that the original PRPD diagram cannot fully represent the PD pattern caused by the corresponding insulation defect (i.e., internal discharge), and thus may not be suitable for transformer insulation diagnosis.

The results in Figure 5.24 prove that the fractal dimension used by the proposed method is capable of identifying and removing stochastic noise. It can also indicate the severity of noise by comparing the signal value in each power cycle. For those signals where noise is located in every power cycle and the signals themselves have similar values of fractal dimension, entropy can further eliminate the noise.

Figure 5.25 presents de-noising results for a PD measurement performed on the PD source model of the corona. From the original signal and its magnification [Figures 5.25(a) and (b)], noise can be observed in most of the power cycles. By contrast, the discharge signals [indicated by red arrows, Figure 5.25(b)] are hardly

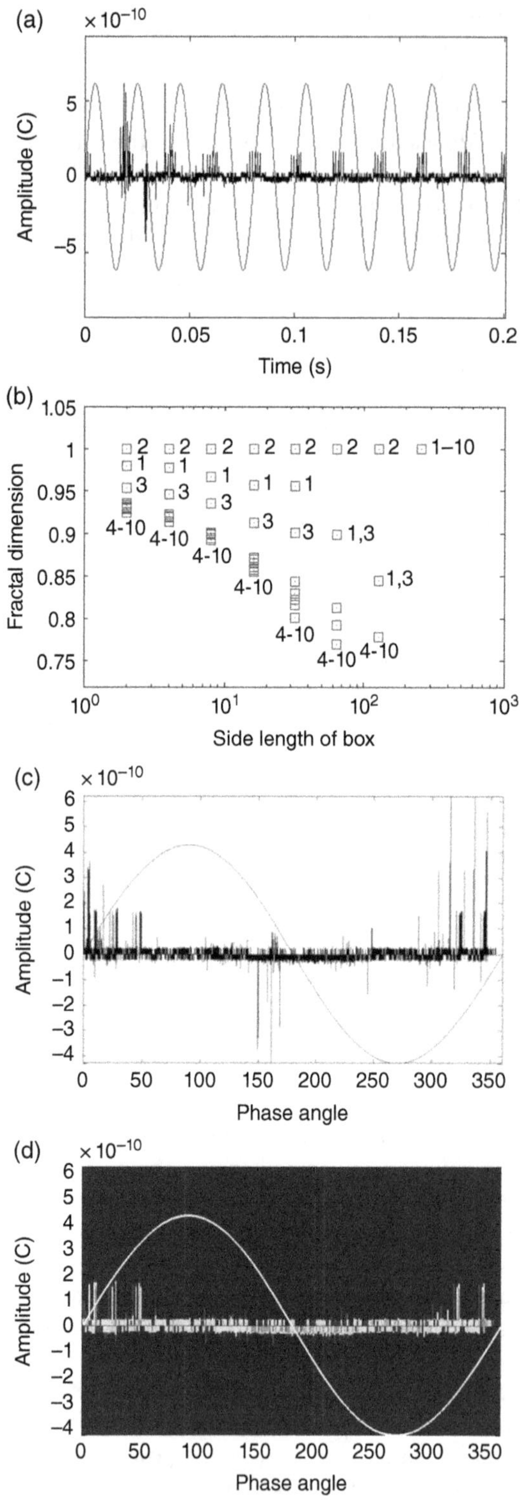

Figure 5.24 De-noising results of internal discharge. (a) Original signal, (b) fractal dimension, (c) original PRPD diagram, and (d) de-noised PRPD diagram.

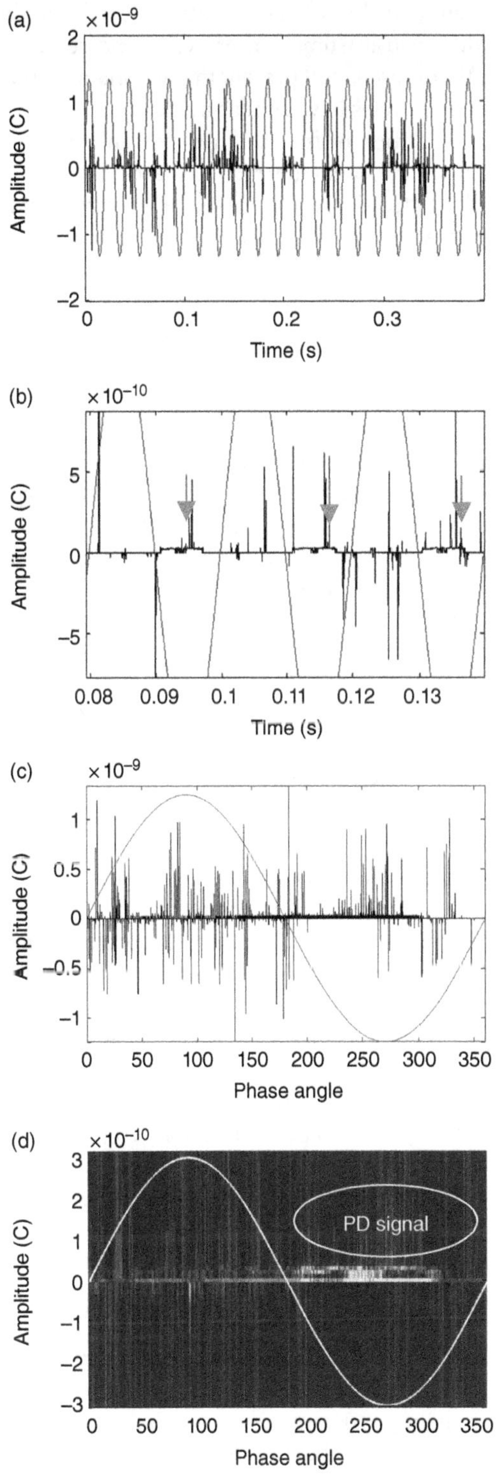

Figure 5.25 De-noising results of corona. (a) Original signal, (b) original signal (magnified), (c) original PRPD diagram, and (d) de-noised PRPD diagram (magnified).

identified due to their relatively small amplitudes. Though some noise can be identified in some power cycles and removed by fractal dimension calculation, other noise is still present in the remaining power cycles (since the signals in these power cycles have similar fractal dimension values). Through calculating entropy in each box of fractal dimension in each cycle and then combining all entropy values into a single cycle, a de-noised PRPD diagram is obtained [Figure 5.25 (d)]. In the figure, the PD signal is clearly identified (the pulses represented in light gray are stochastic noise). However, in the original PRPD diagram without applying the proposed method, the PD signal is submerged in noise [Figure 5.25(c)].

Field measurement and validation. The proposed method was then applied to extract PD signals acquired from a PD measurement on a 5 MVA power transformer at a substation with 22 kV test voltage, as shown in Figure 5.26. It can be seen that the original PD signal is immersed in noise and can hardly be identified [Figures 5.26(a) to (c)]. However, the PD signals can be revealed by using the proposed method [Figure 5.26(d)]. The PD signal located near the negative cycle may be related to the corona, while the PD signal located in both the negative and positive cycles may be generated by internal discharge.

From the above verification of the signal processing techniques on PD signal de-noising, it is showed that a variety of noise can be removed. In the next section, signal processing techniques are applied to separate multiple PD sources that can possibly be embedded in a transformer.

5.5.2 Multiple PD Source Separation

After PD signal extraction (de-noising), multiple PD source separation is performed. In this section, both the TF sparsity map and the conventional TF map (based on equivalent time length and bandwidth) are applied to signals acquired from PD source models and a 100 kVA transformer (10.5 kV/415 V).

Figure 5.27 presents the results of TF sparsity maps and a conventional TF map for separating three PD sources (corona, surface discharge, and discharge in transformer oil). As can be observed from Figure 5.27(a), the conventional TF map is unable to distinguish these three types of PD source. Most of the data points belonging to different types of PD source are merged together. This result indicates that PD pulses cannot be represented accurately by equations (5.20) and (5.21) adopted by the conventional TF map, since some information regarding the shapes of PD pulses is lost. By contrast, the proposed TF sparsity map achieves a very clear separation of the three types of PD source [Figure 5.27(b1)].

In Figure 5.27(b2), the corresponding PRPD diagram indicates that PD pulses in gray are generated by discharge in transformer oil, PD pulses in black are generated by corona, and PD pulses in light gray are generated by surface discharge.

Figure 5.28 shows the results of separating PD sources consisting of corona, surface discharge, and internal discharge. It can be seen from Figure 5.28(a) that the PD sources cannot be clearly distinguished and separated by a conventional TF map. Although three clusters have been formed by DBSCAN, some data points (in light gray) cannot be classified.

By contrast, the three PD sources can clearly be distinguished by the TF sparsity map [Figure 5.28(b1)], in which the data points belonging to different types of PD

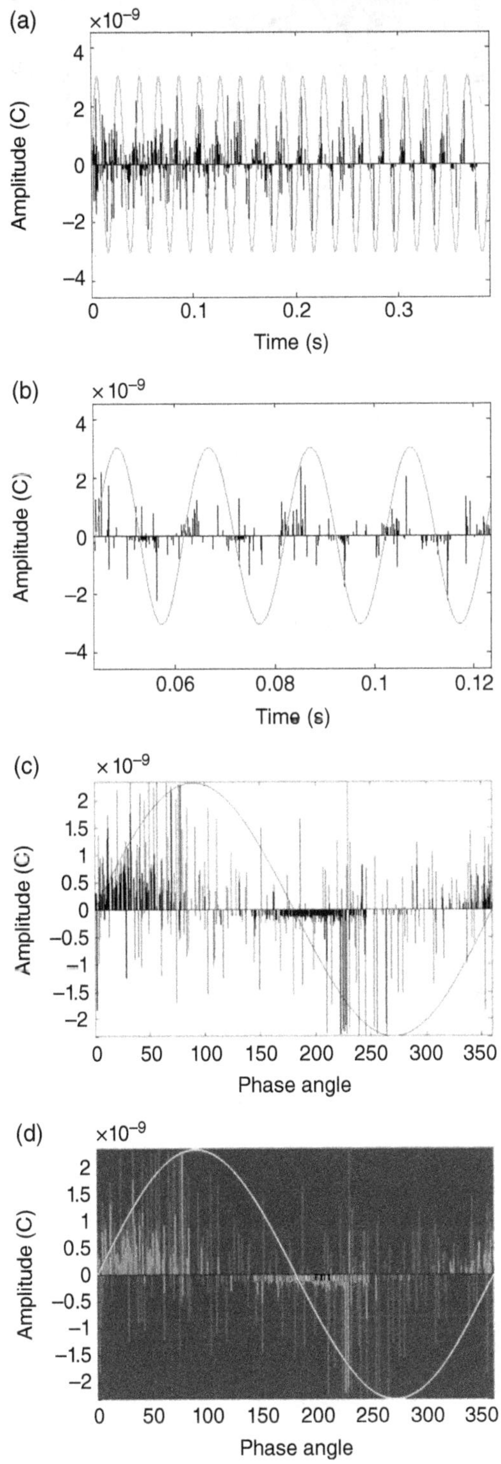

Figure 5.26 De-noising results of PD signals acquired from a transformer. (a) Original signal, (b) original signal (magnified), (c) original PRPD diagram, and (d) de-noised PRPD diagram.

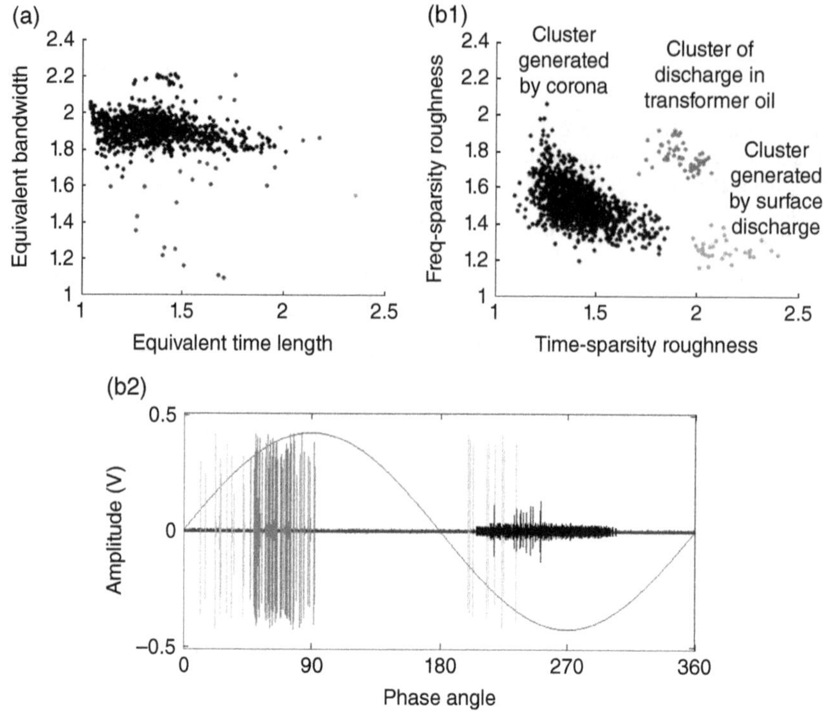

Figure 5.27 Results of multiple PD sources combining corona, surface discharge, and discharge in transformer oil. (a) Result of conventional TF map, (b1) result of TF sparsity map, and (b2) PRPD diagram based on the result of TF sparsity map.

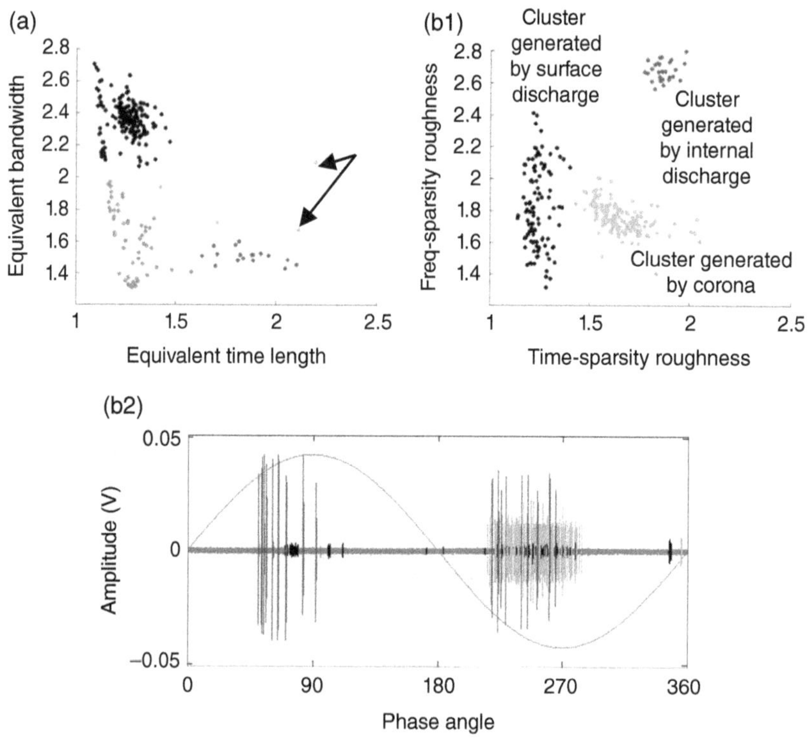

Figure 5.28 Results of multiple PD sources combining corona, surface discharge, and internal discharge. (a) Result of conventional TF map, (b1) result of TF sparsity map, and (b2) PRPD diagram based on the result of TF sparsity map. (*See insert for color representation of the figure.*)

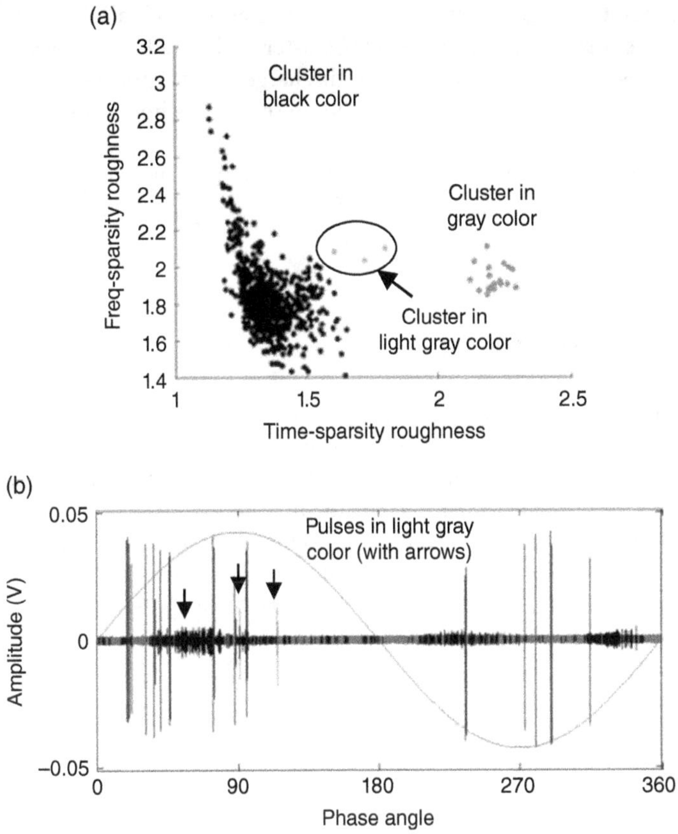

Figure 5.29 Results for a 100 kVA transformer. (a) TF sparsity map and (b) PRPD diagram.

source are well separated. From the PRPD diagram, the PD pulses generated by internal discharge, surface discharge, and corona can be clearly identified. In particular, all three types of PD pulse can be distinguished in the negative power cycle, even if they overlap. Moreover, the TF sparsity map is able to identify the PD pulses generated by surface discharge (in black), even when the amplitudes of these pulses are relatively small and embedded in the PD pulses generated by the corona.

Field measurement and validation. The TF sparsity map has also been applied to analyze signals acquired from an online PD measurement on a 100 kVA transformer (10.5 kV/415 V). Signals were acquired using CT1 clamping on the grounding wire of the transformer. The signals were acquired over 20 power cycles. The results of the 100 kVA transformer are shown in Figure 5.29. From Figure 5.29(a), it is clearly observed that two main clusters are formed on the TF sparsity map. The corresponding distribution of PD pulses shown in the PRPD diagram [Figure 5.29 (b)] may be due to internal and surface discharges. As shown in Figure 5.29(a), a very small cluster on the TF sparsity map between the two clusters (in light gray (circled)) also exists. Since DBSCAN is a density-based clustering method, except for the two clusters in black and gray that are dense enough to form clusters, the data points in light gray are recognized as an individual cluster by DBSCAN. The pulses corresponding to this small cluster are located in the positive cycle, and these pulses may be generated by discharge in transformer oil. In Figure 5.29(b), the signals in dark black are white noise and are not extracted.

The results presented above prove that the TF sparsity map is capable of providing a clear separation of multiple PD sources. After source separation, the next step in PD analysis involves extraction and PD source classification (refer to Figure 5.1). This will be discussed in the next section.

5.6 PD Source Classification

PD source classification is aimed at identifying the types of defect (PD sources) causing discharges in a transformer. Different types of PD source can generate different patterns [77, 78]. The discharge pattern is represented by a set of features, which can be computed using PRPD or directly from measured PD signals. On the basis of these features, it is possible to develop computer algorithms to classify the types of PD source to assist transformer condition assessment. Over the past three decades, considerable efforts have been made to apply pattern recognition algorithms such as genetic algorithms, ANNs, knowledge-based systems, SVMs, and WTs for PD source classification [29, 68, 77–81].

5.6.1 Feature Selection

PD signals after de-noising can be represented as an $N \times D$ dataset $\mathbf{X} = [\mathbf{x}_1, \ldots, \mathbf{x}_N]^T$, where N is the number of PD signals, and each signal is obtained from one PD acquisition and is a D-dimensional vector – i.e., $\mathbf{x}_i = [x_i^1, \ldots, x_i^D]$. The feature selection stage transforms the dataset $\mathbf{X}$ to an $N \times d$ dataset $\mathbf{Y} = [\mathbf{y}_1, \ldots, \mathbf{y}_N]$, $\mathbf{y}_i = [y_i^1, \ldots, y_i^d]$, where $d \ll D$. Though the number of features in dataset $\mathbf{Y}$ is much less than that in dataset $\mathbf{X}$, the features in dataset $\mathbf{Y}$ can still provide essential features to discriminate different PD sources. Feature selection can direct the subsequent pattern recognition algorithm to focus on the most relevant but non-redundant information.

Most feature extraction techniques for PD signal analysis are based on the PRPD diagram [68]. However, the dataset extracted from the PRPD diagram still has considerably high dimensionality. For a PRPD diagram with 200 phase windows, the number of features will be $200 \times 3 = 600$. To deal with the high dimensionality, a number of feature selection approaches including statistical operators, principle component analysis (PCA), kernel principle component analysis (KPCA), and stochastic neighbor embedding (SNE) can be employed to extract representative features from PRPD.

5.6.1.1 Statistical Operators [79]

In this approach, a number of statistical operators are calculated from the three PRPD distributions. The feature set consists of 24 features: (1) skewness and kurtosis for both positive and negative cycles; (2) number of peaks for both positive and negative cycles; (3) asymmetry between positive and negative cycles; and (4) cross-correlation factor of the three PRPD distributions.

5.6.1.2 Principle Component Analysis [79, 82]

PCA maps the original data into a lower-dimensional linear subspace while keeping as much of the variance in the data as possible. PCA solves an eigenvalue problem to find such a linear subspace:

$$\mathrm{cov}(\mathbf{X})\mathbf{M} = \lambda\mathbf{M} \tag{5.37}$$

where cov(**X**) is the covariance matrix of the dataset **X**, λ are the d principal eigenvalues, and **M** is a linear mapping formed by the d principle eigenvectors of cov (**X**). The low-dimensional data representations $\mathbf{y}_i$ of the data points $\mathbf{x}_i$ are computed through a linear mapping $\mathbf{Y} = \mathbf{XM}$. The elements of **Y** (e.g., $\mathbf{y}_i$) form the features set.

5.6.1.3 Kernel PCA [79, 82]

KPCA uses a kernel function to transform the original data into a feature space. The elements of this kernel are in the form

$$K_{ij} = \kappa\left(\mathbf{x}_i, \mathbf{x}_j\right) \tag{5.38}$$

where κ is a kernel function (one commonly used kernel is the Gaussian kernel, refer to Section 5.6.2) and $\mathbf{x}_i$, $\mathbf{x}_j$ are different data points in dataset **X**. KPCA computes the eigenvectors of the covariance matrix $\mathbf{a}_i$ in the above feature space. To obtain the low-dimensional data representation **Y**, the data points in the original dataset **X** are projected onto the eigenvectors of the covariance matrix $\mathbf{a}_i$:

$$\mathbf{y}_i = \left\{ \sum_{j=1}^{N} a_1^{(j)} \kappa\left(\mathbf{x}_i, \mathbf{x}_j\right), \ldots, \sum_{j=1}^{N} a_d^{(j)} \kappa\left(\mathbf{x}_i, \mathbf{x}_j\right) \right\} \tag{5.39}$$

As in PCA, the elements $\mathbf{y}_i$ in dataset **Y** form the features set.

5.6.1.4 Stochastic Neighbor Embedding [79, 82]

SNE is aimed at retaining the pairwise distances between the data points in high-dimensional space. In SNE, the probability p_{ij} that data points $\mathbf{x}_i$ and $\mathbf{x}_j$ are generated by the same Gaussian distribution is computed for all possible pairs of data points in the original dataset **X**. The probabilities q_{ij} of the corresponding data points in low-dimensional space (i.e., $\mathbf{y}_i$, $\mathbf{y}_j$ generated by the same Gaussian distribution) are also computed. These two probability matrices are denoted **P** and **Q**, respectively. SNE minimizes the difference between the probability distributions **P** and **Q** by minimizing the sum of Kullback–Leibler divergences

$$\varphi = \sum_i \sum_j p_{ij} \log \frac{p_{ij}}{q_{ij}} \tag{5.40}$$

As in PCA and KPCA, the elements $\mathbf{y}_i$ form the features set.

Table 5.4 Information on feature datasets

Feature extraction	Information on dataset X	Feature size of dataset X	Feature size of dataset Y
Statistical operator	Phase-resolved PD data in 200 phase windows	600	24
PCA	Phase-resolved PD data in 200 phase windows	600	7
KPCA	Phase-resolved PD data in 200 phase windows	600	7
SNE	Phase-resolved PD data in 200 phase windows	600	7
DWT	PD pulses of 100 power cycles	around 40,000	36

5.6.1.5 Discrete Wavelet Transform [79, 82]

Instead of using the PRPD diagram, DWT utilizes PD signals directly for feature extraction. After being decomposed by DWT, a PD signal is transformed to a set of wavelet coefficients at nine decomposition levels. Accordingly, nine wavelet coefficient distributions are formed. Then, the first four moments (i.e., mean, variance, skewness, and kurtosis) on the distributions of these wavelet coefficients are adopted to form the feature set, which consists of $4 \times 9 = 36$ features [79].

Information on the above datasets **X** and **Y** is presented in Table 5.4.

5.6.2 Pattern Recognition [79]

The pattern recognition algorithm consists of training (learning) and testing (classification) stages. In the training stage, the relationship between data points (e.g., selected features) and their corresponding types of PD source is explored and then approximated using a training database. In the testing stage, a new input data point (not included in the training database) is classified into one type of PD source [79]. Typical pattern recognition algorithms are fuzzy support vector machine (FSVM), radial basis network (RBF), multi-layer perceptron (MLP), Bayesian classifier, k nearest neighbor (kNN), and two-layer (input–output) network. The formulation of FSVM is introduced briefly in the following section. The formulations of RBF, MLP, Bayesian classifier, kNN, and two-layer network can be found in Chapter 4. For a detailed derivation of these algorithms, readers may refer to Refs [82–84].

FSVM is a variant of SVM [84]. In FSVM, each data point is assigned a weight, which describes the degree to which this data point belongs to different types of PD source (classes). These weights are augmented to the original training dataset **Y** in the form $\{\mathbf{y}_k, C_k, \rho_k\}$, where $0 < \rho_k < 1 < 1$ are the weights, determined using an exponentially decaying function [79, 84]

$$\rho_k = \frac{2}{1 + \exp\left(\beta\, d_k^{\text{cen}}\right)}, \beta \in [0,1] \tag{5.41}$$

where $d_k^{\text{cen}} = \mathbf{y}_k - \mathbf{v}_i^{1/2}$ is the Euclidean distance between a data point $\mathbf{y}_k$ and its cluster center $\mathbf{v}_i$, which is the geometrical mean of the data points belonging to the same type of PD source. β determines the steepness of the decay.

The FSVM algorithm converts the input data from its original data space R^m to a higher-dimensional space through a nonlinear function $h = \Phi(x)$. Then, in that space the FSVM searches for an optimal separation hyperplane by solving a quadratic programming (QP) problem [84]

$$\text{Minimize} \quad \frac{1}{2}\mathbf{w}^T\mathbf{w} + G\sum_{k=1}^{N}\rho_k\xi_k \tag{5.42}$$

$$\text{Subject to} \quad z_k\left(\mathbf{w}^T \cdot \Phi(\mathbf{y}_k) + b\right) \geq 1 - \xi_k,\ k = 1, \ldots, N \tag{5.43}$$

$$\text{and} \quad \xi_k \geq 0,\ k = 1, \ldots, N \tag{5.44}$$

where the pair ($\mathbf{w}$, b) defines the separation hyperplane, in which $\mathbf{w}$ is the hyperplane's normal vector while b is the bias. G, the regularization parameter, is used to balance the margin maximization and misclassification; $\xi_k \geq 0$ is the error term due to misclassification.

The dual form of the above QP problem is [84]

$$\text{Maximize} \quad \sum_{k=1}^{N}\alpha_k - \frac{1}{2}\sum_{k=1}^{N}\sum_{j=1}^{N}\alpha_k\alpha_j z_k z_j K\left(\mathbf{y}_k, \mathbf{y}_i\right) \tag{5.45}$$

$$\text{Subject to} \quad \sum_{k=1}^{N}\alpha_k z_k = 0 \tag{5.46}$$

$$\text{and } 0 \leq \alpha_k \leq \rho_k G \tag{5.47}$$

where α_k is the Lagrange multiplier and $K(\mathbf{y}_k, \mathbf{y}_i)$ is the kernel function in the form $K\left(\mathbf{y}_k, \mathbf{y}_i\right) = \Phi\left(\mathbf{y}_k\right)^T \Phi(\mathbf{y}_i)$. The Gaussian kernel $K\left(\mathbf{y}_k, \mathbf{y}_i\right) = e^{-\gamma\|\mathbf{y}_k - \mathbf{y}_i\|^2}$ is adopted in PD source classification, where γ is the variance parameter.

After obtaining the maximized hyperplane, for a new data point $\mathbf{y}_*$ the FSVM can predict its class label z_*:

$$z_* = \operatorname{sgn}\left[\sum_{k=1}^{N}\alpha_k z_k K\left(\mathbf{y}_k, \mathbf{y}_*\right) + b\right] \tag{5.48}$$

5.6.3 Case Studies

5.6.3.1 Feature Selection and Data Preparation [79]

In the case studies, the original dataset is obtained from PD measurements (capacitive measurement system) on a number of different experimental PD source models, which include corona, discharge in oil, surface discharge, internal discharge, and discharge due to floating particles. The database has 1000 data points, with each data point corresponding to one PD acquisition, which is made up of PD pulses of 100 AC cycles. Firstly, the original dataset is converted to the PRPD dataset (i.e., dataset $\mathbf{X}$), which consists of average discharge pulse magnitude, maximum discharge pulse magnitude, and discharge number count in each of the

200 phase windows. Then, the PRPD dataset is processed by feature extraction algorithms to form a feature dataset (i.e., dataset **Y**). Such a feature dataset is then used by pattern recognition algorithms for PD source classification. For feature extraction using DWT, the original dataset is processed directly without converting to PRPD, and the first four moment statistics are calculated to form the feature dataset. Information on the above PRPD dataset and feature dataset can be found in Table 5.4.

To evaluate the pattern recognition algorithms, dataset **Y** is randomly split into two parts: a training dataset (having 70% data points in dataset **Y**) and a testing dataset (having 30% data points in dataset **Y**). The training dataset is used to find the model parameters in FSVM and ANNs. These parameters are: number of neighbors in *k*NN, number of hidden nodes in MLP and RBF, and number of mixture components in Bayesian classifiers, G and γ in FSVM. Tenfold cross-validation on the training dataset is conducted to obtain these parameters [79]. Finally, the trained algorithms make classification on the data points in the testing dataset into one of the PD sources (classes).

5.6.3.1.1 Results

Tables 5.5 to 5.9 summarize the classification results of six pattern recognition algorithms integrated with different feature extraction approaches. Each table presents the overall classification accuracy (%) and the classification accuracy with respect to each type of PD source.

From these tables, it can be seen that FSVM outperforms the other five ANN algorithms. Amongst the ANN algorithms, MLP also attains desirable and consistent classification accuracy. It is also not strange that the two-layer network attains lower classification accuracy, since its simple architecture may not be able to capture the nonlinearity of the dataset. It is interesting to see that *k*NN performs well with the feature extraction approaches PCA, KPCA, and DWT. A similar phenomenon of *k*NN has also been reported in Ref. [85].

The tables also reveal that *k*NN, two-layer network, MLP, RBF, and FSVM integrated with DWT attain higher classification accuracy compared to these algorithms integrated with the other four feature extraction approaches. *k*NN,

Table 5.5 Classification results (%) of algorithms (with statistical operators) [79]

Algorithm	Overall	Class 1	Class 2	Class 3	Class 4	Class 5
*k*NN	90.8	99.4	83.8	86.7	99.5	85.2
Two-layer	90.6	94.7	87.2	86.1	97.6	88.1
RBF	92.5	98.5	87.4	90.1	97.8	89.2
MLP	92.0	98.1	84.8	91.6	99.1	86.6
Bayesian	93.2	97.5	85.6	91.0	99.3	93.3
FSVM	**95.1**	99.5	92.0	92.2	99.4	92.7

Note: Class 1 – corona; Class 2 – discharge in oil; Class 3 – surface discharge; Class 4 – internal discharge; Class 5 – discharge due to floating particle.

Table 5.6 Classification results (%) of algorithms (with DWT) [79]

Algorithm	Overall	Class 1	Class 2	Class 3	Class 4	Class 5
*k*NN	94.2	99.2	91.4	96.0	98.3	85.4
Two-layer	92.6	93.3	80.8	94.2	95.3	91.8
RBF	97.3	98.8	97.5	97.2	98.3	94.3
MLP	98.5	97.9	99.6	99.4	98.2	97.2
Bayesian	92.0	92.8	94.2	87.8	99.7	85.8
FSVM	**98.8**	97.4	99.4	98.9	100.0	98.3

Table 5.7 Classification results (%) of algorithms (with PCA) [79]

Algorithm	Overall	Class 1	Class 2	Class 3	Class 4	Class 5
*k*NN	93.7	100.0	86.4	98.1	100.0	83.3
Two-layer	81.4	100.0	69.7	75.9	100.0	61.0
RBF	91.2	93.9	86.5	91.7	99.8	84.5
MLP	93.6	100.0	83.3	97.8	99.8	81.2
Bayesian	90.1	99.9	88.2	90.6	100.0	74.3
FSVM	**96.5**	99.9	93.4	99.7	99.8	89.1

Table 5.8 Classification results (%) of algorithms (with KPCA) [79]

Algorithm	Overall	Class 1	Class 2	Class 3	Class 4	Class 5
*k*NN	93.9	100.0	90.3	96.0	99.8	82.8
Two-layer	83.2	98.9	74.2	82.8	99.3	59.6
RBF	91.7	96.9	87.8	91.7	100.0	82.0
MLP	93.8	100.0	87.9	99.0	99.5	81.3
Bayesian	92.4	99.9	87.9	95.3	99.1	78.4
FSVM	**96.1**	99.8	92.5	99.7	99.9	88.1

Table 5.9 Classification results (%) of algorithms (with SNE) [79]

Algorithm	Overall	Class 1	Class 2	Class 3	Class 4	Class 5
*k*NN	87.2	97.6	79.4	85.8	93.2	79.9
Two-layer	28.7	43.2	55.1	6.9	34.7	0
RBF	85.2	92.4	76.7	82.5	91.7	83.7
MLP	89.4	99.4	85.7	83.3	99.3	79.3
Bayesian	85.9	94.1	83.8	80.0	94.5	77.0
FSVM	**92.3**	100.0	86.3	91.3	99.2	84.5

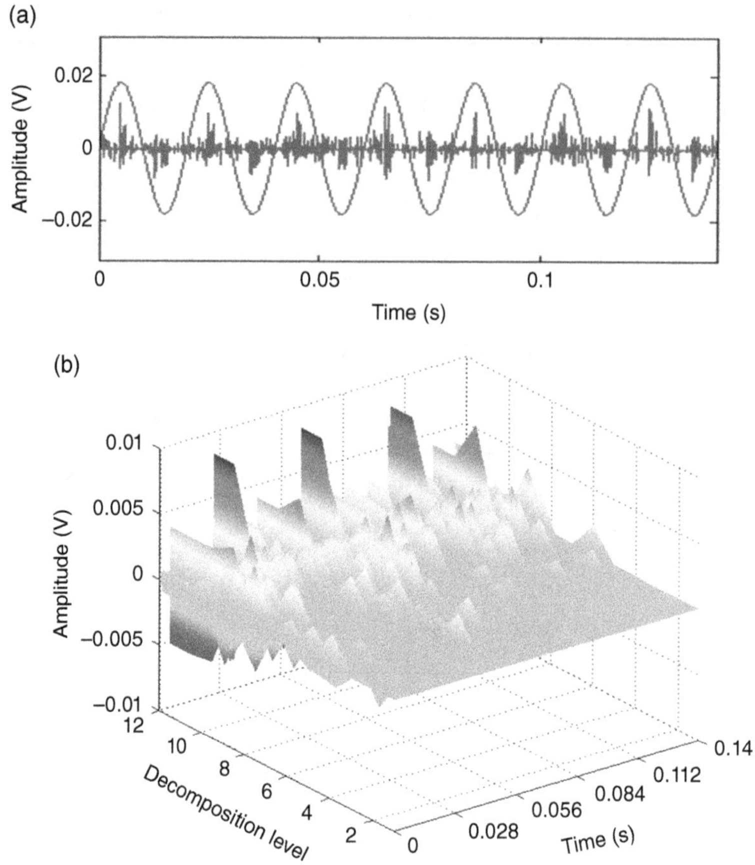

Figure 5.30 PD measurement results of a 10 kVA distribution transformer: (a) measured PD signals; (b) DWT decomposition.

MLP, RBF, Bayesian classifier, and FSVM integrated with KPCA and PCA attain comparable classification accuracy as they are integrated with statistical operators. It is worth mentioning that KPCA and PCA only use seven features, while the statistical operators approach uses 24 features. The pattern recognition algorithms integrated with SNE could not attain desirable classification accuracy. The two-layer network integrated with SNE even fails to make classification. This might be due to the optimization in SNE being trapped into local minima in the cost function [86].

In Figure 5.30, PD patterns of a 10 kVA (10.5 kV/415 V) distribution transformer are presented. An inductive PD measurement system using HFCT was employed for PD measurement of this transformer. The above trained FSVM algorithm (using features extracted by DWT) was employed to recognize the types of PD source inside this transformer. Results indicate that this transformer has 80% possibility of internal discharge, 15% possibility of surface discharge, and 5% possibility of other type discharges.

To achieve the desirable capability of the pattern recognition algorithms in PD source classification, the training database needs to include as many types of PD source as possible. It is also worth mentioning that some other factors also need to be considered when applying the algorithms to PD source classification. For example, the resolution of the PD measurement system has an influence on the overall performance of the algorithms. If the algorithms are trained with PD signals of full resolution and tested on the squashed (distorted) PD signals, then the overall performance of the algorithms will be decreased significantly.

5.7 Acoustic and UHF Methods for PD Signal Detection and Localization

5.7.1 Acoustic PD Measurement for PD Source Localization [87, 88]

5.7.1.1 Sound Wave Propagation

In acoustic PD measurement, a number of acoustic sensors are employed to estimate the location of the PD source inside a transformer. The estimation is made based on the distance between the PD source and the acoustic sensors mounted on the transformer's tank through measuring the time of arrival (TOA) – that is, the elapsed time of sound (generated by discharge) traveling from the PD source to the sensors.

To measure TOA, it is important to identify the starting time instant of the arriving sound waveform acquired by the acoustic sensors. This requires a trigger signal to evoke acoustic sensors to start recording PD events. As shown in Figure 5.31, an HFCT-based PD measurement system can be used to provide such a trigger signal. This is because HFCT measures the electrical signals generated by a PD source, which are much faster than the sound signals generated by the same PD source. Then, the absolute TOA of an acoustic sensor can be obtained by subtracting the arrival time measured by this sensor from the reference time at which the HFCT detects the PD event.

The sound wave generated by a PD source can propagate along different paths inside a transformer and thus there may exist different TOAs [89]. In Figure 5.32, two extreme sound wave travel paths are depicted: (1) the shortest path, in which a sound wave travels only via transformer oil and reaches the acoustic sensors; and (2) the longest path, in which the sound wave first travels in transformer oil (perpendicular to the transformer's tank) and then enters the transformer tank wall and propagates until it reaches the acoustic sensors. Other possible sound wave propagation paths can exist between the above longest and shortest wave paths.

5.7.1.2 Spatial Intersection Technique for Locating PD Source

In the spatial intersection technique, two spheres for each acoustic sensor can be formed. The radii of these two spheres are decided by: (1) the longest distance that the sound wave can travel; and (2) the shortest distance that the sound wave can travel. The sound wave propagation velocity in oil is set to 1400 m/s, and it is set to

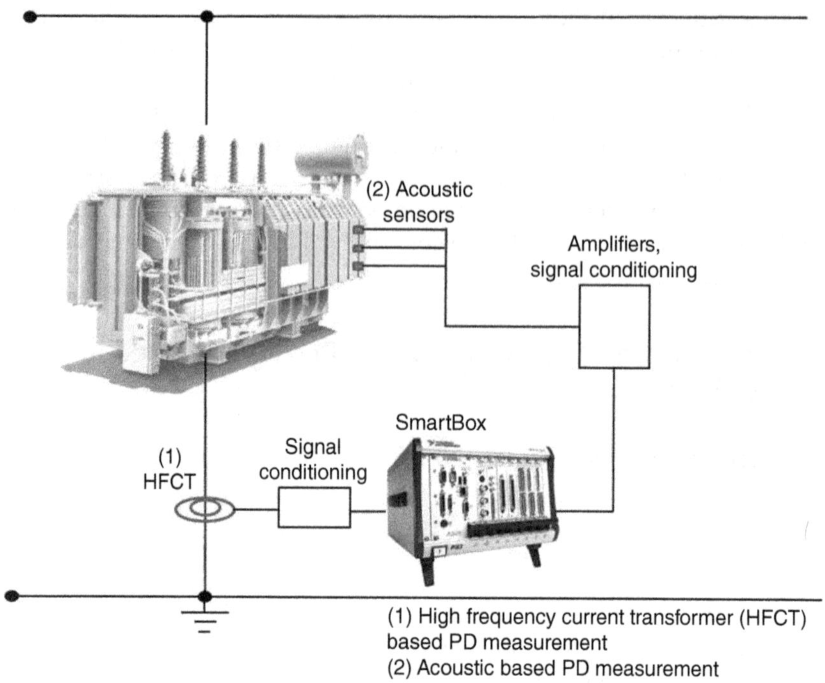

Figure 5.31 Setup of acoustic PD measurement of a transformer.

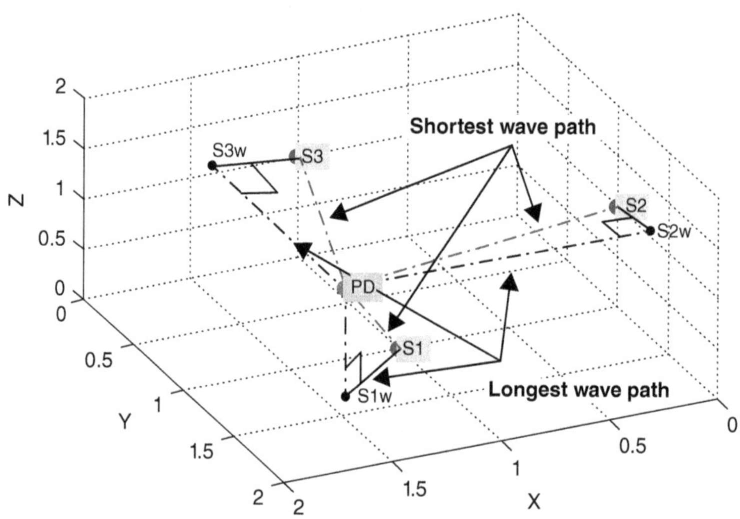

Figure 5.32 Possible sound wave travel paths from PD source to three acoustic sensors.

5900 m/s for longitudinal waves and 3200 m/s for shear waves in the transformer tank wall [89].

Accordingly, each acoustic sensor will have two spheres, one sphere has a small diameter and the other sphere has a large diameter. The intersection area of three acoustic sensors will be the possible location of a PD source.

5.7.1.3 Case Study

The above technique was adopted for localizing a PD source inside a transformer, of which the dimensions are 78 cm in the *X*-axis, 36.2 cm in the *Y*-axis, and 119.4 cm in the *Z*-axis. The transformer and the locations of three acoustic sensors are shown in Figure 5.33. Each AE sensor had an operating frequency range of 50–400 kHz and the output signal of the sensor was amplified by an external preamplifier by 60 dB. The measurement setup was the same as that depicted in Figure 5.31. An internal discharge experimental model was placed inside the transformer tank to generate PD signals.

The graphical visualization is shown in Figure 5.34. Figure 5.34(a) depicts the *X*–*Y*–*Z* 3D view of the intersection area of three acoustic sensors. The projections of the intersection areas on three planes (*X*–*Y*, *Y*–*Z*, and *X*–*Z*) are depicted in Figures 5.34(b)–(d), respectively. In the figures, the gray line denotes the center of the intersection volume by each set of two sensors. Line ① is a central line of an overlapped volume by sensors 1 and 2, line ② by sensors 1 and 3, and line ③ by sensors 2 and 3. The dash line arrow indicates the possible PD location. Among these, line ② by sensors 1 and 3 is the most significant since it appears on all three projection planes.

From the above results, the PD source is estimated at the point $X = 25$ cm, $Y = 18$ cm, $Z = 30$ cm. In Figure 5.35, this point is drawn on the transformer tank as a gray star mark. The error lies within 10% of full length in each axis. The error is thought to be caused mainly by the inaccuracy of sound wave velocity and triggering time. However, the result indicates a reasonably close estimation of the location to the actual PD source.

5.7.2 UHF PD Measurement

Except for the acoustic methods, UHF can also be used for PD source localization. There are different types of UHF sensor that can be used in power transformers.

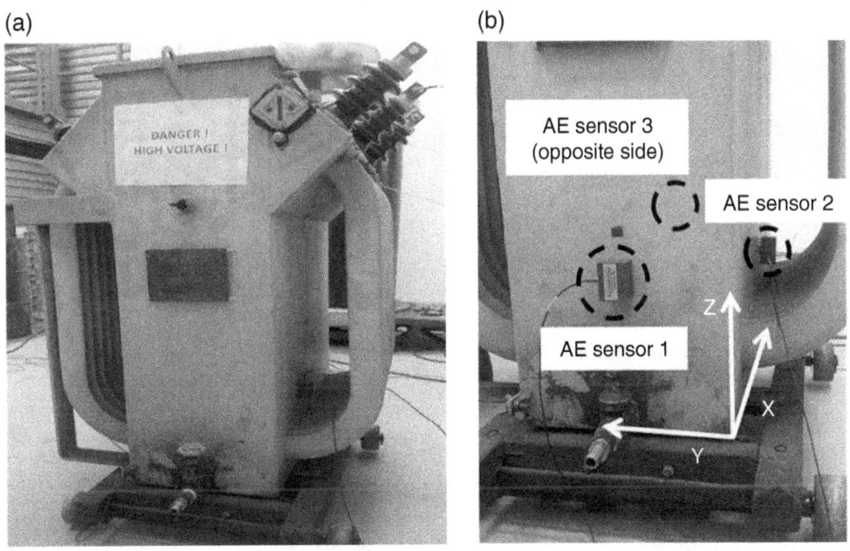

Figure 5.33 Tested transformer tank and AE sensor installation.

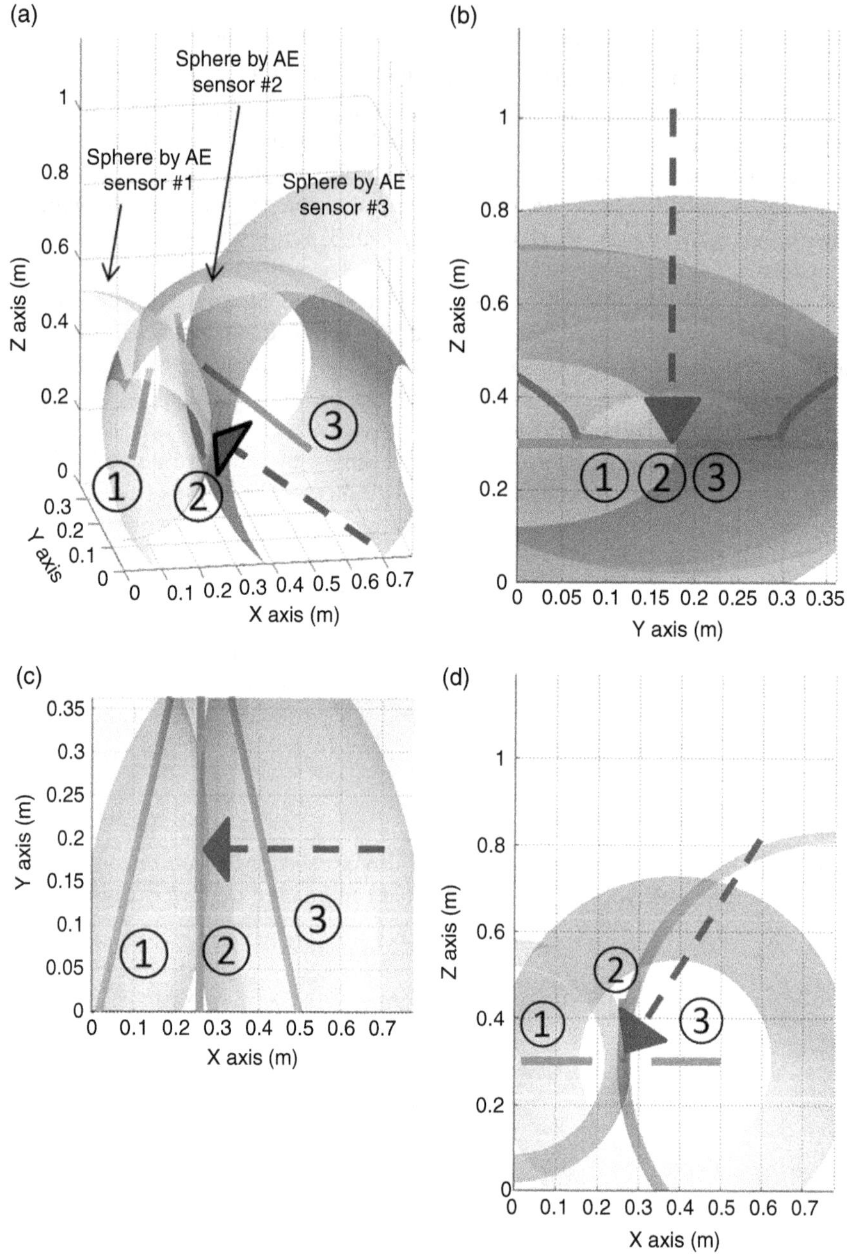

Figure 5.34 The results of spatial intersectional PD localization: (a) *X–Y–Z* axis view; (b) *Y–Z* axis view; (c) *X–Y* axis view; and (d) *X–Z* axis view.

Internal disc couplers are used only for pre-existing locations of transformers specified by manufacturers, while external disc couplers can be installed in existing dielectric windows [90]. An oil-valve probe provides a convenient and online means of installation by putting the sensor into the transformer tank through

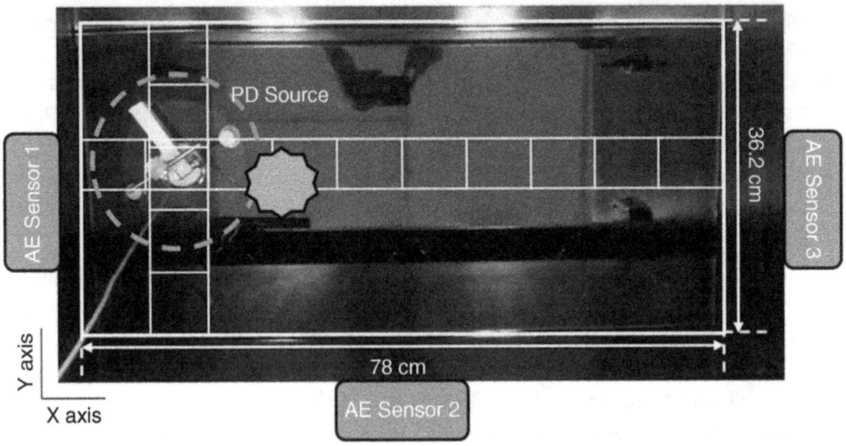

Figure 5.35 The final PD source localization result.

the oil valve. One of the advantages of adopting UHF detection is associated with greater immunity to noise, since the noise in substations is primarily below a few megahertz [26, 91, 92], except for the noise sources from communication systems. A disadvantage of UHF detection is that it is less flexible, since UHF sensors and associated accessories are usually specific to a particular type of equipment [28].

5.8 Summary

In this chapter, a comprehensive PD signal analysis from various PD measurement methods to PD signal de-noising and finally PD source classification and localization has been presented. Several newly developed signal processing techniques have also been detailed. The signal processing methods have been verified by case studies using PD signals acquired from PD source models and substation transformers to indicate their effectiveness on PD signal analysis in practice.

References

1 Jarman P, Hooton R, Walker L, Zhong Q, Ishak T, Wang Z. Transformer life prediction using data from units removed from service and thermal modelling. CIGRE Session 2010, Paris.
2 Picher P, Riendeau S, Gauvin M, Léonard F, Dupont L, Goulet J *et al.* New technologies for monitoring transformer tap-changers and bushings and their integration into a modern IT infrastructure CIGRE Session 2012, Paris.
3 CountryEnergy's Electricity Network Regulatory Proposal 2009–2014, 2008.
4 Grechko O, Kalacheva I. Current trends in the development of in-service monitoring and diagnostic systems for 110–750 kV power transformers. *Applied*

Energy Research: Russian Journal of Fuel, Power, and Heat Systems 1996; 34: 84–97.

5 Lee K, Schneider J. Rockport transient voltage monitoring system: Analysis and simulation of record waveforms. *IEEE Transactions on Power Delivery* 1989; 4: 1794–1805.

6 Bartley W. Analysis of transformer failures. Paper presented at the 36th Annual Conference of the International Association of Engineering Insurers, Stockholm, Sweden, 2003.

7 Minhas M, Reynders J, Klerk PD. Failures in power system transformers and appropriate monitoring techniques. Paper presented at the 11th International Symposium on High Voltage Engineering, London, 1999.

8 Khawaja R. Monitoring of high voltage oil-impregnated transformer insulation. Ph.D. dissertation, University of New South Wales, 2006.

9 Lai K, Phung B, Blackburn T. Application of data mining on partial discharge. Part I: Predictive modelling classification. *IEEE Transactions on Dielectrics and Electrical Insulation* 2010; 17: 846–854.

10 IEEE Std. 60270. High-Voltage Test Techniques – Partial Discharge Measurements, 2000.

11 Ma X, Zhou C, Kemp I. Automated wavelet selection and thresholding for PD detection. *IEEE Electrical Insulation Magazine* 2002; 18: 37–45.

12 Zhou X, Zhou C, Kemp I. An improved methodology for application of wavelet transform to partial discharge measurement denoising. *IEEE Transactions on Dielectrics and Electrical Insulation* 2005; 12: 586–594.

13 Ma X, Zhou C, Kemp I. Interpretation of wavelet analysis and its application in partial discharge detection. *IEEE Transactions on Dielectrics and Electrical Insulation* 2002; 9: 446–457.

14 Zhang H, Blackburn T, Phung B, Sen D. A novel wavelet transform technique for on-line partial discharge measurements. Part 1: WT de-noising algorithm. *IEEE Transactions on Dielectrics and Electrical Insulation* 2007; 14: 3–14.

15 Zhang H. Advanced techniques application of on-line partial discharge detection in power cables. Ph.D. dissertation, University of New South Wales, 2006.

16 Sriram S, Nitin S, Prabhu K, Bastiaans M. Signal denoising techniques for partial discharge measurements. *IEEE Transactions on Dielectrics and Electrical Insulation* 2005; 12: 1182–1191.

17 Cho N, Choi C, Lee S. Adaptive line enhancement by using an IIR lattice notch filter. *IEEE Transactions on Acoustics, Speech, and Signal Processing* 1989; 37: 585–589.

18 Happe S, Kranz H, Krause W. Advanced suppression of stochastic pulse shaped partial discharge disturbances. *IEEE Transactions on Dielectrics and Electrical Insulation* 2005; 12: pp. 265–275.

19 Muhr M, Strehl T, Gulski E, Feser K, Gockenbach E, Hauschild W *et al.* Sensors and sensing used for non-conventional PD detection. CIGRE Session 2006, Paris.

20 Lai K. *Application of data mining techniques for characterisation and recognition of partial discharge*. Ph.D. dissertation, University of New South Wales, 2010.

21 Macedo E, Araújo D, Costa E, Freire R, Lopes W, Torres I *et al.* Wavelet transform processing applied to partial discharge evaluation. Paper presented at the 25th International Congress on Condition Monitoring and Diagnostic Engineering Management, Huddersfield, 2012.

22 WG21.16. Partial Discharge Detection in Installed HV Extrude Cable Systems, 2001.

23 Kheirmand A, Leijon M, Gubanski S. New practices for partial discharge detection and localization in large rotating machines. *IEEE Transactions on Dielectrics and Electrical Insulation* 2003; 10: 1042–1052.

24 Liu Y, Lin F, Zhang Q, Zhong H. Design and construction of a Rogowski coil for measuring wide pulsed current. *IEEE Sensors Journal* 2011; 11: 123–130.

25 Rubio-Serrano J, Rojas-Moreno M, Posada J, Martinez-Tarifa J, Robles G, Garcia-Souto J. Electro-acoustic detection, identification and location of partial discharge sources in oil–paper insulation systems. *IEEE Transactions on Dielectrics and Electrical Insulation* 2012; 19: 1569–1578.

26 Sinaga H. Detection, identification and localization of partial discharges in power transformers using UHF techniques. Ph.D. dissertation, University of New South Wales, 2012.

27 Stone G. Partial discharge diagnostics and electrical equipment insulation condition assessment. *IEEE Transactions on Dielectrics and Electrical Insulation* 2005; 12: 891–903.

28 Chakravorti S, Dey D, Chatterjee B. *Recent Trends in the Condition Monitoring of Transformers: Theory, Implementation and Analysis.* Berlin: Springer, 2013.

29 Jin X. Classification of partial discharges of power transformers with neural network applications. Ph.D. dissertation, University of Southern California, 2000.

30 Su C. *Electromagnetic Transients in Transformer and Rotating Machine Windings.* Pennsylvania, PA: IGI Global, 2013.

31 Chan J, Ma H, Saha T, Ekanayake C. Self-adaptive partial discharge signal de-noising based on ensemble empirical mode decomposition and automatic morphological thresholding. *IEEE Transactions on Dielectrics and Electrical Insulation* 2014; 21: 294–303.

32 Chan J, Ma H, Saha T. Time-frequency sparsity map on automatic partial discharge sources separation for power transformer condition assessment. *IEEE Transactions on Dielectrics and Electrical Insulation* 2015; 22(4): 2271–2283.

33 Mallat S. *A Wavelet Tour of Signal Processing: The sparse way.* New York: Academic Press, 2009.

34 Salomon D, Motta G, Bryant D. *Data Compression: The complete reference,* 4th edn. London: Springer, 2007.

35 Gaouda A, Ei-Hag A, Abdel-Galil T, Salama M, Bartnikas R. On-line detection and measurement of partial discharge signals in a noisy environment. *IEEE Transactions on Dielectrics and Electrical Insulation* 2008; 15: 1162–1173.

36 Li J, Jiang T, Grzybowski S, Cheng C. Scale dependent wavelet selection for de-noising of partial discharge detection. *IEEE Transactions on Dielectrics and Electrical Insulation* 2010; 17: 1705–1714.

37 Zhang H, Blackburn T, Phung B, Sen D. A novel wavelet transform technique for on-line partial discharge measurements. Part 2: On-site noise rejection application. *IEEE Transactions on Dielectrics and Electrical Insulation* 2007; 14: 15–22.

38 Song X, Zhou C, Hepburn D, Zhang G. Second generation wavelet transform for data denoising in PD measurement. *IEEE Transactions on Dielectrics and Electrical Insulation* 2007; 14: 1531–1537.

39 Zhongrong X, Ju T, Caixin S. Application of complex wavelet transform to suppress white noise in GIS UHF PD signals. *IEEE Transactions on Power Delivery* 2007; 22: 1498–1504.
40 Dey D, Chatterjee B, Chakravorti S, Munshi S. Cross-wavelet transform as a new paradigm for feature extraction from noisy partial discharge pulses. *IEEE Transactions on Dielectrics and Electrical Insulation* 2010; 17: 157–166.
41 Chang C, Jin J, Kumar S, Su Q, Hoshino T, Hanai M *et al.* Denoising of partial discharge signals in wavelet packets domain. *IET Science, Measurement and Technology* 2005; 152: 129–140.
42 Huang N, Shen Z, Long S, Wu M, Shih H, Zheng Q *et al.* The empirical mode decomposition and the Hilbert spectrum for nonlinear and non-stationary time series analysis. *Proceedings of the Royal Society London, Series A* 1998; 454: 903–995.
43 Chan J, Tse P. A novel, fast, reliable data transmission algorithm for wireless machine health monitoring. *IEEE Transactions on Reliability* 2009; 58: 295–304.
44 Wu Z, Huang N. Ensemble empirical mode decomposition: A noise-assisted data analysis method. *Advances in Adaptive Data Analysis* 2009; 1: 1–41.
45 Zhang L, Xu J, Yang J, Yang D, Wang D. Multiscale morphology analysis and its application to fault diagnosis. *Mechanical Systems and Signal Processing* 2008; 22: 597–610.
46 Nikolaou N, Antoniadis I. Application of morphological operators as envelope extractors for impulsive-type periodic signals. *Mechanical Systems and Signal Processing* 2003; 17: 1147–1162.
47 Dong Y, Liao M, Zhang X, Wang F. Faults diagnosis of rolling element bearings based on modified morphological method. *Mechanical Systems and Signal Processing* 2011; 25: 1276–1286.
48 Chan J, Ma H, Saha T. Automatic blind equalization and thresholding for partial discharge measurement in power transformer. *IEEE Transactions on Power Delivery* 2014; 29: 1927–1938.
49 Jelonnek B, Boss D, Kammeyer K. Generalized eigenvector algorithm for blind equalization. *Signal Processing* 1997; 61: 237–264.
50 Shalvi O, Weinstein E. New criteria for blind deconvolution of nonminimum phase systems (channels). *IEEE Transactions on Information Theory* 1990; 36: 312–321.
51 Cavallini A, Chen X, Montanari G, Ciani F. Diagnosis of EHV and HV transformers through an innovative partial-discharge-based technique. *IEEE Transactions on Power Delivery* 2010; 25: 814–824.
52 Cavallini A, Montanari G. A new methodology for the identification of PD in electrical apparatus: Properties and applications. *IEEE Transactions on Dielectrics and Electrical Insulation* 2005; 12: 203–215.
53 Contin A, Cavallini A, Montanari G, Pasini G, Puletti F. Digital detection and fuzzy classification of partial discharge signals. *IEEE Transactions on Dielectrics and Electrical Insulation* 2002; 9: 335–348.
54 Cavallini A, Montanari G, Contin A, Puletti F. A new approach to the diagnosis of solid insulation systems based on PD signal inference. *IEEE Electrical Insulation Magazine* 2003; 19: 23–30.
55 Ester M, Kriegel H, Sander J, Xu X. A density-based algorithm for discovering clusters in large spatial databases with noise. Proceedings of the 2nd International Conference on Knowledge Discovery and Data Mining, Portland, OR, 1996.

56 Hao L, Lewin P, Hunter J, Swaffield D, Contin A, Walton C *et al.* Discrimination of multiple PD sources using wavelet decomposition and principal component analysis. *IEEE Transactions on Dielectrics and Electrical Insulation* 2011; 18: 1702–1711.

57 Daszykowski M, Walczak B, Massart D. Looking for natural patterns in data. Part 1. Density-based approach. *Chemometrics and Intelligent Laboratory Systems* 2001; 56: 83–92.

58 Hoof M, Patsch R. Pulse-sequence analysis: A new method for investigating the physics of PD-induced ageing. *IET Science, Measurement and Technology* 1995; 142: 95–101.

59 Kranz H. PD pulse sequence analysis and its relevance for on-site PD defect identification and evaluation. *IEEE Transactions on Dielectrics and Electrical Insulation* 2005; 12: 276–284.

60 Antoni J. The spectral kurtosis: A useful tool for characterising non-stationary signals. *Mechanical Systems and Signal Processing* 2006; 20: 282–307.

61 Antoni J. Fast computation of the kurtogram for the detection of transient faults. *Mechanical Systems and Signal Processing* 2007; 21: 108–124.

62 Randall R. *Vibration-Based Condition Monitoring: Industrial, Aerospace and Automotive Applications: A*. New York: Wiley, 2011.

63 IEEE St. 61198. Mineral Insulating Oils – Methods for the Determination of 2-Furfural and Related Compounds, 1993.

64 Li J, Du Q, Sun C. An improved box-counting method for image fractal dimension estimation. *Pattern Recognition* 2009; 42: 2460–2469.

65 Chan J, Ma H, Saha T, Ekanayake C. Stochastic noise removal on partial discharge measurement for transformer insulation diagnosis. Paper presented at the IEEE Power & Energy Society General Meeting, Washington, D.C., 2014.

66 Li J, Jiang T, Harrison R, Grzybowski S. Recognition of ultra high frequency partial discharge signals using multi-scale features. *IEEE Transactions on Dielectrics and Electrical Insulation* 2012; 19: 1412–1420.

67 Gu F, Chang H, Kuo C. Gas-insulated switchgear PD signal analysis based on Hilbert–Huang transform with fractal parameters enhancement. *IEEE Transactions on Dielectrics and Electrical Insulation* 2013; 20: 1049–1055.

68 Gulski E. Digital analysis of partial discharges. *IEEE Transactions on Dielectrics and Electrical Insulation* 1995; 2: 822–837.

69 Randall R. Applications of spectral kurtosis in machine diagnostics and prognostics. *Key Engineering Materials* 2005; 293/294: 21–32.

70 Wu Z, Huang N. A study of the characteristics of white noise using the empirical mode decomposition method. *Proceedings of the Royal Society London, Series A* 2004; 460: 1597–1611.

71 Li Y, Tse P, Wang X. Recovery of vibration signal based on a super-exponential algorithm. *Journal of Sound and Vibration* 2008; 311: 537–553.

72 Wang D, Tse P. A new blind fault component separation algorithm for a single-channel mechanical signal mixture. *Journal of Sound and Vibration* 2012; 331: 4956–4970.

73 Chan J, Ma H, Saha T. Hybrid method on signal de-noising and representation for online partial discharge monitoring of power transformers at substations. *IET Science, Measurement and Technology* 2015; 9(7): 890–899.

74 Mazzetti C, Mascioli FF, Baldini F, Panella M, Risica R, Bartnikas R. Partial discharge pattern recognition by neuro-fuzzy networks in heat-shrinkable joints

and terminations of XLPE insulated distribution cables. *IEEE Transactions on Power Delivery* 2006; 21: 1035–1044.

75 Venkatesh S, Gopal S. Orthogonal least square center selection technique – A robust scheme for multiple source partial discharge pattern recognition using radial basis probabilistic neural network. *Expert System Applications* 2011; 38: 8978–8989.

76 Wang K, Liao R, Yang L, Li J, Grzybowski S, Hao J. Optimal features selected by NSGA-II for partial discharge pulses separation based on time-frequency representation and matrix decomposition. *IEEE Transactions on Dielectrics and Electrical Insulation* 2013; 20: 825–838.

77 Bartnikas R. Partial discharges: Their mechanism, detection and measurement. *IEEE Transactions on Dielectrics and Electrical Insulation* 2002; 9(5): 763–808.

78 Sahoo NC, Salama MMA, Bartnikas R. Trends in partial discharge pattern classification: A survey. *IEEE Transactions on Dielectrics and Electrical Insulation* 2005; 12(2): 248–264.

79 Ma H, Chan J, Saha T, Ekanayake C. Pattern recognition techniques and their applications for automatic classification of artificial partial discharge sources. *IEEE Transactions on Dielectrics and Electrical Insulation* 2013; 20: 468–478.

80 Chan J, Ma H, Saha T. Partial discharge pattern recognition using multi-scale feature extraction and support vector machine. Paper presented at the IEEE Power & Energy Society General Meeting, Vancouver, Canada, 2013.

81 Gulski E, Kreuger F. Computer-aided recognition of discharge sources. *IEEE Transactions on Electrical Insulation* 1992; 27: 82–92.

82 Bishop CM. *Pattern Recognition and Machine Learning*. Berlin: Springer, 2006.

83 Lin CF, Wang SD. Fuzzy support vector machines. *IEEE Transactions on Neural Networks* 2002; 13: 464–471.

84 Bishop CM. *Neural Networks for Pattern Recognition*. Oxford: Oxford University Press, 1995.

85 Jain AK, Duin RPW, Mao J. Statistical pattern recognition: A review. *IEEE Transactions on Pattern Analysis and Machine Intelligence* 2000; 22: 4–37.

86 Hinton G, Roweis S. Stochastic neighbor embedding. *Advances in Neural Information Processing Systems* 2002; 15: 833–840.

87 Seo J. Advanced signal processing techniques for online partial discharge monitoring and assessment of power transformers. M.Phil. dissertation, University of Queensland, 2014.

88 IEC 62478. High Voltage Test Techniques – Measurement of Partial Discharges by Electromagnetic and Acoustic Methods, draft version.

89 Phung B, Blackburn TR, Liu Z. Acoustic measurements of partial discharge signals. *Journal of Electrical & Electronics Engineering* 2001; 21(1): 41–47.

90 Chen H. Partial discharge identification system for high-voltage power transformers using fractal feature-based extension method. *IET Science, Measurement and Technology* 2012; 7: 77–84.

91 IEEE Std. 60505. Evaluation and Qualification of Electrical Insulation Systems, 2011.

92 Koo J, Jung S, Ryu C, Lee S, Lee B. Identification of insulation defects in gas-insulated switchgear by chaotic analysis of partial discharge. *IET Science, Measurement and Technology* 2010; 4: 115–124.

6

Frequency Response Analysis Interpretation for Winding Deformation of Power Transformers

6.1 Fundamentals of Frequency Response Analysis

Power transformers throughout their lifetime are subjected to various mechanical stresses. If the damage occurs on the internal active parts of a power transformer, then the damaged part needs to be removed from the tank for detailed inspection. This is not warranted, as moisture from the atmosphere could contaminate the transformer oil–paper insulation. Therefore, a common approach is to perform certain transformer diagnostic tests to determine the condition of the internal parts. CIGRE WG A2.26, Technical Brochure 342 provides a very comprehensive list of diagnosis techniques, as shown in Table 6.1 [1].

The mechanical integrity of a power transformer is an important aspect that should be assessed to determine its overall condition. A transformer with minor deformation on its winding could remain in operation without seriously affecting its performance. However, the minor damage can worsen over time, thus it could eventually cause a total failure of the transformer. Generally, there are two factors that can cause mechanical damage on transformer internal components. One is due to high electromagnetic forces (EMFs) from large currents which travel into the transformer winding. The current could originate from a lightning strike on the power line or a fault from external sources (such as single line to earth or equipment failure at the power station). These conditions produce massive EMFs in a short period of time, which are capable of creating instability in the winding structure.

In addition to this, improper design and workmanship of a power transformer could even damage the winding when subjected to various tests performed at the factory, such as a short-circuit withstand test. There can be some damage internally due to shocks during transportation of a transformer from one location to another, or during the installation process. In these situations, any form of damage to the transformer can cause long delays to the entire project.

6.1.1 Mechanical Faults

As mentioned in the previous section, electromagnetic forces could potentially cause damage to the winding. When a current-carrying conductor is located within a magnetic field, a force is created and applied on the conductor. The

Transformer Ageing: Monitoring and Estimation Techniques, First Edition.
Edited by Tapan Kumar Saha and Prithwiraj Purkait.

Table 6.1 Comparison of main electrical diagnostic techniques for winding deformation

Diagnostic technique	Advantages	Disadvantages
Magnetizing (exciting) current	Requires relatively simple equipment.	Not sensitive to winding deformation.
	Can detect core damage.	Measurement strongly affected by core residual magnetism.
Impedance (leakage reactance)	Traditional method currently specified in short-circuit test standards.	Very small changes can be significant.
	Reference (nameplate) values are available.	Limited sensitivity for some failure modes (best for radial deformation).
Frequency response of stray losses	Can be more sensitive than impedance measurement.	Not a standard use in the industry.
	Almost unique to detect short-circuits between parallel strands.	
Winding capacitance	Can be more sensitive than impedance measurements.	Limited sensitivity for some failure modes (best for radial deformation).
	Standard equipment available.	Relevant capacitance may not be measurable (e.g., between series/common/tap windings for autotransformers).
Low voltage impulse (LVI) (time domain)	Recognized as very sensitive.	Specialist equipment required.
		Difficult to achieve repeatability.
		Difficult to interpret.
Frequency response analysis	Better repeatability than LVI with the same sensitivity.	Standardization of techniques required.
	Easier to interpret than LVI (frequency instead of time domain).	Guide to interpretation required.
	Increasing number of users.	

Source: Ref. [1], Table 1.

magnitude of this force, **F**, is given by equation (6.1). Here, **B** is the flux density vector (T), **I** is the current intensity vector (A), and L is the conductor length (m). The equation shows that the electromagnetic forces experienced by the winding are proportional to the winding current and the flux density. In addition, the force can also be proportional to the square of the winding current, since the leakage field can be expressed in terms of winding current.

$$F = IL \times B \tag{6.1}$$

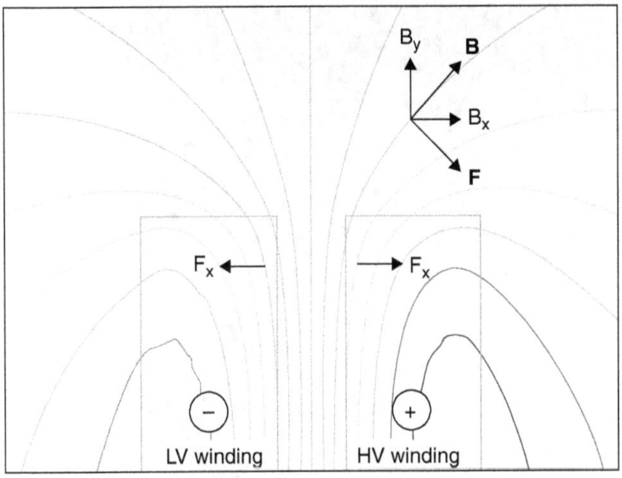

Figure 6.1 Flux lines and force directions on HV and LV windings.

The force direction can easily be determined via Fleming's left-hand rule. As shown in Figure 6.1, the force direction is always perpendicular to the flux direction. While still referring to Figure 6.1, the leakage flux density is composed of two components – the radial component B_x and the axial component B_y. Therefore, based on the left-hand rule, B_x creates an axial component force F_y and B_y creates a radial component force F_x. With this understanding, the radial forces as shown in Figure 6.1 demonstrate that the internal winding (LV) is constantly being pushed inward (compressive stress) while the external winding (HV) is being pushed outward (tensile stress).

Damages due to forces can normally be restricted by appropriate pre-loading of a transformer winding during the assembly process. Very often, the windings are pre-loaded to a pressure value a minimum of five times the calculated axial short-circuit force. Nevertheless, the pre-loaded clamping pressure decreases gradually throughout a transformer's lifetime. This is mainly due to the cellulose insulation material, which reduces in terms of thickness and elasticity from the effects of moisture, temperature, and ageing. If the clamping pressure decreases to a level which short-circuit forces can overcome, this may cause damage to the winding. The axial and radial components of the forces can cause various forms of structural damage on a transformer winding, which is commonly referred to as the failure mode. Buckling of an inner winding caused by such extensive forces is shown in Figure 6.2 [2].

6.1.2 Measuring the Frequency Response

Among the available methods for assessing transformer mechanical condition, FRA is widely recognized as the best tool. FRA is a non-intrusive test, which measures the transfer function response of a transformer winding electrically over a wide frequency range. A sinusoidal voltage V_{input} is injected at one terminal of a winding and the output signal V_{output} is measured at the other terminal. The

Figure 6.2 Buckling of an inner winding. *Source:* Ref. [2].

process is repeated for various frequencies in a predetermined frequency range. This range is typically from 20 Hz to 2 MHz, as suggested in the IEC Standard 60076-18 [3]. The output from this sweep frequency measurement is a transfer function response of a winding, which contains magnitude and phase plots as shown in Figure 6.3.

The magnitude value is calculated from equation (6.2), while the phase value is calculated from equation (6.3). Generally, only the magnitude plot is analyzed, since it is easier to understand and contains more information.

$$\text{Magnitude(dB)} = 20\log_{10}\left(\frac{V_{\text{output}}}{V_{\text{input}}}\right) \tag{6.2}$$

$$\text{Phase}(^\circ) = \tan^{-1}\left(\frac{V_{\text{output}}}{V_{\text{input}}}\right) \tag{6.3}$$

In order to determine if the transformer has suffered any damage, two frequency responses are required. The first response is obtained while the transformer is known to be in a good condition, such as during the commissioning process or at the factory before shipping. This is referred to as the baseline response. The second measurement is performed depending on requirements such as

- for acceptance test
- during commissioning
- relocation of the transformer
- after a fault
- routine inspection
- asset management tool.

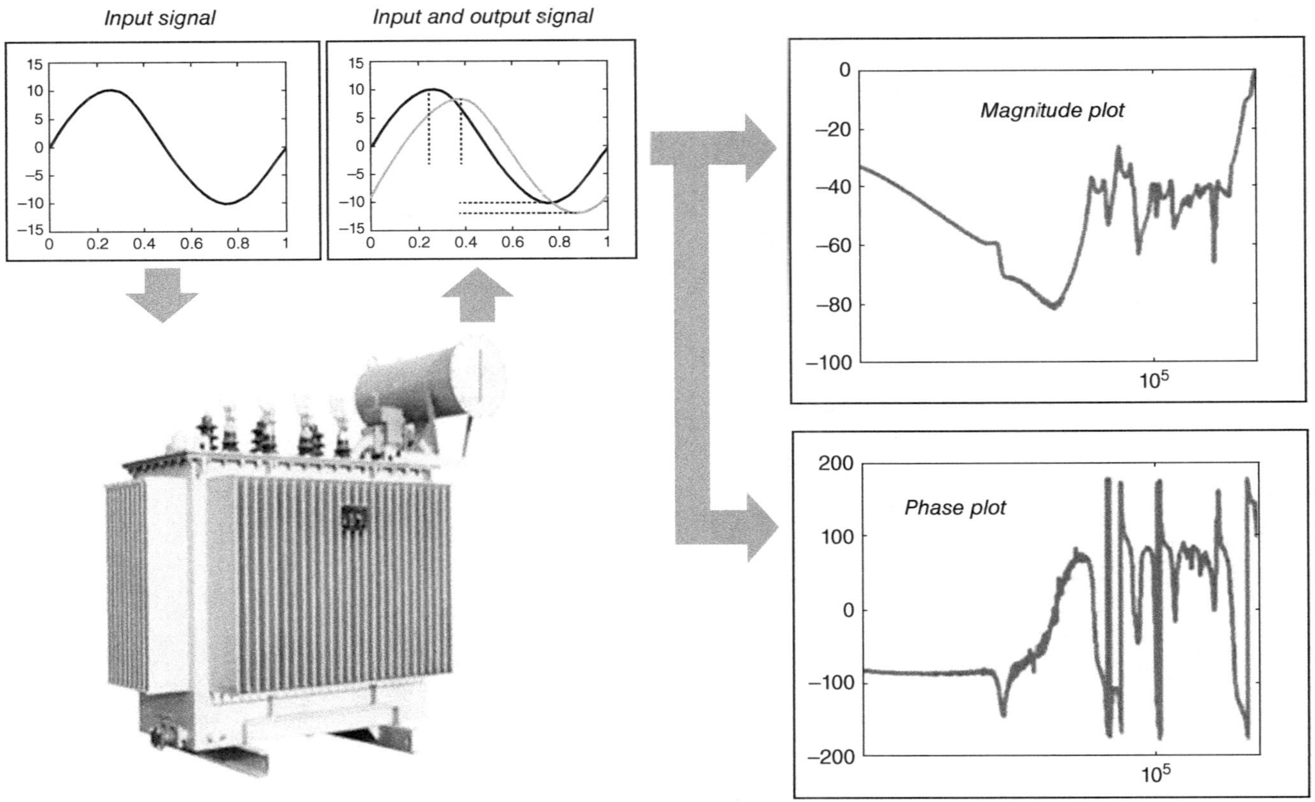

Figure 6.3 FRA measurement signals and typical transfer function plot for a power transformer.

These responses will be compared to observe any difference between them. If there is any variation observed on the latest measurement, it is believed that there has been damage to the winding or core of the transformer. On the contrary, if both responses match perfectly, it can be assumed that the transformer is in the same condition as before. For this reason, FRA is recommended to include as one of the acceptance tests for new transformers to verify if the transformer can withstand the short-circuit test. Subsequently, once the transformer has been shipped to the site, another FRA measurement is conducted during the commissioning process to determine if shocks during transportation have affected the transformer structure. A similar procedure is applied when relocating a transformer to a different site. Besides measuring FRA due to transporting the transformer, another main reason for performing the test is after the transformer has experienced a huge fault current. FRA is performed to determine if the fault has affected the transformer structural integrity. Additionally, FRA measurements can be conducted during routine inspection of the transformer as an asset management tool.

6.1.3 Comparing FRA Results with Other Tests

Normally, it is not adequate to rely entirely on FRA to determine a transformer's mechanical deformation condition. It is suggested to compare the results from FRA with the results from other tests for verification. Reliable analysis can be made by having two or more tests performed on a transformer. Correct analysis is crucial, especially when determining whether the transformer is safe to remain online or should be taken off the grid. Depending on the fault, certain tests can be used to verify or compare to results obtained from FRA, as presented in Table 6.2 [4].

Table 6.2 Suitable power transformer diagnostic test according to faulty condition

	Diagnostic test					
Faulty condition	**Capacitance, dissipation factor**	**Leakage reactance/ short-circuit impedance**	**DC winding resistance**	**Exciting current**	**Transformer turn ratio**	**Frequency response analysis**
Radial winding deformation	×	×				×
Axial winding deformation	×	×				×
Core deformation	×					×
Contact problem on leads			×			×
Short circuit between turns				×	×	×
Open-circuited winding			×			×

Source: Part of the information in this table has been taken Ref. [4].

When comparing the results between FRA and other diagnostic tests to analyze the fault, faults in a power transformer can be arranged into three conditions. The first condition is where the extent of damage on the winding is minimal, thus it is difficult to identify the fault. In this situation, FRA is the key tool to detect the fault as it is sensitive toward winding structure. Deformations under this category are aspects such as radial winding deformation, axial winding elongation (also known as telescoping), and overall or bulk movement. The transformer could operate and remain online, except that the mechanical integrity of the winding has been compromised. In order to verify if the changes in the FRA measurement are due to this damage, additional tests are therefore required. To achieve this, one can measure the winding capacitance and the leakage reactance. These two parameters are influenced by the winding geometry, therefore physical deformation on the winding can be observed from these parameters.

The second condition where the fault severity can be considered in the mid-level is on the core and continuity of the transformer winding. Physical defects in the core and high-contact resistance can be placed under this category. These forms of damages are less complicated to analyze by FRA measurement. This is because the damage mainly affects the low-frequency region of the response, therefore it is quite recognizable. However, if additional tests are needed to verify the FRA results, suitable tests to detect core defects or contact resistance are the exciting current test, DC winding resistance test, and core ground test.

Dielectric failure is the third condition considered a major fault, since the transformer could not then operate as normal. In this instance, FRA is used as the secondary tool for fault diagnosis. Other tests – such as dissolved gas analysis, turns ratio test, and DC winding resistance test – can provide a clear indication compared to FRA. Examples of faults are a short circuit between turns and an open circuit in the winding (as shown in Figure 6.4). In the case of a short circuit, local overheating of strands in the winding can cause damage to the paper insulation and even melt the copper conductor.

Figure 6.4 Shorted parallel strands. *Source:* Ref. [5], Figure 8.

6.1.4 FRA Test Connections

The FRA test configuration is a scheme for making the connection between FRA equipment and transformer terminals to measure the frequency response. There are four different test configurations available, as illustrated in Figure 6.5 for a three-phase transformer.

The end-to-end open-circuit test is performed on one winding, either on the HV or the LV. The input signal is connected to one terminal of the winding, while the output signal is measured from the other end of the same winding. This concept applies for wye and delta-connected windings, and also single-phase windings. The term "open circuit" indicates that the secondary winding of the same phase is left open or floating during the measurement. Additionally, all non-measured windings should be left open-circuited, including windings from other phases or limbs. As shown in Figure 6.6(a), which is the magnitude plot of the response, the typical shape of an open-circuit response will exhibit an apparent anti-resonance (valley) due to the large influence of magnetizing inductance contributed by the laminated core. After the anti-resonance, the response begins to show an increasing magnitude trend which is mainly influenced by the winding capacitance.

The open-circuit test is applied primarily due to the interpretation scheme for the frequency response being generally understood, and has been discussed widely in the literature. Since the test measures only one winding, it allows the user to analyze each winding separately. However, six measurements are required to test all HV and LV windings of a two-winding three-phase transformer.

Similar to the open-circuit test, an end-to-end short-circuit test is conducted on one winding. The only difference between these two is that the two terminals of the secondary winding are connected together for the short-circuit test. Such a connection on the secondary winding will remove the influence of magnetizing inductance of the core from the response. This is helpful to distinguish if the suspected damaged transformer component is the core or something else. The magnitude plot in Figure 6.6(b) is the short-circuit response of the same winding as the response in Figure 6.6(a). Owing to the absence of the core's influence in the response, the main anti-resonance has been shifted to a higher frequency. The response in the low-frequency region is now characterized by only leakage inductance of the winding. The high-frequency response is similar for both open and short-circuit tests, which can be seen on the magnitude and also the phase plot.

In the capacitive inter-winding test, the signal is applied at one terminal of a winding and the output signal is measured at one terminal of the secondary winding. During the measurement, all other terminals are left floating. With this configuration, the capacitance between two windings influences the measurement mainly in the low-frequency region. As can be seen in the magnitude plot of Figure 6.6(c), in the low-frequency region the circuit has very high impedance, which results in the plot beginning at a very low magnitude. The response is highly capacitive, revealed by the positive increment of magnitude, and most of the phases plot a positive value. However, this configuration could not be tested on an autotransformer since series and common windings are not isolated. Moreover, it is believed that this test is less preferable than the end-to-end test due to insufficient understanding of the response. However, the capacitive inter-winding test has a high sensitivity to radial deformation of a winding.

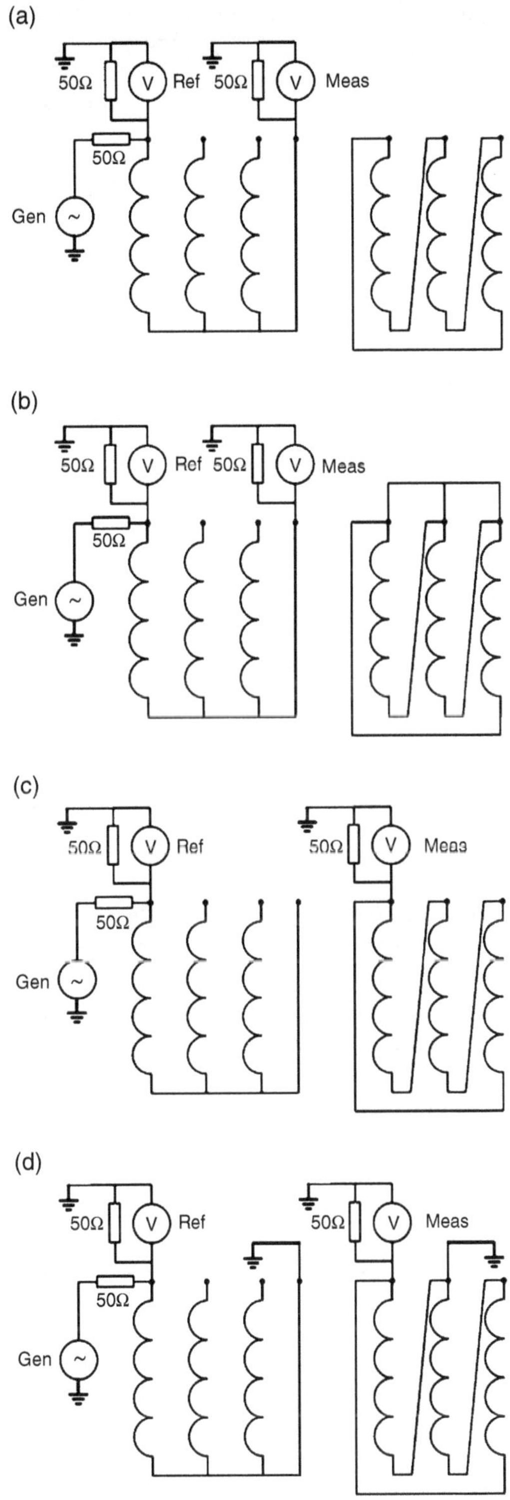

Figure 6.5 FRA test configurations for three-phase transformer: (a) end-to-end open-circuit test; (b) end-to-end short-circuit test; (c) capacitive inter-winding test; (d) inductive inter-winding test.

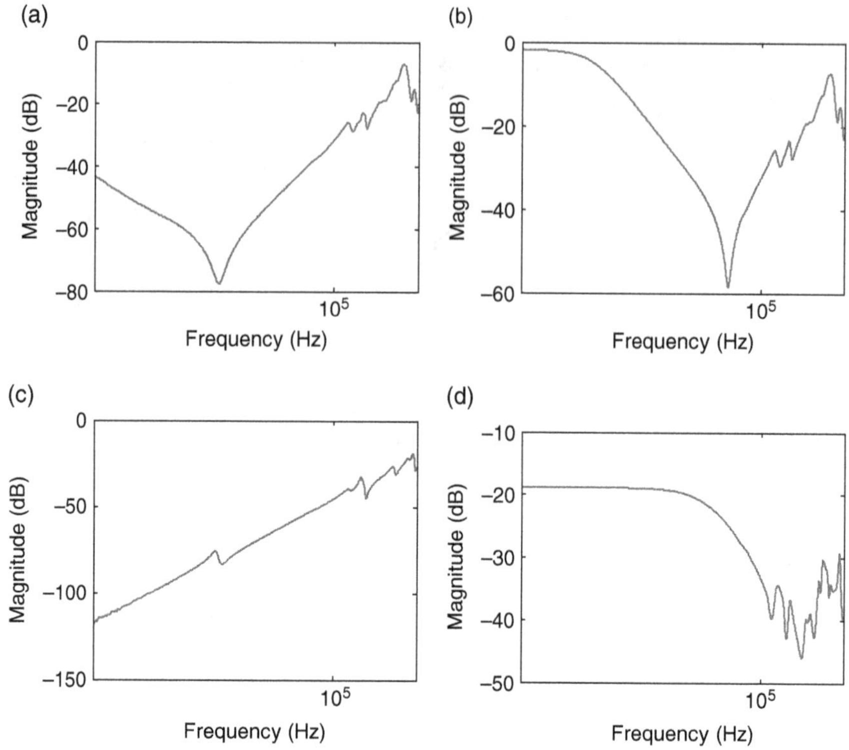

Figure 6.6 Example of measured frequency response from each test configuration: (a) end-to-end open-circuit test; (b) end-to-end short-circuit test; (c) capacitive inter-winding test; (d) inductive inter-winding test.

The fourth test configuration is the inductive inter-winding test, which is similar to the capacitive test except the open terminals of the primary and secondary measured windings are connected to ground. Other terminals of unmeasured windings are left floating to avoid unwanted influences on the response. In this test, the input signal should be supplied at the HV terminal and output is measured at the LV terminal. The response obtained using this test configuration can be used to determine the winding turns ratio from its low-frequency response. Therefore, this could be used as a complementary test for the main turns ratio test (TTR). As can be seen in Figure 6.6(d), the low-frequency response which can be utilized forturns ratio calculation is shown by the flat magnitude region. As in the capacitive inter-winding test, this test is also not widely practiced. The next section will highlight some interpretation schemes for analyzing FRA test results.

6.2 International Practice and Standards

6.2.1 The Chinese Standard DL/T 911/2004 [6]

The Chinese standard on FRA is Chinese Standard DL/T 911/2004, which was published in 2004. It was centrally managed by the Technology Commission

Table 6.3 Relative factor and degree of deformation

Degree of winding deformation	Relative factor R
Severe deformation	$(R_{LF} < 0.6)$
Obvious deformation	$(1.0 > R_{LF} \geq 0.6)$ or $(R_{MF} < 0.6)$
Slight deformation	$(2.0 > R_{LF} \geq 1.0)$ or $(0.6 \leq R_{MF} < 1.0)$
Normal	$(R_{LF} \geq 2.0)$, $(R_{MF} \geq 1.0)$, and $(R_{HF} \geq 0.6)$

for Electric Power Industry and High Voltage Test Technology Standardization, with the cooperation of six institutes. The standard proposed a methodology to compute the relative factor on three frequency sub-bands to determine the degree of winding deformation. The values for relative factor and degree of deformation are given in Table 6.3.

In Table 6.3, R_{LF}, R_{MF}, and R_{HF} are the relative factors for the low (1–100 kHz), medium (100–600 kHz), and high (600–1000 kHz) frequency sub-bands. In this standard, the frequency sub-bands have been predetermined for all transformers. The relative factor for each frequency region is derived from the normalization covariance factor σ, where

$$R_{xy} = \begin{cases} 10 & (1-\sigma) < 10^{-1} \\ -\log_{10}(1-\sigma) & \text{otherwise} \end{cases} \tag{6.4}$$

In this equation, R_{xy} applies for all frequency sub-bands, R_{LF}, R_{MF}, or R_{HF}. The normalization covariance factor is basically derived from

$$\begin{aligned} \sigma(x,y) &= \frac{C_{xy}}{\sqrt{D_x \cdot D_y}} \\ \therefore C_{xy} &= \frac{1}{N}\sum_{k=0}^{N-1}\left[\left(x(k) - \frac{1}{N}\sum_{k=0}^{N-1}x(k)\right)^2 \times \left(y(k) - \frac{1}{N}\sum_{k=0}^{N-1}y(k)\right)^2\right] \\ \therefore D_x &= \frac{1}{N}\sum_{k=0}^{N-1}\left(x(k) - \frac{1}{N}\sum_{k=0}^{N-1}x(k)\right)^2 \\ \therefore D_y &= \frac{1}{N}\sum_{k=0}^{N-1}\left(y(k) - \frac{1}{N}\sum_{k=0}^{N-1}y(k)\right)^2 \end{aligned} \tag{6.5}$$

Take note that this computation only considers the frequency response from 1 kHz to 1 MHz. This is because the relative factor only assesses the main winding condition, and thus the low-frequency response is not included. The literature has applied this method in a few practical cases. The findings show that the actual winding condition, known from inspection, agreed with the result from this method.

6.2.2 CIGRE WG A2.26 Technical Brochure 342, April 2008 [1]

The International Council on Large Electric Systems (CIGRE), under working group A2.26, published an FRA Technical Brochure in April 2008. This brochure is referred to frequently in the literature, along with IEC and IEEE standards published later. The CIGRE brochure provides various practical cases with known actual problems and their measured responses. Important FRA practices are also included, such as the test connections, recommendations on the cable termination, and test equipment requirements. The brochure mentions three methods of obtaining the baseline response for comparing with the latest measured response. These are:

- previous response measured from the same unit;
- measurement on an identical unit (sister transformer);
- measurement on other phases or limbs of the same unit.

The brochure also discusses promising methodologies to assist FRA interpretation. Using statistical indicators such as the correlation coefficient, it is possible to achieve assisted or automatic interpretation. However, the difficulty lies in determining the indicator value or limit that can distinguish between "abnormal" and "normal" transformer conditions. The second methodology suggests to simulate the transformer response based on geometrical parameters. However, the simulated response can have slight deviations from the actual measurement, due to the complexity of the transformer geometry. Nevertheless, this approach is beneficial for simulating a deformed winding response to assist the interpretation. The final methodology provided in the brochure is parameterization of the frequency response based on the transfer function representation. However, due to the complexity of the response, which contains numerous resonances, generating a unique transfer function is a huge challenge.

6.2.3 IEC 60076-18 [3]

This standard was published in 2012 by the International Electrotechnical Commission (IEC). Some content is similar to the CIGRE brochure, such as the terms used, measurement methods, and requirements on measuring equipment. The standard provides a fundamental understanding of the characteristics of the transformer frequency response in every frequency region. Table 6.4 summarizes

Table 6.4 Frequency regions and corresponding influencing factors

Frequency region	Frequency limit	Influencing factors
Low	<2 kHz	Core
Intermediate	2–20 kHz	Interaction between windings
High	20 kHz–1 MHz	Individual winding structure
Very high	>1 MHz (for ratings >72.5 kV) >2 MHz (for ratings ≤72.5 kV)	Measurement set-up and lead (including earthing connection)

the attributes of each frequency region and its influencing factors, based on IEC standard 60076-18 subtopic 4.5 page 15. However, it is vital to note that the limit for each frequency region given in the table is not a generalized value. This is because the frequency limit varies with every transformer due to the variation of design and construction. The standard also discusses several other factors that influence the frequency response. Among these are the influence of the tertiary delta connection, different types of oil insulation, the effect of the DC injection test, and even the effect of temperature on the measurement.

6.2.4 IEEE C57.149 [7]

The IEEE transformer committee of the Power and Energy Society published this standard (IEEE Std. C57.149-2012) on FRA in March 2013. The document serves mainly as a guideline for the application and interpretation of FRA for oil-immersed transformers. Apart from some of the content, which is also available in IEC and CIGRE documents, the IEEE included additional information particularly on test connections and failure modes. Regarding the test connections, the guide listed six types of transformer and the recommended connections to perform FRA measurements. These are summarized in Table 6.5. This table shows that certain transformers would require up to 36 different test connections. However, a minimum number of recommended tests is also suggested, in which these tests are important to provide necessary information to analyze the transformer condition. The minimum recommended test basically consists of open and short-circuit end-to-end tests on each winding. As an example, for a two-winding transformer, six open-circuit tests are required for all LV and HV windings of a three-phase winding. Another three short-circuit tests are required for all HV windings. This is because it has been described specifically in the standard that a short-circuit test can be performed by measuring the HV winding while the LV winding is shorted.

Regarding failure modes, the standard presents measured responses from actual cases along with examination of the response characteristics. This is helpful for guiding an FRA user to analyze the response, since the important and unique features of the response are mentioned. Apart from the common failure modes, other issues are also discussed, such as high contact resistance of connections between windings and bushings and looseness of the winding due to transportation.

Table 6.5 Types of transformer and the number of recommended tests

Transformer type	Minimum recommended tests	Total tests listed
Two-winding transformer	9	15
Three-winding transformer	18	36
Autotransformer without tertiary	9	12
Autotransformer with tertiary	18	33
Autotransformer with buried tertiary	9	18

6.3 Other Interpretation Schemes

6.3.1 Statistical Approach

To analyze and interpret transformer frequency response, a statistical indicator is often applied as an approach to measure the differences between two frequency responses. Rather than only visually comparing the responses, this method can quantify the variation between responses. There are various statistical indicators that have been proposed in the literature for FRA application and interpretation. The most commonly employed one is the correlation coefficient (CC), which is also known as the cross-correlation factor or cross-correlation coefficient in some literature. CC compares two sets of numbers (magnitude in the frequency response) and, if these two are an exact match, CC = 1. However, the value will decrease from 1 as the similarity between two responses decreases. Equation (6.6) defines CC. In this equation, x_i and y_i are the ith values in decibels (dB) from the data set of x and y. $\bar{x}$ and $\bar{y}$ are the mean values from the data set of x and y. N is the size of the data set. Here, x and y are frequency responses from the first and second measurements, respectively. CC has a drawback in that it cannot distinguish between two responses which have similar shapes but are different in magnitude.

$$\mathrm{CC} = \frac{\sum_{i=1}^{N}((x_i-\bar{x})\times(y_i-\bar{y}))}{\sqrt{\sum_{i=1}^{N}(x_i-\bar{x})^2\sum_{i=1}^{N}(y_i-\bar{y})^2}} \tag{6.6}$$

The absolute sum of logarithmic error (ASLE) is another indicator generally mentioned in the literature. ASLE can be calculated according to equation (6.7). The definitions of variables in equation (6.7) are similar to those in equation (6.6). In ASLE, a perfect correlation between two responses will produce 0. The value will increase as the similarity between two responses decreases. As there is no maximum limit for ASLE, it can produce values above 1.

$$\mathrm{ASLE} = \frac{\sum_{i=1}^{N} y_i - x_i}{N} \tag{6.7}$$

Beside CC and ASLE, there are numerous statistical indicators that have been discussed in the literature. However, in practical application, utilities frequently use CC as their main statistical tool. Using CC as a tool to estimate the overall condition of transformer windings is usually accompanied by its range of values. The range value of CC provides an approximation to whether the winding is in good or marginal condition, or should be investigated. To give an example, Table 6.6 shows the range of CC as suggested by Kennedy *et al.* [8].

The range should be used depending on the comparison method, benchmarking, or for phase to phase. Benchmarking is used when the previous response is taken from the same transformer. For this reason, the ranges are much smaller,

Table 6.6 CC range used by a utility according to method of comparison

Explanation	CC		ASLE
	Benchmark	Between side phases or sister unit	For all conditions
Good	0.98–1.00	0.95–1.00	Less than 1.00
Marginal	0.96–0.97	0.90–0.94	1.00–1.70
Monitor/investigate	Less than 0.96	Less than 0.90	Higher than 1.70

such as 0.98 to 1.00 for a good transformer. In contrast, phase-to-phase is when the responses are compared between two outer phase windings in a three-phase transformer or between two sister units of similar phase. In this case, the ranges are slightly wider for each explanation, such as 0.95 to 1.00 for a good transformer. The responses of the outer and center phase windings should not be compared together. This is due to the difference in magnetic flux path between outer and center limbs of the core, whereby the response in the low-frequency region will be different. Fairouz *et al.* [9] also proposed three ranges for ASLE, as given in Table 6.6.

Besides CC, ASLE, and the ranges for estimating the overall condition of a transformer, the relative factor as provided in the Chinese Standard DL 911/2004 is also a form of statistical approach. By dividing the response into three frequency regions (low, medium, and high), the relative factor of each region is computed and referred to Table 6.3 for approximation of the transformer condition. This has been discussed previously.

6.3.2 Other Approaches

Besides analyzing frequency responses using statistical indicators, there are several other methods which have been proposed in the literature. Rahimpour *et al.* [10] proposed a method which is an extension of the use of CC. It is based on certain CC values calculated on three frequency regions to estimate the type of winding fault. Various literature has proposed using neural networks as an approach to distinguish the winding failure mode. Contin *et al.* [11] proposed the use of a fuzzy-logic algorithm to provide the probability of the transformer experiencing four different faults or conditions. These are axial and radial displacement, shorted turns, and residual magnetization. Each of the faults are given three probabilities for good condition, slight deformation, and pronounced damage. The fuzzy-logic approach provides an automated FRA interpretation scheme. Bigdeli *et al.* [12] presented a study which utilized SVM algorithms to distinguish four failure modes: axial winding displacement, radial deformation, gap variation between discs, and short circuit in the winding. In contrast, Yousof *et al.* [13] presented a study using the Nyquist plot to estimate the location of a short circuit between turns. Fairouz *et al.* [9] conducted another study using the Nyquist plot to estimate the severity of winding damage. Axial bending and tilting of conductors were investigated. Equations (6.8) and (6.9) show the minimum real value difference ΔRe_{min} taken from the Nyquist plot to estimate the severity of winding damage WD for axial and tilting damage, respectively:

$$|\Delta Re_{min}| = (0.0002)WD^2 + (0.0018)WD - 0.0016 \quad (6.8)$$

$$|\Delta Re_{min}| = -(0.0008)WD^3 + (0.0068)WD^2 - (0.0061)WD - 0.0002 \quad (6.9)$$

Besides using the Nyquist plot, a polar plot was also applied to analyze the transformer response. Aljohani and Abu-Siada [14] utilized a polar plot to identify and quantify faults in transformer windings. The study investigated several failure modes, such as buckling and axial displacement.

6.4 Case Studies

In this section, three case studies of FRA measurements from power transformers are presented. Each case is analyzed using methods which have been proposed by several references mentioned earlier.

6.4.1 Case Study A

This case study examines a power transformer located at a substation in a university. It has been operating continuously at 80% of full load due to the ongoing construction of a new building. The maintenance team reported that an additional cooling system was required on the transformer due to high loading. Considering the transformer was manufactured in 1996, it was later decided to retrofit the substation with a higher-rating transformer. Before the old transformer was taken out of service, an FRA test was conducted. Results from the end-to-end short-circuit test are provided in Figure 6.7. To compute the statistical indicators for determining the transformer winding condition, a frequency region from 30 kHz to 1 MHz

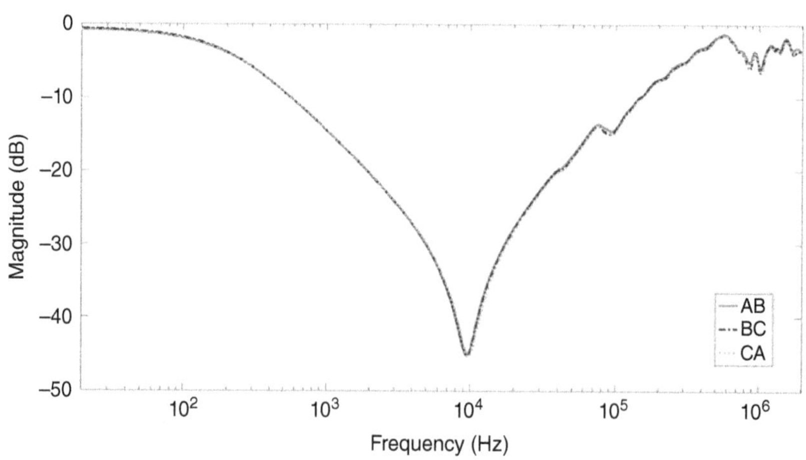

Figure 6.7 End-to-end short-circuit test response of the HV windings.

Table 6.7 Statistical indicators on all phases for the selected frequency region

Statistical indicator		Comparison between phases		
		A vs. B	B vs. C	C vs. A
CC (30 kHz to 1 MHz)		0.9998	0.9997	0.9998
Result		*Good*	*Good*	*Good*
ASLE (30 kHz to 1 MHz)		0.2109	0.1841	0.3435
Result		*Good*	*Good*	*Good*
Relative factor R	R_{LF}	3.81	3.13	2.88
	R_{MF}	4.00	3.94	3.94
	R_{HF}	2.03	2.87	1.87
Result		*Normal*	*Normal*	*Normal*

was selected. The results are given in Table 6.7. It shows a high value of CC and a low value of ASLE, which indicates that the transformer winding is in good condition. This is based on the statistical ranges in Table 6.6. Additionally, the relative factor is also computed as it shows an agreement with the results from CC and ASLE. It was later decided to relocate the transformer to a different substation.

6.4.2 Case Study B

This case study presents responses measured during an inspection on an electric arc furnace (EAF) transformer, used to power arc furnaces in a steelworks. The study indicates that the transformer has been subject to a high level of electrical, mechanical, and thermal stresses. Nevertheless, the transformer was designed to handle such conditions.

Figure 6.8 shows the open circuit responses of the HV windings. Low decibel noises can be observed between 1 kHz and 20 kHz which are suspected to be due to external interference at the site. However, the interference did not affect the responses at other frequency regions thus analysis can still be performed. From the figure, one can determine that the H2-H1 response is taken from the middle phase. This is because the response shape from 1 kHz to 10 kHz is different to the other two phases. Therefore, H1-H3 and H3-H2 are responses from phase A and C respectively. Since the baseline response was not available for this transformer, comparison can only be performed between phases. Statistical indicators are computed between responses from 100 kHz to 700 kHz and the results are given in Table 6.8. All three comparisons in the table show that all windings are in marginal condition or need to be further investigated. This finding is also in agreement with the relative factor, which has estimated that the windings are slightly deformed.

To determine which winding has the best condition among all, the two highest and lowest values of CC and ASLE need to be considered, respectively. Second (phase B vs. C) and third (phase C vs. A) comparisons have the highest CC and lowest ASLE. This suggests that phase C (H3H2) has the best condition, since its response has the lowest variation with respect to the other two responses.

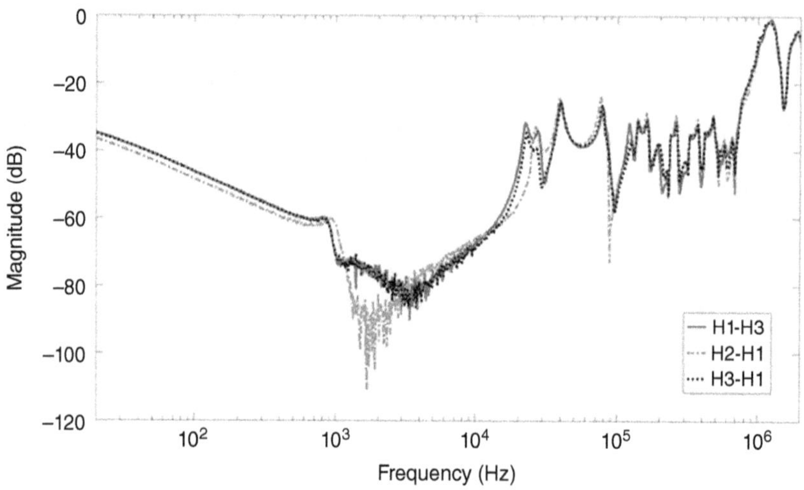

Figure 6.8 End-to-end open-circuit responses of the HV winding with noises at low decibels.

Table 6.8 Statistical indicators on all phases for the selected frequency region

		Comparison between phases		
Statistical indicator		**H1H3 vs. H2H1 (phase A vs. B)**	**H2H1 vs. H3H2 (phase B vs. C)**	**H3H2 vs. H1H3 (phase C vs. A)**
CC (100–700 kHz)		0.8719	0.9398	0.8905
Result		*Investigate*	*Marginal*	*Investigate*
ASLE (100–700 kHz)		2.4146	1.5236	1.9235
Result		*Investigate*	*Marginal*	*Investigate*
Relative factor R	R_{LF}	1.07	1.25	1.67
	R_{MF}	0.88	1.10	0.91
	R_{HF}	2.07	2.22	2.22
Result		*Slight*	*Slight*	*Slight*

Therefore, H3H2 is considered as the reference response. In this case study, the method proposed by Yousof *et al.* using the Nyquist plot is applied. To apply Nyquist plot analysis, all responses are represented in the Nyquist diagram as shown in Figure 6.9. Taking H3H2 as the reference plot, the minimum real value of all plots is observed. The minimum real values of H1H3 and H2H1 are on the left side of H3H2. This is similar to the movement caused by axial bending damage, as presented by Yousof *et al.* Consequently, the absolute difference $|\Delta \mathrm{Re}_{\min}|$ is obtained from the figure to compute the percentage of damage using equation (6.8). The results are given in Table 6.9. Based on the equation, it is estimated that the HV winding of phases A (H1H3) and B (H2H1) are 1% and 2% damaged compared to phase C (H3H2). These estimated percentages are supported by the analysis of the relative factor, which suggests that the windings are slightly damaged.

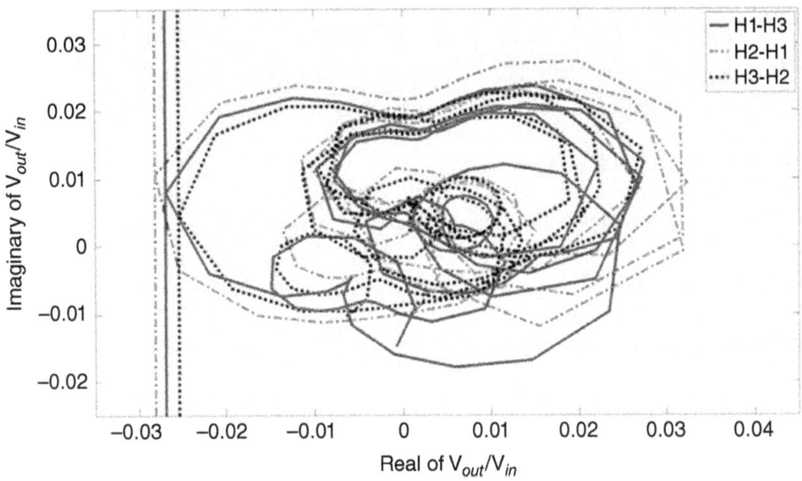

Figure 6.9 Nyquist plot representation of the responses.

Table 6.9 The estimated percentage of winding damage

Comparison between phases	Absolute difference of minimum real value $\|\Delta Re_{min}\|$	Estimated percentage of damage (%)
H3H2 vs. H1H3	0.0015	1
H3H2 vs. H2H1	0.0028	2

This estimation can be improved if the actual baseline response of the transformer is available, which is taken when the winding is in a good or new condition. However, in this case it is not possible, since the transformer itself was manufactured in 1965. Therefore, only careful estimation relying on comparison between responses can be made.

6.4.3 Case Study C

In the third case study, FRA measurement was performed on a 12.5 MVA 33/11 kV power transformer during a routine inspection. The short-circuit responses of the HV winding were measured, as presented in Figure 6.10. Since this is a short-circuit response, the frequency region below 10 kHz is governed mainly by the winding leakage inductance. This is indicated by the negative slope of the response. From the figure, comparison between all phases suggests that there is no problem regarding the winding inductance. The region from 80 kHz to 1 MHz is governed by the winding construction, due to the capacitive trend on the response. CC and ASLE are computed for this frequency region, and the results are given in Table 6.10. CC shows that two comparisons have marginal condition and one is in good condition. In contrast, ASLE shows a slight increase in the warning, where two comparisons suggest investigating the winding and one

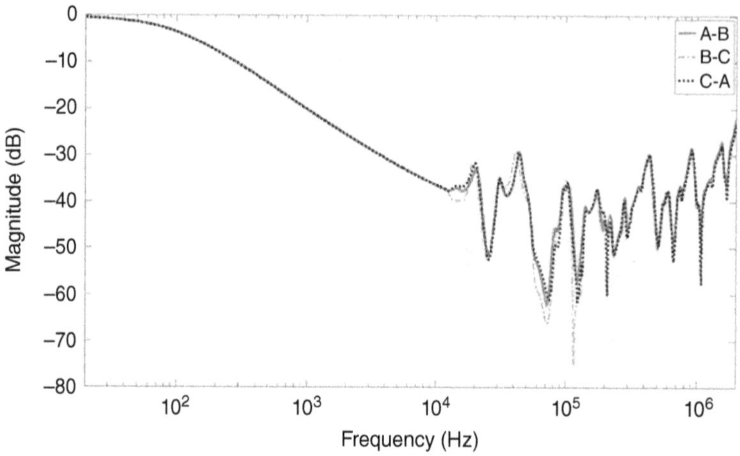

Figure 6.10 End-to-end short-circuit responses of the HV winding.

Table 6.10 Statistical indicators on all phases for the selected frequency region

		Comparison between phases		
Statistical indicator		**AB vs. BC**	**BC vs. CA**	**CA vs. AB**
CC (80 kHz to 1 MHz)		0.91	0.90	0.96
Result		*Marginal*	*Marginal*	*Good*
ASLE (80 kHz to 1 MHz)		1.73	1.77	1.34
Result		*Investigate*	*Investigate*	*Marginal*
Relative factor *R*	R_{LF}	1.78	1.72	2.21
	R_{MF}	0.94	0.91	1.34
	R_{HF}	1.83	1.86	1.85
Result		*Slight*	*Slight*	*Normal*

is in marginal condition. For the relative factor, it produces a similar outcome as for CC.

Since there is no baseline response available for this transformer, as in the previous case, this analysis assumes that one of the three windings has the best condition compared to the other two. To determine this winding, the two highest CC and two lowest ASLE from the comparison are chosen, which are AB vs. BC and CA vs. AB. Response AB appears on both comparisons, thus implying that it has the best winding condition.

All responses are represented in the Nyquist plot shown in Figure 6.11. With the AB plot as the reference plot, its minimum real value Re_{min} is observed. With respect to Re_{min} of the AB plot, the minimum real values of other plots are moving toward the right side of the plot. This movement is similar to the winding deformation caused by conductor tilting, as shown in studies by Yousof *et al.* Therefore,

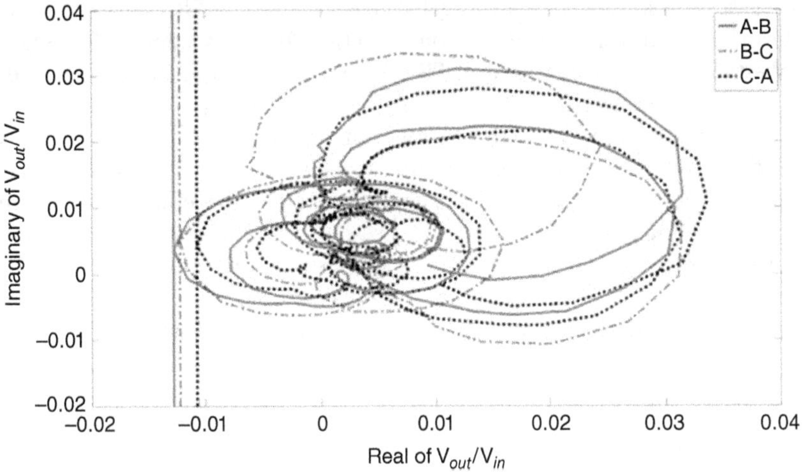

Figure 6.11 Nyquist plot representation of the responses.

Table 6.11 The estimated percentage of winding damage

Comparison between phases	Absolute difference of minimum real value $\lvert\Delta Re_{min}\rvert$	Estimated percentage of damage (%)
AB vs. BC	0.0006	1
AB vs. CA	0.0020	7

the absolute difference $\lvert\Delta Re_{min}\rvert$ is determined and the percentage of damage can be computed from equation (6.9). The results are given in Table 6.11. From this analysis, it is estimated that windings BC and CA are about 1% and 7% damaged, respectively, compared to winding AB.

6.5 Sensitivity of Non-mechanical Factors

Monitoring the condition of oil–paper insulation of a power transformer is an important practice to determine its reliability. There are three notable methods to measure the dielectric response, namely frequency domain spectroscopy (FDS), polarization and depolarization current (PDC), and return voltage measurement (RVM). These techniques operate on the principal of slow polarization of the dielectric material. The most important parameter to be considered here is the moisture content, due to its detrimental effect on the insulation condition. In contrast, FRA is a completely different method, used to assess the transformer mechanical condition. Its main function is to determine whether any deformation has occurred on the active part (windings, leads, and core) of the transformer.

However, several studies have investigated the sensitivity of FRA measurements to determine the transformer insulation condition. In this section, a discussion and investigation of the potential of FRA to assess the insulation condition of power transformers is presented.

The ability of FRA to respond to non-mechanical changes of the transformer is not yet fully understood. It is believed that the earliest study on this issue was done by Abeywickrama *et al.* [15]. They investigated the effect of moisture content, the ageing of pressboard, and the presence of oil on the FRA. Following that, several other research papers have been published which have also attempted to investigate this new FRA application. Various factors regarding the dielectric condition of a transformer have been considered in the literature:

- moisture content
- temperature
- presence of oil.

Regarding the moisture content, Reykherdt and Davydov [16] presented an investigation on the effect of moisture on the transformer FRA response before and after the drying process. The results showed that due to the process, the entire response shifts slightly horizontally toward higher frequencies as the moisture content decreases. In a simulated study by Abeywickrama *et al.* [15], the results showed that high moisture content can shift the resonance to lower frequencies and at the same time dampen its peak. The shifting of resonance and the damping of the peak were governed by real and imaginary parts of the complex permittivity, respectively. Laboratory studies presented by Prevost *et al.* [17] and Krause *et al.* [18] show that the increase of moisture content causes the cellulose insulation to expand, thus increasing the clamping force. Therefore, FRA has the potential to identify the change of clamping forces due to a variation in insulation moisture content.

On the effect of temperature, Reykherdt and Davydov [16] conducted a test on a dry transformer, varying the surrounding temperature while maintaining the moisture content at a certain level. The results indicated that the high temperature could cause the frequency response to shift slightly toward lower frequencies. As discussed by Krause *et al.* [18], temperature also has an impact on clamping forces. Comparing the amount of shift due to temperature and moisture content, the latter is more pronounced. A study by Abeywickrama *et al.* [15] showed results from an oil-filled transformer, indicating that the increase of temperature shifted the resonant frequency while damping its peak. Bagheri *et al.* [19] investigated two cases of temperature effect that were presented on two windings, with and without oil. By increasing the temperature, the results on both winding responses showed an overall response shift toward lower frequencies.

The influence of transformer oil is also found to be sensitive for FRA measurement. This is presented in CIGRE Technical Brochure 342, where two responses of a transformer before and after the oil filling process are compared. The results show that the response above 100 Hz shifts toward lower frequencies with the presence of oil in the system. Yousof *et al.* [20] conducted an FRA sensitivity study of the transformer insulation condition under a complete ageing process. A complete ageing process here denotes the constant effect of operating

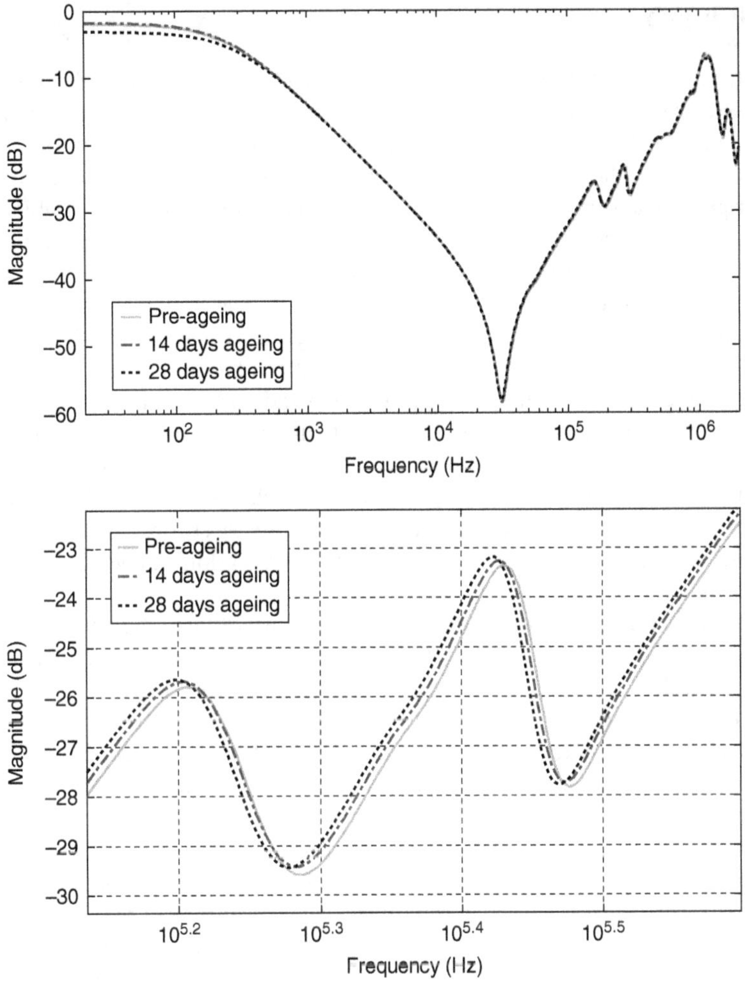

Figure 6.12 Frequency responses of a 16 kVA transformer subject to accelerated ageing process during laboratory study.

temperature and loading condition that contributes to the transformer oil–paper insulation condition as a whole. The study also compared its results with measurements from FDS. Figure 6.12 shows three frequency responses from a 16 kVA transformer which had been subject to an accelerated ageing process. As can be seen in the figure, due to the ageing process (after 14 and 28 days of ageing), the transformer response at the high-frequency region (>30 kHz) shifted slightly toward lower frequencies. Although the variation is fairly small, it shows that FRA measurement could possibly capture the impact caused by the insulation ageing process.

Although several studies have demonstrated the sensitivity of FRA toward non-mechanical parameters in power transformers, the use of FRA to assess these parameters has not been applied seriously in practical measurements.

6.6 Summary

This chapter has presented an overview of the FRA technique, which is a diagnostic tool used primarily for identifying mechanical deformation of transformer windings and core. There are various reasons for conducting an FRA test on a power transformer. Mainly, due to the transportation process, which could disturb the original structure of the transformer and after a large fault current, which could create huge electromagnetic forces, thus possibly damaging the winding. The electromagnetic forces – which comprise axial and radial components – can cause several failure modes on the winding, such as axial displacement, radial deformation, and tilting of conductors. These damages are easily detectable using FRA measurement. Other forms of deformation that can be detected by FRA are damages on the core and localized damage on the winding from overheating. As mentioned in this chapter, to further verify if a fault has occurred on the transformer, it is suggested to employ another diagnostic test. Having two or more tests conducted on the transformer can help to provide a reliable result, and thus an accurate decision can be made. As an example, a fault due to a short circuit within the winding can be identified using FRA and further verification can be done with the transformer turns ratio test.

Four test connections are listed in IEC and IEEE standards. Out of these four, only the end-to-end open and short-circuit tests are used mainly by utilities. This is because the feature of response is fairly well understood from these responses. The low-frequency region is generally governed by the core, the medium-frequency region is governed by interaction between windings, and the high-frequency region is governed by individual winding structure. On the contrary, the other two test connections are not commonly used. The capacitive inter-winding test primarily measures the capacitance between the HV and LV windings. Therefore, the test has high sensitivity to radial deformation of the winding. The inductive inter-winding test measures both HV and LV windings simultaneously. It produces a flat response at the low-frequency region due to a major influence of winding resistance.

The chapter further introduces international standards for FRA measurements and their interpretation schemes. The main standards on FRA interpretations are: The Electric Power Industry Standard of People's Republic of China (2005), CIGRE Technical Brochure 342 (2008), IEC 60076-18 (2012), and IEEE Std. C57.149 (2013). Certain schemes have been practiced widely by utilities, while other schemes are proposed in the literature by researchers. Three case studies were also presented, which use several schemes available to analyze the response. Finally, the chapter discusses non-mechanical factors which are influential on FRA measurement. Although these factors are not the main focus in analyzing the response, it is interesting to understand how the factors could affect the measurement.

References

1 Picher P, Lapworth J, Noonan T, Christian J. Mechanical condition assessment of transformer windings using frequency response analysis. CIGRE WG A2.26, Technical Brochure 342, April 2008.

2 Presentation by Craig L. Stiegemeier: ABB TRES – Transformer Remanufacturing & Engineering Services, 20 August 2013. ABB Red TIE Series – Pomona Transformer Failure Modes.

3 IEC 60076-18. Ed. 1: Power transformers – Part 18, Measurement of frequency response, 2012.

4 Omicron Manual: Diagnostic Testing Solutions for Power Transformers. OMICRON L2206, May 2013.

5 Krüger M, Koch M, Kraetge A, Rethmeier K. New tools for diagnostic measurements on power transformers. International IEEE Conference on Condition Monitoring and Diagnosis, Beijing, China, 21–24 April 2008.

6 Chinese Standard DL/T 911/2004. Frequency Response Analysis on Winding Deformation of Power Transformers. National Development and Reform Commission of the People's Republic of China, 2005.

7 IEEE Std. C57.149-2012. Guide for the Application and Interpretation of Frequency Response Analysis for Oil-Immersed Transformers, 2013.

8 Kennedy GM, McGrail AJ, Lapworth JA. Transformer sweep frequency response analysis (SFRA). Energize, October 2007, pp. 28–33.

9 Fairouz M, Yousof M, Ekanayake C, Saha TK. Frequency response analysis to investigate deformation of transformer winding. *IEEE Transactions on Dielectrics and Electrical Insulation* 2015; 22(4): 2359–2367.

10 Rahimpour E, Jabbari M, Tenbohlen S. Mathematical comparison methods to assess transfer functions of transformers to detect different types of mechanical faults. *IEEE Transactions on Power Delivery* 2010; 25(4): 2544–2555.

11 Contin A, Rabach G, Borghetto J, Nigris MD, Passaglia R, Rizzi G. Frequency-response analysis of power transformers by means of fuzzy tools. *IEEE Transactions on Dielectrics and Electrical Insulation* 2011; 18(3): 900–909.

12 Bigdeli M, Vakilian M, Rahimpour E. Transformer winding faults classification based on transfer function analysis by support vector machine. *IET Electric Power Applications* 2012; 6: 268–276.

13 Yousof MFM, Ekanayake C, Saha TK. Locating inter-disc faults in transformer winding using frequency response analysis. Australasian Universities Power Engineering Conference, Hobart, Australia, 2013, pp. 1–6.

14 Aljohani O, Abu-Siada A. Application of FRA polar plot technique to diagnose internal faults in power transformers. IEEE PES General Meeting, Conference & Exposition, July 2014, pp. 1–5.

15 Abeywickrama KGNB, Serdyuk YV, Gubanski SM. Exploring possibilities for characterization of power transformer insulation by frequency response analysis (FRA). *IEEE Transactions on Power Delivery* 2006; 21(3): 1375–1382.

16 Reykherdt AA, Davydov V. Case studies of factors influencing frequency response analysis measurements and power transformer diagnostics. *IEEE Electrical Insulation Magazine* 2011; 27(1): 22–30.

17 Prevost T, Krause C, Woodcock D. The effects on winding clamping pressure due to changes in moisture, temperature and insulation age. Proceedings of the 67th Annual International Conference of Doble Clients, Boston, MA, 27–31 March, 2000.

18 Krause Ch, Goetz W, Heinrich B. The impact of drying and oil impregnation conditions and of temperature cycles on the clamping force of power transformer

windings. IEEE International Symposium on Electrical Insulation, Boston, MA, 7–10 April 2002.

19 Bagheri M, Phung BT, Blackburn T, Naderian A. Influence of temperature on frequency response analysis of transformer winding. IEEE Electrical Insulation Conference 2013, pp. 40–44.

20 Yousof MFM, Ekanayake C, Saha TK. Examining the ageing of transformer insulation using FRA and FDS techniques. *IEEE Transactions on Dielectrics and Electrical Insulation* 2015; 22(2): 1258–1265.

Further Reading

Bertagnolli G. *The ABB Approach to Short-circuit Duty of Power Transformers*, 3rd edn. ABB Management Services.

Betancourt E, Hernandez C. Power transformer manufacturer's experience with transportation mishandling. Paper presented at the 7th Annual WEIDMANN Diagnostic Solutions Technical Conference, 2008.

Jong-Wook K, ByungKoo P, Seung Cheol J, Sang Woo K, PooGyeon P. Fault diagnosis of a power transformer using an improved frequency-response analysis. *IEEE Transactions on Power Delivery* 2005; 20(1): 169–178.

Kraetge A, Kruger M, Viljoen H, Dierks A. Aspects of the practical application of sweep frequency response analysis (SFRA) on power transformers. Paper presented at the IGRE 6th Southern Africa Regional Conference, Somerset West, 2009.

Ryder SA. Diagnosing transformer faults using frequency response analysis. *IEEE Electrical Insulation Magazine* 2003; 19: 16–22.

Sweetser C, McGrail T. Sweep frequency response analysis transformer applications. Technical paper, Doble Engineering Co., Watertown, MA, 2004.

Transformers: Basics, Maintenance, and Diagnostics. US Department of the Interior Bureau of Reclamation, April 2005.

7
Impact of Moisture and Remaining Life Estimation

7.1 Introduction

The majority of power transformers use paper and pressboard manufactured from wood cellulose and mineral oil as insulation. Kraft paper is wrapped around the copper conductors of the transformer. Pressboard, which is up to several millimeters thick, is used to insulate sections of the winding and to provide mechanical support. Thick sections of wood are used at the base of the windings as support. The cellulosic materials are porous (impregnated with oil); water will migrate into the pores and between the fibers. The water content of cellulose is strongly related to its temperature. Cellulose releases water when heated and adsorbs water during cooling. The cellulosic insulation within a transformer is not at one temperature – that is, the paper wrapped around the conductors and the pressboard adjacent to the flow of warm oil will have a different temperature than the cellulose deep within the thick sections of wood at the bottom of the transformer tank. Consequently, the water content will differ around the transformer. The cyclic loading profile of the transformer also affects the distribution of water as the temperature of the insulation varies. Although the total amount of water within the tank normally changes very slowly, there is a dynamic, continual movement of water around different sections of insulation as the temperature changes (i.e., water migrates from hot to cool sections of cellulose).

Insulation must be kept dry because water, in excessive concentrations, will cause the following problems:

- increase the rate of aging of paper, shortening the life of the insulation;
- failure during transformer overload, where bubbles of water vapor are ejected from the paper into the oil, possibly causing dielectric failure;
- precipitation of water in the oil during transformer de-energization in cold weather, and possible dielectric failure when the unit is subsequently restarted.

During manufacture the windings and leads are dried in a vapor-phase oven to remove water, and the tank is then filled with dried oil. Over time the quantity of water within the insulation generally increases, caused by ingress from the environment and generated by chemical reactions such as oxidation of both mineral

Transformer Ageing: Monitoring and Estimation Techniques, First Edition.
Edited by Tapan Kumar Saha and Prithwiraj Purkait.

oil and cellulose. The electricity utilities sometimes attempt to limit the increase in water by sealing the tank and using desiccant breathers attached to the air intake of the conservator. Not only does this reduce the ingress of water, it also reduces the availability of oxygen.

To provide the reader with a complete overview of water in transformers, this chapter is structured as follows. In Section 7.2, a discussion is given on the techniques used to measure the water content of insulation. In Section 7.3, the effect of water on transformer operation is presented to help the reader understand the problems. In Section 7.4, case studies are shown using measurement data obtained from normally operating power transformers.

7.2 Measurement of Water Content

Within an oil-filled transformer, the majority of water exists within its cellulosic insulation. For instance, the water content of paper (WCP) is usually in the order of a few percent by mass, whereas the water content of mineral oil (WCO) is a maximum of several ten parts per million (ppm). The Karl Fischer (KF) titration method is widely used by the utilities to measure water content. However, a practical problem is that access to the cellulosic insulation of the transformer is limited, so usually only the water content of the oil is measured. Mathematical models must be applied to the KF measurements to ascertain WCP.

Although the total amount of water within the transformer tank changes very slowly over the years, water can migrate around different sections of cellulose dependent on the loading pattern of the transformer. As an example, when running a transformer at rated load, water will migrate out of the hottest paper insulation and be adsorbed by the cooler insulation at the bottom of the transformer. When the load is lowered, there is less of a temperature gradient within the transformer; some of the water adsorbed by the cooler insulation will return to the hotter insulation.

When attempting to determine the remaining life of the cellulosic insulation, it is necessary to measure the water content of the hottest insulation. This produces a challenge, because performing a KF titration on oil sampled from the bottom of the transformer tank is therefore not the best method to determine the life remaining, as will be discussed later in this chapter. Online sensors have been produced, which measure the vapor pressure of water dissolved in oil. These sensors are beneficial, because the movement of water can be tracked in real time.

7.2.1 Karl Fischer Titration

KF titration is an accurate technique to measure the water content of a material or liquid, in mass of water per mass of dry material. KF titration is a chemical-based analysis where a measureable volume of iodine is used to react with the water dissolved in the sample. In the power industry it is very common to extract a sample of oil from a transformer for analysis. Whereas WCP can also be measured, in practice this is not usual because it requires a sample of paper to be removed from the transformer. However, the WCP is dependent on the location where it was removed, because of the temperature gradients within the transformer. Furthermore, once the transformer paper is exposed to the air, it will begin adsorbing water from the atmosphere, adding to the measurement uncertainty.

The temperature of the oil at the sampling point must be recorded accurately, because the water content of oil will be highest when the transformer is hottest, because cellulose desorbs water with rising temperature. IEC 60814 Ed. 2.0 Insulating Liquids – Oil-Impregnated Paper and Pressboard – Determination of Water by Automatic Coulometric Karl Fischer Titration and ASTM D1533 – 12 Standard Test Method for Water in Insulating Liquids by Coulometric Karl Fischer Titration provide comprehensive methodology for moisture measurements in transformer insulation using KF titration.

7.2.2 Vacuum Drying Method

The windings and insulation are usually heated to a temperature between 85 °C and 120 °C using circulated heated dry air, after which a vacuum is applied to remove water. Oil is then injected into the tank to impregnate the cellulose under vacuum. For small transformers the whole unit may be placed into an oven. However, for larger transformers this process could take up to weeks. Therefore, a vapor-phase system is used, where a liquid (such as kerosene) is heated and injected into the tank at low pressure. The kerosene rapidly heats up the insulation and once the desired temperature has been reached, a vacuum is applied to remove the water and kerosene.

Removing water from cellulosic insulation will cause it to shrink. Consequently, the windings may need to be tightened. This operation must be performed as quickly as possible after the dry-out procedure has finished, to minimize the readsorption of water from the air.

7.2.3 Moisture Equilibrium Curves

For solid materials, the relationship between vapor pressure and the mass of dissolved water is described using an adsorption isotherm. The benefit of understanding this relationship is that if the vapor pressure of water dissolved in oil is known, WCP can be estimated. This technique was used by Oommen and Prevost [1], who published Figure 7.1 showing WCP as a function of ppm WCO and temperature. However, this figure is not very reliable for low moisture content and hence Oommen further produced a set of moisture equilibrium charts for the low-moisture region, as shown in Figure 7.2 [2]. Griffin *et al.* (from Doble) produced similar curves, as explained by Du *et al.* [3]. Du *et al.* provided a comprehensive overview of the moisture equilibrium curves and then subsequently explained the validity of these curves under different conditions. They produced MIT curves for water equilibrium in a cellulose–mineral oil system for a wide range of moisture concentrations, up to 800 ppm (see Figure 7.3).

7.2.3.1 Adsorption Isotherms of Solid Materials

The first adsorption isotherm was developed by Langmuir in 1916 [4], and related the fraction of the surface of the material covered by a condensed gas to its relative

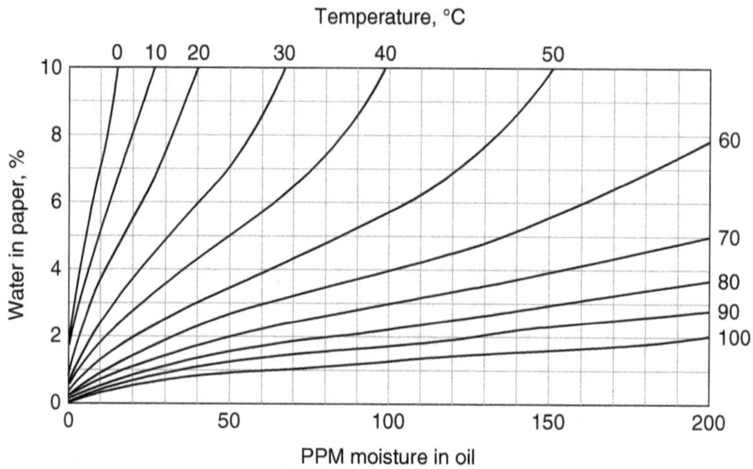

Figure 7.1 Moisture equilibrium charts. *Source:* Ref. [1], Figure 2.

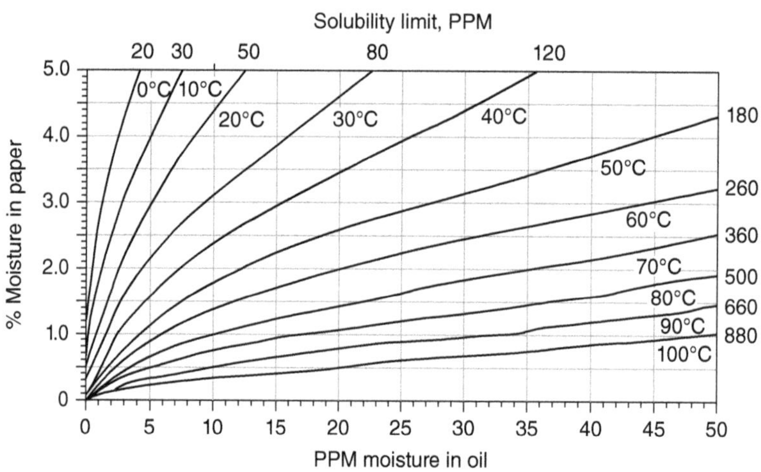

Figure 7.2 Oommen curves for low-moisture region of moisture equilibrium for paper–oil system. *Source:* Ref. [2].

pressure. Langmuir also theorized that the rates of desorption and adsorption were proportional to the surface covered.

Langmuir's model assumed that the gas will form a monolayer on the surface of the solid. This is an incorrect assumption for paper and water, because multiple layers of water can form on top of one another at high humidities. The Langmuir model is only therefore accurate at low humidity (it has been suggested that water only exists as a monolayer at a humidity less than 10% [5]).

The Langmuir model assumes that all sites on a solid that could interact with a gas are the same (i.e., there is no preference which bond with the water first), and that there are no interactions between the adsorbed water molecules. Freundlich

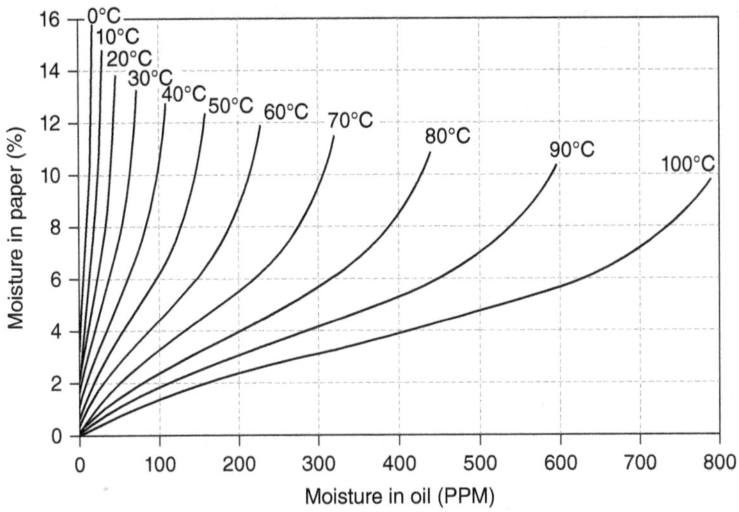

Figure 7.3 MIT-developed curves for water equilibrium in cellulose–mineral oil system for a wide range of moisture concentrations. *Source:* Ref. [3], Figure 10.

noted that these assumptions are incorrect, and so developed an improved model [4]. Given in equation (7.1), the water content C is equal to a coefficient K, multiplied by the gas pressure P raised to a power f:

$$C = K \times P^f \tag{7.1}$$

The Freundlich isotherm was used by Fessler to model the water content of transformer paper. Fessler assumed that coefficient K could be related to temperature using an Arrhenius equation (7.2), where A and B are empirical constants and T is the temperature (K):

$$K = A \times e^{B/T} \tag{7.2}$$

Fessler *et al.* [6] fitted Freundlich isotherms to measurement data provided by various researchers, given in Table 7.1. There can be uncertainty in measurements, which were minimized by determining one set of coefficients that fitted the data best. It is notable that only data up to 2% water content was considered; previous work had suggested that at a WCP of higher than 4% the isotherm models lose accuracy [2].

Other isotherms exist, such as the BET, GAB, and Dent models [4]. Researchers have attempted to produce an isotherm which can be used to accurately model the water content of paper cellulose over a wide range of humidity. However, since the paper used in a transformer is kept relatively dry, the accuracy of a model only has to be considered in the low-humidity region.

7.2.3.2 Vapor Pressure of Water Dissolved in Oil

When oil and pressboard are in thermal and moisture equilibrium, the vapor pressure of the water dissolved in the oil will equal that of the water adsorbed by the

Table 7.1 Freundlich isotherm coefficients calculated by Fessler *et al.* [6]

	A	*B*	*f*
Piper	1.477E-7	4952.6	0.7016
Houtz	6.728E-9	6225.5	0.7813
Ewart	8.086E-7	4211.6	0.6722
Oommen	6.081E-7	4430.1	0.6742
Fessler	**2.173E-7**	**4725.6**	**0.6685**

pressboard. One common method to determine the vapor pressure of water-in-oil is to measure its water activity.

Water activity A_w is the ratio of the vapor pressure of water divided by the vapor pressure above pure water at the same temperature in equilibrium. Mathematically, this relationship is expressed as in equation (7.3), where WCO is the concentration of water dissolved in oil and S_T is the solubility of the oil to water at that temperature. The "≈" symbol denotes that the relationship is approximate, because liquids do not behave exactly as ideal fluids. This ratio is numerically very similar to the ratio of the mass of water in the oil divided by its solubility at that temperature, because Henry's law of vapor pressure above dilute mixtures applies.

$$A_w = \frac{P_v}{P_{o(T)}} \approx \frac{\text{WCO}}{S_T} \tag{7.3}$$

Figure 7.4 shows a realistic plot of the variation in WCO as A_w changes. In this experiment, water was slowly added to a volume of oil. The A_w of the oil was measured, and a KF titration was used to measure the WCO. According to this relationship, a water activity of 1 is measured when the oil is fully saturated with water.

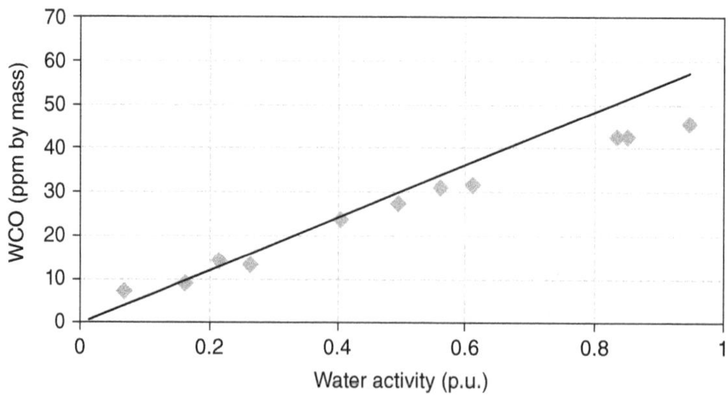

Figure 7.4 Plot of water activity and WCO for mineral oil. The black line shows the ideal relationship between water activity and WCO when using equation (7.3).

Table 7.2 Comparison of *A* and *B* oil solubility coefficients for mineral oil, both measured during this investigation and published by other researchers

	A	*B*
New mineral oil		
Oommen [7]	7.42	1670
Shell [7]	7.09	1567
Doble [7]	7.3	1630
Hyvolt II [8]	7.06	1570
Diala S4-ZX-I [9]	6.8	1500
Nynas Libra [8]	6.5	1400
Nynas Nytro 4000 X [9]	6.5	1400

If A_w is to be calculated from ppm, the solubility of water in the oil must be known. This is calculated using equation (7.4), where $S_{(T)}$ is the solubility of water in the oil at the given temperature, A and B are coefficients specific to the oil and its quality, and T is the temperature (K):

$$S_{(T)} = 10^{A-B/T} \tag{7.4}$$

The constants A and B can be determined graphically from plotting the linear relationship log $S_{(T)}$ against $1/T$. Suggested coefficients for new mineral oil are given in Table 7.2. In practice, using these coefficients provides very similar results.

Over time, the molecular state of the oil and paper change, caused by chemical reactions occurring with oxygen, acids, and water. In general, oil will dissolve more water as it degrades. Consequently, the A and B solubility coefficients will change to reflect this higher affinity to water. To ascertain the degree of effect of ageing on the A and B solubility coefficients, mineral oil from several service-aged transformers was analyzed (as shown in Table 7.3). The difference in solubility of water for the most degraded oil is shown in Figure 7.5.

Table 7.3 Comparison of *A* and *B* oil solubility coefficients for service-aged mineral oil (values used from Ref. [8])

Service-aged mineral oil	*A*	*B*
From a 1990 8 MVA transformer containing naphthenic mineral oil	7.63	1732
From a 1982 22.5 MVA transformer containing Shell Diala S	6.58	1329
From a 1971 225 MVA transformer containing Shell Diala AX	6.54	1349
From a 1952 10 MVA transformer containing naphthenic mineral oil	6.35	1262
From a 1928 4 MVA transformer containing uninhibited mineral oil	6.47	1325

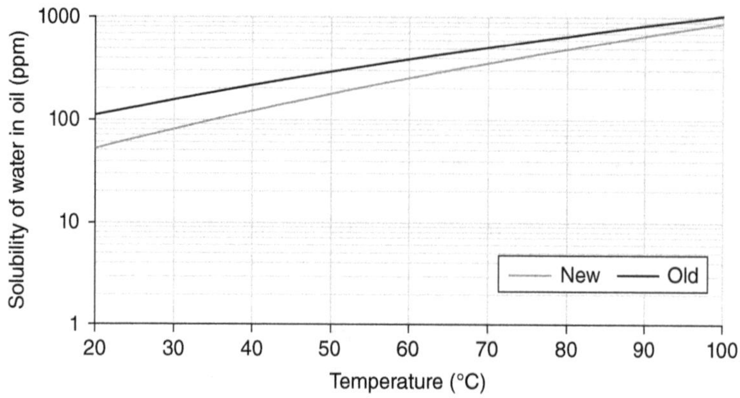

Figure 7.5 Comparison of solubility for differently degraded mineral oils.

7.2.4 Changes in Cellulose Adsorption Isotherms During Ageing

Investigations have shown that paper adsorbs less water as it degrades [10]. This is in contrast to oil, which dissolves more water as it degrades. The water content of new and aged paper at two levels of relative humidity has been investigated, with the results given in Table 7.4.

The condition of aged paper was quantified by its degree of polymerization (DP). A DP of 200 is usually taken as being near the end of life of the paper, because at this point it is brittle. Table 7.4 gives the water content of paper measured by KF titration, according to standard IEC 60641-2. Aged paper adsorbs less water than when new. This change in adsorption is not taken into account by the Fessler isotherm, and thus the paper in an old transformer will be drier than that of the calculation.

For this experiment, strips of paper were placed into a desiccator with the air relative humidity controlled to a high and low level. The low relative humidity (11.3%) was created using a saturated lithium chloride solution, whereas the high relative humidity (43.2%) was that of ambient air. The temperature of the room was carefully controlled (22 ± 1 °C) to prevent differences in water uptake.

7.2.4.1 Uncertainties in Method

Possible inaccuracies in using the adsorption equilibrium curves with KF titration are discussed in an IEEE standard [11]. An analysis of the possible degree of

Table 7.4 Comparison of paper water content at room temperature [10]

Relative humidity: Paper condition	DP	11.3%	43.2%
		Water content of paper(%)	
New	1182	2.9	6.5
Aged	245	2.2	4.9

Table 7.5 Assumptions implied in the use of cellulose adsorption isotherms in conjunction with KF titration, and the resulting inaccuracies if the assumptions are invalid [12]

Assumption	Estimated inaccuracy (%)
1. Solubility of water in oil is the same for all oils and does not change with age.	±100
2. WCO is uniformly distributed throughout the transformer oil.	±15
3. WCP is uniformly distributed throughout the cellulosic insulation structure.	±20
4. Moisture equilibrium between oil and cellulose paper, and between oil and pressboard, is readily attainable.	±20
5. The time required to reach moisture equilibrium between oil and insulation is no longer than the time required to reach temperature equilibrium throughout the transformer.	±15
6. Cellulose adsorption and desorption isotherms are described by the same equations.	±20
7. The same adsorption isotherm is valid for all types of cellulose used in the construction of a transformer.	±15
Total	> ± 200

inaccuracy was given by Ward *et al.* [12], who argued that the potential inaccuracy in using moisture equilibrium isotherms could be up to 200%. These assumptions are listed in Table 7.5. A more precise estimation of WCP can be made using water activity probes, as discussed in the next section.

7.2.5 Water Activity Measurement

A number of uncertainties in using moisture equilibrium curves are given in Table 7.5. The advantages of using water activity probes are as follows:

- They measure the vapor pressure of water dissolved in oil directly, which is not affected by changes in the solubility of water in oil.
- It is not important for the insulation to be in moisture equilibrium.
- Algorithms can be applied to determine the distribution of water in the areas of interest (such as hotspot areas).

Water activity can be measured using a capacitive sensor probe, such as shown in Figure 7.6. These probes are immersed into the oil of a transformer. Accurate measurement requires that the polymer strip is in thermodynamic equilibrium with the surrounding fluid. Despite the fact that the temperature within a transformer is usually changing continuously due to fluctuating loads, thermodynamic equilibrium between the probe and the oil will be reached provided the time required for the probe strip to equilibrate with the oil is short compared with the time within which the transformer internal conditions change significantly.

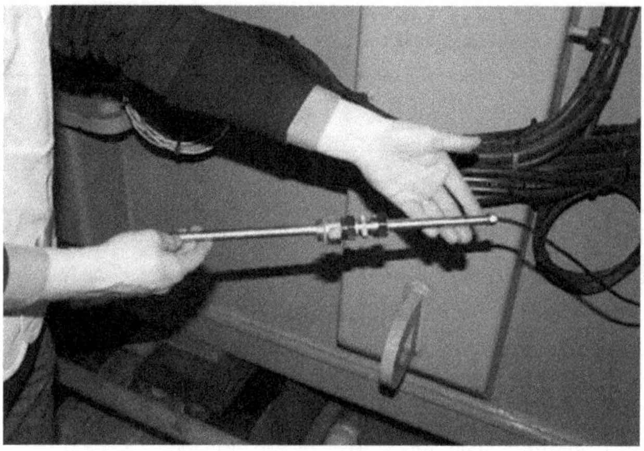

Figure 7.6 Example water activity probe. *Source:* Ref. [13], Figure 1. (*See insert for color representation of the figure.*)

The probe must be inserted into flowing oil, preferably at the top of the transformer (because the difference in temperature between the winding and the measured oil is lowest). The response time of such a probe immersed in transformer oil has been found to be in the order of minutes, which is much faster than the change in temperature or moisture within the transformer.

The vapor pressure of pure water, $P_{o(T)}$ is calculated from temperature. Buck [14] published an equation to calculate $P_{o(T)}$ [equation (7.5)], where T is temperature (°C). The Buck equation expresses pressure in terms of hecto-pascals (hPa), whereas the Fessler equation requires pressure in terms of atmospheres (atm). A conversion factor of 0.00098 must be used in the calculation of equation (7.5) to change between units.

$$P_{o(T)} = 6.1121 \times e^{\frac{17.502 \times T}{240.97 + T}} \tag{7.5}$$

Vapor pressure P_v is therefore calculated by rearranging equation (7.3), substituting equation (7.5) into equation (7.3), and applying the conversion factor. This is shown in equation (7.6):

$$P_v = 0.00603 \times e^{\left(\frac{17.502 \times T}{240.97 + T}\right)} \times A_w \tag{7.6}$$

A source of error when using this calculation is that the temperature of the oil (measured by the probe) can be slightly less than the temperature of the paper wrapped around the windings of the transformer. The degree of difference is dependent on load, cooling mode, and design. As paper is heated it releases water, and thus it can be expected that paper at the top of the windings (where the transformer insulation is usually hottest) will have a lower water content than that calculated using the temperature of the oil. Given that one use of the water calculation is to estimate the remaining life of the paper insulation, the effect of this error on the WCP calculation is that the transformer will appear to have less life remaining. In the next section a discussion is given on how to correct for temperature gradients.

7.2.6 Correcting WCP Measurement for Temperature Gradients when Using Water Activity Sensors

Ideally, the water activity sensor should be mounted at the top of the transformer, in the flow of warm oil (which minimizes the temperature difference between the oil and the winding). This can be performed by installing the sensor into the main tank itself, or into the pipe carrying warm oil to the heat exchanger. For a transformer designed in accordance with the IEC 60076 series of standards, at rated load there can be a 15 K temperature gradient between the hotspot area of the transformer and the oil at the top of the transformer tank.

The utilities have also occasionally installed water activity sensors into the flow of oil at the bottom of the tank, and thus there will be a greater temperature difference between the oil at that point and the top of the winding. The oil flows in a loop around the transformer through the winding structure, out the top of the tank, through the heat exchanger, returning to the tank. If the temperature of the warm oil is known, the temperature of the cool oil can be adjusted to its value before it reaches the heat exchanger.

A summary of this temperature correction is given in the following bullet points.

- The cooling system of a transformer is closed, therefore the mass of water dissolved in the oil remains constant after the given volume of oil has left the top of the transformer tank until it enters the bottom of the tank. There is no cellulose to adsorb or desorb water in the cooling system.
- The vapor pressure of water dissolved in oil is a function of both the mass of water and its temperature. Consequently, if the mass of water is known, then its vapor pressure at a different temperature can be calculated.

As an example of the accuracy of this method, two water activity probes were installed on a 275 MVA power transformer by inserting them into the pipes carrying oil to and from the radiator bank. One probe was installed into the pipe carrying warm oil, the other into the pipe carrying cool oil back to the tank.

The water activity of the cool oil was recalculated using the temperature of the warm oil monitored by the sensor at the top of the tank.

The Fessler equation was used to calculate WCP using the water activity data from the probe mounted in the top pipe, and calculated using measurements from the probe in the cool oil adjusted for temperature. As can be seen in Figure 7.7, the difference in WCP is very small. This infers that to appropriately recalculate WCP, the winding temperature must be known.

7.2.7 Summary

In this section the techniques used to monitor the water content of paper have been discussed. The moisture equilibrium method has limitations, which can be minimized using online water activity probes. In the next section, a discussion will be presented on how water affects the operation of a transformer.

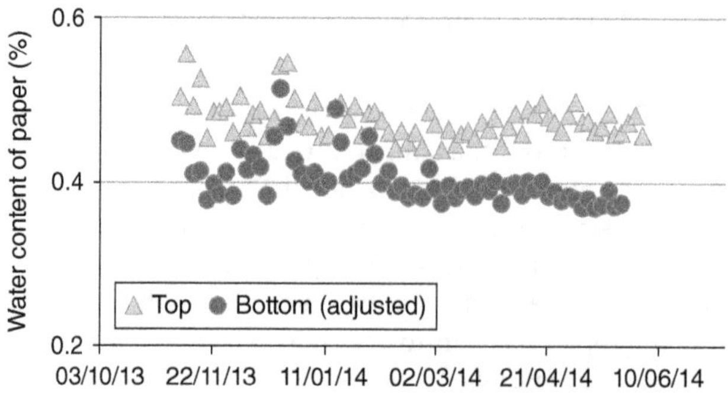

Figure 7.7 Comparison of measuring WCP using probes at different locations in transformer, corrected for temperature difference. *Source:* Ref. [13], Figure 10. (*See insert for color representation of the figure.*)

7.3 Consequence of Wet Insulation on Transformer Operation

In this section, two detrimental effects of water are presented. The first is from an operational viewpoint. During network contingency events (e.g., short term emergency loading), the temperature of transformer insulation can reach in excess of 150 °C. If the cellulosic insulation is too wet, bubbles of water will be expelled into the oil, possibly causing dielectric failure. The second is from an asset management perspective, where the remaining life of the insulation is reduced by chemical reactions occurring with water. Mathematical models are available to determine insulation life.

7.3.1 During Overload Conditions

Cellulosic insulation desorbs water as it is heated. If the temperature of paper is too high, the resulting desorbed water can form bubbles on the surface of the insulation, which then detach, floating into the oil. The dielectric strength of a bubble is lower than that of the oil, therefore possibly causing an arc between the turns of a winding.

7.3.1.1 Bubble Formation

Paper is a thin material, thus most of the water it contains exists on its surface. The pressure of the oil (hydrostatic) and air (atmospheric) bearing down on the paper prevents formation of bubbles. If the transformer is either free breathing, or has a flexible diaphragm in the conservator, the atmospheric pressure will still bear on the oil. However, if the tank is sealed, the air pressure will be dependent on the internal conditions (e.g., whether the oil was vacuumed, or there was a gas blanket, etc.).

The internal pressure of a bubble, P_{int}, is equal to the external pressure, P_{ext}, plus a term related to its surface tension, γ, and radius, r. Both sides of equation (7.7) are equal for a bubble at rest. If the P_{int} term increases past its value at rest, the bubble will expand:

$$P_{int} = P_{ext} + \frac{2\gamma}{r} \tag{7.7}$$

The external pressure is related to atmospheric and hydrostatic pressures [equation (7.8)]. The pressure due to the atmosphere is larger than the hydrostatic pressure. Atmospheric pressure is around 101 kPa. Hydrostatic pressure is equal to the density of the fluid, ρ, multiplied by the acceleration due to gravity, g, multiplied by the height of the fluid above the point of interest, h [equation (7.9)]. For standard mineral oil having a density of 900 kg/m^3 with several meters of fluid above (typical for a transformer with a conservator mounted on top), the hydrostatic pressure is about 30–40 kPa.

$$P_{ext} = P_{atmospheric} + P_{hydrostatic} \tag{7.8}$$

where

$$P_{hydrostatic} = \rho g h \tag{7.9}$$

Bubbles tend to develop first in cavities in the solid insulation. They grow until eventually they detach from the surface and float into the oil.

Various researchers have investigated bubbling under varying conditions of aged oil and paper. Koch and Tenbohlen [15] proposed the empirical equation (7.10), where A and B are coefficients given in Table 7.6. Figure 7.8 later shows a comparison of work performed by two different research groups, where Koch and Tenbohlen's equation is compared to that of a research group based at Monash University [16]. Both sets of measurements are very similar. However, given the temperature uncertainty within a transformer, it may be desired to include a safety margin in the coefficients. Measurement data in Figure 7.6 can be approximated using equation (7.11), where the calculated bubbling inception temperatures have been lowered by 10 °C as a safety margin. A utility could use a greater safety margin by subtracting this margin from 160 in equation (11).

$$T = A \times e^{B \times WCP} \tag{7.10}$$

$$T = -31.07 \times \ln(\mathrm{WCP}) + 160 \tag{7.11}$$

Table 7.6 ***A*** **and** ***B*** **coefficients proposed by Koch and Tenbohlen [15]**

Oil	Paper	*A*	*B*
New Shell Diala D	New Kraft paper	195.5	−0.11186
New Shell Diala D	New thermally upgraded Kraft paper	237.7	−0.13718
Service-aged Shell K6SX	Aged thermally degraded Kraft paper	178	−0.07338

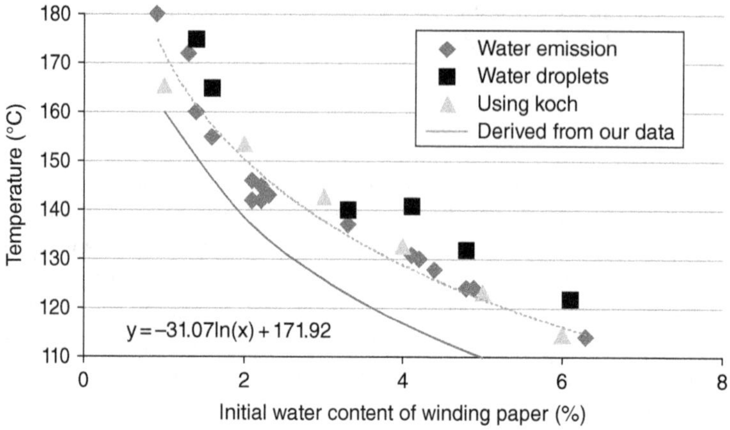

Figure 7.8 Bubbling inception temperature. The triangle dots are calculated from equation (7.10). The orange trace [equation (7.11)] has a 10 °C safety margin added – the original line of best fit included 171.92, however, this was reduced to 160 to take into account a 10 °C margin of safety. 'Water emission' and 'Water droplets' taken from Ref. [16], Koch data from Ref. [15].

7.3.2 Effect of Water on Life Expectancy of Transformer Insulation

The usability of a transformer is dependent on the condition of its cellulosic insulation. Other parts of a transformer can be replaced or changed over its life, whereas the paper cannot be easily replaced or refurbished and rewinding the transformer is expensive. Over the years, research has been performed to determine equations to model the degradation of cellulose. Providing accurate tools for life estimation would be highly valuable for a utility. Not only would they be able to effectively plan for replacement units, but they could track the effectiveness of their asset management strategy.

Cellulosic paper insulation fails when it becomes too brittle, and if it tears or lifts from the surface of the conductor it is no longer an effective insulator. The chemical reactions which attack the cellulose also generate water, further increasing the rate of degradation.

Efforts to produce equations to estimate the life expectancy of paper have been conducted by both the IEEE and IEC, focusing on the temperature of the insulation. These models are based on the assumption that for approximately every 6 °C increase, the rate of ageing is doubled. However, no consideration is given to the effects of water, oxygen, and oil acids, all of which degrade the paper. Oxygen can either degrade the cellulose directly, or indirectly when the oxygen reacts with the mineral oil-generating acids. Together with water, these acids then react with the cellulose. Newer algorithms have been proposed by the authors of this book, which use WCP and whether or not the transformer tank is sealed from the air. Research found that when paper has a water content of 3% and is immersed in oxygenated oil, it will age nearly 40 times faster than when the paper is dry in a sealed atmosphere.

7.3.3 Cellulose Ageing

A cellulose molecule is a long-chain polymer comprising a large number of glucose rings. As mentioned before, for the Kraft paper inside a newly manufactured transformer, the average number of glucose rings per molecule (DP) is about 1000. Water, oxygen, and oil acids steadily react with the cellulose, breaking down the number of rings per molecule. As the degree of polymerization decreases, the paper becomes more brittle. A DP of 200 is usually taken as being the end of life, because the tensile strength of the paper has fallen to around a quarter of its original value.

The key reaction modes affecting cellulose are as follows.

- Oxidation, involving atmospheric oxygen which enters the tank either through an unsealed conservator or through leaks (e.g., aged seals).
- Hydrolysis, which is the reaction between water and paper. The water is either sourced from the atmosphere or generated from chemical reactions. Oxygen will react with the oil, generating water. Acids, which are generated from the oxidation of oil, are also required by the hydrolysis reaction. Consequently, oxidation speeds up hydrolysis.
- Pyrolysis, which involves temperature.

7.3.4 Modeling End of Life

Emsley and Stevens [17] have reported that most data on cellulose ageing can be modeled by a pseudo-zero-order rate law. This is represented by the Ekenstam equation (7.12), where DP is the average degree of polymerization per molecular chain at the initial value (DP_0) and after time $t(DP_t)$, k is the ageing rate, and t is time:

$$\frac{1}{DP_t} - \frac{1}{DP_0} = kt \tag{7.12}$$

The equation for the rate of ageing k, shown in equation (7.13), is based on the Arrhenius relationship [17]. The activation energy E_a is the minimum energy, in terms of J/mol, required to start the reaction. As the temperature increases, more molecules have the required energy to react and thus the overall rate increases. The term A is in units of 1/time (h). The term R is the ideal gas constant, 8.314 J mol^{-1} K^{-1} and temperature T is in kelvin:

$$k = A \times e^{\frac{-E_a}{RT}} \tag{7.13}$$

Substituting equation (7.12) into equation (7.13) produces equation (7.14). The time required for the DP to fall to a given value can therefore be calculated by rearranging this equation:

$$\frac{1}{DP_t} - \frac{1}{DP_0} = t \times A \times e^{\frac{-E_a}{RT}} \tag{7.14}$$

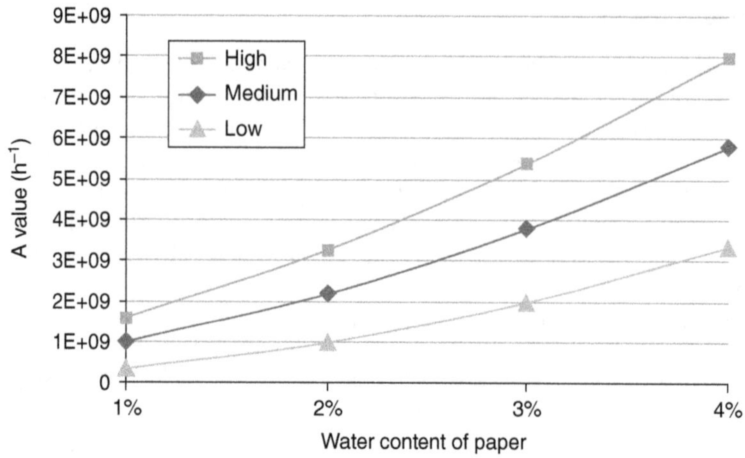

Figure 7.9 Dependency of *A* value on oxygen content of oil and water content of paper. Low oxygen is <7000 ppm, medium oxygen 7000–14,000 ppm, and high oxygen 16,500–25,000 ppm. A high oxygen level is typical for a free-breathing transformer. *Source:* Ref. [26], Figure 1.

Each chemical reaction has different E_a and A values. A previous research group took the activation energy of Kraft paper to be 111 kJ/mol for both oxidation and hydrolysis, with the justification that these reactions are both always present to some degree [18]. Rate k approximately doubles for every 6 °C increase in temperature, due to the $e^{-E_a/RT}$ term of the equation.

The A value is dependent on the oxygen content of the oil and the water content of the cellulose. Within the context of this work (specifically the temperature range), it is not dependent on temperature. The effect of water and oxygen on the A value was investigated thoroughly by Lelekakis *et al.* [19–21], and is shown in Figure 7.9 and given in equations (7.15)–(7.17), where w is the paper water content expressed as a decimal fraction.

For $O_2 < 7000$ ppm:

$$A = (1.78 \times 10^{12})w^2 + (1.10 \times 10^{10})w + 5.28 \times 10^7 \tag{7.15}$$

For 7000 ppm $< O_2 <$ 14,000 ppm:

$$A = (2.07 \times 10^{12})w^2 + (5.61 \times 10^{10})w + 2.31 \times 10^8 \tag{7.16}$$

For 16,500 ppm $< O_2 <$ 25,000 ppm:

$$A = (2.29 \times 10^{12})w^2 + (9.78 \times 10^{10})w + 3.86 \times 10^8 \tag{7.17}$$

7.3.5 Paper Life Tables

Equation (7.14) was used to estimate the life of paper at various temperatures and water content, with both low and high oxygen. These estimates are shown in Table 7.7 (appropriate for a sealed transformer) and Table 7.8 (for a free-breathing transformer). Given that a transformer is expected to last around

Table 7.7 Estimated life, in years, of paper aged in low oxygen

	WCP (%)									
	0.5	**1**	**1.5**	**2**	**2.5**	**3**	**3.5**	**4**	**4.5**	**5**
50 °C	2630	1170	650	400	277	200	150	120	95	79
60 °C	760	340	187	117	80	58	44	34	27	22
70 °C	236	105	58	36	25	18	13	10	8	7
80 °C	78	35	19	12	8	6	4	3	2	2

Table 7.8 Estimated life, in years, of paper aged in high oxygen

	WCP (%)									
	0.5	**1**	**1.5**	**2**	**2.5**	**3**	**3.5**	**4**	**4.5**	**5**
50 °C	429	251	169	122	93	74	60	50	42	36
60 °C	124	72	48	35	27	21	17	14	12	10
70 °C	38	22	15	11	8	6	5	4	3	3
80 °C	12	7	5	3	2	2	1	1	1	1

40 to 50 years, its average temperature should not be allowed to rise above 60–70 °C, because this is when the insulation will age much more quickly.

7.4 Calculating Water Content of Paper in an Operating Transformer

Given the problems that excessive water causes for transformer operation, it is desirable to measure it within the tank. In Section 7.2, the key assumption was that oil and cellulose were in thermodynamic equilibrium; the water within the system had reached an ultimate partition between oil and cellulose. This is an incorrect assumption for operating transformers, for the following reasons.

- The transformer load (and therefore the temperature) generally cycles on a daily basis, far faster than the diffusion time constants, thus the distribution of water does not reach an ultimate partition.
- The transformer insulation is not at one uniform temperature. The water content of the cellulosic insulation therefore differs.

7.4.1 Extending the Fessler Equation to Non-equilibrium Conditions

One of the assumptions implied in the use of cellulose adsorption isotherms (Table 7.5) is that moisture equilibrium between oil and paper, and oil and pressboard, is easily obtainable. The ultimate partition reached by water is strongly dependent on temperature. Given that the load – and therefore the temperature – of a transformer is continually changing, it is unlikely that during normal cyclic loading, enough time is given for moisture equilibrium to be reached. Investigations have been performed on transformers whose insulation was not in moisture equilibrium. When using online water activity probes, average measurements provide accurate determinations. The Fessler equation (7.18) was then extended to quasi-equilibrium conditions [22]:

$$\mathrm{WCP} = 2.173 \times 10^{-7} \times P_{\mathrm{v}}^{0.6685} \times e^{\frac{4725.6}{T}} \tag{7.18}$$

The exchange of water between the oil and cellulose is dependent on the vapor pressure gradient between the oil and cellulose. The rate of flow of water R is calculated using Fick's equation (7.19), where D_{p} is the diffusion constant and $\frac{dP_{\mathrm{v}}}{dx}$ is the difference in vapor pressure between the surface of the cellulose and the adjacent oil volume (its gradient):

$$R = -D_{\mathrm{p}} \times \frac{dP_{\mathrm{v}}}{dx} \tag{7.19}$$

The change in water concentration measured in the oil (WCO) is given by equation (7.20), where k is a proportionality coefficient dependent on the volume of oil:

$$\frac{d\mathrm{WCO}}{dt} = kR \tag{7.20}$$

The relationship between the vapor pressure of water and its mass differs depending on whether the water is adsorbed by the cellulose, or dissolved in the oil. For instance, during conditions of equilibrium, the water content of paper can be several percent by mass of water, whereas in oil only tens of ppm of water is present.

According to equation (7.20), when the change in water content equals zero, the rate of flow between the two media is also zero, and therefore there is zero vapor pressure gradient. This condition arises when the insulation media are in thermodynamic equilibrium.

Changing temperature will affect the vapor pressure of water. However, its degree of effect differs depending on whether the oil is dissolved in the oil or adsorbed to the surface of the cellulose. For instance, when referring to the equations below, it can be seen that the effect of temperature on the vapor pressure of water is quite different for the two types of insulation [equations (7.21) and (7.22)]. The general consequence is that with rising temperature, the vapor pressure of water adsorbed into the paper increases faster than the vapor pressure of water dissolved in the oil. This means that while the temperature is increasing, there is a tendency for water to migrate out of the cellulose into the oil:

$$P_{\mathrm{v}}(\text{paper}) = 9.2683 \times 10^{9} \times C_{\text{paper}}^{1.4959} \times e^{\frac{-7069}{T}} \tag{7.21}$$

$$P_{\mathrm{v}}(\text{oil}) = \frac{C_{\text{oil}}}{10^{\left(A - \frac{B}{T}\right)}} \times 0.00603 \times e^{\left(\frac{17.502 \times T}{240.97 + T}\right)} \tag{7.22}$$

During thermodynamic equilibrium the water content of oil can be measured and the vapor pressure of water calculated. Since there is zero vapor pressure gradient between the oil and paper, the calculated vapor pressure can also be used in the Fessler equation to calculate water content.

The net rate of water flow is momentarily zero when the temperature is either at its local maximum or minimum. Since the direction of flow reverses, either into or out of the cellulose, the flow must have momentarily been zero. At this point the vapor pressure gradient must have also been zero. Thus, the insulation media are, for a short period of time, in a similar state as when in equilibrium; the Fessler equation can therefore be used. This investigation [22] showed that the WCP determined using this method was very similar to that calculated using the long-term average.

7.4.2 Impact of Top Winding Temperature

The WCP calculation must be corrected because the paper on the winding is usually at a higher temperature than that of the oil at the measured point. Given that cellulose releases water as it becomes warmer, the general consequence is that if the calculation is not corrected, it will over-estimate WCP. This leads to the life expectancy of the paper appearing lower.

The principle of mass conservation is used for the correction. The WCO is assumed to stay constant after the oil has flowed past the top of the winding until it reaches the water activity probe. The temperature of the oil then has to be corrected to reflect the temperature of the paper wrapped around the top of the winding, and the water activity recalculated at the winding temperature. The temperature of the winding and hotspot is either measured, or is calculated using IEC 60076/7 thermal model [23]. The vapor pressure of that concentration of water is then recalculated for that temperature. The steps are as follows.

1) Calculate WCO from water activity and temperature:

$$C_{\text{oil}} = A_{\text{W}} \times 10^{\left(A - \frac{D}{T+273}\right)} \tag{7.23}$$

2) Calculate vapor pressure of this mass concentration of water at the winding temperature $T_{\text{w}} + 273$:

$$P_{\text{v}} = \frac{C_{\text{oil}}}{10^{\left(A - \frac{B}{T_{\text{W}}+273}\right)}} \times 0.00603 \times e^{\left(\frac{17.502 \times T_{\text{W}}}{240.97 + T_{\text{W}}}\right)} \tag{7.24}$$

where T is in degrees C.

The recalculated vapor pressure, along with the hotter temperature, is then used with the Fessler equation (7.18).

7.5 Case Studies

Algorithms to determine the water content of paper described in the above sections were investigated by applying them to power transformers operated by several Australian electricity utilities. Dielectric response and oil KF measurements

Table 7.9 Transformers investigated during study

	TR1/TR2	TR3
Rating	12.5 MVA 33/11 kV	25 MVA 110/11 kV
Year of manufacture	1981	1985
	Free breathing + desiccator	Sealed + desiccator
Comments	Nearing end of expected 40-year life	Mid-life transformer undergoing refurbishment

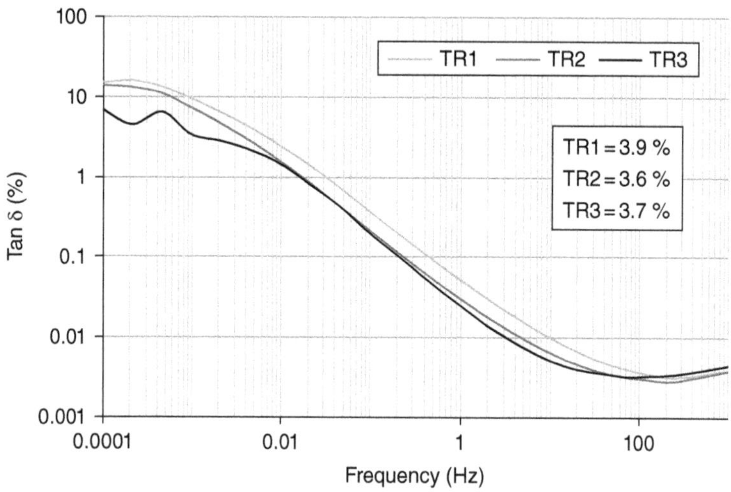

Figure 7.10 Dielectric response of transformers. *Source:* Ref. [27], Figure 11.

were also performed on these transformers to aid the comparison of the different methods. These transformers all operate in zone substations and their ratings are given in Table 7.9.

FDD response measurements were carried out on these power transformers. In some cases this would take nearly a day by the time the unit was isolated from the busbars within the substation. The traces for the three transformers are shown in Figure 7.10. For a dielectric response trace:

- around 0.001 Hz, tan δ increases with increasing water content;
- between 0.01 and 0.1 Hz, tan δ is affected by the insulation geometry;
- between 0.1 and 10 Hz, tan δ increases with increasing electrical conductivity of the oil;
- around 100 Hz, tan δ increases with increasing water content.

7.5.1 Analysis of Data for Transformers TR1, TR2

These transformers operate as part of a trio in a substation, where the load is shared nearly evenly between them. The substation is designed to be able to operate with only one transformer, for instance if maintenance were being carried out

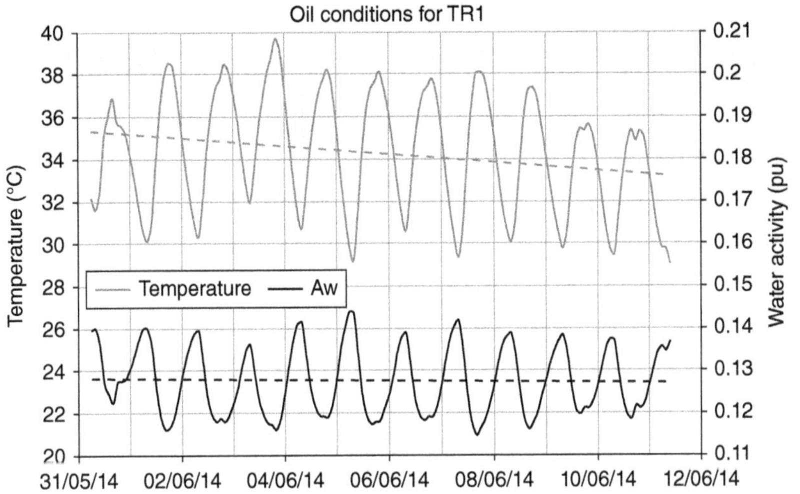

Figure 7.11 Temperature and water activity of oil in TR1. *Source:* Ref. [27], Figure 12.

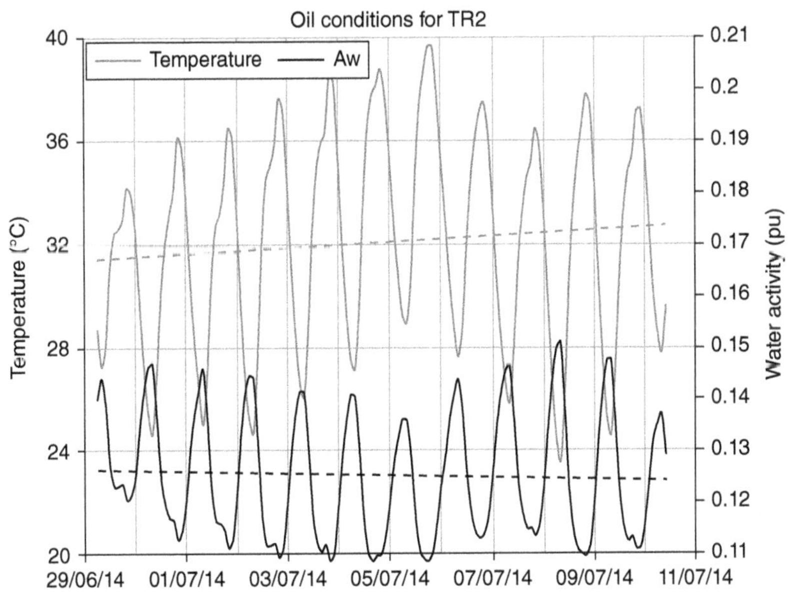

Figure 7.12 Temperature and water activity of oil in TR2. *Source:* Ref. [26], Figure 13.

on one and the second one failed, supply would not be interrupted. Consequently, the maximum load of each unit during normal operation is fairly low. One water activity probe was installed per transformer, through a valve at the top of the tank. The measurements made by this probe are shown in Figures 7.11 and 7.12.

Figures 7.13 and 7.14 are created by applying equations (7.6) and (7.18) to calculate vapor pressure and then WCP. The average WCP was determined for TR1 to be around 3.8% (Figure 7.13). This is close to the dielectric response

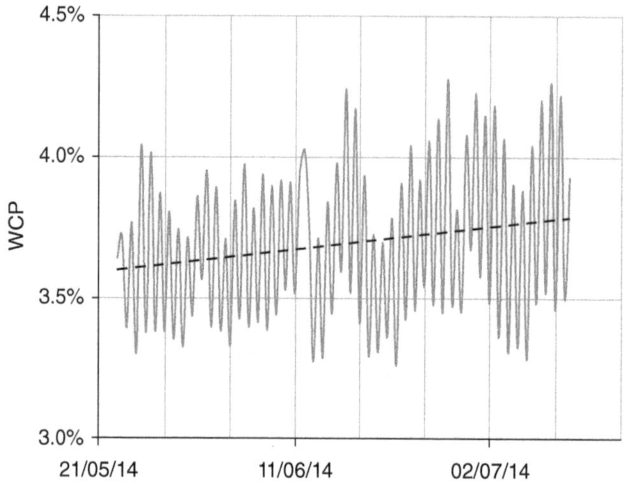

Figure 7.13 WCP calculated using water activity data for TR1. *Source:* Ref. [28], Figure 6.

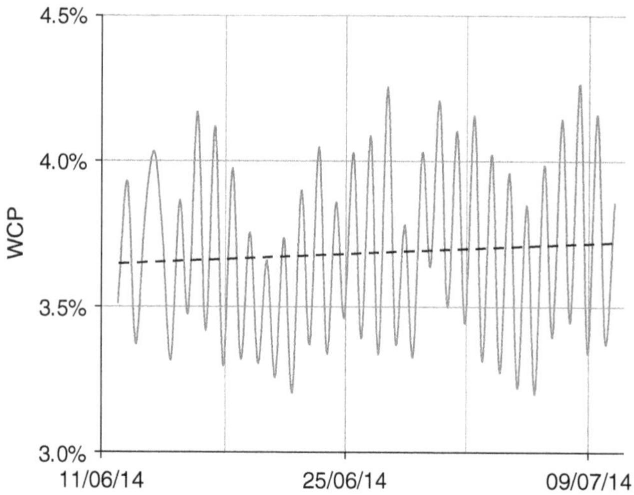

Figure 7.14 WCP calculated using water activity data for TR2. *Source:* Ref. [28], Figure 7.

measurement of 3.9%. For TR2, the water content of paper was determined from the water activity probe to be around 3.7% (Figure 7.14). This is reasonably close to the dielectric response measurement of around 3.6%. One source of small variability was noted in the dielectric response measurement; small changes in temperature during the measurement will affect its value.

7.5.2 Comparison of WCP Using Water Activity and KF Titration

WCO was measured using KF titration. Two sets of oil solubility coefficients were used in the calculation for the paper water content. The first set is for new oil ($A = 7.42$, $B = 1670$), the second set is from a similarly aged transformer

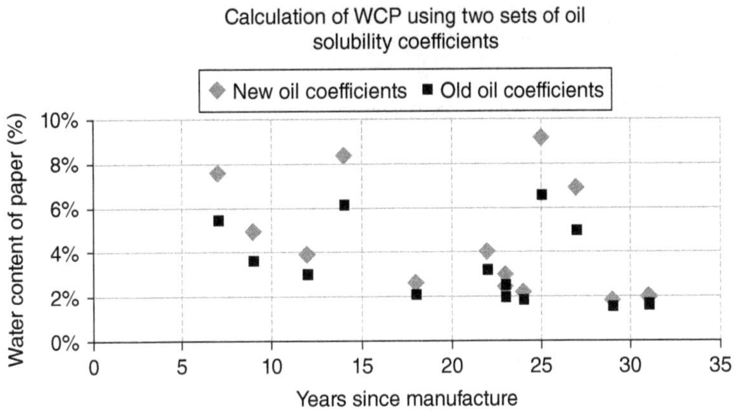

Figure 7.15 TR1 calculation of WCP from WCO.

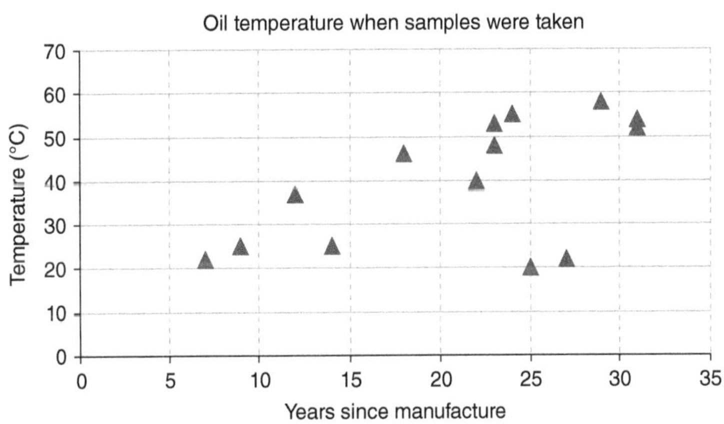

Figure 7.16 TR1 oil temperature on sampling.

($A = 6.54$, $B = 1349$). Using two sets of coefficients demonstrates the uncertainty in WCP caused by the condition of the oil.

For TR1, the WCP calculated using WCO and solubility coefficients is shown in Figure 7.15 and Figure 7.17 for TR2. The oil temperature at the time of sampling is shown in Figures 7.16 and 7.18 for TR1 and TR2, respectively. Given that paper becomes less hygroscopic with age, the calculated paper water content is likely to be lower than determined. It should be noted that these calculations are inaccurate, because the transformer insulation media were probably not in thermodynamic equilibrium. For instance, the samples collected when the oil was near 20 °C show an abnormally high paper water content. Consequently, oil should only be collected when the transformer is in a warm condition.

Since paper adsorbs less water when aged, the degree of polymerization was estimated for TR1 and TR2 from the oil furan content (using some empirical relationship between furan in oil and DP as described in Chapter 2), and is shown in Figures 7.19 and 7.20. For both units, the DP is calculated to be around 300. It

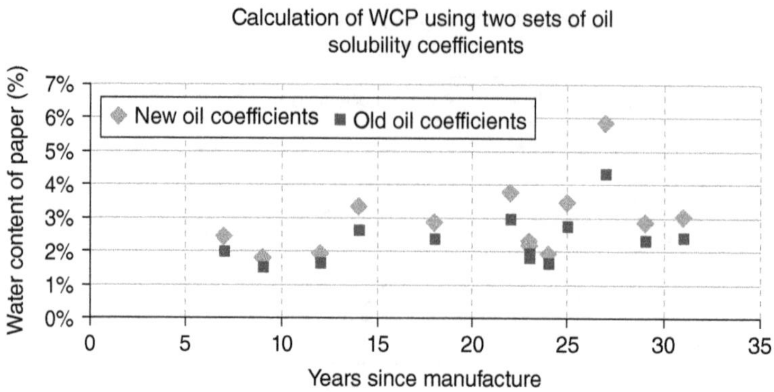

Figure 7.17 TR2 calculation of WCP from WCO.

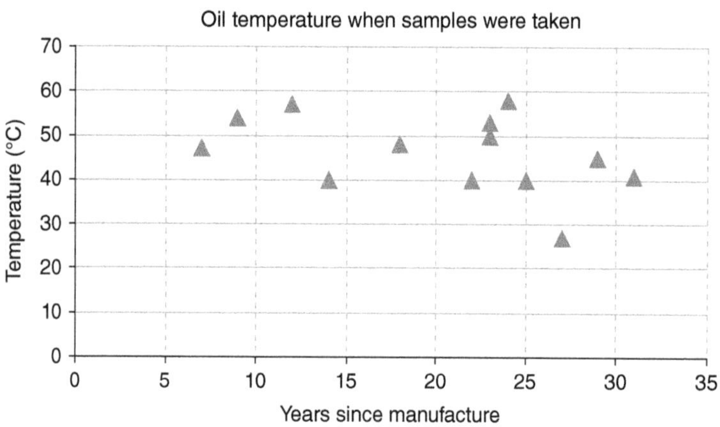

Figure 7.18 TR2 oil temperature on sampling.

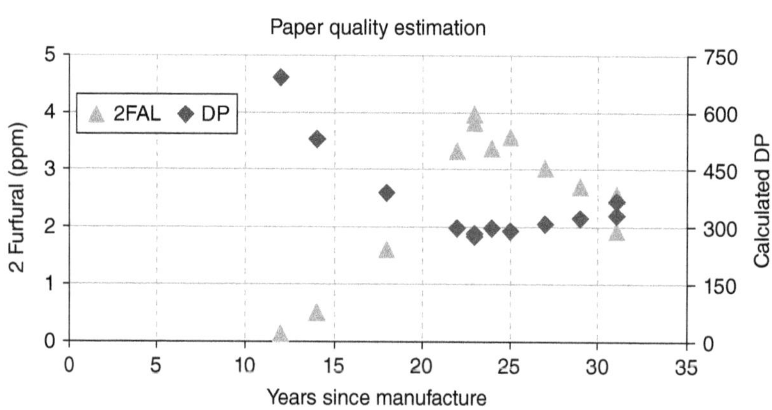

Figure 7.19 TR1 paper quality estimation. *Source:* Ref. [27], Figure 8.

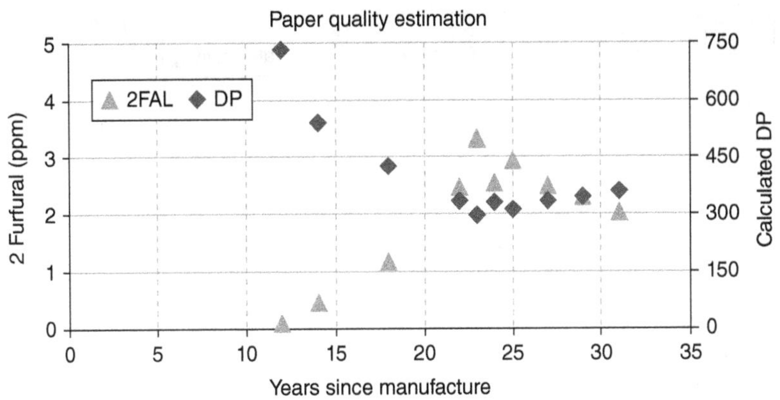

Figure 7.20 TR2 paper quality estimation. *Source:* Ref. [27], Figure 9.

should be noted that this method is not highly reliable, because paper does not age uniformly in a transformer. An observation is that the furan level appears to fall at around 23–25 years. In 2004, at the 23rd year of operation, the cooling of these transformers was upgraded. The fall in furan is therefore likely to be due to the transformer now running cooler and so less furans are produced, and the furan/oil partition changing so a higher concentration of furan is adsorbed by the paper. The paper within these transformers is indeed degraded, and thus the true water content of paper is very likely to be slightly less than that calculated.

7.5.3 Transformer TR3

This unit had a sealing-bag in its conservator to prevent the ingress of air and water into the oil. There were no valves in the pipes to and from the radiator, so the water activity probe was installed directly into the tank at the top.

While the data was being collected from the probe one morning, the oil within the transformer was noted as being cool (only 25 °C). This implies that its load was reasonably low. Figure 7.21 shows the temperature and water activity measurements. As can be observed, there is a slow fall in temperature, caused by the weather moving toward the southern winter. As the temperature of the cellulose reduces, it adsorbs more water from the oil and so the water activity of the oil also falls.

Table 7.10 shows the averages for temperature and water activity used in the calculation for paper water content. Both give a similar water content. The equations for the trend-lines of both temperature and water activity in Figure 7.21 were used to plot long-term trends for paper water content, as shown in Figure 7.22. The average WCP is 2.9%.

The WCP should normally be recalculated for the winding temperature, which results in the water content decreasing. However, as will be seen in Figure 7.23, the temperature of the winding is very similar to that of the oil. Therefore, any correction to WCP is minimal.

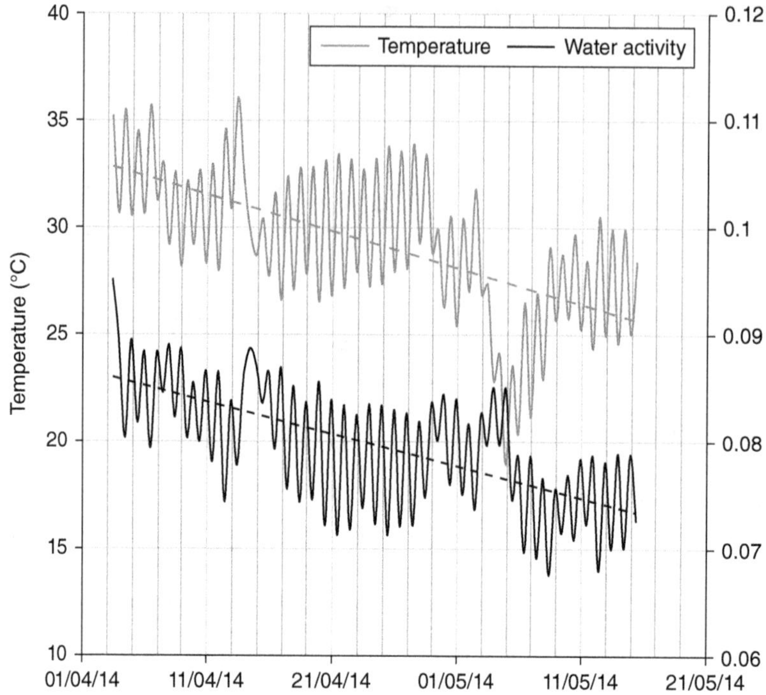

Figure 7.21 Temperature and water activity of the oil. *Source:* Ref. [27], Figure 14.

Table 7.10 Conditions used in calculations, and the estimated water content of paper

Description	Temperature (°C)	Water activity (pu)	WCP (%)
Average using 12-hourly samples	29.2	0.080	2.9

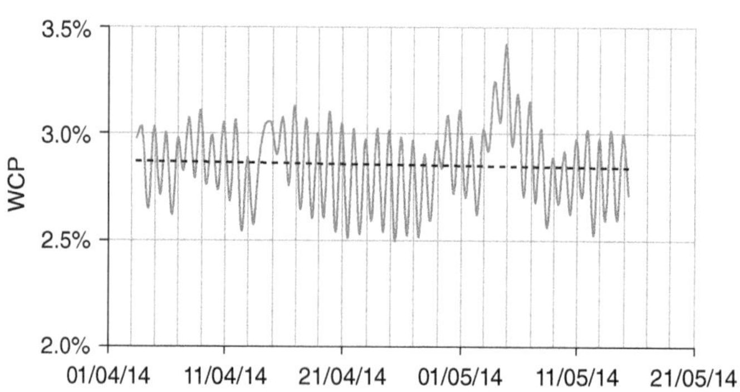

Figure 7.22 WCP for TR3.

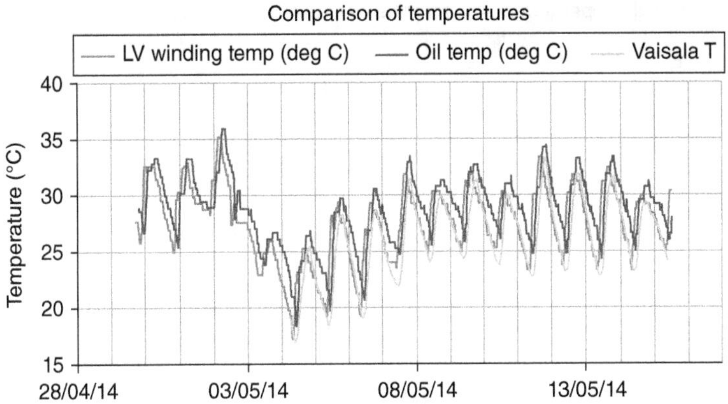

Figure 7.23 Comparison of temperature measurements in transformer.

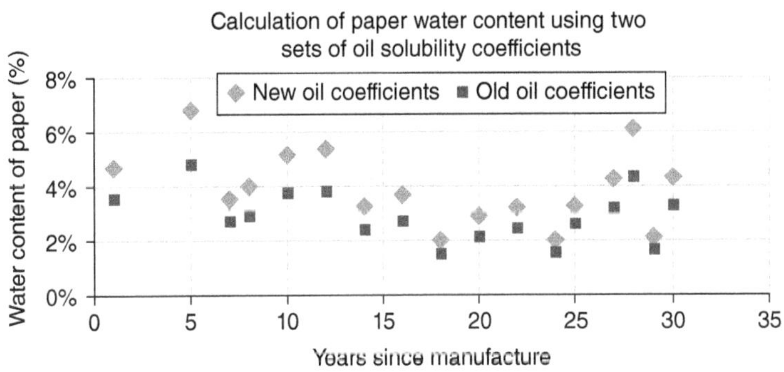

Figure 7.24 Calculation of WCP using oil measurements.

As can be seen from Figure 7.23, the oil temperature tends to vary slightly higher (1 or 2 °C) than the LV winding. The temperature measured by the water activity probe is practically the same as that of the winding. Consequently, no temperature correction of the water activity measurement was performed.

The oil water content, measured periodically using KF titration, was used to calculate the water content of paper, as shown by the data in Figures 7.24 and 7.25. The WCP measurements are somewhat scattered, probably due to the transformer insulation not being in thermodynamic equilibrium. The effect of using different oil solubility coefficients can also give an error of up to one percentage point.

7.5.3.1 Comparison with Dielectric Response Measurements

Frequency domain dielectric response equipment Omicron Dirana [24] was used to measure WCP; the trace is shown in Figure 7.10. The WCP of the unit was estimated to have been 3.7% in October 2013, approximately six months before the water activity probe was installed. After drying, water will migrate from the cellulose into the oil until a new partition is reached. More data was downloaded

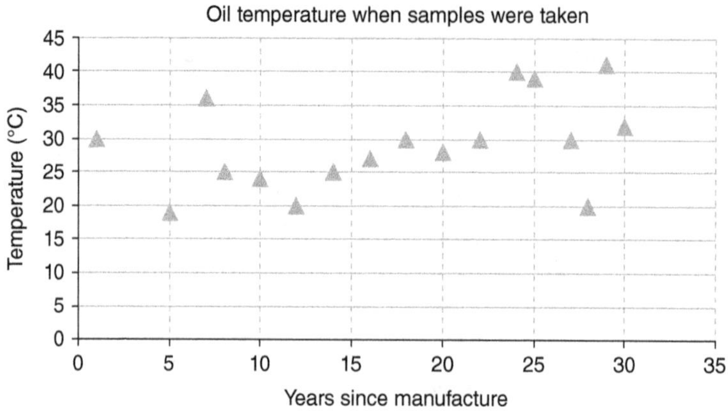

Figure 7.25 Temperature of oil when sampled.

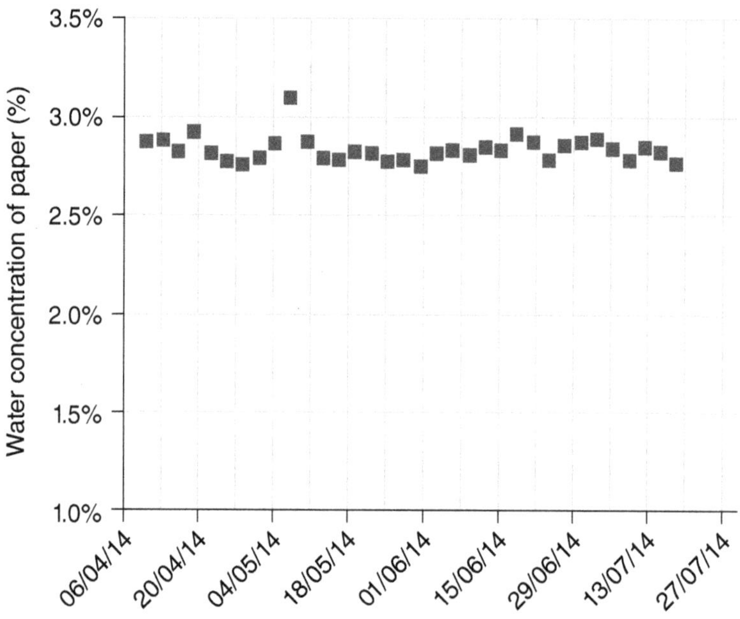

Figure 7.26 Long-term WCP for TR3. *Source:* Ref. [27], Figure 15.

from the water activity probe to determine whether the ultimate water partition had been reached, or whether the water content of the oil was still rising while returning to a level reflective of 3.7% water content of paper. As can be seen in Figure 7.26, the water content of paper appears stable, it is not increasing, and thus the water partitioning is likely to have reached its long-term equilibrium level. The difference in paper water content determined from both dielectric response and water activity measurement is therefore likely to have been caused by the drying.

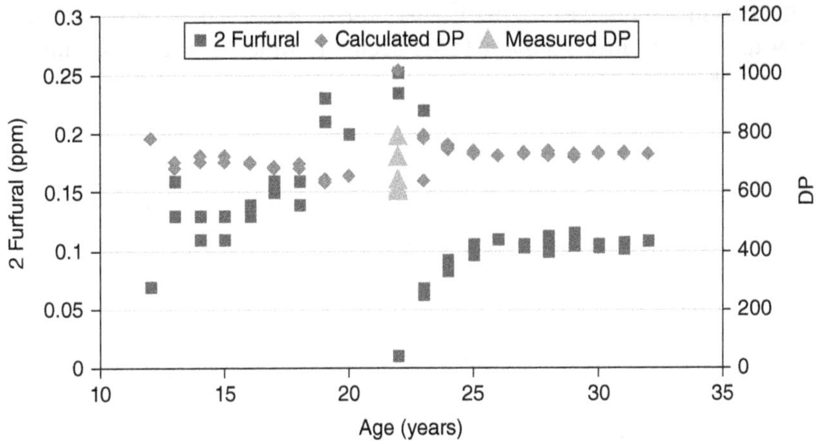

Figure 7.27 Analysis of oil furan content, calculated and measured DP. *Source:* Ref. [27], Figure 10.

7.5.3.2 Condition of Paper Insulation

Figure 7.27 shows the DP of the paper within the transformer. The triangles show the DP for samples of paper, which had been removed from the transformer and measured in a laboratory. The diamonds show the DP calculated using the concentration of 2-furfural dissolved in the oil and the Stebbins method [25]. There is good agreement between the DP measurements and those calculated.

7.6 Summary

As has been discussed, water within the transformer tank is detrimental because it reduces the remaining life of the paper insulation, and decreases the maximum operating temperature that the transformer can run at in contingency operation. Insulating paper which has a water content of 4% will age around six times quicker (if the oil's oxygen content is the same) than when it has a water content of only 1%. The temperature at which bubbling occurs for paper will fall from around 170–180 °C for 1% to 130 °C for 4%, which is a concern because during emergency overload conditions the paper can reach higher temperatures [23]. If the water content of insulation is known, then the utility can factor this into their asset replacement and operation strategies.

Measuring the water content of transformer insulation is challenging because firstly, most of the water is adsorbed by the blocks of cellulose and secondly, the distribution of water through the oil and solid insulation of a transformer is dynamic as it is continuously changing. Online techniques to measure the water content of transformer insulation are desirable because other methods, such as dielectric response, require that the unit is taken offline and electrically isolated from the network. This is not required when installing water activity probes.

Processing data from water activity probes to determine the water content of paper requires the use of adsorption isotherms of cellulose, a well-established method which has been used in the asset managing industry for decades, along with oil KF titrations. An advantage of water activity probes compared with the

KF oil titration method is that the insulation does not need to be in moisture equilibrium for the measurement to be accurate, which is a key assumption when using KF titrations.

References

1 Oommen TV, Prevost TA. Cellulose insulation in oil-filled power transformers. Part II: Maintaining insulation integrity and life. *IEEE Electrical Insulation Magazine* 2006; 22: 5–14.

2 Oommen TV. Moisture equilibrium in paper–oil insulation systems. Proceedings of 16th Electrical/Electronics Insulation Conference, Chicago, IL, October 3–6, 1983, pp. 162–166.

3 Du Y, Zahn M, Lesieutre BC, Mamishev AV, Lindgren SR. Moisture equilibrium in transformer paper–oil systems. *IEEE Electrical Insulation Magazine* 1999; 15(1): 11–20.

4 Laidler K, Meiser J. *Physical Chemistry*, 2nd ed. Montreal, QC: Houghton Mifflin, 1995.

5 Time B. Hygroscopic moisture transport in wood. Doctoral dissertation, Norwegian University of Science and Technology, 1998.

6 Fessler WA, Rouse O, McNutt WJ, Compton OR. A refined mathematical model for prediction of bubble evolution in transformers. *IEEE Transactions on Power Delivery* 1989; 4(1): 391–404.

7 Du Y. *Measurements and Modelling of Moisture in Transformer Insulation.* Saarbrücken: VDM Verlag Dr. Müller, 2009.

8 Martin D, Saha T, Perkasa C, Lelekakis N, Gradnik T. Fundamental concepts of using water activity probes to assess transformer insulation water content. *IEEE Electrical Insulation Magazine* 2016; 32(3): 9–16.

9 Gradnik T, Čuček B, Končan-Gradnik M. Temperature and chemical impact on determination of water content in dielectric liquids by capacitive moisture sensors. IEEE International Conference on Dielectric Liquids, Slovenia, 2014, pp. 381–384.

10 Lelekakis N, Wijaya J, Martin D, Saha T, Susa D, Krause C. Aging rate of Grade 3 presspaper insulation used in power transformer. *IEEE Transactions on Dielectrics and Electrical Insulation* 2014; 21(5): 2355–2362.

11 IEEE Std. C57.106-2006. Guide for Acceptance and Maintenance of Insulating Oil in Equipment, 2006.

12 Ward BW, Oommen TV, Thompson JA. Moisture estimation in transformer insulation. Technical presentation at the IEEE/PES Transformers Committee, Spring 2004 Meeting.

13 Martin D, Saha T, Gray T, Wyper K. Determining water in transformer paper insulation: Effect of measuring oil water activity at two different locations. *IEEE Electrical Insulation Magazine* 2015; 31(3): 17–23.

14 Buck AL. New equations for computing vapour pressure and enhancement factor. *Journal of Applied Meteorology* 1981; 20: 1527–1532.

15 Koch M, Tenbohlen S. Evolution of bubbles in oil–paper insulation influenced by material quality and ageing. *Electric Power Applications* 2011; 5: 168–174.

16 Perkasa CY, Lelekakis N, Wijaya J, Martin D. Investigating bubble formation in vegetable and mineral oil impregnated transformer paper insulation systems. Australasian Universities Power Engineering Conference, Indonesia, 2012.
17 Emsley AM, Stevens GC. Kinetics and mechanisms of the low temperature degradation of cellulose. *Cellulose* 1994; 1: 26–56.
18 Lundgaard L, Hansen W, Linhjell D, Painter TJ. Aging of oil-impregnated paper in power transformers. *IEEE Transactions on Power Delivery* 2004; 19(1): 230–239.
19 Lelekakis N, Guo W, Martin D, Wijaya J, Susa D. A field study of ageing in paper–oil insulation systems. *IEEE Electrical Insulation Magazine* 2012; 28(1): 12–19.
20 Lelekakis N, Martin D, Wijaya J. Ageing rate of paper insulation used in power transformers. Part 1: Oil/paper system with low oxygen concentration. *IEEE Transactions on Dielectrics and Electrical Insulation* 2012; 19(6): 1999–2008.
21 Lelekakis N, Martin D, Wijaya J. Ageing rate of paper insulation used in power transformers. Part 2: Oil/paper system with medium and high oxygen concentration. *IEEE Transactions on Dielectrics and Electrical Insulation* 2012; 19(6): 2009–2017.
22 Martin D, Perkasa C, Lelekakis N. Measuring paper water content of transformers: A new approach using cellulose isotherms in non-equilibrium conditions. *IEEE Transactions on Power Delivery* 2013; 28(3): 1433–1439.
23 IEC Standard 60076-7. Power Transformers – Part 7: Loading Guide for Oil-Immersed Power Transformers, 2005.
24 DIRANA, Dielectric Response Analysis and Moisture in Oil–Paper Dielectrics Brochure. Available at: www.omicron.at/fileadmin/user_upload/pdf/literature/DIRANA-Brochure-ENU.pdf.
25 Stebbins RD, Myers DS, Shkolnik AB. Furanic compounds in dielectric liquid samples. International Conference on Properties and Applications of Dielectric Materials, Japan, 2003, pp. 921–969.
26 Martin D, Cui Y, Ekanayake C, Ma H, Saha T. An updated model to determine the life remaining of transformer insulation. *IEEE Transactions on Power Delivery* 2015; 30(1): 395–402.
27 Martin D, Saha T, Dee R, Buckley G, Chinnarajan S, Caldwell G et al. Determining water in transformer paper insulation: Analyzing aging transformers. *IEEE Electrical Insulation Magazine* 2015; 31(5): 23–32.
28 Martin D, Saha T, Krause O, Buckley G, Chinnarajan S, Dee R, and Russell G. Improving the determination of water content of power transformer insulation paper near the end of its functional life, 2016 Australasian Universities Power Engineering Conference (AUPEC), pp. 1–6, Australia, 2016.

Further Reading

Emsley AM, Xiao X, Heywood RJ, Ali M. Degradation of cellulosic insulation in power transformers. Part 3: Effects of oxygen and water on ageing in oil. IET Science, *Measurement and Technology* 2000; 147: 115–119.

Heathcote M. *The J & P Transformer Book*, 1998.
IEC 60641-2 (Ed. 2.0). Pressboard and presspaper for electrical purposes – Part 2: Methods of tests, 2004.
International Food Information Service. *Dictionary of Food Science and Technology*, 2nd edn. West Sussex, UK: Wiley-Blackwell, 2009.
Martin D, Lelekakis N, Ekanayake C, Saha T, Ma H. Improving measurement techniques of power transformer insulation: A study of the intermolecular interactions between water and vegetable oil based dielectrics. IEEE Conference on Electrical Insulation and Dielectric Phenomena, Shenzhen, China, October 20–23, 2013.
Perkasa C, Lelekakis N, Czaszejko T, Wijaya J, Martin D. A comparison of the formation of bubbles and water droplets in vegetable and mineral oil impregnated transformer paper. *IEEE Transactions on Dielectrics and Electrical Insulation* 2014; 21(5): 2111–2118.

8

Biodegradable Oils and their Impact on Paper Ageing

8.1 Introduction

Mineral oils are still the preferred choice for insulation and cooling purposes in liquid-filled transformers due to their low cost and excellent dielectric properties [1]. A lower fire point is a major disadvantage associated with typical mineral insulating oils, which increases the risk of a subsequent fire in the case of a transformer failure. Moreover, a high fire risk associated with mineral oil results in increasing the lifetime cost of a transformer, mainly due to additional works required for fire mitigation. As a consequence, less flammable insulating oils – such as high molecular weight hydrocarbons (HMWH) and silicon fluids – were introduced in the early 1980s as an alternative to typical mineral oil for improving the fire safety of transformers used in fire-sensitive areas and applications. Today, several thousand transformers filled with silicon and HMWH insulating oils are in service. However, these insulating liquids – including typical mineral and less flammable insulating oils – create environmental problems, mainly due to their poor biodegradable characteristics. Tightening of environmental regulations and the large costs associated with cleaning up an oil spill from downed transformers have forced utilities to use environmentally friendly insulating fluids with high fire safety characteristics. As a result, synthetic-based polyol ester with improved biodegradable properties was introduced since 1970's. The high cost associated with synthetic esters compared to other less flammable fluids hampered their wide application and limited their use for traction and mobile transformers and speciality applications. Therefore, in the early 1990s, immense interest was focused on the use of edible vegetable oils (natural esters) as an environmentally friendly insulating liquid. Natural esters are not only readily biodegradable, but also have excellent fire safety and dielectric properties. By the late 1990s, several types of natural ester-based insulating oils – such as FR3 and Biotemp – were commercialized and their application for new transformers and retrofilling of existing mineral oil filled units has grown substantially over the last two decades. There are some advantages of natural ester insulating oils over other less flammable fluids – for example, they are essentially nontoxic when consumed. Higher oxidation susceptibility is the major disadvantage of natural esters being utilized as insulating oil. Therefore, natural ester insulating oils are recommended for use only in sealed-type transformers. Synthetic and natural esters are collectively known as biodegradable oils. Even with proven environmentally friendly and

Transformer Ageing: Monitoring and Estimation Techniques, First Edition.
Edited by Tapan Kumar Saha and Prithwiraj Purkait.

fire-resistant properties, a lack of field experience and technical knowledge hampers the wide application of biodegradable oil in power transformers. The main purpose of this chapter is to provide a better understanding of the chemistry of biodegradable oil, including chemical composition and degradation mechanisms. The application of existing chemical and electrical-based condition monitoring tools for biodegradable oil-filled transformers will be discussed in this chapter and will be of value to those dealing with biodegradable oil-filled transformers.

8.2 Comparison of the Properties of Insulating Liquids with Mineral Oil

8.2.1 Chemistry of Insulating Oil

Mineral insulating oils are produced from crude petroleum. The quality and composition of crude petroleum extracted from different geographical regions in the world are quite different. However, crude oil is subjected to a series of chemical and physical treatments to produce acceptable insulating oils, which comply with certain specifications provided in international standards such as IEC 60296 and IEEE Guide C57.106 [2–4].

There are a certain amount of olefins–alkenes and heterocyclic compounds containing oxygen, nitrogen, and sulfur in crude oil. These compounds make oil oxidation susceptible [5]. Moreover, some sulfur-containing compounds including sulfides, disulfides, and mercaptans contribute to corrosion of copper conductors [5, 6]. Therefore, these unnecessary contaminants are selectively removed during the refining process to produce transformer oil with acceptable quality. The overall refining process could include acid treatment, solvent extraction, dewaxing, hydrogen treatment, or combinations of these methods [7].

Refined mineral oils are a complex mixture of hydrocarbons, which mainly have three different types of molecule: paraffinic, naphthenic, and aromatic. Figure 8.1 shows the general molecular structures of these three hydrocarbons, such that paraffinic molecules have a straight or branched chain structure whilst naphthenic and aromatic molecules possess cyclic structures. There are two main types of mineral insulating oil: paraffinic and naphthenic. Paraffinic oils are produced from a crude oil, which contains a substantial quantity of naturally occurring n-paraffin molecules. In contrast, crude oil with a very low concentration of naturally occurring n-paraffin is used for manufacturing naphthenic oil [9]. HMWHs are also a type of paraffinic mineral oil, which mainly contain saturated hydrocarbon compounds with long straight-chain structures.

8.2.2 Natural Esters

Natural ester-based insulating liquids are commonly produced from soy, sunflower, and rapeseed, mainly due to their abundant supply, lower cost, and anticipated characteristics for use as a heat transfer fluid and insulant in transformers [9, 10]. Raw vegetable oils contain a myriad of unnecessary compounds, such as gums, wax, volatile compounds, and free fatty acids. These substances show an

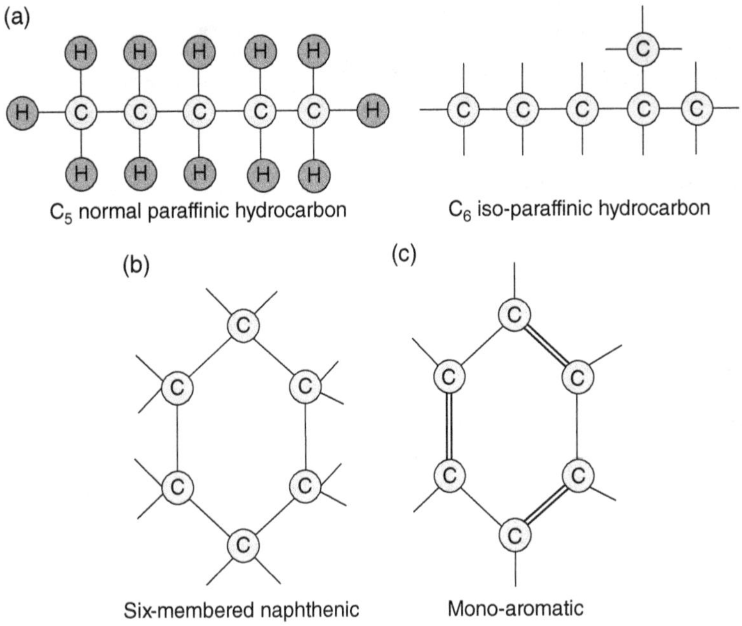

Figure 8.1 Molecular structures of different hydrocarbons in mineral insulating oil [6, 8]. *Source:* Part (a): Ref. [8], Figure 3.1.

adverse impact on the appearance and performance of the oil. Therefore, raw oil is subjected to a series of chemical and physical processes – refining, bleaching, deodorization, and winterizing – to selectively remove unwanted substances and improve the quality standard (appearance and chemical stability) [11–13]. A small amount of nontoxic additives is included in processed oil to decrease the pour point and increase oxidative stability. IEEE Std. C57.147-2008 recommends tests and evaluation procedures as well as criteria and methods of maintenance for natural ester-based transformer insulating oils [14].

Natural esters consist mainly of triglycerides, in which the molecular structure has three fatty acid molecules chemically linked to the glycerol backbone, as shown in Figure 8.2. Diverse physical and chemical characteristics of natural esters derived from different sources are caused mainly by their different fatty acid compositions. Table 8.1 provides typical fatty acid compositions of oil produced from natural seeds: soy, sunflower, and rapeseed. Fatty acids have a linear chain structure with an even number of carbon atoms (typically in the range from 8 to 22) and ending with a carboxylic group [17]. The degree of unsaturation of these fatty acid groups typically varies from 0 to 4. Stearic acid is a saturated acid which is present in natural esters in small concentrations [10]. Oleic acid is the most available mono-unsaturated fatty acid in vegetable oil. Linoleic and linolenic acids are good examples of di and tri-unsaturated fatty acids that can be found in natural esters.

The physicochemical properties of natural esters – such as viscosity, pour point, and oxidation stability – are determined solely by their fatty acid composition. Saturated fatty acid is a chemically stable compound. However, an increase of saturated fatty acid content in oil causes an increase of both viscosity and pour point.

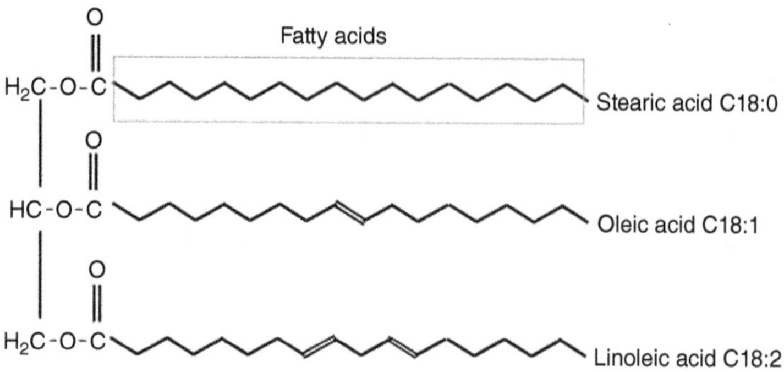

Figure 8.2 Molecular structure of triglycerides [15].

Table 8.1 Fatty acids composition of different vegetable oils [11, 15, 16]

Type of oil	Palmitic C16:0 + stearic acid C18:0 (%)	Oleic acid C18:1 (%)	Linoleic acid C18:2 (%)	Linolenic acid C18:3 (%)
HO* sunflower	5	86	6	0.1
HO soy	3–5	75–85	2–5	2–5
HO rapeseed	4–6	75–85	6–10	3
Typical sunflower	10	19	65	—
Typical soy	15	24	54	7

* HO refers to high oleic.

In contrast, increases of poly-unsaturated fatty acid composition in oil result in lower pour point and viscosity. However, poly-unsaturated fatty acids are highly susceptible to oxidation. The relative oxidation susceptibility of saturated to mono to di to tri-unsaturated C18 (18 carbon atoms) fatty acids is roughly 1:10:100:200 [11]. Therefore, an increase of unsaturated fatty acid composition reduces the chemical stability of the oil significantly. Therefore, it is very important to maintain the balance in the degree of saturation of fatty acids when selecting a base fluid for producing insulating oil. High oleic natural esters which contain over 80% oleic acid are the preferred choice in manufacturing insulating oil. Biotemp and FR3 are two different commercially available natural ester-based insulating liquids, which are being produced respectively from high oleic sunflower and soy oil. In general, a high oleic concentration is not naturally occurring, and thereby special breeding and gene manipulation techniques are used to produce high oleic oil [11].

8.2.3 Synthetic Esters

Chemical reaction between polyols (molecules with more than one alcohol group) and synthetic or natural carboxyl acids yields a synthetic ester whose molecular structure has multiple saturated fatty acid groups (mostly two, three, or four)

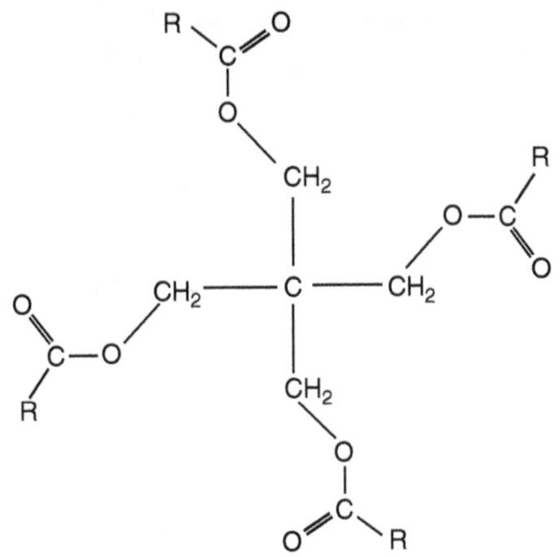

Figure 8.3 Generic molecular structure of synthetic ester [17].

connected to the polyol backbone. There are seven main types of synthetic ester: diester, phthalate, trimellitate, pyromellitate, dimer acid ester, polyols, and polyoleates [17]. Pentaerythritol ester is an example of a commercially available synthetic ester insulating oil, which is produced from a branched mono-acid containing 5 to 18 carbon atoms and alcohol pentaerythritol [9, 17]. The molecular structure of pentaerythritol synthetic ester, which has four fatty acid groups connected to a central polyol backbone, is shown in Figure 8.3. Since synthetic ester insulating oil molecules only contain saturated fatty acids, it possesses excellent oxidation stability compared to both mineral and natural esters. Detailed technical specifications of new synthetic ester insulating oils are available in IEC 61099 [18] and an in-service maintenance guide published as IEC 61203 [19].

8.3 Ageing of Biodegradable Oil

8.3.1 Degradation of Natural Esters

Oxidation is the major degradation mechanism of natural esters (NEs), and it is a radical chain reaction involving three major stages of initiation, propagation, and termination, as shown in Figure 8.4. Metals such as tin, iron, and copper act as an effective catalyst in the oxidation of NEs, whereas lead shows very little impact. Since the strength of hydrogen–carbon bonds next to the carbon–carbon double bond sites in fatty acid groups is weaker, a free radical is easily formed in NEs by eliminating a hydrogen atom from the methylene group next to a double bond [16, 20]. This could be caused by intense heat or a high electric field. It has been reported in the literature [20] that vegetable oil containing poly-unsaturated fatty acids auto-oxidizes even at room temperature, but mono-unsaturated fatty acids (oleic acids) oxidize only at high temperature. The oxidative stability of poly-unsaturated acids in an organic solvent is inversely proportional to the number of unsaturated carbon double bond sites [21]. Therefore, it is clear that the degree of unsaturation is the major factor determining the oxidation stability of NEs. In

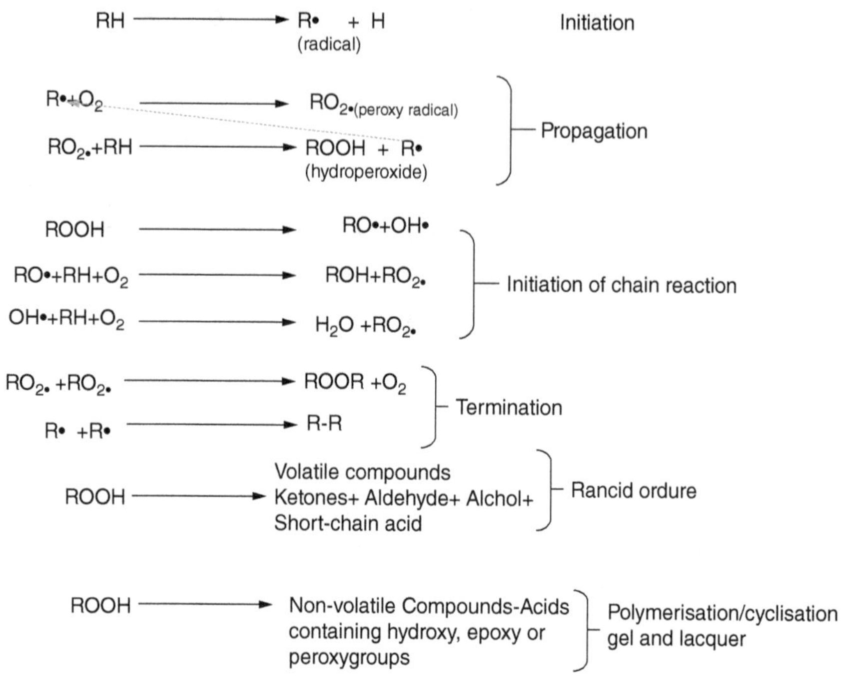

Figure 8.4 Major step of oxidative degradation of natural ester-based insulating oil [20–22].

addition, temperature is another factor which determines the reaction rate of oxidative degradation.

As shown in Figure 8.4, the reaction among free radical, oxygen, and oil molecules results in formation of a triglyceride hydroperoxide and initiation of another radical propagating the oxidative reaction [20, 23]. Triglyceride hydroperoxides break down to form more radicals, leading to a chain reaction. A myriad of non-volatile and volatile compounds – including smaller oxygen-containing by-products such as alcohols, aldehydes (octanal and nonanal with some heptanal), ketones (2, 4-heptadienal), and high molecular acids – are produced by decomposition of triglyceride hydroperoxides [9, 20, 22]. In the termination reaction, peroxy radicals $\left(RO_2^o\right)$ are converted into the stable high molecular compound through dimerization without converting to triglyceride hydroperoxides. At the last stage of oxidation, secondary non-volatile substances of oxidation are subjected to cyclization and polymerization processes, leading to the formation of high molecular weight compounds including gel and lacquer [20]. Concurrently, the viscosity of NEs increases measurably and degrades the cooling capability of oil.

In the case of NEs, hydrolysis is another degradation mechanism which occurs concurrently with oxidation. This is an autocatalytic reaction, because free fatty acid molecules themselves accelerate the hydrolysis reaction [16]. There are three major steps in the hydrolysis reaction in vegetable oil, which are also reversible, as shown in Figure 8.5 [24]. In general, hydrolytic degradation largely increases the acidic level of NEs over time. Temperature is the main factor which determines

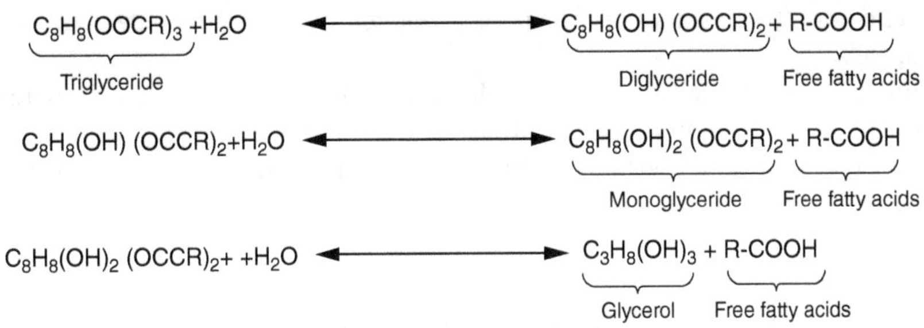

Figure 8.5 Major steps of NE hydrolysis degradation. *Source:* Ref. [24].

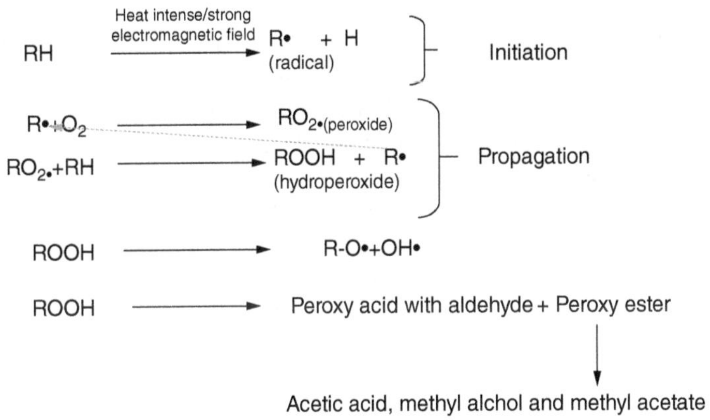

Figure 8.6 Major steps of SE oxidative degradation.

the rate of all chemical reactions. Thus, a large increase in the acidity of natural ester insulating oil under overheating conditions due to pronounced hydrolytic degradation can be expected.

8.3.2 Degradation of Synthetic Esters

Synthetic esters show excellent oxidation stability over mineral and natural ester insulating liquids. However, synthetic esters oxidize slowly at high temperature, accompanied by an evolution of gases (CO, CO_2, and H_2). Thermo-oxidative degradation of synthetic ester oil is also a free radical chain mechanism. At the initial step, α-acylhydroperoxide is formed in a way similar to the oxidation of hydrocarbons [25]. This intermediate by-product then decomposes via two routes, one the usual degradation into free radicals via O–O bond splitting and the other involving hydroperoxide decomposition into both peroxy acid (together with an aldehyde) and so-called peroxy ester [25]. Liquid-phase oxidation of these peroxy esters yields low molecular carboxyl acid (acetic), methyl alcohol, and methyl acetate. Figure 8.6 illustrates the major steps of synthetic ester oxidative degradation. As explained in Ref. [25], it is assumed that α-acylhydroperoxide is produced in synthetic ester (SE) in a way similar to the production of hydroperoxide in hydrocarbon oils.

Synthetic esters are also subjected to hydrolysis degradation at high temperatures. The acids generated via hydrolysis of SEs are shorter-chain acids. These acids are more aggressive than those from NEs and are more corrosive at elevated temperatures.

8.4 Comparison of Physiochemical and Dielectric Properties

8.4.1 General

Table 8.2 presents the desired properties of commercially available synthetic and natural ester insulating liquids in comparison to mineral oil. Fire resistance and environmentally friendly characteristics (e.g., biodegradability and reduced carbon footprint) of synthetic and natural ester insulating oils are promoted to increase their demand as transformer insulant. Moreover, the use of synthetic and natural ester insulating oils in transformers gives advantages such as reduced maintenance and lifecycle costs, longer insulation lifetime, and higher loading capacity [26].

The higher specific gravity of synthetic ester and the higher pour point of natural ester are the main unfavorable characteristics, which could cause problems in low-temperature applications. The specific gravity of synthetic ester is in between that of

Table 8.2 Comparison of typical properties of commercially available insulating liquids [15]

Properties	Units	Test method	Typical mineral oil	Typical BIOTEMP®	TypicalFR3	MIDEL 7131 synthetic ester
Fire point	°C	ASTM D92	180	360	362	322
Pour point	°C	ASTM D92	-50	−15	−21	-50
Specific gravity	g/ml	ASTM D1298 @ 20 °C	0.88	0.914	<0.92	0.97
Kinematic viscosity	cSt	ASTM D445 @ 40 °C	12	42	34	29
Relative permittivity			2.2	3.2	3.2	3.2
Oxidation stability	% per mass	Sludge after 72 h	<0.1	0.01	solid	non
Neutralization number (acidity)	mg KOH/g	ASTM D974	—	0.03	0.02	<0.015
		IEC 61099 9.11	—	<0.03	—	<0.03
Biodegradability	%	CEC L-33-A-93	—	99	>99	30
		EEC Standard 79/831	83	—	—	—

water and ice, and thereby in a condition where supersaturation has occurred, ice particles would float on top of the oil, leading to a catastrophic failure. Natural ester insulating oil in a transformer which is operating at subzero ambient temperature could solidify and the transformer might experience thermal runaway and eventual damage or failure. Rapp *et al.* [27] investigated the cooling behavior of ester insulating oil near the pour point such that a test transformer filled with natural ester was operated at rated load under a controlled ambient temperature of −30 °C and the temperature rises at the top of the oil, core steel, secondary and primary windings were recorded. They reported that there was no indication of abnormal temperature excursion and the transformer operated well below the rated temperature. However, it is worth taking recommendations from oil manufacturers regarding low-temperature application of synthetic and natural ester insulating oils.

8.4.2 Moisture Solubility

Mineral oils are nonpolar liquids which have few or no polar molecules. Therefore, mineral oils show low affinity for water, leading to low moisture solubility. However, a substantial increase in moisture solubility in mineral oils occurs with ageing. Natural and synthetic ester insulating oils are slightly polar liquids and the ester group (COOR) in their molecular structure has a greater tendency to make hydrogen bonds. This behavior results in higher moisture solubility in both natural and synthetic esters compared to mineral oil. Figure 8.7(a) compares the moisture saturation level of unaged natural and synthetic esters with new mineral insulating oil. It indicates that synthetic ester has the highest moisture solubility, followed by natural ester. The moisture solubility of all insulating liquids increasing exponentially with temperature and moisture solubility at a temperature T (K) of mineral oil, natural ester, and synthetic ester can be explained using equations (8.1), (8.2), and (8.3), respectively [14]. The equations corresponding to mineral oil and synthetic ester have been obtained from data given in Refs [14, 28]. One of the major advantages of high moisture solubility of ester oil is that there will be no free-water formation in an ester-filled power transformer in the event of rapid cool down.

$$\text{Mineral oil (ppm)} = 10^{\left(-\frac{1566}{T}+7.08\right)} \tag{8.1}$$

$$\text{Natural ester (ppm)} = 10^{\left(-\frac{802}{T}+5.708\right)} \tag{8.2}$$

$$\text{Synthetic ester (ppm)} = 10^{\left(-\frac{654}{T}+5.614\right)} \tag{8.3}$$

Moisture in cellulose insulation is detrimental and results in decreasing transformer life. Therefore, accurate estimation of water content in cellulose insulation is of utmost importance in order to maximize transformer life through appropriate maintenance actions such as field drying. An equilibrium chart is a widely utilized technique to estimate the moisture content of transformer solid insulation by measuring the moisture content in oil. This technique has been well established for mineral oil-based transformer insulation systems over the past few decades, and the accuracy of this method has improved. However, high moisture

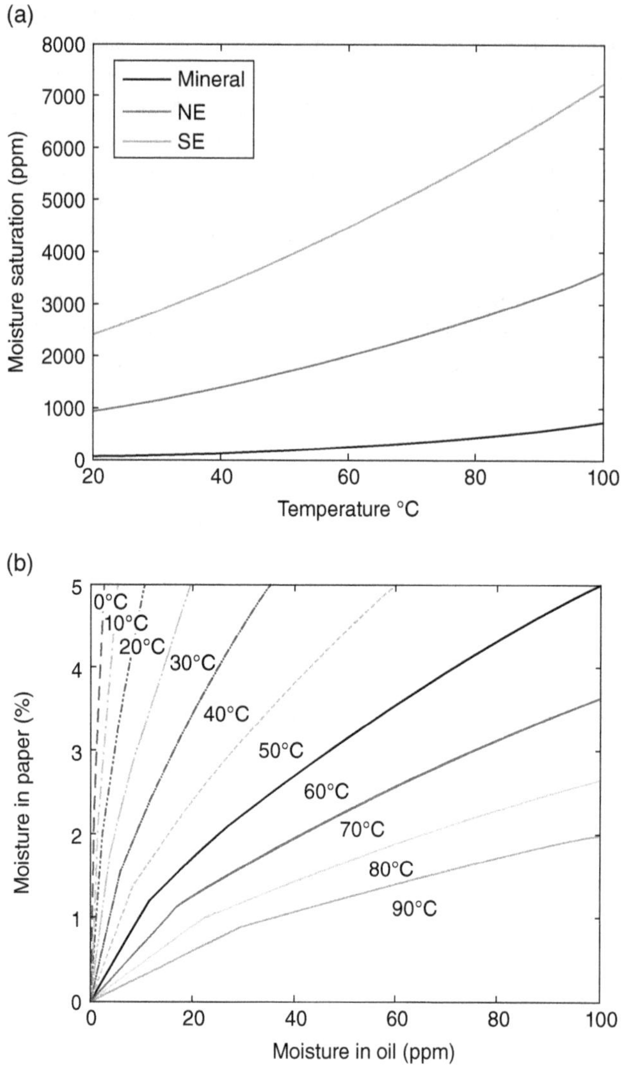

Figure 8.7 (a) Moisture saturation graph for new natural and synthetic ester insulating liquids in comparison with typical new mineral oil [14, 28–30]. (b) Equilibrium chart for mineral oil. (c) Equilibrium chart for natural ester. (d) Equilibrium chart for synthetic ester.

solubility of esters hampers the utilization of existing equilibrium charts, and new charts should be developed for natural and synthetic ester insulating oils individually.

The principle of constructing a moisture equilibrium chart is that relative moisture saturations for adjacent materials are the same, and thereby the moisture isotherms of cellulose and oil are combined by equating relative moisture saturation [29]. The same principle is used in moisture equilibrium charts for mineral, synthetic ester, and natural ester insulating oils, formed using equations (8.4) to (8.7). The relative humidity (water activity) of air [RH(%)] and saturated vapor pressure (P_s) can be calculated using equations (8.4) and (8.5), respectively [30].

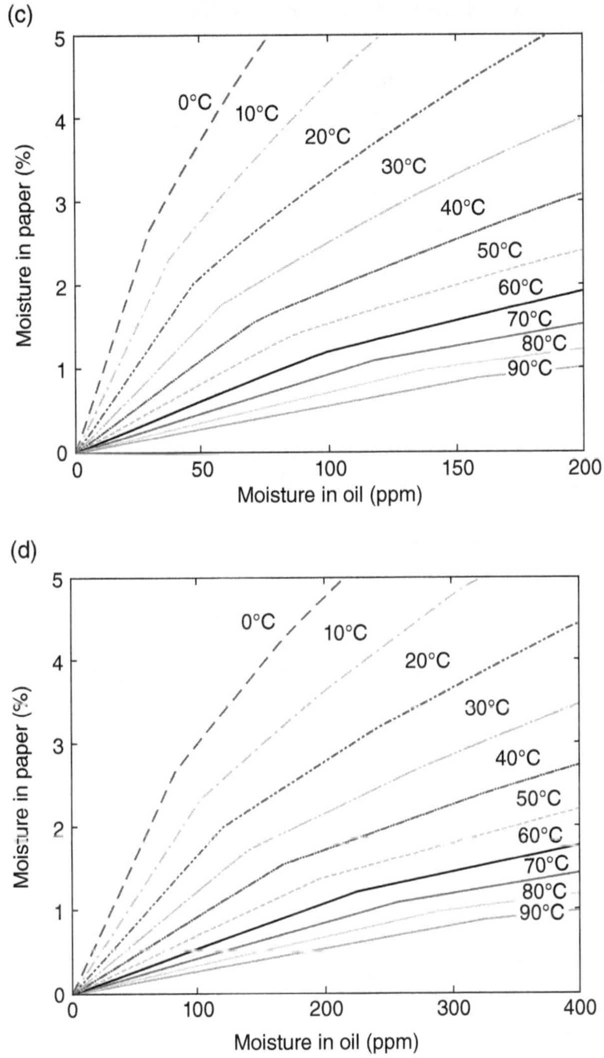

Figure 8.7 *(Continued)*

$$RH(\%) = \frac{p}{P_s} \times 100\% \tag{8.4}$$

$$P_s\,(\text{pa}) = 611 \times \exp\left(\frac{17.28 \times T}{T + 237.3}\right) \tag{8.5}$$

where p is water vapor pressure and T is temperature (°C). The Fessler equation (8.6) gives the water–paper equilibrium formula, and moisture in oil can be related to relative humidity using equation (8.7).

$$\text{Moisture in paper}\,(\text{g}\,H_2O/\text{g}\,\text{paper}) = 2.173 \times 10^{-7} \times p^{0.6685} \exp^{\left(\frac{4725.6}{T+273.15}\right)} \tag{8.6}$$

$$\text{Moisture in oil (ppm)} = \text{Moisture solubility (ppm)} \times \text{RH}(\%) \qquad (8.7)$$

8.4.3 Oxidation Stability and Biodegradability

Natural esters have less resistance to oxidation than both mineral and synthetic oils. Therefore, the application of natural ester-based insulating fluids has been limited to sealed-type units and their use in free-breathing transformers and other equipment (e.g., bladderless conservator design) is not generally recommended [14]. To ensure an extended operational life with good dielectric and cooling characteristics, long-term exposure of natural ester insulating oils to air should be avoided. Thin films of natural ester oil occur naturally in an untanked core and coils, hoses, and fittings essentially oxidize and form a polymerized liquid layer when such surfaces contact with air [10, 14]. Therefore, practical measures should be taken to minimize the time that such surfaces are exposed to air. Excellent oxidation stability of synthetic esters compared to both natural ester and mineral insulating oils allows them to be used in free-breathing transformers even at extreme temperature conditions.

Biodegradability is an essential property for an insulating oil to comply with environmental regulations. The biodegradability is an index that reflects the degree to which fluid is metabolized by naturally occurring microbes in soil or water courses, in the event of a spillage or leak [9]. Mineral and silicon insulating oils show resistance to biodegradation. In contrast, both synthetic and natural esters are generally classified as readily biodegradable substances. The CECL-33-A-93 test is a widely used technique to test the biodegradability of lubricating oil, which is also applicable for insulating oil. Figure 8.8 compares the biodegradability test results for both types of ester and mineral oil.

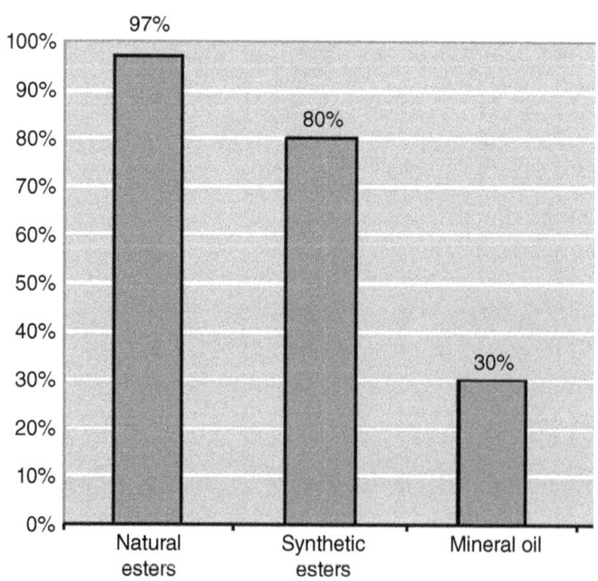

Figure 8.8 Biodegradability of synthetic ester insulating liquids in comparison with mineral oil. *Source:* Ref. [31].

In addition to reducing environmental impacts, fast biodegradation characteristics of synthetic and natural ester give economic benefits to end users such as simple, low-cost spill remediation. However, in the case of a large spill on soil or water, it is better to contact the oil manufacturer for instructions on a cleaning procedure. IEEE Std. C57.147-2008 [14] provides basic remedial actions for cleaning in the case of a natural ester spill, and suggests using a bioremediation catalyst such as fertilization and seeding to speed up biodegradation [14].

8.4.4 Advantages of Higher Fire Point of Ester Liquids

As listed in Table 8.2, both types of ester liquid have a fire point greater than 300 °C and are classified as K-class less flammable fluids. These less flammable oils are ideal for transformers located at strategically important places such as subway tunnels, ships, nuclear power stations, and offshore wind farms. The application of less flammable insulating oils in transformers not only increases fire safety, but also gives substantial economic benefits to end users by reducing overall lifetime costs. Use of ester liquids potentially reduces insurance premiums and essentially eliminates the requirement of costly fire mitigation systems. In general, maintaining an adequate space between adjacent transformers and nearby infrastructure is recommended for mineral oil-filled transformers to eliminate the spread of fire and reduce collateral damage.

There are several international standards which should be followed by utilities when installing electrical equipment with potential fire risks. IEC Std. 61936-1 is one of the widely utilized guidelines, allowing reduced spacing and fire protection for both indoor and outdoor transformers when they are filled with a K-class fluid in such a way that these transformers can be placed much closer to the building without fire walls [32]. Thus, the installation costs of ester-filled transformers are substantially lower compared with mineral oil-filled transformers. The building-to-transformer and equipment-to-equipment separation distances given in FM Global Property and Loss Prevention Data Sheets are listed in Tables 8.3 and 8.4, respectively [33].

Table 8.3 Separation distance between outdoor fluid-insulated transformer and building

			Horizontal distance (m)			
Liquid	**Approved transformer or equivalent**	**Liquid volume (m^3)**	**2 h Fire resistance construction**	**Non-combustible construction**	**Combustible construction**	**Vertical distance (m)**
Less flammable	Yes	N/A	0.9	0.9	0.9	1.5
	No	<38	1.5	1.5	7.6	7.6
		>38	4.6	4.6	15.2	15.2
Mineral oil, unapproved	N/A	<1.9	1.5	4.6	7.6	7.6
		1.9–19	4.6	7.6	15.2	15.2
		>19	7.6	15.2	30.5	30.5

Source: Ref. [33], Table 2.a.

Table 8.4 Outdoor fluid-insulated transformer equipment distance

Liquid	Approved transformer or equivalent	Fluid volume (m^3)	Distance (m)
Less flammable	Yes	N/A	0.9
	No	<38	1.5
		>38	7.6
Mineral, unapproved fluid	N/A	<1.9	1.5
		1.9–19	7.6
		>19	15.2

Source: Ref. [33], Table 2.b.

8.4.5 Influence of Higher Viscosity of Ester Liquids

As listed in Table 8.2, viscosity values of ester insulating liquids are almost three times greater than those of a typical mineral oil. However, their difference diminishes with temperature [9]. Viscosity is a parameter which reflects the fluidity of a liquid, such that a liquid with low viscosity has higher fluidity and vice versa. As a result of high viscosity, ester insulating fluids flow slowly through winding ducts, leading to a reduced heat transfer capability by convection. This effect is more significant for naturally cooled transformers. In contrast, higher thermal conductivity of ester oil shows a positive impact on their cooling characteristics and compensates for the effect of high viscosity to some extent. However, design changes – such as increasing the size of oil ducts – will still be required for an ester-filled transformer to get the same cooling performance as conventional mineral oil-insulated transformers. Once a transformer designed for typical mineral oil is filled with ester fluid, it could expect higher operating temperatures. This section will report the results of some studies performed to investigate the temperature rise of ester-filled transformers.

Based on the results of heat run tests performed by transformer manufacturers, CIGRE Brochure 436 [9] reported that the temperature rise in natural ester-filled transformers up to 5 MVA is 2–3 °C greater than an identical mineral oil-filled unit. It has been reported in Ref. [34] that the operating temperature of a transformer designed for conventional mineral oil will be 4–5 °C higher after retrofilling with FR3 oil. However, a similar temperature rise can be expected for units equipped with FOA or FOAF cooling systems. Moore [35] ran a thermal design program using properties of natural ester insulating liquids to calculate the temperature rise for dozens of existing transformer designs. He reported that, when a design program uses properties of natural esters, the maximum temperature rise is 21 °C greater than that using properties of mineral oil. Pukel *et al.* [36] simulated the temperature distribution of the disc windings of a transformer when it is filled with mineral oil, synthetic ester oil, and natural ester oil. This study revealed that the hotspot temperatures of synthetic and natural ester-filled transformers are 7 °C and 10 °C, respectively, higher than that of the mineral oil-filled transformer under similar conditions. Figure 8.9 shows the results of heat run tests

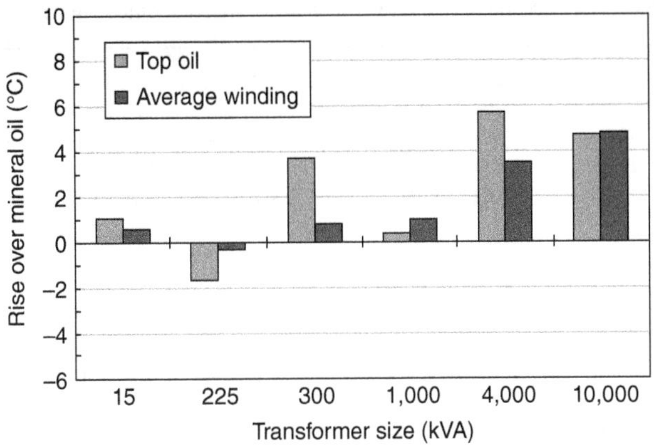

Figure 8.9 Average winding temperature rise and top oil differentials between mineral oil and natural ester. *Source:* Ref. [37], Figure 11.

for transformers with a range of power ratings from 15 kVA to 10 MVA. Overall, one can comment that natural ester-filled transformers operate at a little higher temperature than mineral oil-filled transformers and minor modifications in cooling system design could compensate for this effect.

A longer time required for the impregnation process in ester insulating liquids is one problem experienced by transformer manufacturers. Generally, this process is performed under vacuum conditions and is principally governed by so-called capillary action, which depends on the viscosity of the impregnate liquid. The impregnation process occurs slowly when it uses fluid with a high viscosity. Researchers reported that the rate of fluid impregnation was similar in esters at 60 °C compared to mineral oil at 20 °C [36]. Accordingly, the impregnation process with ester oil should be performed at a higher temperature than that with mineral oil.

8.5 Guide for Assessing the Quality of Ester Liquids in Service Unit

Several test methods are utilized on insulating liquids to determine their appropriateness for further use in a service unit and whether corrective action is necessary. Color and appearance, kinematic viscosity, neutralization value, interfacial tension, breakdown voltage, and dielectric dissipation factor are the most widely used parameters for analyzing the insulation quality of transformer oil. Change of density may also be an indicator for contamination with other liquids and materials. IEC 61203-1992, IEEE Std. C57.106-2006 and IEEE Std. C57.147-2008 [14] have been tailored to provide guidelines classifying the condition of new and service-aged synthetic ester, mineral oil, and natural ester insulting liquids, respectively, based on the above parameters. See Table 8.5 for an overview.

In the case of mineral insulating oil, increasing the neutralization number mainly reflects the degree of deterioration of oil under oxidizing conditions.

Table 8.5 Overview of limiting values of dielectric and chemical parameters of insulating oils

Oil condition	Parameter	Mineral oil[a]	Synthetic ester[b,c]	Natural ester[a,d]
New	Acidity (mg KOH/g)	0.03 (ASTM D974)	0.03 BS EN 61099	0.06 (ASTM D974)
	DDF (%)	0.05 @ 25 °C	3 @ 90 °C	0.2 @ 25 °C
	Color	<0.5 (Clear)	<200 ISO 2211	1.0 (ASTM D1500)
	Breakdown voltage (kV) ASTM D1816 – 1 mm gap	>35 (1 mm) ASTM D1816	>45 (2.5 mm) IEC 60156	>35 (1 mm) ASTM D1816
Service aged	Acidity (mg KOH/g)	0.2 (Voltage class <69 kV)	2	0.3
	DDF (%) @ 25 °C (ASTM D924)	0.5	1	0.5[a] 3[d] (provisional)
	Color (ASTM D1500)	4	Not dark in color No acrid odor	1.5[d] (provisional)
	Breakdown voltage (kV) ASTM D1816 – 1 mm gap	>30 (Voltage class >230 kV)	>30 (Voltage class <35 kV)	>30 (Voltage class >230 kV)

[a] See IEEE Std. C57.152-2013.
[b] See BS EN 61099 (1992).
[c] See BS EN 61203 (1994).
[d] See IEEE Std. C57.147-2008.

Short-chain acids produced by degradation of mineral oil are detrimental to an insulation system and create some problems by oxidizing metallic parts. However, a rise in neutralization number of synthetic and natural ester insulating oils does not merely represent oxidizing conditions. Hydrolysis degradation of esters yields acids and gives rise to a neutralization number. Since the degradation process of cellulosic insulation releases water, the hydrolysis reaction in ester oil in a transformer could become pronounced with time. Synthetic ester produces mainly short-chain acids and these acids accelerate the ageing process of paper insulation and create corrosion on copper conductors. However, a hydrolysis reaction in natural ester produces long-chain fatty acids which are not harmful to insulation systems or other components in transformers.

Since the chemical activity of acids generated in different insulating oils is not similar, different limiting values for the neutralization number have been proposed for in-service mineral, synthetic, and natural ester insulating oils. Synthetic ester insulating oil with satisfactory conditions for further use in electrical equipment with a voltage rating of 35 kV or lower should be less than 2 mg KOH/g [38]. The acceptable maximum acidity values for in-service mineral and natural ester oils of <69 kV voltage class are 0.2 mg KOH/g [4] and 0.3 mg KOH/g [39], respectively. The proposed limiting neutralization value of natural ester oil is arguable, because it is much lower than the limiting value provided for synthetic ester, although short-chain acids generated by degradation of synthetic ester are more reactive and detrimental than long-chain acids coming from the degradation of natural ester. Moreover, acidity value measurement results of several laboratory studies – including the case study presented in this section – point out that the acidity of natural ester insulating oil can increase far beyond the specified limiting value, even after being aged in a sealed environment with minimum oxygen [22, 23, 40, 41]. In addition, the results of some studies confirmed better ageing performance of cellulosic material in natural ester with a high acidity level rather than in a typical mineral oil [41, 42]. This suggests considering the behavior of natural ester insulating liquids when revising international standards corresponding to natural ester insulating oils in the future.

The color of oil is a primary parameter which reflects the degree of degradation and possible contamination of oil during use. A numeric value based on international color standards (ASTM D1500) is generally used in expressing color changes. New mineral oil has a clear appearance, with its color number in the range 0 to 0.5. A dramatic increase in color number indicates abnormal degradation in oil. Severely aged mineral oil with a dark brown color has a color number greater than 7. Such oil is not suitable for further use as an insulant [39]. Synthetic ester with a clear appearance, without being excessively dark in color, is considered as an insulant with acceptable condition for further use [38]. Natural ester color changes have not yet been standardized for quality assurance. IEEE Std. C57.147-2008 has provided provisional limiting values (1.5) for color number for service-aged natural ester insulating oils [14]. The case study presented here reveals that there is no significant color change in natural ester insulating oils after ageing of 2800 h in a sealed environment with minimal oxygen supply and dry pressboard insulation. On the contrary, a dramatic change in color number occurs even after 1150 h of ageing when it was aged with pressboard having 2% moisture. However, viscosity remains almost the same as new oil, indicating that there is no severe oxidation in the oil. The color change could be due to a higher solubility of decay products in natural ester, because vegetable oil ageing by-products are less carbonaceous [43]. Moisture cannot be excluded completely from a transformer, as it is a by-product of paper degradation. Therefore, rapid color change in natural ester oils could be expected in service conditions, compared to mineral oil. However, for a conclusive interpretation, further investigation will be required to understand the color change of natural ester insulating oil.

The DDF of oil is used as an indication of changes in quality resulting from contamination and deterioration in service or as a result of handling [39]. As per IEEE Std. C57.152-2013, the limiting value for DDF of both service-aged mineral oil and natural ester insulating liquids is 0.005 at 25 °C. British Std. EN 61203 specified a value of 0.01 as the acceptable maximum DDF for service-aged synthetic esters. However, it is to be noted that natural ester inherently has a higher DDF than mineral oil. The results of several studies point out that the DDF of natural ester increases far beyond the limiting value specified in standards, even when aged under an oxygen-free environment [44–46]. The DDF values measured at different temperatures in the literature have been normalized to 25 °C using 0.4 eV activation energy. Table 8.6 provides the results of a case study performed using

Table 8.6 Measured oil parameter under two different ageing conditions [47, 48]

Ageing condition	Parameter	Mineral	SE	NEA	NEB
Oil aged with dry pressboard for 2800 h	Acidity (mg KOH/g)	0.01	0.81	0.59	0.23
	DDF (%) @ 25 °C	0.09	0.67	0.2	0.3
	Color	1	<0.5	1.5	1.5
	Viscosity change (% of initial value)	0	0	0	0
Oil aged with wet pressboard (2%) for 1152 h	Acidity (mg KOH/g)	0.02	—	2	0.5
	DDF (%) @ 25 °C	0.04	—	0.52	0.7
	Color (ASTM D1500)	1	—	4	2.5
	Viscosity change (% of initial value)	0	—	0	0

sunflower and soybean-based natural ester insulating oils under controlled conditions in a laboratory of the University of Queensland. It appears clearly that both natural esters used in this study possess DDF values greater than the limiting value after 1150 h of ageing when the ageing system includes wet pressboard [48]. 0.5% is a disputable limiting value for the DDF of service-aged natural ester insulating oil, and it may be necessary to consider this in revising the corresponding international standards in the future. In contrast, the provisional limiting value of 0.03 provided in IEEE Std. C57.147-2008 [14] seems to be more appropriate for diagnostic purposes.

8.6 Case Studies

8.6.1 Case Study 1

Four different types of commercially available insulating oil – including one type of mineral oil, synthetic ester (SE), sunflower oil-based natural ester (NEA), and soybean oil-based natural ester (NEB) – were aged by the authors of this book under a minimum oxygen supply (under a nitrogen cushion) at 120 °C in the presence of dry pressboard insulation and copper conductors. Oil properties were measured after 2800 h of ageing. Another set of oil samples from mineral, NEA, and NEB were aged with pressboard insulation having 2% moisture for 1150 h. Apart from moisture in the pressboard, all other conditions were similar. The measured oil properties, including acidity, DDF, color, and viscosity, are as listed in Table 8.6.

In the case of oil aged with dry pressboard insulation, the DDF of synthetic ester and acidity of NEA exceed the limiting values. The acidity and color of NEA aged with wet pressboard insulation are exceptionally high. The DDF and acidity of NEB oil aged with wet pressboard are marginally higher than the limiting value.

Though NEA oil with a higher color number has a cloudy appearance, there was no indication of sludge formation. Overall, there is no change in viscosity of any of the oils, indicating that their cooling performance is the same as that of new oil. In the case of a transformer, an increase of moisture in paper insulation to about 2% after 10–15 years of operation cannot be avoided. In such a situation, a rapid increase in acidity and color of natural ester insulating oil will be expected. Therefore, the interpretation scheme based on acidity value, DDF, viscosity, and dielectric breakdown strength will be more appropriate in making decisions for maintenance actions such as oil reclaiming and removal of dissolved water for natural ester-based insulating oil.

8.6.2 Degradation of Cellulosic Insulation in Ester Liquids

The thermal stability of solid insulation (cellulosic materials) in insulating oil is a very important factor, because common failure modes of electrical transformers are associated with degradation of cellulosic insulation. Moreover, constraints on the loading capability of transformers are primarily determined by the thermal stability of solid insulation. In the last 15 years, the ageing behavior of typical cellulosic insulation materials in natural ester-based insulating liquids has been investigated extensively and also compared to typical mineral insulating oil [36, 49–51].

Liao *et al.* [42] performed a sealed-tube accelerated ageing experiment to understand the ageing behavior of typical Kraft insulation paper and thermally upgraded paper in both commercially available natural ester (BIOTEMP®) and conventional mineral insulating oils. Their study identified that both Kraft and thermally upgraded paper show better thermal stability in natural ester than in mineral oil. The results of this study show that the levelling off DP (LODP) of both types of paper insulation aged in natural ester is about 100 to 400 times higher than that of paper insulation aged in mineral oil. Lijun *et al.* [52, 53] compared the ageing behavior of typical Kraft paper insulation in a commercially available natural ester insulating liquid (BIOTEMP®) with that of conventional naphthenic-based mineral oil through an accelerated ageing experiment performed at three different temperatures. Their study also reveals that natural esters slow down the ageing of Kraft paper, leading to a longer life expectancy. Coulibaly *et al.* [41] confirmed experimentally that cellulosic insulation materials in natural and synthetic esters age at a slower rate than in mineral oil for both oxygen-rich and low-oxygen environments.

Cargill's brochure for Envirotemp™ FR3 [54] fluid presents a lifetime curve for thermally upgraded Kraft insulation paper in natural ester insulating oil compared with that provided in the IEEE loading guide [55] for mineral insulating oil. IEEE Std. C57.91 has defined the relationship of insulation deterioration with time and temperature, with an exponential function which gives 1 per unit lifetime (65,000 h based on 50% retained tensile strength of insulation) at 110 °C. As shown in Figure 8.10, an FR3-filled transformer has a 20 °C temperature advantage over a conventional mineral oil-filled unit. This means that a natural ester-filled transformer can be operated 20 °C warmer than an equivalent mineral oil-insulated unit, while maintaining the same life expectancy. Similar behavior has been

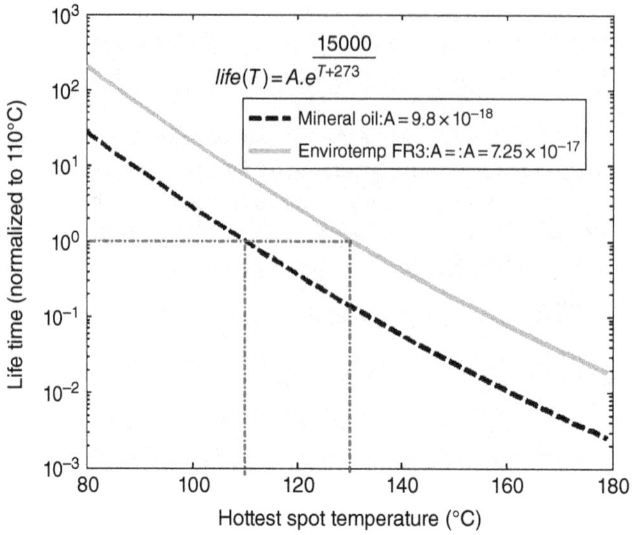

Figure 8.10 Life of thermally upgraded insulation in mineral and natural ester. *Source:* Ref. [54].

reported for sunflower oil-based natural ester insulating oil manufactured by ABB (Biotemp) [56].

8.6.3 Case Study 2

Typical Kraft pressboard insulation samples were aged by the authors of this book at 120 °C in three different glass ampoules which contained Shell Diala mineral oil, NEA, and NEB. Ageing ampoules also included copper conductors and core steel. The mass ratio between cellulose insulation, oil, and copper was 1:10:1. The headspaces of the ageing ampoules were filled with dry nitrogen. In addition to that, sets of wet pressboard insulation samples with 2.1%, 2.4%, and 2.0% moisture content were also aged under similar conditions in Shell Diala mineral oil, NEA, and NEB, respectively. The ageing condition of pressboard insulation was assessed using DP. The method given in IEC Standard 60450-2004 was utilized to measure the average DP. The results shown in Figure 8.11 indicate that cellulosic insulation could have a longer lifetime in natural ester insulating liquids than in conventional mineral oil. Moreover, when the moisture content in cellulose insulation is high, this advantage is more significant.

Overall, one could make a generalized statement such that natural ester insulating oil shows resistance to thermal ageing of cellulosic insulation materials and confirms the longer life expectancy and high overloading capability of natural ester-filled transformers. Most studies claim that this advantage of natural ester insulating liquids is mainly due to their high moisture solubility. Natural esters attract more water from paper insulation than mineral oil, causing them to slow down the hydrolytic degradation of cellulosic insulation. This hypothesis is also valid for the protective behavior of synthetic ester on cellulose insulation [41]. Moreover, hydrolysis reaction in ester consumes free water, causing a shielding

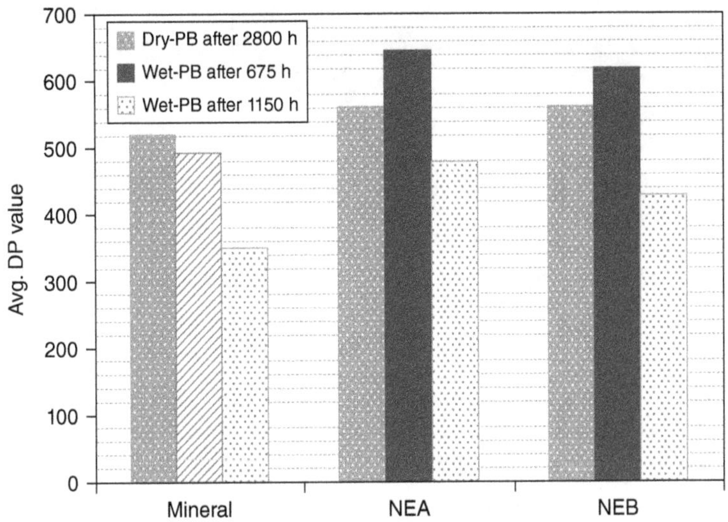

Figure 8.11 DP measurement results of case study [47, 57].

effect, which reduces the interaction between water and PB [42]. This could also be the reason for a decrease in hydrolytic degradation of cellulose insulation in ester insulating liquids. Hydrolysis of natural ester produces long-chain fatty acids. These fatty acids are chemically bonded onto the C6-hydroxyls groups of cellulose molecules, which make them more stable [53]. This could also be a reason for better thermal stability of cellulosic material in natural ester insulating oil.

8.7 Use of 2-FAL as an Ageing Indicator in Ester-Based Insulation Systems

Thermal degradation of cellulosic insulation in electrical equipment such as transformers produces a class of furanic compounds which can be detectable in oil. Moreover, 2-FAL can generally be found in its highest concentration in oil, because of the fact that production of furanic compounds due to degradation of pressboard insulation in decreasing order is 2-FAL > 5-HMF > 2-FOL [58]. In addition, 2-FAL is a thermally stable compound. Thereby, dissolved 2-FAL concentration in transformer oil has been used as an indicator of paper insulation degradation for decades. However, there is not enough discussion on the use of 2-FAL as an ageing indicator in ester-based insulation systems. This section will summarize the results of some laboratory investigations into 2-FAL analysis of ester-based insulation systems.

McShane *et al.* [49] compared the ageing behavior of Kraft insulation paper in mineral and FR3 insulating oils through an accelerated sealed-tube ageing experiment. This study identified that 2-FAL concentration dissolved in FR3 oil is several orders of magnitude less than that dissolved in mineral oil at a similar ageing

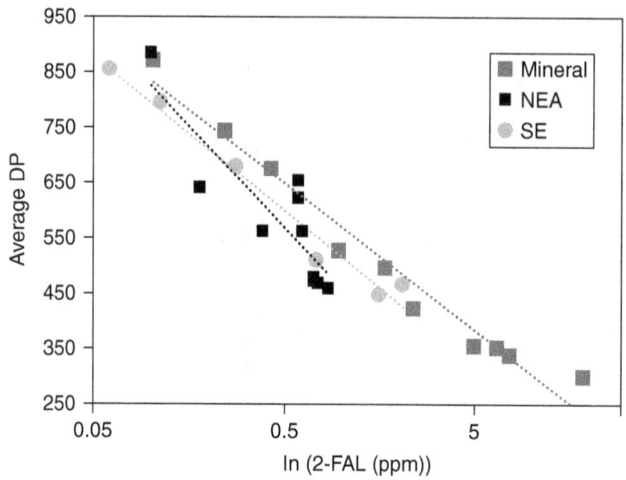

Figure 8.12 Relationship between DP and 2-FAL concentration of oil [47].

condition. Junru *et al.* [50] also observed a similar behavior, whereby 2-FAL concentration dissolved in camellia oil (natural ester) is several orders of magnitude less than that in mineral oil for a similar degree of ageing of paper insulation in both types of oil. In Refs [42, 59], the ageing behavior of both Kraft and thermally upgraded paper insulation in Biotemp insulating oil has been investigated and compared to that in typical mineral insulating oil. The results of both studies indicate that dissolved 2-FAL concentration in mineral oil is several orders of magnitude greater than that in Biotemp. In addition, Figure 8.12 presents the results of our study to correlate detected 2-FAL concentration in three different types of insulating oil [Shell Diala mineral oil, NEA, and MIDEL 7131 (synthetic ester)] to the DP value of pressboard insulation aged in corresponding oils.

It appears clearly in Figure 8.12 that ln(2-FAL) and DP values of pressboard insulation aged in mineral oil and synthetic ester can be represented using a liner relationship with good accuracy. On the contrary, it is difficult to establish a linear relationship between ln(2-FAL) and DP for the NEA–pressboard system with reasonable accuracy. In addition, detected 2-FAL concentration in NEA is relatively low compared to that in both mineral and synthetic ester oils. Thereby, one could propose that a relatively lower concentration of 2-FAL can be detected in natural ester insulating oil than in typical mineral oil for a similar degree of paper insulation ageing. Anti-oxidative additives used in natural ester insulating oil could react with furanic compounds, giving a low concentration of 2-FAL in natural ester oil. In addition, in an oil–paper system, 2-FAL can be lost due to thermal and hydrolytic decomposition, and this mechanism is more pronounced in an acidic environment [60]. Over thermal ageing, acidity of natural ester increases to a large value and this may also be partly accountable for detecting a low concentration of 2-FAL in aged natural ester oil. Overall it can be suggested that the use of 2-FAL as an ageing indicator of paper insulation in natural ester-based insulation could be problematic.

8.8 Stray Gas Generation in Ester Insulation Liquids

8.8.1 Gassing Behavior of Ester Oil

Stray gases are formed in insulating oil as a consequence of high electric field stress or local overheating of oil due to abnormal operating conditions such as overloading, local electric discharges, and excessive conductor resistance [61]. Free radicals are generated through breaking of carbon–carbon and carbon–hydrogen bonds under the influence of intense heat or high electric field, and they are then combined with each other to form different types of stray gases – namely hydrogen (H_2), methane (CH_4), ethane (C_2H_6), etc.

Ester insulating liquids possess completely different chemical structures compared to mineral insulating oil. This leads to a quantitatively different stray gassing behavior for ester insulating oil, but fortunately gas composition is the same for all types of oil. Different types of chemical bonds in ester oil, such as C = O, C = C, and C–H allylic bonds, give this quantitative difference [62]. A notable difference is that esters yield a large amount of carbon oxide (CO and CO_2) due to the breakdown of the carbonyl group (COO) [11, 61]. In natural ester insulating oil, a radical can easily be formed by eliminating a hydrogen atom in allylic position (next to a C = C bond), because the energy required for this mechanism is low – 100, 75, and 5 kcal/mol, respectively, for a hydrogen atom in saturated, allylic, and doubly allylic methylene group [21]. This could also lead to a different gassing behavior for natural ester insulating oils derived from different sources such as sunflower, soy, and rapeseed. IEEE Std. C57.155-2014 (draft) reported that production of methane, ethane, and ethylene in esters is greater due to a low-temperature overheating and ester fluids containing linolenic acid generates ethane even under non-fault conditions. Ester insulating oils show different gas solubility characteristics compared to mineral oil, and gas solubility constants depend on the type of ester too. This makes analyzing and interpreting dissolved gases in ester oils more complicated. This section will present three case studies to enhance the understanding of gassing behavior of different insulating oils. The use of a DGA interpretation scheme for ester insulation will be discussed at the end of the section.

8.8.1.1 Case Study A: Simulated Low-Energy Discharge

A standard transformer oil breakdown tester (100 kV, 2 kVA) was used to create electrical breakdowns in the oil. Firstly, 500 ml of degassed oil was poured into the test cell of the breakdown tester. The top cover was then closed and high-vacuum grease was applied to all the joints. Gas leaks from the oil are thus avoided. Then, 100 breakdown tests were performed on the oil sample. The time difference between two consecutive breakdowns was set to 2 min. After each breakdown, the oil samples were stirred for 30 s to ensure that the fault gases were distributed homogenously. Since the breakdown current had been limited using the internal circuit, the electrical breakdowns created in the oil were considered to be low-energy discharges. These low-energy discharge faults were simulated for four different oil types: Shell Diala mineral oil, NEA, NEB, and MIDEL 7131 synthetic ester. The dissolved gases in the oil after 100 breakdowns were analyzed using

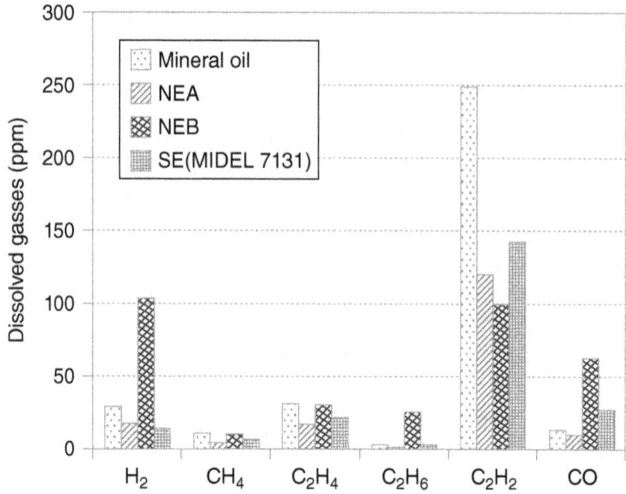

Figure 8.13 DGA results after 100 electrical breakdowns. *Source:* Ref. [63].

the method prescribed in ASTM D3612-02. The results from this work, conducted by the authors of this book, are presented in Figure 8.13.

This figure indicates that under low-energy electrical arcing conditions, gases produced in all types of oil are mainly acetylene (C_2H_2) but in NEB, an additional significant quantity of hydrogen was also produced. A notable difference in stray gases production under low-energy discharge is that the generation of C_2H_2 in mineral oil is significantly greater compared to ester fluids. Ethylene, hydrogen, and methane become the second, third, and fourth highest hydrocarbon gases, respectively, detected in mineral oil and moreover, almost the same gas generation patterns can be observed in NEA, NEB, and MIDEL 7131. However, in the case of NEB, one could also assume that hydrogen and acetylene are equally produced under low-energy discharge conditions. The stray gases generation pattern observed in this case study shows good agreement to some extent with the results in Refs [64, 65], indicating the applicability of these results for diagnostic purposes.

8.8.1.2 Case Study B: Simulated Thermal Fault at 150 °C

250 ml of degassed oil was poured into a specially designed glass bottle. Then, air in the headspace was pumped out and the headspace was filled with dry nitrogen. High-vacuum grease was applied to all the ground glass joints to ensure a fully sealed environment. Thus, one could assume that there was no gas leak in the system. The oil-filled glass bottle was then placed in an aluminum heater block and the temperature of the heater block was set to 150 °C (±2 °C). Ageing was performed over 72 h at this temperature, and DGA was performed on two aged oil samples. The average is reported in Figure 8.14. This thermal fault was simulated on all types of oil considered in case study A.

Figure 8.14(a) indicates that carbon monoxide is the key combustible gas produced in synthetic ester insulating oil under a thermal overheating condition at

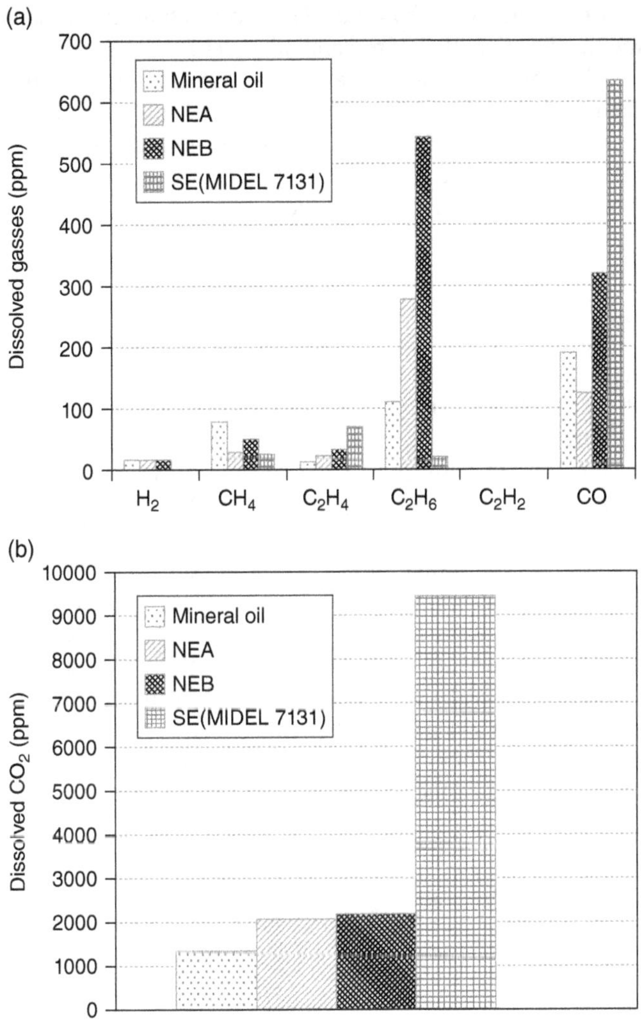

Figure 8.14 Stray gas in oil after ageing over 72 h at 150 °C: (a) combustible gases; (b) CO_2.

150 °C followed by ethylene methane and ethane, respectively (and they exist in lesser quantities compared to carbon monoxide). One could assume that mineral oil produces mainly carbon monoxide, ethane, and methane under low-temperature overheating, with a maximum concentration of carbon monoxide followed by ethane and methane. In the case of natural esters, ethane remains the main combustible gas under the simulated thermal fault and, in addition, a substantial concentration of carbon monoxide was also detected. If one compares the concentration of ethane in NEB with that detected in the other three oils, it clearly proves that the production of ethane in NEB is greater by several orders of magnitude. Such behavior for soybean oil-based insulating liquids has also been reported in IEEE Std. C57.155-2014 (draft). As shown in Figure 8.14(b), carbon dioxide is produced abundantly in synthetic ester oil under thermal overheating

conditions at 150 °C. Moreover, the carbon dioxide concentration in synthetic ester is almost six times greater than that of mineral oil under similar conditions. This behavior indicates that the rate of carbon dioxide production from synthetic ester could exceed that from paper degradation, leading to a misinterpretation of DGA results with cellulose involvement [61]. Though the concentration of carbon dioxide in both types of natural ester oil under overheating conditions at 150 °C is greater than mineral oil, it is not as high as synthetic ester.

8.8.1.3 Case Study C: Simulated Thermal Fault at 120 °C with Similar Proportion of Paper Insulation Copper Conductors in Power Transformer

DGA was performed on an oil sample taken from case study 1 after thermal ageing over 2800 h. The results are presented in Figure 8.15.

This figure shows that a very low quantity of stray gases is produced in synthetic ester due to thermal ageing at 120 °C. An almost similar amount of methane and ethane gases is produced in synthetic ester, and this statement is true for hydrogen and ethylene also. A notable fact is that ethane production in NEB is abundantly high, which remains the main stray gas in NEB for a thermal fault followed by methane. A nearly similar gassing pattern has been observed for mineral oil and NEA. However, total hydrocarbon gases detected in NEB is larger than all other oils. Figures 8.14 and 8.15 show clearly that the pattern of hydrocarbon gases generation in mineral oil, NEA, and NEB is similar for both simulated faults at 120 °C and 150 °C, indicating that both faults are in a similar category. In contrast, the hydrocarbon gas generation pattern in synthetic ester under an overheating condition at 120 °C and 150 °C is dissimilar to some extent, because no hydrogen has been detected in synthetic ester oil after overheating at 150 °C.

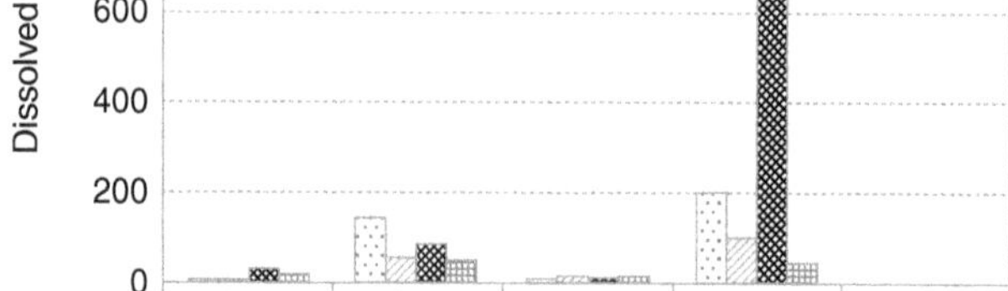

	H_2	CH_4	C_2H_4	C_2H_6	C_2H_2
Mineral oil	6	147	9	201	0
NEA	4	56	19	101	0
NEB	32	87	14	1256	0
SE(MIDEL 7131)	20	50	18	46	0

Figure 8.15 Combustible gas dissolved after ageing for 2800 h with pressboard and copper [47, 57].

It is worthwhile mentioning that no acetylene is produced in ester insulating oil due to an overheating condition at 120 °C and 150 °C.

8.8.2 Interpretation of the DGA Data of Ester Oil

Duval's triangle, IEC gas ratio, and IEEE key gas methods are the most widely utilized interpretation tools used to identify a possible fault occurring in a transformer based on DGA ester oil data. Duval's triangle technique has recently been extended for ester-based insulating oil by considering the different gassing behaviors [66, 67]. The differences in new Duval's triangle for synthetic ester and two different natural esters (FR3 and Biotemp) occur primarily at the boundary between T1 and T3 thermal fault zones and zones D1 (partial discharge) and D2 (low-energy discharge). The insulating fluid subcommittee of the IEEE transformers committee has revised Table 1 in IEEE Std. C57.104, which gives the threshold value for dissolved gases of a transformer with no previous sample history, to comply with a gas formation pattern that is likely to occur when ester fluids are used in transformers. This section will investigate the applicability of the revised Duval's triangle techniques, IEC gas ratio, and IEEE key gas methods for interpretation of DGA provided in the above case studies and Refs [61, 65, 68]. Here, low-energy discharge faults for oil samples from each type of oil after 50, 75, and 100 breakdowns will be considered. In the case of thermal faults at 120 °C, DGA results after 1280 h, 1960 h, and 2800 h of ageing for each type of oil will be considered.

DGA data from case studies A to C and some of the literature was interpreted using three different techniques. The results are presented in Tables 8.7 and 8.8.

Table 8.7 Correctly identified faults for DGA results of case studies A to C

Oil type	Fault condition	Duval	IEC 60599 [69]	IEEE key gas
Mineral	Low-energy discharge	3 of 3	3 of 3	N/A
	120 °C thermal faults	3 of 3	3 of 3	N/A
	150 °C thermal faults	2 of 2	2 of 2	N/A
NEA	Low-energy discharge	3 of 3	3 of 3	N/A
	120 °C thermal faults	3 of 3	3 of 3	N/A
	150 °C thermal faults	2 of 2	2 of 2	N/A
NEB	Low-energy discharge	2 of 3	1 of 3	N/A
	120 °C thermal faults	3 of 3	3 of 3	N/A
	150 °C thermal faults	2 of 2	2 of 2	N/A
MIDEL 7131	Low-energy discharge	3 of 3	3 of 3	N/A
	120 °C thermal faults	3 of 3	1 of 3	2 of 3 overheating cellulose
	150 °C thermal faults	0 of 2	0 of 2	N/A

Table 8.8 Correctly identified faults for DGA results in Refs [61, 65, 68]

Oil type	Fault condition	Duval	IEC 60599	IEEE key gas
FR3	Low-energy discharge, cold corona-type discharge, 150 °C and 200 °C thermal faults [65]	4 of 7	3 of 6	2 of 6
MIDEL 7131	Low-energy discharge, cold corona-type discharge, 150 °C and 200 °C thermal faults [65]	5 of 7	6 of 6	3 of 6
Soybean	Overheating of oil at 200, 300, 500 °C (Table E.1 of Ref. [61])	2 of 3	2 of 3	1 of 3 overheating of oil
	Electrical arching Table E.2 Case study 1 and 2 [61]	2 of 3	2of 3	2 of 3

Based on the fault diagnosis results shown in Table 8.7, one could conclude that extended Duval's triangle and IEC gas ratio methods interpret the DGA results of all types of oil considered in this study with reasonable accuracy. In contrast, the IEEE key gas method possesses the lowest accuracy in diagnosing faults in all types of insulating oil system with DGA data. The IEEE key gas method diagnoses the DGA results of synthetic ester oil aged at 120 °C as overheating of the cellulose insulation. This is caused by production of a large concentration of carbon monoxide due to overheating of synthetic ester oil. By considering the gas generation pattern of soybean-based insulating oil considered in case studies B and C under low-temperature thermal overheating, it is suggested to use ethane as a key indicator of thermal faults in combination with ethylene in equipment using a soybean-type insulating oil.

8.9 Analyzing of Moisture in Solid Insulation of an Ester-Filled Transformer Using FDS Methods

Moisture determination of transformer solid insulation using FDS is based on a comparison of the transformer's dielectric response to a modeled dielectric response. The model response is obtained by considering the relative volume proportion of oil and paper in the insulation system [70]. The accuracy of this method depends on the database used to derive the modeled response. The database characterizes primarily the dependence of the FDS of transformer solid insulation material (cellulose) on moisture, temperature, and ageing conditions. Most of the commercial FDS-based moisture diagnosis tools use a well-characterized database for mineral oil–paper insulation system. These databases consist of measurements on oil-impregnated paper insulation material at various moisture and ageing conditions. In general, the temperature dependence of dielectric response of mineral oil-impregnated paper insulation and mineral oil is, respectively, characterized with activation energy 0.9 eV and 0.4 eV. However, it is still controversial whether the influence of moisture, temperature, and ageing on the dielectric response of ester-impregnated paper insulation is similar to a mineral oil–paper system. The following case study will outline the major issues in the use of existing tools for evaluating moisture content in paper insulation of an ester-filled transformer.

8.9.1 Case Study D

The dielectric responses as dissipation factor vs. frequency were measured for dry and moderately wet pressboard specimens, impregnated with four different types of insulating oil: Shell Diala (mineral oil), NEA, NEB, and MIDEL 7131. Figure 8.16 shows the response corresponding to dry and moderately wet samples,

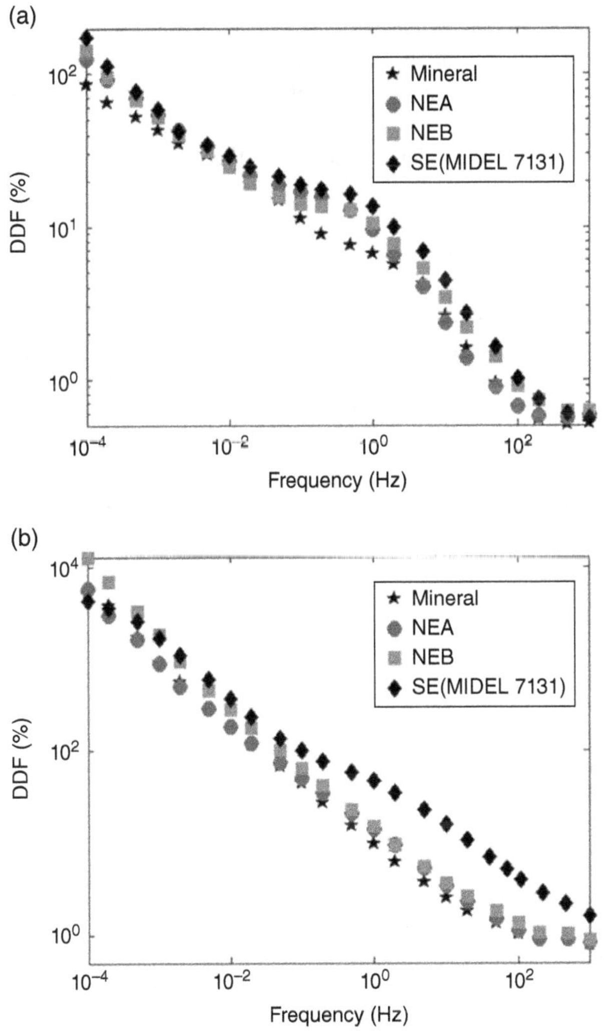

Figure 8.16 Comparison of FDS for different types of oil-impregnated unaged pressboard insulation at 55 °C: (a) mineral 0.8%, NEA 0.8%, NEB 0.8%, SE (MIDEL 7131) 0.6%; (b) mineral 2.4%, NEA 2.1%, NEB 2.7%, SE (MIDEL 7131) 3%.

Table 8.9 Activation energy for moderately wet pressboard insulation

Impregnated oil type	Activation energy (eV)
Mineral (Shell Diala)	1.05
NEA	0.96
NEB	1
SE (MIDEL 7131)	0.9

respectively. It is worth mentioning that the corresponding mineral oil, NEA, NEB, and synthetic ester oils possess DC conductivities of 2.5, 27, 20, and 37 pS/m, respectively, at 55 °C.

Figure 8.16(a) shows that the dielectric response of ester-impregnated dry pressboard qualitatively resembles that of mineral oil-impregnated pressboard, but ester-impregnated pressboard possesses a slightly higher DDF throughout the observed frequency domain. These results show good agreement with the experimental data provided in Ref. [71]. Based on the results shown in Figure 8.16(b), one could assume that the DDF of moderately wet ester-impregnated pressboard insulation is also greater than that of mineral oil-impregnated pressboard, but this cannot be confirmed due to their difference in moisture content. Overall, one can conclude that ester-impregnated cellulosic insulation seems to have a higher dielectric loss due to the higher conductivity of ester oils. However, for a conclusive interpretation, further investigation will be required to understand the influence of oil conductivity on the dielectric response. In order to characterize the temperature dependence of FDS of ester-impregnated cellulose insulation, measurements were carried out at 35, 55, and 75 °C. Then, activation energy values were calculated for all types of oil-impregnated pressboard insulation, as listed in Table 8.9. The results indicate that the temperature dependence of natural ester-impregnated cellulosic insulation resembles that of mineral oil-impregnated cellulosic insulation. However, the calculated activation energy for the dielectric response of synthetic ester-impregnated pressboard insulation is lower than that for other types of oil-impregnated pressboard insulation.

Thermal ageing of oil–paper insulation produces a myriad of conductive and polar by-products, which are not similar for different oil–paper combinations. In this case study, the frequency domain dielectric response of aged pressboard insulation impregnated with mineral oil, NEA, and NEB was investigated. The results in Figure 8.17 show that NE–paper insulation systems produce a larger amount of conductive and polar substances, giving a higher DDF factor for aged NE-impregnated pressboard insulation compared to that of aged mineral oil-impregnated pressboard insulation at similar moisture contents. This effect is more substantial at low frequencies for wet insulation, such that the DDF of aged NE-impregnated pressboard insulation is greater even with about 0.7% lower moisture content than mineral oil-impregnated pressboard insulation.

Overall, one could deduce conclusively that ester–paper insulation possesses higher conductive and dielectric losses than a mineral oil–paper system. Accordingly, if the algorithm used in the FDS-based moisture estimating tool does not consider this behavior, it could lead to overestimation of moisture in ester-based

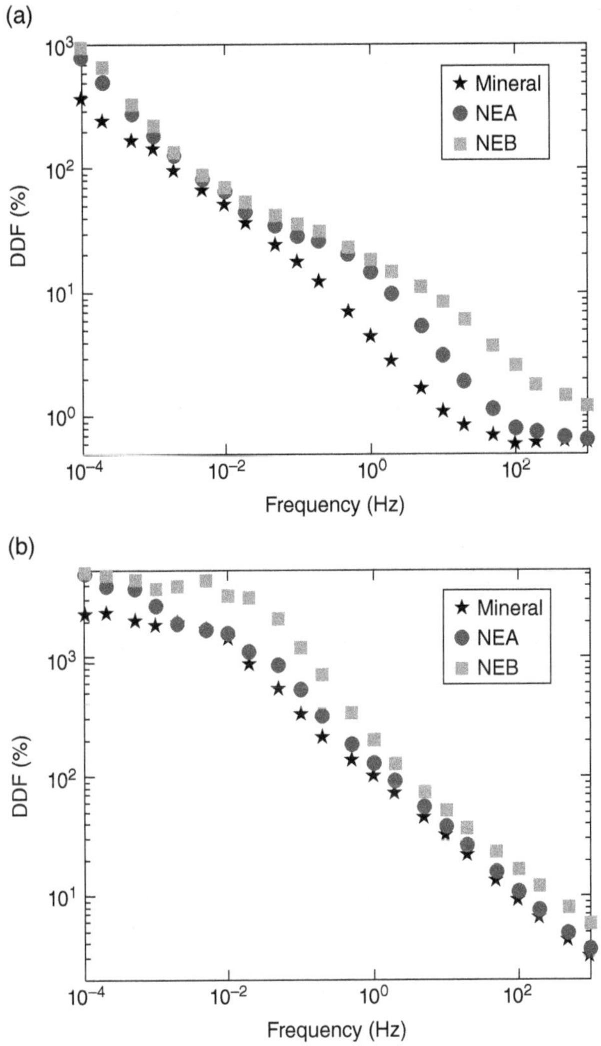

Figure 8.17 Comparison of FDS for different types of aged oil-impregnated pressboard insulation: (a) mineral 0.8%, NEA 0.8%, NEB 0.8%; (b) mineral 3.6%, NEA 2.8%, NEB 2.9%.

insulation systems. In order to provide quantitative understanding of this effect, simulated frequency domain dielectric responses of an ester-filled transformer with aged and unaged paper insulation were analyzed using the commercial tool MODS [72]. This is an FDS-based tool, used to estimate the moisture content in transformer solid insulation. Simulated responses were obtained using a so-called X–Y model, selecting 20% value for both X and Y parameters. To index the sample, M, UA, and A are used to denote mineral oil, unaged sample, and aged sample. The last numbers, 1 and 2, represent the corresponding dry and wet samples.

Figure 8.18 compares the measured moisture content in pressboard insulation using KF titration with that derived from MODS using simulated frequency domain dielectric response data. The estimated moisture content of mineral

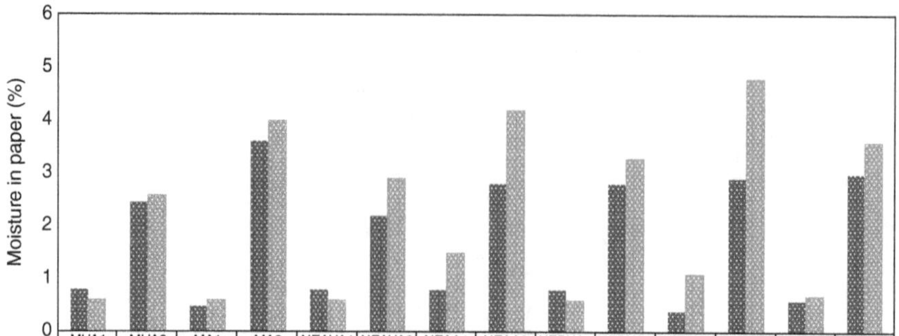

	MUA1	MUA2	MA1	MA2	NEAUA1	NEAUA2	NEAA1	NEAA2	NEBUA1	NEBUA2	NEBA1	NEBA2	SEUA1	SEUA2
KFT	0.8	2.45	0.5	3.6	0.8	2.2	0.8	2.8	0.8	2.8	0.4	2.9	0.6	3
MODS	0.6	2.6	0.6	4	0.6	2.9	1.5	4.2	0.6	3.3	1.1	4.8	0.7	3.6

Figure 8.18 Comparison of moisture analysis of paper insulation using KFT and FDS-based tool (MODS).

oil-impregnated pressboard insulation using MODS shows good agreement with that obtained using KF titration, showing the applicability of this method for both new and aged mineral oil–paper insulation systems of transformers. Moreover, it seems that the MODS tool can predict the moisture in ester-impregnated unaged dry insulation with good accuracy. In contrast, the dielectric response-based tool overestimates the moisture level of both moderately wet and thermally aged ester-impregnated insulation. This effect is more substantial for moderately wet aged insulation, such that the estimated moisture levels of samples NEAA2 and NEBA2 are 1.7% and 1.6% higher than the actual value measured using the KFT method, respectively. This behavior underlines that application of existing FDS-based moisture diagnosis tools for ester–paper insulation systems could lead to unnecessary costly maintenance action. Therefore, in the case of assessing moisture in solid insulation of an ester-filled transformer, it is recommended not to rely completely on FDS-based commercialized tools, which have been developed for mineral oil–paper insulation systems. It is suggested to use equilibrium charts proposed for ester–paper systems as a complementary validation.

Natural ester-filled transformers are now half way through their life, and some synthetic ester-filled units have been operating over 30 years. It is time to monitor closely the health condition of such units, particularly the moisture content in solid insulation. Accordingly, it is an urgent requirement to quantitatively analyze the influence of moisture, temperature, and ageing on the dielectric response behavior of ester-impregnated insulation; the findings should then be incorporated into existing dielectric response-based diagnosis tools.

8.10 Field Experience in Condition Monitoring of Natural Ester-Filled Transformers

Though the population of natural ester-filled transformers is over 600,000, including more than 25,000 medium and large power transformers, there is limited public domain information available in respect of the diagnosis of such

	ppm	09.10.2009	09.09.2008
H_2	Hydrogen	51	77
CH_4	Methane	3	2
C_2H_6	Ethane	29	12
C_2H_4	Ethylene	1	< 1
C_2H_2	Acetylene	< 1	< 1
C_3H_8	Propane	1	< 1
C_3H_6	Propylene	< 1	< 1
CO	Carbon monoxide	60	62
CO_2	Carbon dioxide	344	228
O_2	Oxygen	862	950
N_2	Nitrogen	4291	4555

Property	Standard	09.10.2009	19.09.2008	Unit
Colour	ISO 2049	0,5	0,5	
Appearance	VDE 0370	clear	clear	
Neutralisation value	IEC 62021–1	<0.01	0,01	mg KOH/g
Breakodwn voltage	IEC 60156	47	46	kV
Dissipation factor at 50 Hz	IEC 60247	0,097	0,056	
Water content	DIN 51777	12	23	mg H_2O/kg
Interfacial tension	ASTM D971	16	17	mNm
Viscosity at 40°C	ISO 3104	34 3	4	mm^2/s
2-Furfurylalcohol (2-FOL)	IEC 61198	0,054 <	0,05	mg/kg
2-Furfural (2-FAL)	IEC 61198	<0,05 <	0,05	mg/kg

Figure 8.19 DGA and physicochemical analysis of natural esters. *Source:* Ref. [73], Figures 9 and 10.

transformers, possibly due to confidentiality. This section will summarize some field experiences in diagnosing natural ester-filled transformers.

Figure 8.19 depicts oil-based diagnosis results of a 40 MVA, 110 kV natural ester-filled network transformer, manufactured in 2008 [73]. The DGA results in this figure indicate that the only fault gas with an increasing trend is ethane, which remains as the key stray gas in natural ester. Moreover, similar to the results of our case studies B and C, it confirms the suitability of ethane as a marker for thermal and oxidative stress in natural ester fluids. Physicochemical test results including appearance, neutralization number, DDF, and viscosity indicate that there is no severe degradation in oil in the transformer after a year. This confirms the ability to attain reliable oxidation protection for natural ester insulating oil through a suitable design.

McShane *et al.* [37] have periodically analyzed the key properties of natural ester insulating oil of single-phase pole and several three-phase pad-mounted

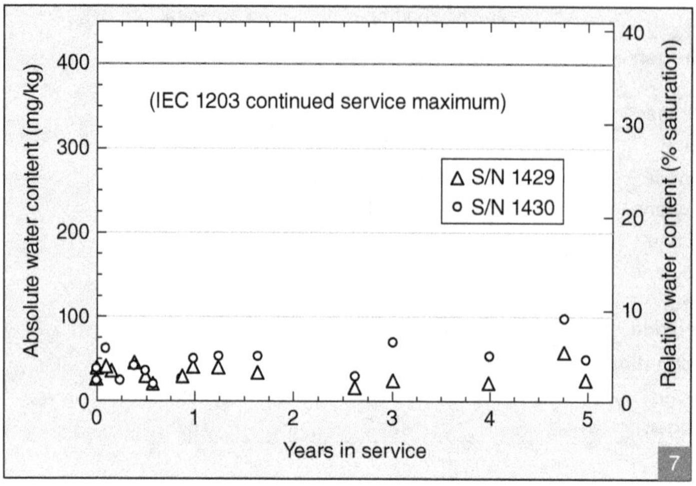

Moisture content of prototype transformers. The IEC limit for in-service synthetic esters is shown as a dashed line.

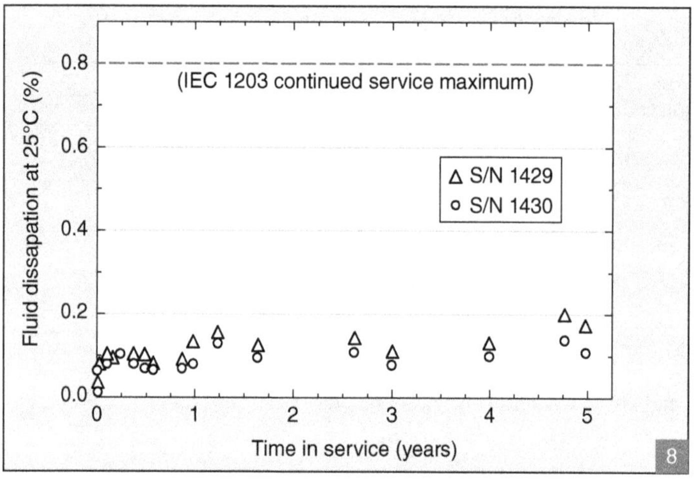

Dissipation factor of prototype transformers. The IEC limit for in-service synthetic esters is shown as a dashed line.

Figure 8.20 Change of oil properties over ageing. *Source:* Ref. 8.20, Figures 7–10.

units. The results have been compared with the standard acceptance limits of service-aged synthetic ester insulating oil.

Figure 8.20 indicates that there is no significant change in key parameters of natural ester insulating oil even after 5 years of operation. Moreover, moisture content, dielectric dissipation factor, and acid number remained well below the limiting value defined for service-aged synthetic ester insulating oil. However, the dielectric breakdown voltage is much higher than the acceptance value for continued use of aged oil. Overall, these positive results confirm better stability and performance of natural ester insulating oil as a transformer oil. In

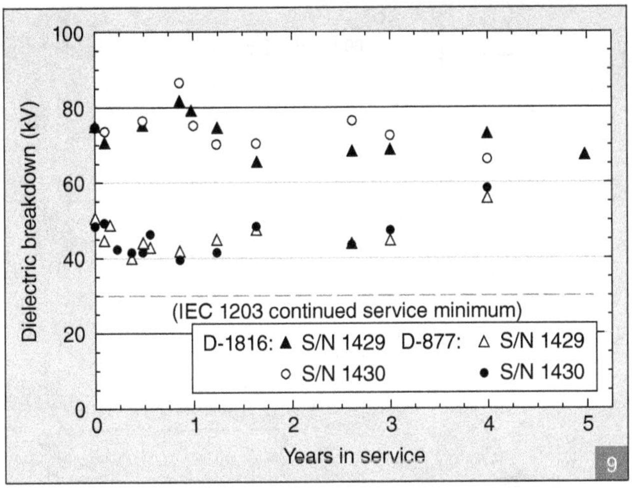

Dielectric breakdown strength of prototype transformers. The IEC limit for in-service synthetic esters is shown as a dashed line.

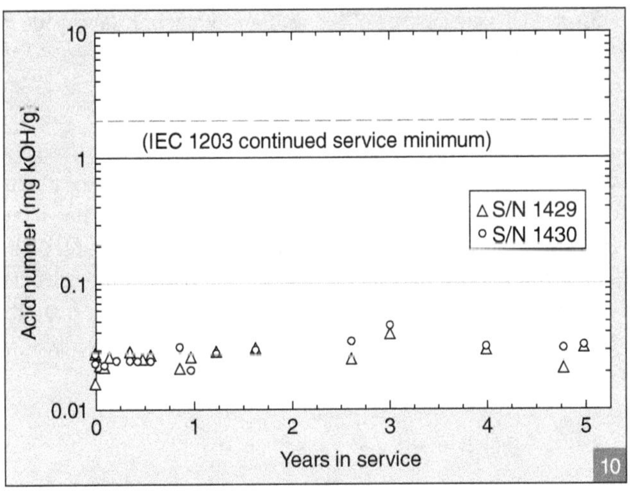

Neutralization (acid) number of prototype transformers. The IEC limit for in-service synthetic esters is shown as a dashed line.

Figure 8.20 (*Continued*)

Ref. [74], McShane has reported that the acidity value of a natural ester oil sample from a field transformer is about 0.5 mg KOH/g after 1 year of service. This value is higher than the acceptance acidity level of 0.3 mg KOH/g defined in IEEE Std. C57.147-2008 [14] for service-aged natural ester insulating oil. In contrast, other parameters including DDF and moisture content were at a satisfactory level. These results underline the requirement for revision of the acceptance standard value of acidity level of service-aged natural ester insulating oil.

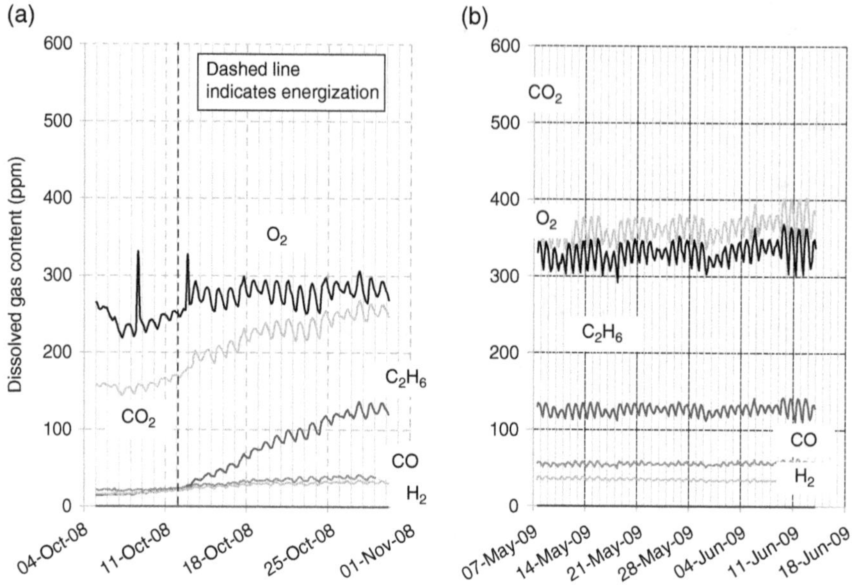

Figure 8.21 Plot of gases dissolved in vegetable oil of Sydney transformer unit: (a) change in gases around the time of energization; (b) gases after a 6-month settling in period. *Source:* Ref. [43], Figure 5.

Martin *et al.* [43] have periodically analyzed the gases dissolved in oil of a natural ester-filled transformer located in Sydney, Australia (Figure 8.21). They have identified that ethane is the key fault gas dissolved in natural ester insulating oil in transformers [43]. Ethane is mainly produced by a reaction between linolenic acid and oxygen. Thereby, ethane can be used as an indicator for natural ester insulating oil oxidation and to trigger an early warning before oil quality is affected.

8.11 Summary

Application of ester-based insulating oils improves the fire safety of transformers. In addition, ester insulating oils show a protective behavior to typical cellulosic insulation material used in power transformers, increasing the life expectancy and overloading capability. In a way similar to the degradation of mineral oil, the thermo-oxidative process is mainly responsible for ageing of ester oil. In addition, ester oil can be subjected to hydrolytic degradation in the presence of moisture. However, the ageing behavior of vegetable oil-based insulating oil has not yet been studied comprehensively. Thereby, application of common oil diagnostic methods such as DDF, neutralization number, color, and interfacial tension for quality assurance of field-aged vegetable oil-based insulating oils has still not been standardized. Moreover, not enough data is available to fully characterize the fault gas generation patterns in ester insulating oils. Thereby, standards corresponding to analysis and interpretation of dissolved fault gases in ester-based insulating oils have not been completely tailored yet. Overall, the case studies presented in this chapter underline the requirement for revisions of international standards used for quality assurance of mineral insulating oil to comply with ageing and gassing behavior of ester insulating oils. Modern moisture determination of a transformer's solid insulation is based on the comparison of modeled dielectric response with the measured response of the transformer. The existing FDS-based moisture determination tool has been built considering the effect of moisture, temperature, and ageing on the dielectric response behavior of mineral oil-impregnated cellulose paper insulation. Case study D presented in this study shows that the use of existing FDS-based tools on a vegetable oil–paper insulation system may result in overestimation of moisture content in paper insulation. Thereby, we propose using the equilibrium chart technique for moisture estimation validation, along with a need to further develop the FDS method for ester-based oil-filled transformer insulation systems.

References

1 Valagro BY, Hong LC. Vegetable oils as substitute for mineral insulating oils in medium-voltage equipments. D1-202-CIGRE Session, Paris, 2004.
2 IEC Std. 60296-2003. Fluids for Electrotechnical Applications – Unused Mineral Insulating Oils for Transformers and Switchgear.
3 ASTM Std. D3487-09. Standard Specification for Mineral Insulating Oil Used in Electrical Apparatus.
4 IEEE Std. C57.106-2006 (Revision of IEEE Std. C57.106-2002). Guide for Acceptance and Maintenance of Insulating Oil in Equipment.
5 Rouse TO. Mineral insulating oil in transformers. *IEEE Electrical Insulation Magazine* 1998; 14: 6–16.
6 Heathcote MJ. Basin materials. *J & P Transformer Book – A Practical Technology of the Power Transformer*, 13th edn. Amsterdam: Elsevier, 2007.
7 Du Y, Zahn M, Lesieutre BC, Mamishev AV, Lindgren SR. Moisture equilibrium in transformer paper–oil systems. *IEEE Electrical Insulation Magazine* 1999; 15: 11–20.
8 Hodges PKB. Mineral base oils and oxidation stability. *Hydraulic Fluids*. Amsterdam: Elsevier, 1996.
9 CIGRE Brochure 436. Experience in Service with New Insulating Liquids, 2010.
10 Viertel J, Ohlsson K, Singha S. Thermal aging and degradation of thin films of natural ester dielectric liquids. IEEE International Conference on Dielectric Liquids 2011, pp. 1–4.
11 Oommen TV. Vegetable oils for liquid-filled transformers. *IEEE Electrical Insulation Magazine* 2002; 18: 6–11.
12 Smith FH. Vegetable oil refining. *Journal of the American Oil Chemists' Society* 1956; 33: 473–476.
13 Environmental Technology Verification Report ABB inc. Biotemp vegetable oil based insulating dielectric fluid, June 2002.
14 IEEE Std. C57.147-2008. Guide for Acceptance and Maintenance of Natural Ester Fluids in Transformers.

15 Milledge R. Introduction to BIOTEMP: The superior, biodegradable, high fire point dielectric insulating fluid. ABB seminar presentation, Australia, 2011.

16 Tenbohlen S, Koch M. Aging performance and moisture solubility of vegetable oils for power transformers. *IEEE Transactions on Power Delivery* 2010; 25: 825–830.

17 McShane CP. Relative properties of the new combustion resistant vegetable oil based dielectric coolants. Industry Applications Society 47th Annual Conference on Petroleum and Chemical Industry 2000, pp. 265–272.

18 IEC Std. 61099-2010. Insulating Liquids – Specifications for Unused Synthetic Organic Esters for Electrical Purposes.

19 IEC Std. 61203. Synthetic Organic Esters for Electrical Purposes – Guide for Maintenance of Transformer Esters in Equipment.

20 Fox NJ, Stachowiak GW. Vegetable oil-based lubricants – A review of oxidation. *Tribology International* 2007; 40: 1035–1046.

21 Gunstone FD. *The Chemistry of Oil and Fat.* Oxford: Blackwell, 2004.

22 Wilhelm HM, Tulio L, Jasinski R, Almeida G. Aging markers for in-service natural ester-based insulating fluids. *IEEE Transactions on Dielectrics and Electrical Insulation* 2011; 18: 714–719.

23 Ciuriuc A, Vihacencu MS, Dumitran LM, Notingher PV. Comparative study on power transformers vegetable and mineral oil ageing. International Conference on Applied and Theoretical Electricity 2012, pp. 1–6.

24 Rooney D, Weatherley LR. The effect of reaction conditions upon lipase catalysed hydrolysis of high oleate sunflower oil in a stirred liquid–liquid reactor. *Journal of Process Biochemistry* 2001; 36: 947–953.

25 Bakunin VN, Parenago OP. Mechanism of thermo-oxidative degradation of polyol ester lubricants. *Journal of Synthetic Lubrication* 1992; 9: 127–143.

26 Cargill's Brochure. Envirotemp™ Dielectric Fluids overview, 2014.

27 Rapp KJ, Gauger GA, Luksich J. Behavior of ester dielectric fluids near the pour point. Conference on Electrical Insulation and Dielectric Phenomena, Austin, TX, October 17–20, 1999, pp. 459–462.

28 Koch M, Tenbohlen S, Stirl T. Diagnostic application of moisture equilibrium for power transformers. *IEEE Transactions on Power Delivery* 2010; 25: 2574–2581.

29 Oommen TV. Moisture equilibrium charts for transformer insulation drying practice. *IEEE Transactions on Power Apparatus and Systems* 1984; PAS-103: 3062–3067.

30 Abtew W, Melesse A. Vapor pressure calculation methods. *Evaporation and Evapotranspiration.* Amsterdam: Springer, 2013, pp. 53–62.

31 Wheeler M. Insulating fluids: An overview of alternative dielectric fluids used in transformers. Life of a Transformer Seminar, ABB Doble, 2012.

32 Lashbrook M, Kuhn M. The use of ester transformer fluids for increased fire safety and reduced costs. CIGRE Session A2-210, Paris, 2012.

33 FM Global. Property Loss Prevention Data Sheets, 5–4 Transformers, Section 2.3.1.1.1 Separation Distance, May 2003.

34 Cargill Service Information R 2040. Dielectric fluids – Guide for retrofilling power class transformers >7500 kVA, January 2013.

35 Moore SP. Some considerations for new and retrofill applications of natural ester dielectric fluids in medium and large power transformers revisited. IEEE Power & Energy Society General Meeting 2009, pp. 1–7.

36 Pukel GJ, Schwarz R, Baumann F, Muhr HM, Eberhardt R, Wieser B, Chu D. Power transformers with environmentally friendly and low flammability ester liquids. CIGRE Session A2-210, Paris, 2012.

37 McShane CP. Vegetable-oil-based dielectric coolants. *IEEE Industry Applications Magazine* 2002; 8: 34–41.

38 British Std. BS EN 61203:1995, IEC 61203:1992. Synthetic Organic Esters for Electrical Purposes. Guide for Maintenance of Transformer Esters in Equipment, 1995.

39 IEEE Std. C57.152-2013. Guide for Diagnostic Field Testing of Fluid-Filled Power Transformers, Regulators, and Reactors, 2013, pp. 1–121.

40 Jung-Il J, Jung-Sik A, Chan-Su H. Accelerated aging effects of mineral and vegetable transformer oils on medium voltage power transformers. *IEEE Transactions on Dielectrics and Electrical Insulation* 2012; 19: 156–161.

41 Coulibaly ML, Perrier C, Marugan M, Beroual A. Aging behavior of cellulosic materials in presence of mineral oil and ester liquids under various conditions. *IEEE Transactions on Dielectrics and Electrical Insulation* 2013; 20: 1971–1976.

42 Liao R, Liang S, Sun C, Yang L, Sun H. A comparative study of thermal aging of transformer insulation paper impregnated in natural ester and in mineral oil. *European Transactions on Electrical Power* 2010; 20: 518–533.

43 Martin D, Ekanayake C, Saha T, Ma H, Lelekakis N, Williams K. State of the art review on managing vegetable oil filled transformers. Colloquium CIGRE Study Committee B3, Brisbane, Australia, September 2013.

44 Singha S, Asano R, Frimpong G, Claiborne C, Cherry D. Comparative aging characteristics between a high oleic natural ester dielectric liquid and mineral oil. *IEEE Transactions on Dielectrics and Electrical Insulation* 2014; 21: 149–158.

45 Martin D, Guo W, Lelekakis N, Heyward N. Using a remote system to study the thermal properties of a vegetable oil filled power transformer: How does operation differ from mineral oil. IEEE PES Conference on Innovative Smart Grid Technologies, Asia, 2011, pp. 1–5.

46 Duart J, Bates LC. Aging of high temperature insulation systems with alternative fluids. IEEE International Symposium on Electrical Insulation 2010, pp. 1–5.

47 Bandara K, Ekanayake C, Saha TK, Kumar P. Understanding the ageing aspects of natural ester-based insulation. *IEEE Transactions on Dielectrics and Electrical Insulation* 2016; 23(1): 246–257.

48 Bandara K, Ekanayake C, Saha TK. Comparative study to understand behaviour of natural ester as transformer insulating liquid. IEEE International Conference on Industrial and Information Systems, Sri Lanka, 2015.

49 McShane CP, Rapp KJ, Corkran JL, Gauger GA, Luksich J. Aging of Kraft paper in natural ester dielectric fluid. Proceedings of 14th IEEE International Conference on Dielectric Liquids, 2002, pp. 173–177.

50 Junru X, Jian L, Zhaotao Z. Influence of water content on the aging performance of natural ester–paper insulation. IEEE International Conference on High Voltage Engineering and Application, 2012, pp. 663–666.

51 Stockton DP, Bland JR, McClanahan T, Wilson J, Harris DL, McShane P. Natural ester transformer fluids: Safety, reliability and environmental performance. IEEE International Conference on Petroleum and Chemical Industries, 2007, pp. 1–7.

52 Lijun Y, Ruijin L, Sun C, Jianguo Y, Mengzhao Z. Influence of vegetable oil on the thermal aging rate of Kraft paper and its mechanism. IEEE International Conference on High Voltage Engineering and Application, 2010, pp. 381–384.

53 Lijun Y, Ruijin L, Sun C, Mengzhao Z. Influence of vegetable oil on the thermal aging of transformer paper and its mechanism. *IEEE Transactions on Dielectrics and Electrical Insulation* 2011; 18: 692–700.

54 Cargill's Brochure R 2110-FR3. Fluid Loading Guide A and B Factors for Envirotemp FR3 Fluid and Thermally Upgraded Kraft Insulation, April 2013.

55 IEEE Std. C57.91-2011 (Revision of IEEE Std. C57.91-1995). Redline Guide for Loading Mineral-Oil-Immersed Transformers and Step-Voltage Regulators, 2012, pp. 1–172.

56 Frimpong GK, Oommen TV, Asano R. A survey of aging characteristics of cellulose insulation in natural ester and mineral oil. *IEEE Electrical Insulation Magazine* 2011; 27: 36–48.

57 Bandara K, Ekanayake C, Saha TK. Compare the performance of natural ester with synthetic ester as transformer insulating oil. IEEE International Conference on Properties and Applications of Dielectric Materials, Sydney, Australia, 2015.

58 Levchik S, Scheirs J, Camino G, Tumiatti W, Avidano M. Depolymerization processes in the thermal degradation of cellulosic paper insulation in electrical transformers. *Journal of Polymer Degradation and Stability* 1998; 61: 507–511.

59 Lijun Y, Ruijin L, Sun C, Huigang S. Study on the influence of natural ester on thermal ageing characteristics of oil–paper in power transformer. IEEE International Conference on High Voltage Engineering and Application, 2008, pp. 437–440.

60 Emsley AM, Stevens GC. Review of chemical indicators of degradation of cellulosic electrical paper insulation in oil-filled transformers. *IEE Proceedings on Science, Measurement and Technology* 1994; 141: 324–334.

61 IEEE PC57.155/D5. Draft Guide for Interpretation of Gases Generated in Natural Ester and Synthetic Ester Immersed Transformers, 2014, pp. 1–63.

62 Yu L, Jian L, Zhaotao Z. Gases dissolved in natural ester fluids under thermal faults in transformers. IEEE International Symposium on Electrical Insulation, 2012, pp. 223–226.

63 Bandara K, Ekanayake C, Saha TK. Comparative study for understanding the behaviour of natural ester with mineral oil as transformer insulating liquid. IEEE Conference on Electrical Insulation and Dielectric Phenomena, 2014.

64 Perrier C, Marugan M, Beroual A. DGA comparison between ester and mineral oils. *IEEE Transactions on Dielectric Insulation* 2012; 19: 1609–1614.

65 Imad UK, Zhongdong W, Cotton I, Northcote S. Dissolved gas analysis of alternative fluids for power transformers. *IEEE Electrical Insulation Magazine* 2007; 23: 5–14.

66 Duval M. The Duval triangle for load tap changers, non-mineral oils and low temperature faults in transformers. *IEEE Electrical Insulation Magazine* 2008; 24: 22–29.

67 CIGRE Brochure 443-2010. DGA in Non-Mineral Oils and Load Tap Changers and Improved DGA Diagnosis Criteria, 2010.

68 Zhongdong W, Xiao Y, Jinping H, Hinshaw JV, Noakhes J. Fault gas generation in natural-ester fluid under localized thermal faults. *IEEE Electrical Insulation Magazine* 2012; 28: 45–56.

69 IEC Std. 60599-2007. Mineral Oil Impregnated Electric Equipment – Service Guide to the Interpretation of Dissolved and Free Gases Analysis.

70 Koch M, Kruger M. A fast and reliable dielectric diagnostic method to determine moisture in power transformers. International Conference on Condition Monitoring and Diagnosis 2008, pp. 467–470.

71 Hao J, Liao R, Ma Z, Yang L. Influence of natural ester on frequency dielectric response of impregnated insulation pressboard. *IET Science, Measurement & Technology* 2012; 6: 403–411.

72 User Manual for Insulation Diagnostic System IDAX 350, Megger Sweden AB, 2012.

73 Atanasova-Hoehlein I, Hammer Th, Schaefer M. Diagnostic markers for oxidation condition of mineral oil and ester insulating fluids. CIGRE Session D1-213, Paris, 2010.

74 McShane CP. Relative properties of the new combustion-resist vegetable-oil-based dielectric coolants for distribution and power transformers. *IEEE Transactions on Industry Applications* 2001; 37: 1132–1139.

9

Smart Transformer Condition Monitoring and Diagnosis

9.1 Introduction

Power transformers have a significant impact on the reliable delivery of electricity and their operation is extensively complicated, involving electrical, magnetic, and mechanical operations. Moreover, electrical utilities around the world are currently facing an increasingly aged population of power transformers in their networks. Therefore, in order to ensure reliable operation of the electricity grid, transformer condition must be monitored and evaluated continuously.

Over the past three decades, many diagnostic techniques have been introduced for the condition monitoring of transformers [1–5]. As already explained in this book, DGA is the most established chemical technique to identify any incipient faults in transformers [5, 6]. Furan measurement on oil samples by high-performance liquid chromatography (HPLC) provides an average ageing condition of transformer solid insulation [7]. Polarization-based electrical measurements including PDC measurement and FDS on transformer insulation have been used to estimate moisture content and ageing in transformer solid insulation [8, 9]. Temperature has a significant influence on insulation ageing, and research has been performed to calculate the hotspot temperature of a transformer based on oil temperatures using continuous temperature recording of oil and loading history [10]. Winding displacement is a key concern for in-service aged transformers, and identification of winding deformation/displacement reduces the risk of sudden transformer failure. Frequency response analysis has been developed to detect the altered electrical parameters of the winding due to winding deformation/displacement [11].

To accurately provide dynamic behavior of a transformer's condition, PD measurement has been carried out either through measurement impedance installed at the bushing tap adapter or through a high-frequency current transducer (HFCT) clamped on the transformer grounding cable or through an ultra-high frequency (UHF) sensor inserted in the oil drain valves or through an acoustic emission (AE) sensor mounted on the transformer tank [12]. Major challenges of PD measurement are noise suppression, multiple PD source separation and recognition, and PD source localization. A number of PD analysis techniques based on advanced signal processing and pattern recognition techniques have been developed to deal with these challenges [13–19].

Transformer Ageing: Monitoring and Estimation Techniques, First Edition.
Edited by Tapan Kumar Saha and Prithwiraj Purkait.

Moisture content in transformers has numerous adverse effects on their cellulose insulation. Accurate online evaluation of moisture content in a transformer is of great concern for utilities. Moisture in oil can be measured directly through a moisture-in-oil sensor and then moisture in cellulose can be estimated through equilibrium charts [20]. This approach assumes an equilibrium state of moisture inside a transformer, which is almost impossible for an operational transformer since it is normally subjected to continuous fluctuations of loading, temperature, and other conditions. Some researchers investigated moisture diffusion in transformers using Fick's second law and determined the diffusion coefficients of cellulose [21]. However, the diffusion coefficients were estimated by conducting laboratory experiments on pressboard/paper samples. This may introduce some level of discrepancy when these diffusion coefficients are applied to transformers constructed with different cellulose materials and having a heterogeneous temperature distribution. Recently, some researchers have developed the cellulose isothermal method, which estimates moisture content in transformer cellulose insulation from the online moisture-in-oil measurement [22–24]. However, this method may only provide accurate moisture estimation for cellulose insulation which is in close proximity to the sensor. Therefore, transformer moisture estimation should be conducted based on a comprehensive approach, which combines online moisture-in-oil measurement, multi-physics modeling, and theoretical derivations [25].

In the previous chapters of this book, theories, implementations, and interpretations of a variety of condition monitoring techniques have been presented. However, each of these techniques normally focuses on evaluating one aspect of transformer condition and may not be able to provide an overall "picture" of the health status of a transformer. Moreover, due to the complexity of transformer structure and its degradation mechanism, interpreting measurement data and evaluating transformer condition is a challenging task, especially for non-experts. Therefore, it is necessary to develop an integrated condition monitoring and diagnosis system, which can dynamically coordinate and automatically collect measurements data from various online sensor measurements, integrate data and information from offline measurements and historic data, and finally provide an overall condition assessment of the transformer.

In this chapter, an intelligent framework for transformer condition monitoring and diagnosis is presented. A number of signal and data analysis techniques, which have been implemented within the proposed framework, are also described and there may be some overlap with previous chapters. The techniques are: (1) digital signal processing, which is concerned with de-noising and extracting signals acquired by online sensor measurements; (2) pattern recognition, which focuses on identifying transformer fault types; (3) data fusion, which integrates results from different sources to determine transformer condition; and (4) determining the health index of a transformer insulation system. A variety of case studies are provided in this chapter to demonstrate the applications of these techniques. Moreover, a hardware and software platform for implementing the proposed intelligent transformer condition monitoring and diagnosis framework and deploying the above signal and data analysis techniques is also presented.

9.2 Intelligent Framework for Transformer Condition Monitoring and Diagnosis

The interpretation of the data obtained from transformer condition monitoring techniques is non-trivial; this is particularly so for non-experts in utilities. It is necessary to develop intelligent techniques to process the collected data automatically, and subsequently transform them into useful knowledge regarding the condition of the transformer and also convey such knowledge to engineers and managers in utilities wherever and whenever needed. A variety of efforts have been made to develop intelligent techniques for monitoring and diagnosis of the overall condition of a transformer. For example, CIGRE developed a number of recommendations on applying artificial intelligence (AI) to process large volumes of measurement data, extract useful information, and subsequently make condition assessment on a transformer [26–29]. Attempts have also been made to calculate a health index for a transformer to provide an indication of its overall health condition [2, 30–32]. However, there is still a lack of a common framework, which properly defines the requirements and procedures of applying different signal processing and pattern recognition techniques to process transformer measurement data and make condition assessment of the transformer.

In Figure 9.1, we present an intelligent framework for transformer condition monitoring and diagnosis. This framework incorporates a number of key components including sensor deployment, data acquisition, signal processing, pattern recognition, data fusion, etc., which are based on the authors' research and development works conducted over the past five years [4].

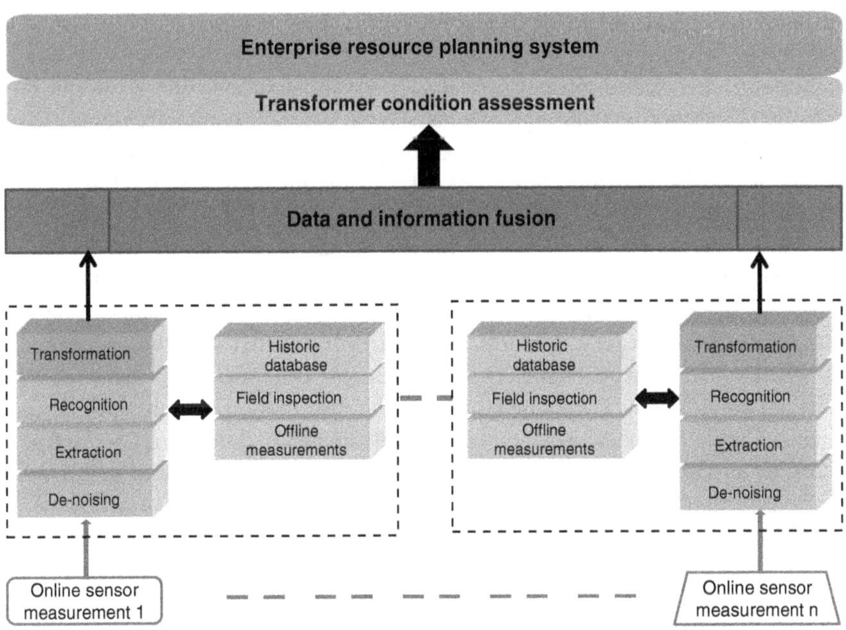

Figure 9.1 Intelligent framework for power transformer condition monitoring and diagnosis.

Table 9.1 Typical online sensor-based measurement and offline measurement for transformer condition monitoring and diagnosis

Online sensor-based transformer condition monitoring	Current and voltage measurement Moisture-in-oil and temperature PD (HFCT, UHF, IEC 60270) Acoustic/vibration measurement	**Doesn't cause disruption on transformer's normal operation and provides dynamic information on transformer condition**
Offline techniques for transformer condition monitoring	FDS PDC measurement FRA Oil characteristic test (including DGA)	**Can provide more focused and specific information on a particular aspect of transformer condition**

The first component of the framework is an integration of various online sensors to provide dynamic information on transformer condition. Necessary data acquisition hardware, which can acquire sensor-measured signals and then transmit these signals through communication peripherals for further data processing, need to be incorporated. A number of offline measurements are also included in the first component to complement the online measurement. Table 9.1 presents a variety of online sensor-based measurements and offline measurements, which have been widely adopted for transformer condition monitoring and have been discussed in previous chapters.

For transformer condition monitoring in a substation environment, online sensor-acquired signals are normally highly coupled with interference and noise. This interference and noise can significantly affect the sensitivity and reliability of sensor measurements, and consequently has reverse impacts on properly interpreting measurement data and subsequently identifying different types of fault occurring in a transformer. Therefore, the second component of the proposed framework is a suite of signal processing algorithms, which can remove noise from the original measured signals.

With the advanced sensors and high-speed data acquisition systems, it is relatively easy to acquire and access raw data in a more straightforward manner. However, the huge volume of raw data may consist of redundant and irrelevant information. Therefore, it is necessary to perform feature extraction, which aims to select representative characteristics (features) from raw data. The selected features are embedded with more condensed information, which will be more suitable and manageable for further information processing to support transformer condition diagnosis. Feature extraction forms the third component of the proposed framework.

After feature extraction, a pattern recognition algorithm is adopted as the fourth component in the proposed intelligent framework. It is tasked with classifying different types of fault that may occur in a transformer tank. The major advantage of pattern recognition algorithms is the learning from data (i.e., they make use of a historic database consisting of previous measurement data and corresponding fault diagnosis results to obtain knowledge regarding the correlation between the measurement data and transformer condition). Subsequent condition assessment of the transformer of interest can be performed on the basis of such knowledge.

The overall dynamic condition of a transformer can only be evaluated by combining every piece of available data and information. Within the proposed intelligent framework, a data fusion component is incorporated, which integrates the pattern recognition results of individual measurements to estimate the probability of different types of fault that might occur in a transformer. Moreover, a quantitative health index is calculated to provide an indication of the overall health condition of a transformer.

To deploy signal processing and pattern recognition algorithms for transformer condition monitoring and diagnosis within the above intelligent framework, a configurable architecture (termed as "SmartBox" in the rest of this chapter) can be adopted. It acts as both a hardware and software development platform. The SmartBox consists of signal conditioning circuits and a high-speed digitizer, which can acquire signals from sensors during online condition monitoring of a transformer. The SmartBox can also execute signal processing algorithms and pattern recognition algorithms for signal de-noising, feature extraction, and fault pattern recognition.

The following sections discuss key issues of development, implementation, and deployment of a number of representative techniques within the proposed intelligent condition monitoring and diagnosis framework. Case studies are provided to demonstrate the applicability of these techniques in transformer condition monitoring and diagnosis.

9.3 Advanced Signal Processing for Transformer Condition Monitoring

9.3.1 Signal De-noising in Transformer Condition Monitoring

During onsite measurements of transformers in a substation environment, sensor-measured signals can be highly coupled with extensive noise and interference. Some major types of noise are as follows [15–18].

1) Discrete spectral interference (DSI), generated by communication systems and power electronics.
2) White noise, which is normally produced by measurement systems and the ambient environment.
3) Repetitive pulses, which are generated by power electronics or other periodic switching.
4) Stochastic impulsive interferences, from infrequent switching operation or lighting, arcing and corona from other high-voltage apparatus which can be coupled to the transformer under test.

To effectively extract signals from noise and interference (i.e., signal de-noising), a number of criteria can be adopted [17]:

1) Occurrence time instances of signals, for example, the repetitive pulse-type noise may only appear at some specific time instances and thus they can be removed by time domain filtering techniques.

2) Frequency spectrum of signals, for example, the DSI-type noise may only appear at some discrete frequencies and can be removed by frequency domain techniques such as discrete Fourier transform (DFT) and discrete wavelet transform (DWT).
3) Time frequency characteristics of signals, for example, DWT and empirical model decomposition (EMD) separate sensor signals and noise signals in both time and frequency domains at multiple resolutions.
4) Phase-resolved diagram of applied AC voltage, for example, white noise may span a whole AC cycle of the applied voltage while PD signals may only appear at some phase locations of the AC cycle.

In Chapter 5, a variety of signal processing techniques were presented for de-noising signals from different types of noise and interference [4, 14–19]. In the following section a newly developed multi-scale thresholding wavelet transform algorithm is utilized to demonstrate the application of advanced signal processing techniques to de-noise sensor-measured signals for transformer condition monitoring.

9.3.2 Case Study on Signal De-noising

As discussed in Chapter 5, the wavelet transform has become a de-facto tool for PD signal de-noising. The conventional wavelet transform performs PD signal de-noising in the following three steps: (1) selecting an appropriate mother wavelet from a mother wavelet bank to decompose measured signals into a series of coefficients at different decomposition levels corresponding to different frequency bands; (2) thresholding the above wavelet coefficients for the purpose of discarding wavelet coefficients associated with noise while keeping those associated with PD signals; and (3) reconstructing PD signals using the retained wavelet coefficients.

In practical PD measurements of a transformer, however, it is difficult to find a mother wavelet having a high level of similarities (in shape) with all PD signals that may be generated by different types of PD source [14, 17]. As a consequence, the above conventional wavelet transform may also discard some PD signals. Moreover, noise-related wavelet coefficients may have larger magnitudes than PD signal-related wavelet coefficients. As such, PD signals and noise can be merged in the reconstructed signal in wavelet transform. Therefore, thresholding based on a single amplitude value may not be able to effectively de-noise PD signals.

To deal with the limitations of conventional wavelet transform using single amplitude-based thresholding, a multi-scale thresholding wavelet transform has been developed [17]. Figure 9.2 illustrates the major steps involved in multi-scale thresholding. Firstly, a signal cumulative distribution is formed by sorting the signals according to their absolute amplitudes. Since an inductive PD measurement system is adopted, the signal amplitude is in millivolts (mV). Then, the base value for determining multi-scale thresholds (named as the "baseline threshold") is decided. For an individual decomposition level, the baseline threshold is calculated as the ratio between the average magnitudes of wavelet coefficients and the number of wavelet coefficients at this decomposition level.

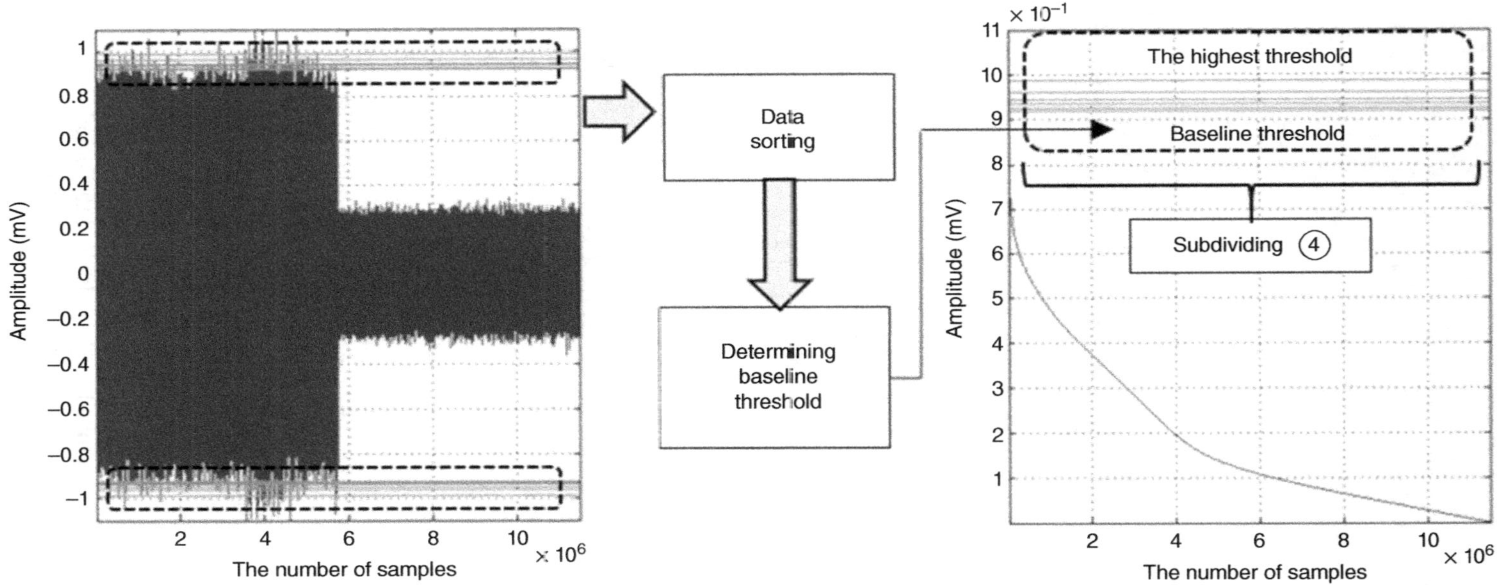

Figure 9.2 Multi-scale thresholding. The horizontal axis denotes the cumulative number of signals and the vertical axis denotes the amplitude of signals (mV).

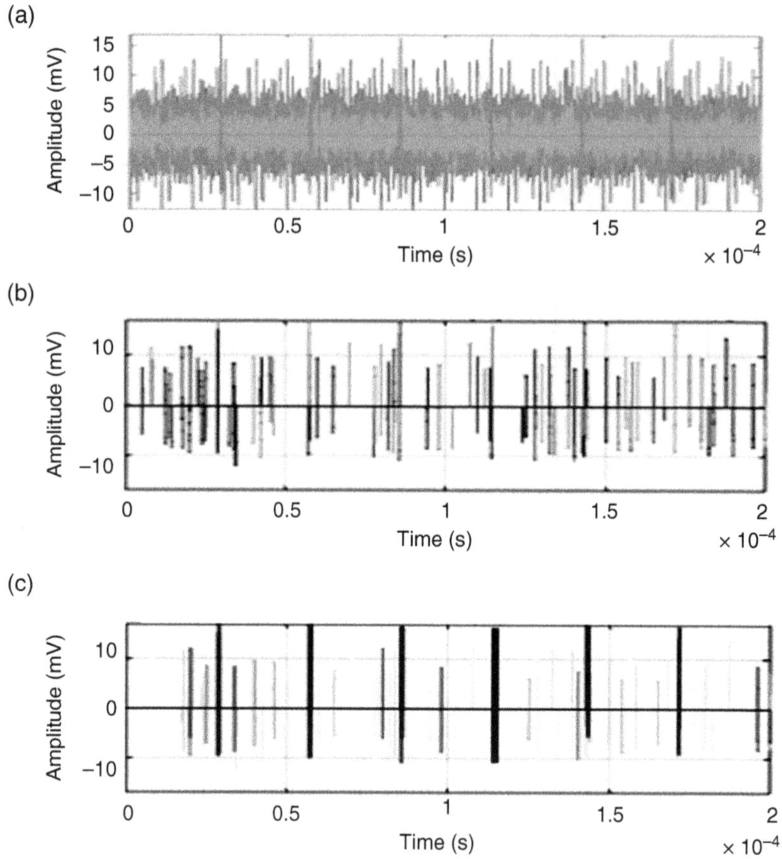

Figure 9.3 Results on PD signal de-noising using simulated signals: (a) simulated noise-corrupted PD signals (SNR = −18.6 dB); (b) de-noised PD signals by single-amplitude-based thresholding wavelet transform; (c) de-noised PD signals by multi-scale thresholding wavelet transform (signals having higher probability indexes are depicted in the darkest shade).

Once the baseline threshold is decided, multi-scale thresholds are set by dividing the wavelet coefficient level using quantile-based segmentation [17].

To select an appropriate mother wavelet for multi-scale thresholding wavelet transform, the correlation coefficients between the candidate mother wavelet and the wavelet coefficients remaining after multi-scale thresholding is computed. To decide the number of decomposition levels, the normalized energy of signals reconstructed after the multi-scale thresholding is calculated. A probability index can finally be computed for an extracted signal based on its recurrence rate over all thresholds at all decomposition levels [17]. This probability index provides an indication of the signal's likelihood of being generated by PD events.

Figure 9.3 illustrates the application of the multi-scale thresholding method for de-noising PD signals which are coupled with DSI and white noise. In the figure, there exist six PD signals embedded in noise and the signal-to-noise ratio (SNR) is −18.6 dB [Figure 9.3(a)]. The PD signal de-noising results obtained by conventional single-amplitude-based thresholding and the new multi-scale thresholding

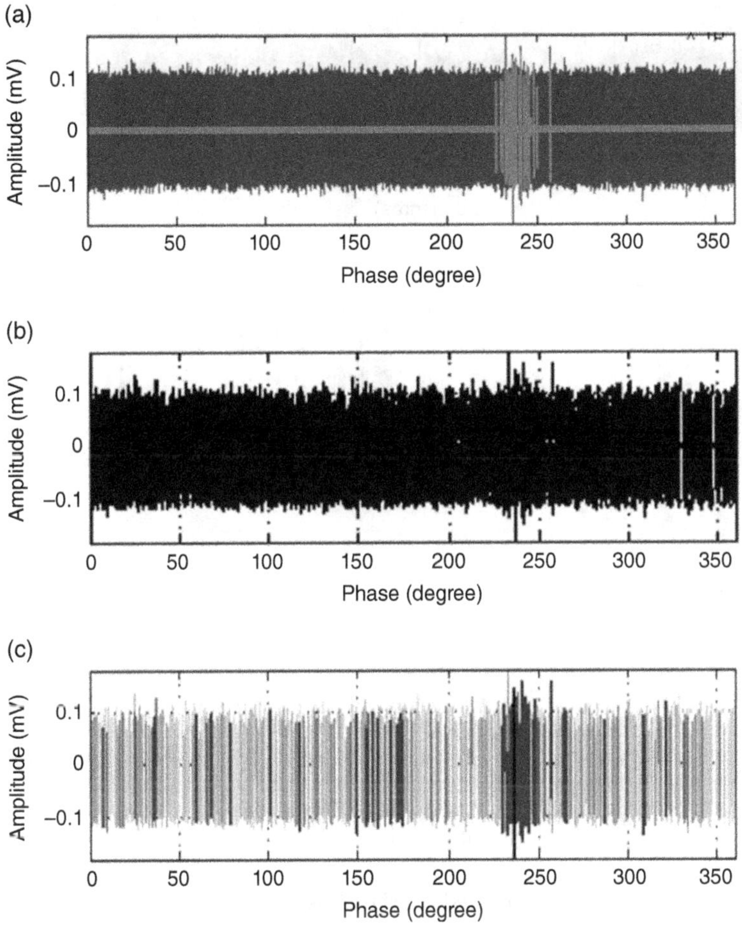

Figure 9.4 Results of de-noising PD signals obtained from experimental model of corona. (a) noise-corrupted PD signals (PD signals are in red, noise-corrupted signals are in blue) with SNR = −28; (b) de-noised PD signals by single-amplitude-based thresholding wavelet transform; (c) de-noised PD signals by multi-scale thresholding wavelet transform (signals having higher probability indexes are depicted in a darker color).

wavelet transforms are shown in Figures 9.3(b) and 9.3(c). In Figure 9.3(c), the signals are depicted in different shades based on their probability indexes (the possibilities of being generated by PD events); signals having higher probability indexes are depicted in the darkest shade. It can be seen from Figure 9.3(c) that the six PD signals have the highest probability indexes. They are significantly separable from noise and can readily be extracted (i.e., removing noise). In contrast, as shown in Figure 9.3(b), the wavelet transform using single-amplitude-based thresholding failed to extract the six PD signals.

Figure 9.4 presents the results of de-noising PD signals acquired from an experimental PD corona model. Since PD measurements were conducted in the laboratory environment, the noise level was rather low. DSI and white noise were added to the originally acquired signals (SNR = −28). The phase-resolved PD

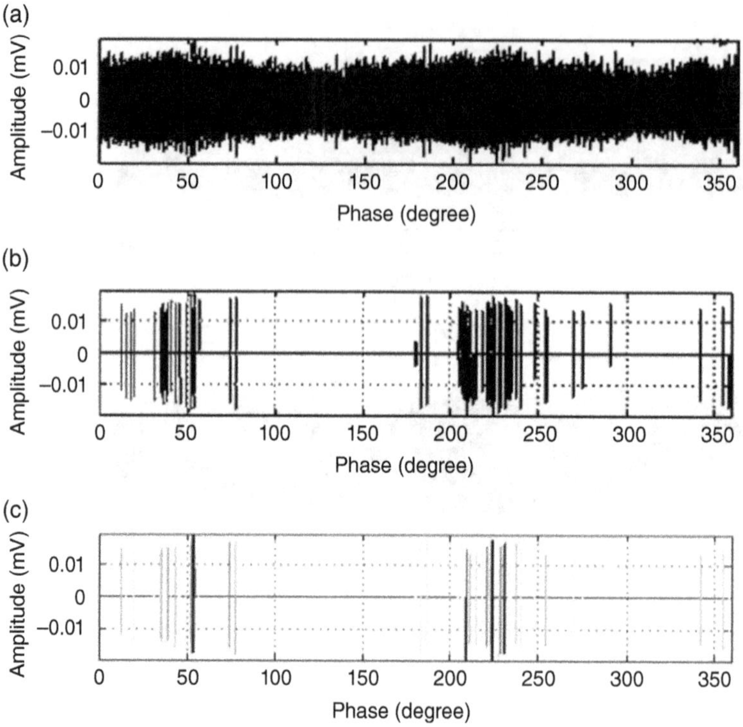

Figure 9.5 Results of de-noising PD signals obtained from a 10 MVA (33 kV/11 kV) substation transformer: (a) measured signals in 25 AC cycles; (b) de-noised PD signals by single-amplitude-based thresholding wavelet transform; (c) de-noised PD signals by multi-scale thresholding wavelet transform (signals having higher probability indexes are depicted in a darker color).

(PRPD) diagram of the noise-corrupted PD signals is shown in Figure 9.4(a). It can be seen that the PD signals were totally embedded in the noise. It is not unexpected that the conventional single-thresholding wavelet transform failed to de-noise the PD signals [Figure 9.4(b)]. However, the multi-scale thresholding wavelet transform successfully separated corona discharges from noise [Figure 9.4(c)]. The PD signals can easily be identified in Figure 9.4(c), in which the PD signals have the highest probability indexes (the highest possibility of being generated by a PD event).

Figure 9.5 shows the results of extracting PD signals from PD measurement on a 10 MVA transformer (33 kV/11 kV). It can be seen that the wavelet transform with multi-scale thresholding can effectively de-noise PD signals, while the conventional single-threshold wavelet transform cannot effectively de-noise PD signals. The results presented in Figures 9.3, 9.4, and 9.5 demonstrate that, by adopting multi-scale thresholds at each decomposition level in the wavelet transform and using a probability indexes for the extracted signals, it is possible to enhance the wavelet transform's capability for separating PD signals and noise in time and frequency at multiple resolutions.

It is worth mentioning that successful signal de-noising depends largely on the differences between signals and noise in the criteria described in Section 9.3.1

[17]. If a signal and a signal generated by noise exhibit the same properties in all of these criteria, it would be difficult to differentiate and separate these two signals. Moreover, it also needs mentioning that it is difficult for a single technique to remove all types of noise that may be encountered at a substation environment. A suite of combinations of different signal processing algorithms might be required for de-noising signals which are heavily coupled with different types of noise. A number of other de-noising techniques have also been presented previously.

9.4 Feature Extraction and Pattern Recognition for Transformer Diagnosis

9.4.1 Mathematical Formulation of Feature Extraction and Pattern Recognition

After signal de-noising, the next step within the intelligent framework for transformer condition monitoring and diagnosis involves feature extraction and pattern recognition, aimed at making a diagnosis on a transformer's condition, especially identifying types of fault occurring inside a transformer. The feature extraction and pattern recognition can be formulated as follows.

The de-noised measurement signal (data) can be represented in the form of an $N \times D$ dataset $\mathbf{X} = [\mathbf{x}_1, \ldots, \mathbf{x}_N]$, where $\mathbf{x}_i = [x_i^1, \ldots, x_i^D]$ is a D-dimensional vector, with each dimension regarded as a feature in which a particular characteristic of the measured signal is embedded. However, the dimensionality of the dataset $\mathbf{X}$ may be very large. Thus, feature extraction is performed on the dataset $\mathbf{X}$ to transform it into an $N \times d$ dataset $\mathbf{Y} = [\mathbf{y}_1, \ldots, \mathbf{y}_N]$, where $\mathbf{y}_i = [y_i^1, \ldots, y_i^d]$ and $d < < D$. Though a significant portion of the features are removed from the dataset $\mathbf{X}$, the remaining features in dataset $\mathbf{Y}$ can still provide essential characteristics for discriminating different types of fault [13, 33].

Each record in dataset $\mathbf{Y}$ can be categorized into one of K independent classes, $C_k \in \{1, \ldots, K\}$, with each class corresponding to one type (class) of transformer fault. The task of pattern recognition is firstly to learn the correlation between records (i.e., features in dataset $\mathbf{Y}$) and the corresponding types of fault occurring in transformers (i.e., exploring the function of $f: \mathbf{y}_j \rightarrow C_k$, $j = 1, \ldots, N$, $C_k \in \{1, \ldots, K\}$. Then, pattern recognition is used to identify the fault types of transformers, which are not included in the above dataset $\mathbf{Y}$.

In the following sections we present two case studies, one is the classification of types of transformer insulation defect causing discharges inside the transformer and the other is the determination of the health index level of the transformer insulation system. They are presented to illustrate the key steps of feature extraction and pattern recognition for transformer diagnosis.

9.4.2 Case Study on Discharge Source Classification

As mentioned in Chapter 5, different types of PD source can generate different discharge patterns, which can be represented by a PRPD dataset. However, the PRPD dataset has considerably high dimensionality. For a PD dataset of 200 phase

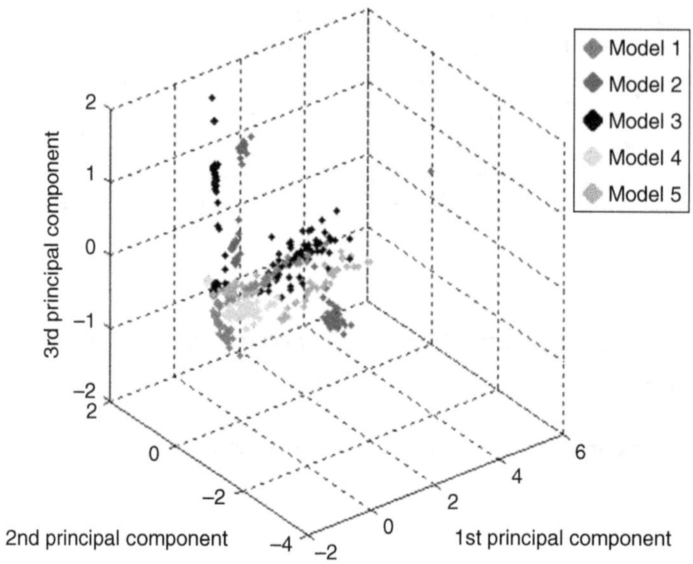

Figure 9.6 PCA visualizations for DWT feature extraction on PD dataset (Model 1 – corona, Model 2 – discharge in oil, Model 3 – surface discharge, Model 4 – internal discharge, Model 5 – discharge due to floating particles).

windows (1.8° per window), there will be $200 \times 3 = 600$ features per data point. Therefore, instead of using the PRPD PD data, DWT is adopted to extract features from the original measured PD signals. In the DWT extraction approach, a mother wavelet first decomposes the measured signal into nine levels. After thresholding, the first four moments (mean, variance, skewness, and kurtosis) are calculated from the distribution of wavelet coefficients in a nine-level decomposition. These four moments are the extracted features. As a result, 36 features ($4 \times 9 = 36$) are used to represent each data point [13, 14].

In this case study, a training database having 1000 records for five different PD sources (internal discharge, surface discharge, discharge in oil, discharge in air, and discharge due to floating particles) has been constructed using experimental PD models [13, 34]. Each record consists of the above 36 features and corresponding types of PD source. Therefore, the **Y** dataset has dimension 1000×36. This dataset can be visualized using PCA, as shown in Figure 9.6 [13]. It can be seen that data points (records) belonging to different PD sources have a certain degree of overlap. To carry out PD source classification on this dataset, algorithms with nonlinear discriminability – such as ANNs and SVM algorithms – need to be employed.

An SVM algorithm is adopted as the pattern recognition tool for PD source classification. The input to the SVM algorithm is the above 36 features extracted by DWT. The training and testing procedure of the SVM algorithm have been described in Chapter 5 and are not repeated here. Different types of PD source can be co-existent in a transformer. The SVM algorithm can estimate the probability of existence of each type of PD source. Figure 9.7 depicts the test cell configurations and corresponding patterns of internal discharge, discharge due to

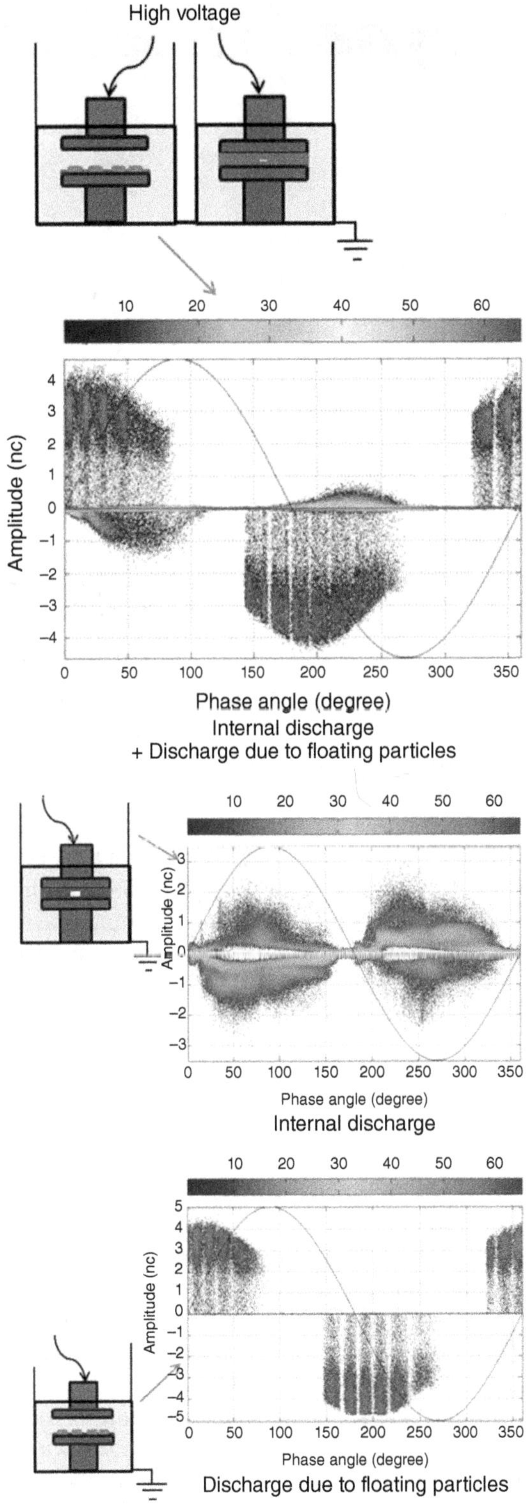

Figure 9.7 Multiple PD sources obtained from internal discharge and discharge due to floating particles.

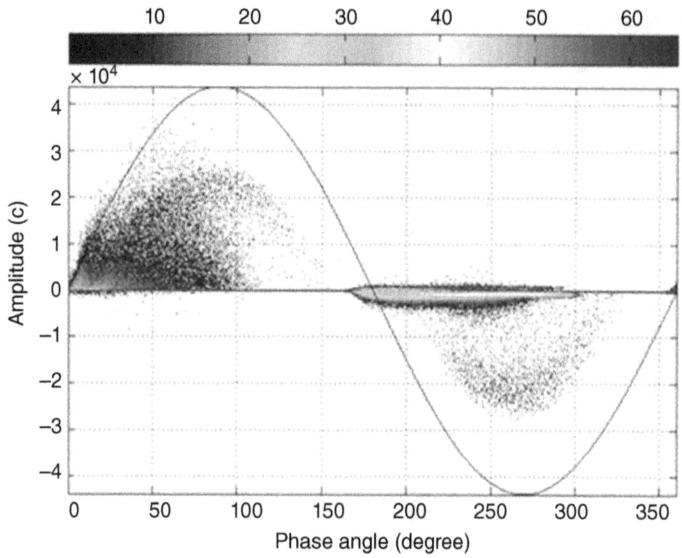

Figure 9.8 PRPD diagram of a 5 MVA power transformer in a substation.

floating particles, and multiple discharges attributed to the simultaneous occurrence of these two types of discharge. The SVM algorithm indicates that the probability of discharge due to floating and internal discharges is 67% and 17%, respectively [34]. This complies with the observation obtained from Figure 9.7 (top graph) that the pattern of the multiple PD sources contains the components of both internal discharge and discharge due to floating particles.

Figure 9.8 presents the PD pattern of a 5 MVA power transformer obtained at the test voltage of 22 kV. The above SVM algorithm (using features extracted by DWT) indicates that this transformer may have a 54% possibility of having internal discharges, a 34% possibility of having surface discharges, and a 12% possibility of having corona [34].

9.4.3 Case Study on Transformer Insulation Health Index Level Classification

Transformer oil tests including acidity, BDV, DDF, 2-FAL, moisture, and dissolved gases (C_2H_4, C_2H_2, C_2H_6, CH_4, H_2, CO, and CO_2) have been conducted in utilities as one of the major tools for evaluating transformer insulation. However, the information obtained from different types of oil test (termed "oil characteristics" in this chapter) may have different degrees of significance in revealing the condition of transformer insulation. This case study demonstrates a novel feature selection technique, which can be integrated into a pattern recognition algorithm to improve its performance in determining transformer insulation health index level using oil characteristics [30, 33].

In this case study, a feature selection method named minimum redundancy/maximum relevance (mRMR) is integrated with the SVM algorithm to determine the health index level of transformer insulation. This feature selection technique chooses a subset of the most informative features from the obtained oil characteristics and its advantages are two-fold. It can improve transformer diagnosis efficiency (in terms of computation time and data storage) because only a subset of oil characteristics is used. In addition, the selected features as inputs to a pattern recognition algorithm (e.g., SVM) can assist the algorithm to consistently achieve desirable performance by focusing on the most relevant but non-redundant information.

The mRMR selects a feature subset by using the maximal relevance and minimal redundancy criteria [33]. In the context of maximal relevance, the selected features y_i have the largest mutual information $I(y_i; c)$ with respect to target values c (e.g., health index levels). However, the combination of several features with largest mutual information may not always lead to desirable accuracy in determining the health index level for a transformer insulation system. This might be due to the extent of redundant information amongst these features. The approach to deal with this situation is to select the features with minimal redundancy while still maintaining maximum relevance [33, 35].

The mRMR first searches for a set of features S within m features which jointly have the largest dependency on the target value c [33, 35]:

$$\max D(S,c),\quad \text{where}\quad D(S,c) = I(\{y_i, i = 1, \ldots, m\}; c) \tag{9.1}$$

Assuming the set of $m-1$ features, S_{m-1}, has been found, then the mth feature can be added if it leads to the largest increase in the mutual information $I(S_m; c)$ [33, 35]:

$$I(S_m;c) = \int \ldots \int p(y_1,\ldots,y_m,c)\log\frac{p(y_1,\ldots,y_m,c)}{p(y_1,\ldots,y_m)p(c)}dy_1\ldots dy_m dc \tag{9.2}$$

To estimate $p(y_1, \ldots, y_m)$ and $p(y_1, \ldots, y_m, c)$ in equation (9.2), mRMR averages all mutual information values and then searches for features satisfying

$$\max\ D(S,c),\quad \text{where}\ D(S,c) = \frac{1}{|S|}\sum_{y_i \in S} I(y_i;c) \tag{9.3}$$

To prevent the redundant features being included in equation (9.3), mRMR introduces the minimal redundancy condition

$$\min R(S),\quad \text{where} R(S) = \frac{1}{|S|^2}\sum_{y_i, y_j \in S} I(y_i, y_j) \tag{9.4}$$

By combining the maximal relevance and minimal redundancy criteria, feature selection is now equivalent to solving one of the following optimization problems [33, 35]:

$$\max_{y_j \in y - S_{m-1}} \left\{ I(y_j;c) - \frac{1}{m-1}\sum_{y_i \in S_{m-1}} I(y_j;y_i) \right\} \tag{9.5}$$

$$\max_{y_j \in y - S_{m-1}} \left\{ I(y_j;c) \Big/ \left[\frac{1}{m-1}\sum_{y_i \in S_{m-1}} I(y_j;y_i)\right] \right\} \tag{9.6}$$

Table 9.2 Oil characteristics database for demonstrating mRMR feature selection technique

Health index level	Transformer insulation condition	Number of transformers
1	Excellent	80
2	Good	50
3	Fair	21
4	Poor	30

Table 9.3 Feature subsets with different number of features

Number of features	MID search scheme	MIQ search scheme
7	7, 9, 4, 11, 6, 2, 5	7, 11, 4, 9, 2, 5, 6
8	7, 9, 4, 11, 6, 2, 5, 3	7, 11, 4, 9, 2, 5, 6, 3
9	7, 9, 4, 11, 6, 2, 5, 3, 10	7, 11, 4, 9, 2, 5, 6, 3, 8
10	7, 9, 4, 11, 6, 2, 5, 3, 10, 1	7, 11, 4, 9, 2, 5, 6, 3, 8, 12
11	7, 9, 4, 11, 6, 2, 5, 3, 10, 1, 12	7, 11, 4, 9, 2, 5, 6, 3, 8, 12, 1

1: CO, 2: CO_2, 3: H_2, 4: CH_4, 5: C_2H_4, 6: C_2H_6, 7: C_2H_2, 8: H_2O, 9: acidity, 10: BDV, 11: 2-FAL, 12: DDF.

Equations (9.5) and (9.6) are named mutual information difference (MID) and mutual information quotient (MIQ), respectively.

A database consisting of the oil characteristics of 181 transformers (provided by a utility company) is used to demonstrate the application of mRMR in assisting the SVM algorithm to classify the health index level of transformer insulation systems. Table 9.2 presents the record distribution of this database. In this database, the transformer insulation health index levels are categorized into four different levels, including excellent, good, fair, and poor condition. The procedure for determining the health index level for transformer insulation in this table will be detailed in Section 9.6.

The mRMR-based subset selection adds a feature at each step. Table 9.3 presents the selected feature subsets (with a different number of features) obtained by applying the mRMR method on the oil characteristics database shown in Table 9.1. In the process of finding optimal features, the feature variables in the dataset are first normalized to have zero mean and unit variance. Both MID [equation (9.5)] and MIQ [equation (9.6)] search schemes are adopted.

Records in the database shown in Table 9.2 will be split into two parts: a training part consisting of 70% records and a testing part consisting of 30% records. After mRMR feature selection, the selected features will be fed into an SVM algorithm. The SVM will be trained using the 70% records. After training, the SVM algorithm predicts the health index levels for the transformers whose records are not included in the training [33]. The prediction accuracy of the above integration

Table 9.4 Prediction result obtained by mRMR and SVM algorithm (using MID search scheme, results in %)

m	Average predication accuracy	Average predication accuracy of each health index level			
		Health index 1	Health index 2	Health index 3	Health index 4
7	95	95	90	94	93
8	95	98	89	94	96
9	95	98	89	93	96
10	95	99	90	92	95
11	94	98	91	83	94
12	94	98	91	86	93

Table 9.5 Prediction result obtained by mRMR and SVM algorithm (using MIQ search scheme, results in %)

m	Average predication accuracy	Average predication accuracy of each health index level			
		Health index 1	Health index 2	Health index 3	Health index 4
7	94	95	90	94	93
8	95	98	89	94	96
9	93	96	88	90	95
10	94	98	91	86	92
11	94	98	94	82	94
12	94	98	91	86	93

of mRMR and SVM is summarized in Tables 9.4 and 9.5. In both tables, m is the number of features in the selected feature subsets corresponding to the features determined in Table 9.3. The case of $m = 12$ refers to the original SVM algorithm without integrating mRMR feature selection (i.e., using all 12 features).

It can be observed from Tables 9.4 and 9.5 that the prediction accuracies of the SVM algorithm integrated with mRMR are in the range of 94% to 95% (except for 93% by MIQ with nine features), which are slightly higher than or equal to that of the SVM algorithm without integration with mRMR ($m = 12$ in both tables, 94%). This indicates that the SVM algorithm integrated with mRMR feature selection can attain slightly better or equivalent prediction accuracy compared to the original SVM at lower computational cost, since it uses fewer features.

Tables 9.4 and 9.5 also show the predication accuracies for different health index levels. For health index level one, the algorithm achieves the highest classification accuracy (above 95%). This may be due to the fact that the number of transformers in health index level one is dominant (44%) in the database (refer to Table 9.2). The predication accuracy for the transformers in health index level one is not significantly affected by the number variations of selected features. The algorithm also attains reasonable predication accuracy (around 90%) for the transformers with health index level two and four. It is interesting to observe that for health index level three, the algorithm with fewer features (7, 8, and 9) attained better predication accuracy compared to the original SVM algorithm (12 features). For example, the algorithm with 7 and 8 features attained 94% accuracy while the original SVM algorithm with 12 features only attained 86% accuracy in both tables. This might be due to the mRMR's capability to select the most informative but non-redundant features.

A potential application of feature selection techniques is for the condition assessment of a whole power transformer unit, which may include the transformer main body (insulation, winding, and core), bushings, on-load tap changer (OLTC), and cooling systems. In such a situation, the data can include that obtained from various chemical, mechanical, and electrical tests, different online and offline measurements, historic databases, and information on transformer operating conditions. Consequently, the database may consist of a larger amount of data with very high dimensionality (i.e., number of features). Such a training database could lead to high computational cost and low efficiency of intelligent algorithms in evaluating transformer conditions. Feature selection techniques would be very useful to deal with such a large amount of data with high dimensionality.

9.5 Data and Information Fusion for Transformer Diagnosis

Owing to the complicated construction of transformers, inaccurate interpretation of measurement results, extensive coupling of noises, and interference, it is almost impossible for a single technique to make an explicit assessment of the overall health condition of a transformer. An effective and reliable condition assessment should be made based on all available evidence of the transformer, including both online and offline measurements, historic database regarding past records of the transformer and other similar transformers, industry standards, utility practices, and expert experience.

Information fusion aggregates every piece of information collected from various sources [36]. In an information fusion system, several sources may perceive the same aspect of a physical phenomenon, but with a different fidelity. Integrating information from different sources to make inferences on the physical phenomenon can reduce overall uncertainty and increase the accuracy of the diagnosis results. Moreover, an individual source may only acquire a fragment or a piece of a feature (characteristic) regarding the physical phenomenon. Information fusion allows the features to be perceived in an integrated and meaningful way, which was previously impossible, by making use of the information from each individual source operating separately.

In transformer condition assessment, the inference (interpretation) is made based on the information collected from multiple sources (DGA, PDC, PD, oil analysis, etc.). For each individual measurement, there might be several inference schemes available for interpretation. For example, the DGA diagnosis can be made through conventional ratio-based schemes, expert experience, utility practice, and AI techniques such as various ANNs and SVM and its variants. In the following two sections, Bayesian fusion and Dempster-Shafer fusion are applied for transformer condition assessment using measurement results obtained from multiple sources.

9.5.1 Bayesian Fusion for Integrating Information from Multiple Information Sources

By combining the latest measurements made by each individual information source and a priori knowledge about the probability of different fault types (calculated from a historic dataset), Bayesian fusion computes the a posteriori probability for each fault type. It is assumed that two independent information sources will make measurements and subsequently diagnose the same physical phenomenon [37]. For the first source, its diagnosis result is denoted as S_1^1, which is computed using the current measurement z_1^1 and a priori knowledge S_0^1. Similarly, for the second source, its diagnosis result is denoted as S_1^2, which is computed using its current measurement z_1^2 and a priori knowledge S_0^2. x denotes one type of fault; the task is to assign a probability to x given the latest measurements obtained from the above two information sources – $\left(x\middle|S_1^1\,S_1^2\right)$. Using Bayes' rule, this can be expressed as

$$\begin{aligned}\left(x\middle|S_1^1S_1^2\right) &= \left(x\middle|z_1^1z_1^2S_0^1S_0^2\right)\\ &= \frac{\left(z_1^1\,z_1^2\middle|x,\,S_0^1\,S_0^2\right)\left(x\middle|\,S_0^1\,S_0^2\right)}{\left(z_1^1\,z_1^2\middle|\,S_0^1\,S_0^2\right)}\\ &= \frac{\left(z_1^1\middle|x,\,S_0^1\right)\left(z_1^2\middle|x,\,S_0^2\right)\left(x\middle|\,S_0^1\,S_0^2\right)}{\left(z_1^1\,z_1^2\middle|\,S_0^1\,S_0^2.\right)}\end{aligned} \tag{9.7}$$

In equation (9.7), measurement independence at different sources is assumed. This leads to the simplification of $\left(z_1^1z_1^2\middle|x,S_0^1S_0^2\right)=\left(z_1^1\middle|x,S_0^1\right)\left(z_1^2\middle|x,S_0^2\right)$. Applying Bayes' rule again and swapping measurement z and state x in the first two terms of the numerator of equation (9.7), the final form of fusing the information can be obtained [36]:

$$\begin{aligned}\left(x\middle|S_1^1S_1^2\right) &= \frac{\left(x\middle|\,S_1^1\right)\left(z_1^1\middle|\,S_0^1\right)}{\left(x\middle|\,S_0^1\right)}\cdot\frac{\left(x\middle|\,S_1^2\right)\left(z_1^2\middle|\,S_0^2\right)}{\left(x\middle|\,S_0^2\right)}\frac{\left(x\middle|\,S_0^1\,S_0^2\right)}{\left(z_1^1\,z_1^2\middle|\,S_0^1\,S_0^2\right)}\\ &= \frac{\left(x\middle|\,S_1^1\right)\left(x\middle|\,S_1^2\right)\left(x\middle|\,S_0^1\,S_0^2\right)}{\left(x\middle|\,S_0^1\right)\left(x\middle|\,S_0^2\right)}\end{aligned} \tag{9.8}$$

In equation (9.8), the terms $\left(x\middle|S_1^1\right)$ and $\left(x\middle|S_1^2\right)$ represent the contribution of current measurements at two sources. The terms $\left(x\middle|S_0^1\right)$ and $\left(x\middle|S_1^2\right)$ represent the a priori knowledge about the probability distribution of x at two sources. The term $\left(x\middle|S_0^1S_0^2\right)$ is the fault diagnosis result of the previous time step, which is provided by the fusion algorithm. From equation (9.8), it can be seen that Bayesian fusion incorporates not only the current measurements but also the previous knowledge (i.e., historic data and information) from multiple sources. This will greatly improve the accuracy of fault classification. For more than two information sources, equation (9.8) can readily be extended. For example, for three sources, equation (9.8) simply becomes

$$\left(x \middle| S_1^1 S_1^2 S_1^3\right) = \frac{\left(x \middle| S_1^1\right)\left(x \middle| S_1^2\right)\left(x \middle| S_1^3\right)\left(x \middle| S_0^1\, S_0^2\;\; S_0^3\right)}{\left(x \middle| S_0^1\right)\left(x \middle| S_0^2\right)\left(x \middle| S_0^3\right)}$$

The following case study is on DGA interpretation of a 200 MVA three-winding transformer (made in 1972). DGA tests were conducted on two sets of oil samples: sample A (collected from the transformer oil valve, bottom sample) and sample E (collected from Buchholz, Buchholz sample). The DGA results of these two samples are analyzed using SOM clustering and an SVM algorithm. In both SOM and SVM, five gas concentrations (C_2H_2, C_2H_4, CH_4, H_2, and C_2H_6 by volume) are used as features that are fed into the algorithms. A total of 390 DGA records (with known transformer condition) provided by a utility company is used to form a database for training the SOM and SVM algorithm. The trained SOM and SVM are then applied to recognize normal operation and fault conditions using the DGA results on the above two oil samples. Figure 9.9 presents the visualization of SOM clustering results. Table 9.6 summarizes the SVM classification results and also provides Bayesian fusion results (i.e., combining SVM results from the analysis on two samples).

It can be seen from Figure 9.9 that the DGA test results of samples A and E indicate this transformer is either in normal operating condition, or has a discharge fault, or has a low to medium-temperature thermal fault, or has a high-temperature thermal fault. SVM classification results shown in Table 9.6 indicate that this transformer has around 67% (sample A 67%, sample E 68%) chance of being in normal operating condition, 13% (sample A 14%, sample E 12%) chance of having a discharge fault, 8% chance of having a low to medium-range thermal fault, and around 11% (sample A 11%, sample E 12%) chance of having a high-range thermal fault. The Bayesian fusion algorithm is then applied to combine the diagnosis results of samples A and E. In the computation, the a priori probability in equation (9.8) is computed using the above-mentioned training database. The a priori probability for normal operating condition, discharge fault, low to medium-range thermal fault, and high-range thermal fault are 48%, 8%, 24%, and 8%, respectively.

It can be seen from Table 9.6 that Bayesian fusion reinforces the diagnosis results that this substation transformer is in normal operating condition and with some chance of having a discharge and high-range thermal fault. However, the chance of having a low to medium-range thermal fault as calculated by the Bayesian fusion algorithm is much lower than that calculated by the SVM algorithm. This is due to the introduction of the a priori probability in the computation. Bayesian fusion provides a means to combine the diagnosis results from different sources and also incorporate prior knowledge regarding the transformer condition.

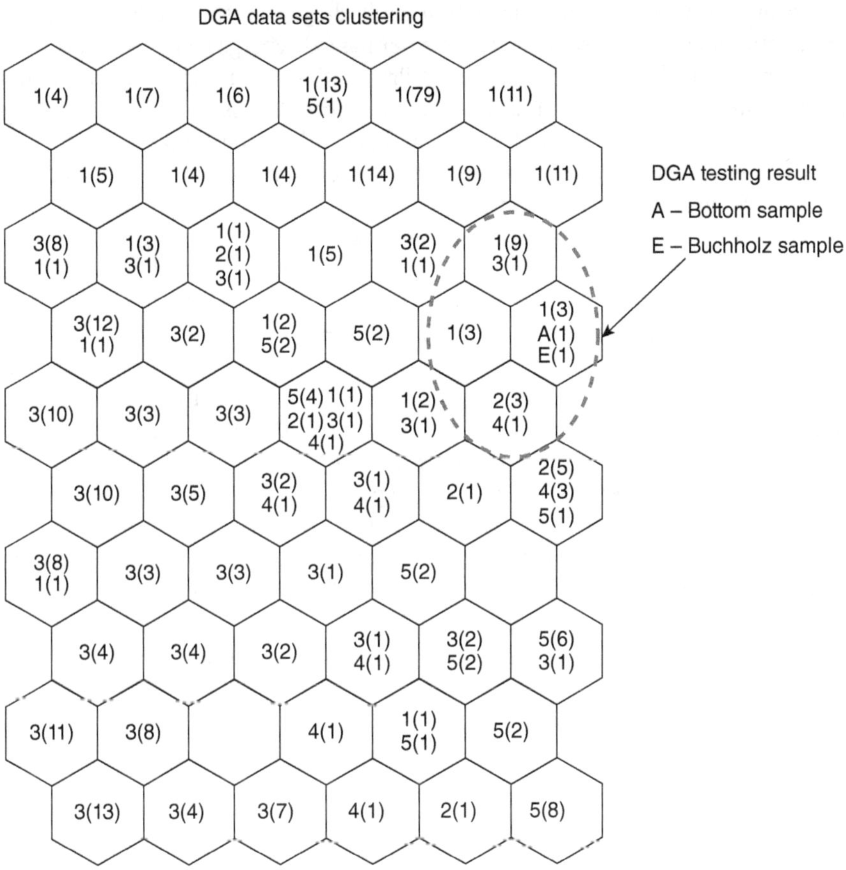

Figure 9.9 SOM clustering results on two sets of samples: sample A (collected from bottom) and sample E (collected from Buchholz). The transformer conditions include 1 – normal operating condition, 2 – discharge fault, 3 – lower to medium-range thermal fault, 4 – high-range thermal fault, and 5 – partial discharge. (The numbers in brackets are the number of data points with that fault type in the training dataset, which consists of 390 DGA records with known transformer condition provided by a utility company).

Table 9.6 Bayesian fusion of DGA test

Oil sample	SOM results	SVM classification
A	1, 2, 3, 4	[**67%** 14% 8% 11%]
E	1, 2, 3, 4	[**68%** 12% 8% 12%]
	Bayesian fusion result	[**65%** 20% 1% 14%]

9.5.2 Dempster–Shafer Fusion

Dempster–Shafer fusion differs from Bayesian fusion in that it can assign a measure of belief to propositions, which might be a single hypothesis or a combination of several hypotheses containing overlapping or even conflicting hypotheses [38].

For example, in transformer inception fault diagnosis, there might be three propositions: (1) Proposition 1 – transformer has a discharge fault; (2) Proposition 2 – transformer has a thermal fault; (3) Proposition 3 – transformer has a discharge fault or thermal fault. It can be seen that both Propositions 1 and 2 overlap Proposition 3, however, Proposition 1 conflicts with Proposition2.

In a Dempster–Shafer fusion system, a set of n mutually exclusive and exhaustive propositions is denoted $\Theta = \{A_1, A_2, \ldots, A_n\}$ (named the *frame of discernment*). These propositions are the causes that would explain observable evidence. Accordingly, there will be 2^n general propositions (including both single and combined hypotheses). For the above transformer example, the Dempster–Shafer theory will assign one of the possibilities $\Theta = \{A_1, A_2, \{A_1, A_2\}, \emptyset\}$ to the transformer condition, corresponding to: (1) transformer has a discharge fault; (2) transformer has a thermal fault; (3) transformer has either a discharge or a thermal fault; (4) transformer has neither discharge nor thermal fault.

In Dempster–Shafer theory, the contribution of each information source S^i to the frame of discernment Θ is described by a probability mass function, named the *basic probability assignment (BPA)* of S^i and denoted m^i [38]. The BPA cannot be treated as the probability in the classical sense. This is because BPA can be assigned not only to a single hypothesis (as for the classical probability) but also to combined hypotheses. Therefore, BPA is defined such that $m^i(\Theta) \leq 1$ and $\sum m^i(\Theta) = 1$. For the above transformer example, the BPA is $m^i(A_1)$, $m^i(A_2)$, and $m^i(A_{1\vee}A_2)$. Here, v denotes the Boolean expression OR.

After assigning BPA for the hypothesis at each information source (e.g., sensor measurement, expert diagnosis, etc.), Dempster's rule of combination is then adopted for data fusion. For each proposition in the frame of discernment Θ, it combines source S^i's BPA m^i and source S^j's BPA m^j [38]:

$$\left(m^i \oplus m^j\right)(A) = \frac{\sum_{E_i \cap E_j = A} m^i(E_i) m^j(E_j)}{1 - \sum_{E_i \cap E_j = \varphi} m^i(E_i) m^j(E_j)} \tag{9.9}$$

where E_i and E_j are evidences (i.e., diagnosis results) at sources S^i and S^j. As in Bayesian fusion, the above rule can readily be extended to include more sources.

In the Dempster–Shafer theory, two new functions – Belief (*Bel*) and plausibility (*Pl*) – are used. *Bel* is defined as the sum of the probability mass for a proposition:

$$Bel(A) = \sum_{E_k \subseteq A} m(E_k) \tag{9.10}$$

Apparently, *Bel* combines all pieces of evidence that support proposition A. If A is a single proposition (i.e., $A = A_i$), then *Bel*(A) is simply the probability of A_i. If A is a combined proposition (e.g., $A = A_1 \vee A_2$), then *Bel*(A) is the sum of probability mass contributing to all elements of A:

$$Bel(A_1 \vee A_2) = m_i(A_1) + m_i(A_2) + m_i(A_1 \vee A_2) \tag{9.11}$$

The difference between $m(A)$ and $Bel(A)$ is that $m(A)$ measures the assignment of belief only to A, while $Bel(A)$ measures the total assignment of belief to A and all its subsets.

The plausibility function is defined as the sum of the BPAs of all sets that intersect A, i.e. $E_k \cap A \neq \emptyset$:

$$Pl(A) = \sum_{\forall E_k\, E_k \cap A \neq \emptyset} m(E_k) \tag{9.12}$$

The relationship between Bel and Pl is

$$\begin{aligned} &Pl(A) = 1 - Bel(\neg \mathrm{A}) \\ &Pl(A) \geq Bel(A) \end{aligned} \tag{9.13}$$

In equation (9.13), ($\neg A$) denotes the negation of hypothesis A. The "belief" and "plausibility" can be regarded as the lower and upper bound of the probability range [38].

The major difference between Bayesian fusion and Dempster–Shafer fusion is that Bayesian fusion works with the classically defined probabilities (i.e., how often an event will happen if an experiment is performed a large number of times). In contrast, Dempster–Shafer fusion considers the way humans give judgement after making a measurement of a physical phenomenon. Dempster–Shafer fusion includes an "unknown type" state, which indicates that there is no definite diagnosis. For example, in transformer fault diagnosis the claim that "the transformer might be either in normal operating condition or have a discharge fault, or have a thermal fault" can be regarded as the "unknown type" state. This might be the state of a human's knowledge about a transformer on some occasions.

In the case study presented in Section 9.5.1, Bayesian fusion was used to make transformer fault diagnosis by integrating pattern recognition algorithms' interpretation results, obtained on oil samples from different locations of the transformer. In utilities, DGA test data sets are normally evaluated based on utility practices, expert knowledge, and conventional ratio-based schemes. As human evaluation will introduce the specific "unknown type" state, Dempster–Shafer fusion is applied. This combines the diagnosis results obtained from the Bayesian fusion algorithm (Table 9.6) and from utility experts.

Table 9.7 shows the mass function of Dempster–Shafer fusion. In this table, the mass functions of the four transformer conditions are assigned according to the above Bayesian fusion result (the last row of Table 9.5). One extra condition is also assigned with the mass function, which corresponds to the unknown type state (i.e., the transformer condition might be in normal operating condition, or has a discharge fault, or has a lower to medium-range thermal fault, or has a high-range thermal fault). For the utility expert diagnosis result, the mass functions are assigned as follows: normal operating condition (60%), discharge fault (30%), unknown (10%) [36].

It can be seen from Table 9.7 that the Dempster–Shafer fusion further reinforces the diagnosis results that the transformer is in normal operating condition (82.7%) and with some chance of having a discharge fault (14.6%). Therefore, Dempster–Shafer fusion provides a means to combine the objective diagnosis

Table 9.7 Dempster–Shafer fusion result

Condition	Bayesian fusion result	Utility expert's result	Fused masses
Normal	65%	60%	82.7%
Discharge	20%	30%	14.6%
L-thermal	1%	0	0.2%
H-thermal	14%	0	2.5%
Unknown	0	10%	0
Total mass	100%	100%	100%

Note: The Bayesian fusion result is taken from Table 9.6. The utility's result is based on the conventional ratio-based method, expert's diagnosis, and utility's practice. L-thermal and H-thermal refer to low to medium and high-range thermal fault, respectively.

results (e.g., by SOM and SVM algorithm) and the subjective diagnosis results (e.g., by expert experience and utility practice).

From the case studies shown in Tables 9.6 and 9.7, it can be concluded that adopting data and information fusion techniques for transformer condition assessment will be an advanced feature in an intelligent transformer assessment system, given that it integrates all evidence from multiple sources, historic data, and subjective expert's experience and utility practices. It can also be a component in determining the transformer health index, as discussed in the next section.

9.6 Determining Health Index for Transformer Insulation System [30, 43]

9.6.1 Introduction

During the past 10 years, attempts have been made to calculate a health index for a power transformer to provide an indication of its health condition [2, 30, 38–41]. Various online and offline measurements have been performed on transformers. A quantitative health index was then computed based on relevant industrial standard evaluations. Such an approach requires an experienced engineer processing the data and arriving at an evaluation of the condition of the transformer. However, due to the nature of the power transformer construction and the complexity of its operation, the above approach for transformer condition assessment is non-trivial; this is particularly so for non-experts. Therefore, it is necessary to develop intelligent techniques to process the collected oil test records automatically, and subsequently compute a health index for a transformer's insulation system.

This section presents the application of pattern recognition techniques to determine the health index for a transformer oil–paper insulation system using oil characteristics (oil test). Pattern recognition-based health index prediction is

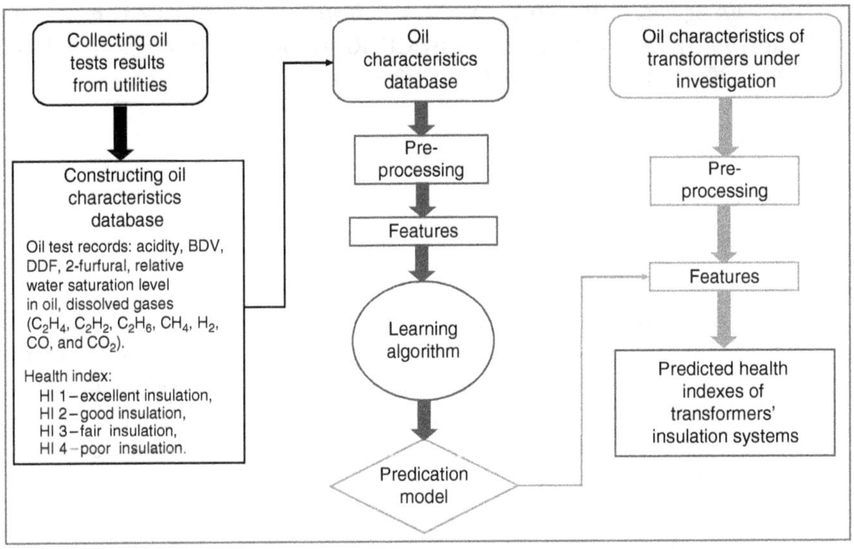

Figure 9.10 Conceptual diagram of pattern recognition-based transformer health index prediction [38].

an automatic computer-based process, depicted in Figure 9.10. It includes three steps [38]:

1) *Construction of a training database.* A database is constructed using oil test records collected from utilities. The records included in the database are: acidity, BDV, DDF, 2-FAL, dissolved gases (C_2H_4, C_2H_2, C_2H_6, CH_4, H_2, CO, and CO_2), and relative water saturation (RWS) level in oil. RWS is the ratio between the measured water content in oil and the saturated water content in oil at the temperature of oil sample collection. The higher RWS value indicates the higher possibility of forming free water in transformer oil. A comprehensive approach is developed to determine health indexes for the transformers in this database (refer to Section 9.6.2).
2) *Training of pattern recognition algorithm.* Using the above database, an algorithm can learn the relationships between oil characteristics and health indexes of transformer insulation.
3) *Prediction of health index.* After training, the pattern recognition algorithm can be used to predict health indexes for transformers under investigation (they are not included in the training database), based on their oil characteristics.

9.6.2 Construction of Training Database

A pattern recognition algorithm can be used to approximate the relationship between the oil characteristics and the health index of the transformer insulation. This algorithm relies on a model trained using a database of oil characteristics. A comprehensive approach for constructing the oil characteristics database is as follows [30].

a) *Determining health index (HI_a) based on industry standards.* Based on industry standards (IEEE and IEC standards, CIGRE recommendations), both DGA factor (DGAF) and oil quality factor (OQF) are defined [2, 28] as

$$\mathrm{DGAF} = \sum_{i=1}^{7} S_i \times W_i \Big/ \sum_{i=1}^{7} W_i \tag{9.14}$$

$$\mathrm{OQF} = \sum_{j=1}^{4} S_j \times W_j \Big/ \sum_{j=1}^{4} W_j \tag{9.15}$$

where i and j denote dissolved gases and other oil characteristics (DDF, BDV, relative water saturation level in oil, and acidity), respectively. S_i and S_j are the scores determined from the test results. W_i and W_j are the weighting of each individual test [2, 30]. The paper insulation factor (PIF) is defined based on the amount of 2-furfural. Finally, DGAF, OQF, and PIF are combined and normalized into the health index $\mathrm{HI_a}$ as

$$\mathrm{HI_a} = \sum_{j=1}^{3} F_i \times K_i \Big/ \sum_{j=1}^{3} K_i \tag{9.16}$$

where F_i are values of DGAF, OQF, and PIF. A method to determine K_i values is suggested in Ref. [2].

b) *Determining health index (HI_b) based on expert judgement.* In addition to oil test records, comments from utility experts are taken into consideration by the health index. The utility experts have access to further information regarding their transformers, such as loading history and operating conditions. Their comments on transformer condition include: (1) transformer condition is acceptable; (2) further investigation required, repeat oil tests every three months for one year; (3) DGA caution, repeat oil tests daily for seven consecutive days; and (4) DGA danger, abnormal or serious situation. Based on these comments, the condition of transformer insulation is quantified into four health index levels (i.e., health index $\mathrm{HI_b}$).

c) *Determining health index (HI_c) based on Duval's triangle.* This takes into account the dissolved gas content of the oil. According to the faults identified by Duval's triangle, the corresponding transformer insulation condition is quantified into health index $\mathrm{HI_c}$ [30].

d) *Determining health index (HI_d) based on a fuzzy c-means (FCM) algorithm.* Owing to the complicated ageing mechanism of the insulation system, imprecision of the test equipment, and limitation of expert judgements on the oil test data, a certain degree of inaccuracy and uncertainty may exist when the health index for each oil characteristic record is decided using the database. As a result, whereas some oil characteristic records can be fully assigned to one of the health index levels, others may not be assigned exactly to one of the health index levels. To deal with such a problem, a computer algorithm was developed based on an FCM algorithm. FCM has the advantage of assigning an oil characteristic record with a weight of belonging to a particular health index. It helps to take the above uncertainties into account in database construction [42].

e) *Assigning final health index and forming the database.* After obtaining the above four sets of health index (HI_a, HI_b, HI_c, and HI_d), this task combines them into one final health index. Since the four methods compute a health index based on different criteria, discrepancies may exist. Therefore, the majority vote strategy is adopted to decide the final health indexes of insulation systems for the transformers included in the training database [30].

The above comprehensive approach can be demonstrated with the following database. In database construction, oil test records of 618 transformers collected from several utilities are used. Each record consists of the concentrations of dissolved gases (C_2H_4, C_2H_2, C_2H_6, CH_4, H_2, CO, and CO_2) and test results of acidity, BDV, DDF, 2-furfural, and RWS level in oil. As an example, Table 9.8 shows oil test records for eight out of the 618 transformers.

By using the above comprehensive approach, the health indexes for 618 transformer insulation systems are calculated. Table 9.9 presents the results of the eight transformers which are shown in Table 9.8. In Table 9.9, the insulation conditions of transformers are quantified into four health index levels: excellent (HI = 1), good (HI = 2), fair (HI = 3), and poor (HI = 4).

It can be seen from Table 9.9 that the health indexes ranked by four assessment criteria are consistent for transformer 4. For example, FCM assigns transformer 4 to health index level 3 with a high weight of 96%. Duval's triangle, expert judgement, and industry standard evaluation also determine transformer 3 to have health index level 3. Thus, the final health index for transformer 3 is HI 3 (fair condition of insulation). For the other seven transformers, there exist some discrepancies amongst different criteria. For example, FCM assigns transformer 7 to health index level 3 (88% possibility in HI 3 and 12% possibility in HI 2); however, the other three criteria assign transformer 7 to health index level 4. By majority vote, transformer 7 is finally assigned to HI 4 (poor condition of insulation).

9.6.3 Predication of Health Index for Transformer Insulation System

The next step is to train the pattern recognition algorithms using the database constructed in the previous section. After training, a pattern recognition algorithm can build up a model to approximate the relationship between the oil characteristics and the health index of the transformer insulation. The model can then predict the health indexes of insulation systems for transformers of interest (named "testing transformers"). It needs to be mentioned that (1) these testing transformers are not included in the oil characteristics database; (2) only oil test results of these transformers are available; and (3) the health indexes of these transformers will be predicted by the pattern recognition algorithm, which has been trained by the training database.

Several pattern recognition algorithms were trained using the database described in Section 9.6.2, which consists of 618 transformers. The algorithms we investigated are those based on SVM, RBF, and *k*NN. Descriptions of these algorithms are given in Chapter 4 and in the literature [40–44, 46, 47]. After the above three algorithms were trained, they were used to predict the health index of a further 264 transformers, which were not included in the training

Table 9.8 Oil test record of eight transformers digested from historic records [38]

No.	C_2H_4 (ppm)	C_2H_2 (ppm)	C_2H_6 (ppm)	CH_4 (ppm)	H_2 (ppm)	CO (ppm)	CO_2 (ppm)	Acidity (mg KOH/g)	Water (ppm)	Temp* (°C)	RWS* (%)	BDV (kV)	DDF (%)	Furfural (ppm)
1	16	1	73	45	14	300	3720	0.01	2	33	2.2	51	0.036	0.01
2	9	1	3	1	10	46	170	0.01	4	56	1.8	65	0.009	0
3	1	0.1	0.3	1.5	38.1	76.1	1555	0.01	8	52	4.2	29	0.020	0.01
4	3	1	60	98	20	640	8050	0.02	2	46	1.3	74	0.164	0.01
5	4	1	38	80	12	1580	5100	0.03	8	55	3.8	41	0.003	0.02
6	2	1	115	735	5840	225	2260	0.04	9	63	3.2	27	0.114	0.01
7	1	1	81	545	7970	215	1580	0.02	7	56	3.1	49	0.183	0
8	2	1	73	535	8750	220	1770	0.04	4	42	3.0	32	0.103	0.01

* Temp is the temperature when oil samples were collected. RWS is calculated based on the saturation level of transformer oil at different temperatures as given in Ref. [45].

Table 9.9 Examples of determining health index of transformer insulation for the construction of oil characteristics database [38]

	Health index based on standard						FCM					
No.	DGAF	OQF	PIF	HI_a	Expert's judgment HI_b	Duval triangle HI_c	M1	M2	M3	M4	HI_d	HI
1	2	3	1	3	1	1	0	85%	15%	0	2	1
2	1	2	1	2	1	1	0	2%	98%	0	3	1
3	1	3	1	2	2	1	0	97%	3%	0	2	2
4	3	2	1	3	3	3	0	4%	96%	0	3	3
5	3	2	1	3	4	3	0	96%	4%	0	2	3
6	4	3	1	4	4	4	0	68%	32%	0	2	4
7	4	3	1	4	4	4	0	12%	88%	0	3	4
8	4	3	1	4	3	4	0	92%	8%	0	2	4

Note: HI_b is based on the utility expert's comments: 1 – acceptable; 2 – further investigation required, sample three monthly for one year; 3 – DGA caution, resample daily for seven consecutive days; 4 – DGA danger, abnormal or serious situation.

Table 9.10 Examples of health index of transformer insulation systems [38]

Transformer	Probability of each health index level	Predicted health index
TX1	[**93%** 2% 4% 1%]	HI 1 (Excellent condition of insulation)
TX2	[**94%** 4% 1% 1%]	HI 1 (Excellent condition of insulation)
TX3	[**86%** 5% 7% 2%]	HI 1 (Excellent condition of insulation)
TX4	[**80%** 4% 3% 13%]	HI 1 (Excellent condition of insulation)
TX5	[3% **93%** 2% 2%]	HI 2 (Good condition of insulation)
TX6	[8% **88%** 4% 0%]	HI 2 (Good condition of insulation)
TX7	[17% **75%** 3% 5%]	HI 2 (Good condition of insulation)
TX8	[1% 8% **87%** 4%]	HI 3 (Fair condition of insulation)
TX9	[1% 6% **90%** 3%]	HI 3 (Fair condition of insulation)
TX10	[1% 10% 1% **88%**]	HI 4 (Critical condition of insulation)

database. Table 9.10 shows the predicted insulation health index of 10 transformers amongst these 264 transformers.

A comparison between the algorithm predicted health index and the manually determined health index using the methods presented in Section 9.6.2 is shown in Table 9.11. The prediction accuracy is the ratio between the number of transformers with correctly predicted health indexes and the total number of testing transformers (i.e., 264). As shown in this table (left column, without using data preprocessing and boosting), the prediction accuracy of HI 1 and 2 is lower than

Table 9.11 Comparison of prediction accuracy (%) on health index of three algorithms [38]

	Without data preprocessing and boosting					With data preprocessing and boosting				
Algorithm	HI 1	HI 2	HI 3	HI 4	Overall	HI 1	HI 2	HI 3	HI 4	Overall
*k*NN	78	57	100	90	82	100	90	100	96	97
RBF	84	59	97	94	84	100	100	100	98	99
SVM	59	77	100	97	82	100	97	100	97	98

that of HI 3 and 4. For example, the SVM algorithm failed to reliably predict HI 1 (59% accuracy), while the *k*NN and RBF algorithms failed to reliably predict HI 2 (around 58% accuracy). This is because there aren't enough samples of HI 1 and HI 2 in the training database. Samples of HI levels 3 and 4 outnumber those of HI levels 1 and 2. If trained by such an imbalanced database, the algorithms may generate a biased model, which is in favor of the health index levels with a larger number of records and provides lower prediction accuracy on the health index levels with a smaller number of records [42].

To deal with the above class imbalance database, some data processing approaches can be adopted. For example, oversampling (preprocessing) and bootstrap techniques are implemented to statistically equalize the record numbers in the original oil characteristics database [38, 41, 42]. After applying these data processing methods, the oil characteristics database is balanced with evenly distributed records. Table 9.10 (right column, using data preprocessing and boosting) shows the prediction accuracy of the algorithms integrating data preprocessing and boosting. It can be seen that data preprocessing and boosting can improve the performance of pattern recognition algorithms significantly. The overall prediction accuracy of three pattern recognition algorithms is improved to above 97% from around 83%.

9.6.4 Discussions on Using HI to Describe Overall Condition of Transformers

In the above sections, intelligent algorithms were applied to compute the health index for a transformer's insulation system. Recently, efforts have also been made to apply a health index to quantify the overall condition of a transformer [26, 48–51]. Accordingly, the health index is defined as "one overall indicator describing the condition status of the transformer that the decision maker can use to define and prioritize the required action" in operation, maintenance, and replacement [26]. Such comprehensive health indexing can be applied to life management of a fleet of transformers. When a transformer in a fleet is considered, firstly its health index is calculated and then an impact index is also calculated, which reflects the transformer's importance to the network – such as reliability, failure impact on the whole network, feasibility of replacement, etc. [26]. The combination of these two indexes can be used to rank the transformer in the fleet. In deriving and making use of the health index, the following two issues need to be considered.

1) *Data quality issues.* Any algorithm's reliability is highly dependent on the quality of input data (sensor measurement data, historic records, test results, interpretation, etc.). There may exist missing data, erroneous data, and redundant data. Thus, there should be a mechanism in the health index approach to compensate for the impact of this data, as well as to provide a confidence level to account for uncertainties due to this data. Also, the generalization ability of algorithms for calculating the health index needs to be improved (i.e., algorithms trained on a "local" database can be applied "globally").
2) *A single-value health index may not be enough in assisting decision making for transformer replacement, refurbishment, and maintenance.* In determining the health index, one should not only rely on the mathematical weighting alone, but also consider the criticality of each indicator to the failure probability of the transformer as a whole. Moreover, the worst-case approach may be required (i.e., using the critical defects next to the weighting result).

To make the health index a practical tool, further research and development work with collaboration between academia and practicians is much needed.

9.7 SmartBox for Smart Transformer [4]

To implement and deploy various condition monitoring and diagnosis techniques, a hardware and software system (named SmartBox) has been developed, as shown in Figure 9.11(a). The SmartBox consists of a main processing board (functioning as a normal computer), high-speed data acquisition (DAQ) boards, and multi-functional I/O. SmartBox acquires signals from online sensors during online condition monitoring of transformers and executes signal processing and pattern recognition algorithms for signal de-noising, feature extraction, and fault pattern recognition. Laboratory measurements and field measurements have been performed to investigate the most suitable configurations for the SmartBox, including sampling rate, sensitivity, bandwidth, and peak detection mode. Algorithms for real-time signal acquisition, advanced signal processing, feature extraction, pattern recognition, data fusion, and health index predication have been implemented [Figure 9.11(b)].

An online PD (both electric and acoustic measurements) monitoring and diagnosis system has been deployed successfully in SmartBox and tested for field PD measurement of transformers. As shown in Figure 9.11(a), it uses an HFCT to measure the magnitude of PD pulses in peak detection mode with low sampling rate. After being triggered by HFCT upon detecting PD signals at a predefined threshold level, the system changes to a high sampling rate mode for continuous signal acquisition. Then it executes a suite of advanced signal processing and pattern recognition algorithms to accomplish the tasks of PD signal de-noising and extraction and PD source separation and identification. Meanwhile, immediately after HFCT detection of PD signals, three AE sensors will also be evoked to take measurements that will be used to determine the positions of the PD sources inside the transformer tank. It needs to be mentioned that other types of PD sensor – such as the UHF PD probe inserted at a transformer oil valve and RLC

(a)

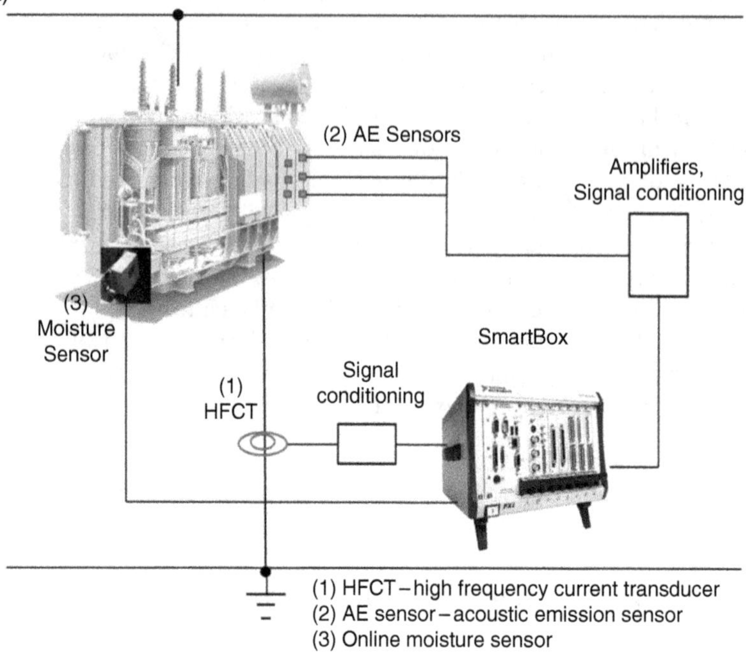

(b)

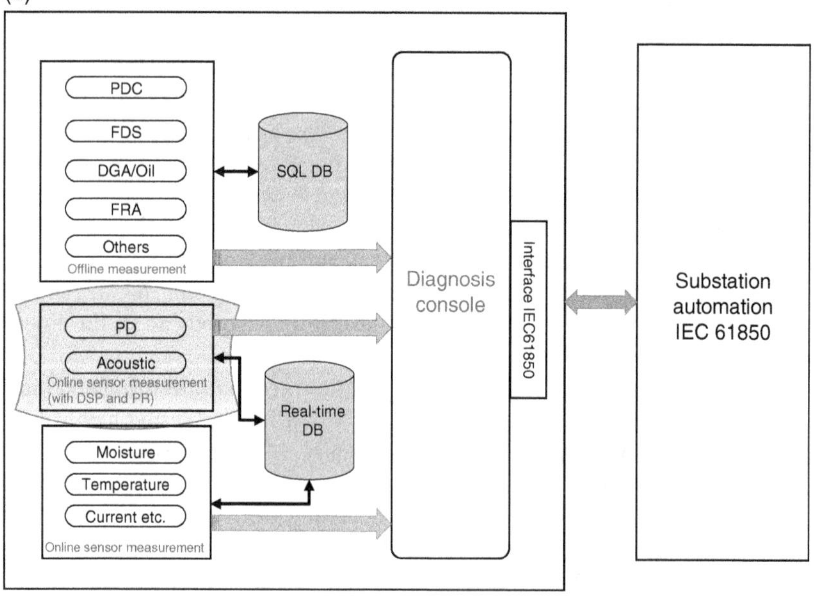

Figure 9.11 SmartBox – configurable hardware and software platform: (a) hardware modules; (b) system software architecture. (*See insert for color representation of the figure.*)

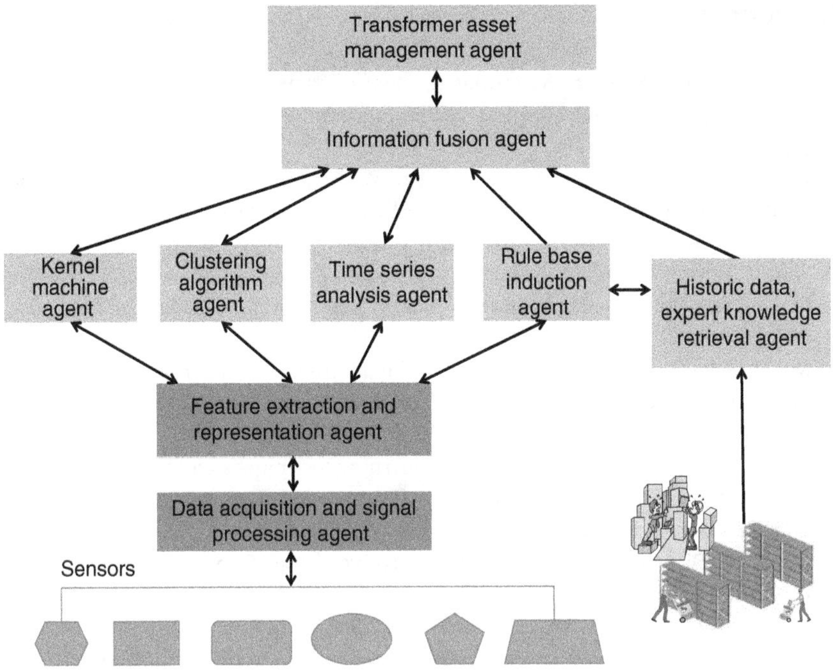

Figure 9.12 Software platform based on a multi-agent system for transformer condition monitoring and assessment.

measuring impedance PD detector installed at the transformer bushing taps – can also be installed without changing the hardware configuration of the SmartBox.

As shown in Figure 9.11, a moisture-in-oil sensor is also installed. This can provide oil temperature and water content in the transformer oil. With the proper transformer modeling, it is possible to derive water content in the transformer paper/pressboard [22, 25].

To provide a platform for deploying various algorithms, a multi-agent system (MAS) can be adopted as shown in Figure 9.12. The MAS can provide system extensibility to include newly installed sensors and the latest signal processing and pattern recognition algorithms as they become available.

MAS systems encompass attributes required to automate condition monitoring and diagnosis applications [52, 53]. With integration and interaction with pattern recognition algorithms, each intelligent agent in a MAS system can learn from data for decision making. Also, through historical data analysis using pattern recognition algorithms, these agents can enhance their capability to handle uncertainties, which is important for practical intelligent condition monitoring and assessment of power transformers in a substation environment.

References

1 Xiang Z, Gockenbach E. Asset-management of transformers based on condition monitoring and standard diagnosis. *IEEE Electrical Insulation Magazine* 2008; 24 (4): 26–40.

2 Jahromi A, Piercy R, Cress S, Service J, Fan W. An approach to power transformer asset management using health index. *IEEE Electrical Insulation Magazine* 2009; 25(2): 20–34.

3 Saha T. Review of modern diagnostic techniques for assessing insulation condition in aged transformers. *IEEE Transactions on Dielectrics and Electrical Insulation* 2003; 10(5): 903–917.

4 Ma H, Saha T, Ekanayake C, Martin D. Smart transformer for smart grid – intelligent framework and techniques for power transformer asset management. *IEEE Transactions on Smart Grid* 2015; 6(2): 1026–1034.

5 Ma H, Saha T, Ekanayake C. Statistical learning techniques and their applications for condition assessment of power transformer. *IEEE Transactions on Dielectrics and Electrical Insulation* 2012; 19(2): 481–489.

6 Duval M. Dissolved gas analysis: It can save your transformer. *IEEE Electrical Insulation Magazine* 1989; 5(6): 22–27.

7 Unsworth J, Mitchell F. Degradation of electrical insulating paper monitored with high performance liquid chromatography. *IEEE Transactions on Electrical Insulation* 1990; 25(4): 737–746.

8 Saha T, Purkait P. Investigation of an expert system for the condition assessment of transformer insulation based on dielectric response measurements. *IEEE Transactions on Power Delivery* 2004; 19(3): 1127–1134.

9 Ekanayake C, Gubanski S, Graczkowski A, Walczak K. Frequency response of oil impregnated pressboard and paper samples for estimating moisture in transformer insulation. *IEEE Transactions on Power Delivery* 2006; 21(3): 1309–1317.

10 Cui Y, Ma H, Saha T. Transformer hot spot temperature prediction using a hybrid algorithm of support vector regression and information granulation. Proceedings of the 7th IEEE PES Asia-Pacific Power and Energy Engineering Conference (APPEEC), Brisbane, Australia, November 15–18, 2015.

11 Yousof M, Ekanayake C, Saha T. Frequency response analysis to investigate deformation of transformer winding. *IEEE Transactions on Dielectrics and Electrical Insulation* 2015; 22(4): 2359–2367.

12 Kreuger FH. *Partial Discharge Detection in HV Equipment.* London: Butterworth, 1989.

13 Ma H, Chan J, Saha T, Ekanayake C. Pattern recognition techniques and their applications for automatic partial discharge source classification. *IEEE Transactions on Dielectrics and Electrical Insulation* 2013; 20(2): 468–478.

14 Chan J, Ma H, Saha T, Ekanayake C. Self-adaptive partial discharge signal extraction based on ensemble empirical mode decomposition and optimal morphological thresholding. *IEEE Transactions on Dielectrics and Electrical Insulation* 2014; 21(1): 294–303.

15 Chan J, Ma H, Saha T. Automatic blind equalization and morphological thresholding for partial discharge signal de-noising. *IEEE Transactions on Power Delivery* 2014; 29(4): 1927–1938.

16 Chan J, Ma H, Saha T. Time-frequency sparsity map on automatic partial discharge sources separation for power transformer condition assessment. *IEEE Transactions on Dielectrics and Electrical Insulation* 2015; 22(4): 2271–2283.
17 Seo J, Ma H, Saha T. Probabilistic wavelet transform for partial discharge measurement of transformer. *IEEE Transactions on Dielectrics and Electrical Insulation* 2015; 22(2): 1105–1117.
18 Chan J, Ma H, Saha T. Hybrid method on signal de-noising and representation for online partial discharge monitoring of power transformers at substations. *IET Science, Measurement & Technology* 2015; 9(7): 890–899.
19 Seo J, Ma H, Saha T. A novel signal extraction technique for online partial discharge (PD) measurements of transformers. *International Transactions on Electrical Energy Systems* 2015; 26(5): 1032–1048.
20 Du Y, Zahn M, Lesieutre BC, Mamishev AV, Lindgren SR. Moisture equilibrium in transformer paper–oil systems. *IEEE Electrical Insulation Magazine* 1999; 15: 11–20.
21 Garcia B, Burgos JC, Alonso AM, Sanz J. A moisture-in-oil model for power transformer monitoring - Part I: Theoretical foundation. *IEEE Transactions on Power Delivery* 2005; 20(2): 1417–1422.
22 Martin D, Perkasa C, Lelekakis N. Measuring paper water content of transformers: A new approach using cellulose isotherms in non-equilibrium conditions. *IEEE Transactions on Power Delivery* 2013; 28(3): 1433–1439.
23 Martin D, Saha T, Dee R, Buckley G, Chinnarajan S, Caldwell G *et al.* Determining water in transformer paper insulation: Analysing aging transformers. *IEEE Electrical Insulation Magazine* 2015; 31(5): 23–32.
24 Martin D, Saha T, Gray T, Wyper K. Determining water in transformer paper insulation: Effect of measuring oil water activity at two different locations. *IEEE Electrical Insulation Magazine* 2015; 31(3): 18–25.
25 Cui Y, Ma H, Saha T, Ekanayake C. Understanding moisture dynamic in transformer and its effects on dielectric response of transformer insulation. *IEEE Transactions on Power Delivery* 2015; 30(5): 2195–2204.
26 CIGRE WG A2.44. Guide on Transformer Intelligent Condition Monitoring (TICM) Systems, 2015.
27 CIGRE WG D1.11.04. Data Mining Techniques and Applications in the Power Transmission Field, 2006.
28 CIGRE WG A2.27. Recommendations for Condition Monitoring and Condition Assessment Facilities for Transformers, 2008.
29 CIGRE WG D1.17. Generic Guidelines for Life Time Condition Assessment of HV Assets and Related Knowledge Rules, 2010.
30 Ashkezari A, Ma H, Saha T, Ekanayake C. Application of fuzzy support vector machine for determining the health index of insulation system of in-service power transformer. *IEEE Transactions on Dielectrics and Electrical Insulation* 2013; 20 (3): 965–973.
31 Abu-Elanien AEB, Salama MMA, Ibrahim M. Calculation of a health index for oil-immersed transformers rated under 69 kV using fuzzy logic. *IEEE Transactions on Power Delivery* 2012; 27(4): 2029–2036.
32 Morais D, Bencz E, Rolim J. Condition assessment of power transformers through the integration of historical and on-line data in a multi-agent system. *Journal of Control, Automation and Electrical Systems* 2014; 25(1): 93–102.

33 Ashkezari A, Ma H, Saha T, Cui Y. Investigation of feature selection techniques for power transformer insulation condition assessment. *IEEE Transactions on Dielectrics and Electrical Insulation* 2014; 21(2): 836–844.

34 Ma H, Seo J, Saha T, Chan J, Martin D. Automatic partial discharge sources classification of power transformers using pattern recognition techniques. roceedings of IEEE Conference on Electrical Insulation and Dielectric Phenomena, Shenzhen, China, October 20–23, 2013.

35 Peng H, Long F, Ding C. Feature selection based on mutual information criteria of max-dependency, max-relevance, and min-redundancy. *IEEE Transactions on Pattern Analysis and Machine Intelligence* 2005; 27(8): 1226–1238.

36 Ma H, Saha T, Ekanayake C. Predictive learning and information fusion for condition assessment of power transformer. Proceedings of the IEEE Power and Energy Society General Meeting, Detroit, MI, July 24–28, 2011.

37 Hall DL, McMullen SAH. *Mathematical Techniques in Multi Sensor Data Fusion*, 2nd edn. London: Artech House, 2004.

38 Koks D, Challa S. An Introduction to Bayesian and Dempster–Shafer Data Fusion. Australian Government Defence Science and Technology Science Technical Report DSTO-TR-1436, 2005 (approved for public release).

39 Ghosh K, Ng YS, Srinivasan R. Evaluation of decision fusion strategies for effective collaboration among heterogeneous fault diagnostic methods. *Computers & Chemical Engineering* 2011; 35(2): 342–355.

40 Saha T, Ma H, Ekanayake C, Martin D. Pattern recognition techniques for determining the health index of oil–paper insulation of in-service power transformers. CIGRE 2014 Session, Paris.

41 Cui Y, Ma H, Saha T. Pattern recognition techniques for power transformer insulation diagnosis: A comparative study. Part 1: Framework, literature, and illustration. *International Transactions on Electrical Energy Systems* 2015; 25(10): 2247–2259.

42 Cui Y, Ma H, Saha T. Pattern recognition techniques for power transformer insulation diagnosis: A comparative study. Part II: Implementation, case studies, and statistical analysis. *International Transactions on Electrical Energy Systems* 2015; 25(10): 2260–2274.

43 Cui Y, Ma H, Saha T. Power transformer insulation diagnosis using oil characteristics by implementing SMOTEBoost. *IEEE Transactions on Dielectrics and Electrical Insulation* 2014; 21(5): 2363–2373.

44 Ma H, Ekanayake C, Saha T. Power transformer fault diagnosis under measurement originated uncertainties. *IEEE Transactions on Dielectrics and Electrical Insulation* 2012; 19(6): 1982–1990.

45 Shukla P, Sood YR, Jarial RK. Experimental evaluation of water content in transformer oil. *International Journal of Innovative Research in Science, Engineering and Technology* 2013; 2(1): 284–291.

46 Bishop CM. *Pattern Recognition and Machine Learning*. Berlin: Springer, 2006.

47 Jain AK, Duin RPW, Mao J. Statistical pattern recognition: A review. *IEEE Transactions on Pattern Analysis and Machine Intelligence* 2000; 22(1): 4–37.

48 Picher P, Boudreau J-F, Manga A, Rajotte C, Tardif C, Bizier G *et al.* Use of health index and reliability data for transformer condition assessment and fleet ranking. CIGRE 2014 Session, Paris.

49 Németh B, Vörös CS, Csépes G. Health index as one of the best practice for condition assessment of transformers and substation equipment – Hungarian experience. CIGRE 2014 Session, Paris.

50 Scatiggio F, Fraioli A, Iuliani V, Pompili M. Health index: The TERNA's practical approach for transformers fleet management. CIGRE 2014 Session, Paris.

51 Vermeer M, Wetzer J. Transformer health and risk indexing, including data quality management. CIGRE 2014 Session, Paris.

52 Weyns D, Uhrmacher AM (eds). *Multi-Agent Systems Simulation and Applications*. Boca Raton, FL: CRC Press, 2009.

53 Cao LB (ed). *Data Mining and Multi-Agent Integration*. Berlin: Springer, 2009.

10

Conclusions and Future Research

The properties of oil–paper insulation systems used in transformers have been introduced in this book. A comprehensive discussion on the ageing of oil–paper insulation systems has been provided, in the context of transformer operating conditions. For better understanding, thermal, oxidative, and hydrolytic ageing effects on insulation materials are discussed in the early part of the book, to relate these ageing effects to transformer condition monitoring methodologies. The introductory part of the book provides a foundation for understanding a number of chemical and electrical-based diagnostics, and also helps the reader to relate the diagnosis with the cause of insulation ageing.

Among the chemical-based techniques, dissolved gas analysis (DGA), furan analysis, and degree of polymerization (DP) are the most commonly used diagnosis techniques. Many interpretation techniques exist in the literature, and the use of their relevant international standards is highly recommended. Though DGA, furan, tensile strength, and DP measurements are still being used extensively by the utilities, in recent years new chemical-based diagnostic methods – such as molecular weight measurements by gel permeation chromatography (GPC), UV spectroscopy, and Fourier transform infrared (FTIR) spectroscopy methods – have been trialed by many researchers. Some of these have shown significant advancements in knowledge for insulation diagnosis. Unfortunately, due to the complexity of measurements/interpretation schemes and cost, many of them have not become mainstream diagnostic tools to date.

Classical insulation resistance, power frequency dissipation factor, and polarization index measurements are still used by many utilities as electrical techniques for insulation diagnosis. Since the early 1990s, new dielectric response-based diagnostic tests have been in wide use, based on either time or frequency domain measurements. Time domain dielectric response diagnosis techniques, which include polarization–depolarization current (PDC) and recovery voltage measurements (RVM), are explained from theoretical to practical points of view. Comprehensive interpretation schemes of PDC and RVM are presented, along with impacts of moisture and temperature ageing and insulation geometry on the interpretation of their measurement data. Frequency domain dielectric spectroscopy (FDS) has also been used widely by many researchers and engineers

Transformer Ageing: Monitoring and Estimation Techniques, First Edition.
Edited by Tapan Kumar Saha and Prithwiraj Purkait.

in utilities. A comparison between time and frequency domain measurements is analyzed and reported. Several attempts have been made to reduce the measurement time of dielectric response diagnostics. Some commercial equipment has been developed to address this issue by combining time and frequency domain measurements. Standardization of insulation diagnosis using dielectric response measurements still needs to be addressed in future research.

DGA being the most commonly used tool for insulation assessment, several standards are available from the IEC and IEEE and also several guidelines from CIGRE. Utilities very often use interpretation schemes such as the key gas concentration method, Doernenburg ratio, Rogers ratio, IEC ratio, and Duval's triangle. Our focus in this book has been on intelligent diagnostic techniques for power transformer insulation diagnosis using DGA and oil property test data by a number of pattern recognition algorithms. A general pattern recognition application framework is proposed involving the process of oil characteristics database preparation, training/testing datasets formation, cross-validation, testing, and evaluation in applying a pattern recognition algorithm for diagnosing transformer insulation. An unbalanced training dataset, which is very often biased toward healthy transformer condition, can be overcome by a hybrid of SMOTEBoost and bootstrap algorithm. This has been proposed to enhance pattern recognition algorithms to consistently achieve desirable classification accuracy in power transformer diagnosis using DGA and oil property data. Based on a number of case studies with several intelligent diagnostic tools, it is found that the proposed hybrid approach can improve the generalization capability of pattern recognition algorithms. Further validation is warranted for field DGA data for transformers with known fault conditions.

PD monitoring to detect incipient faults is discussed in detail, with a focus on PD signal de-noising and PD source classification tools for better interpretation of online PD measurements of transformers at substations. Wavelet transform (WT), empirical mode decomposition (EMD), ensemble EMD (EEMD), blind equalization (BE), and mathematical morphology (MM) methods are proposed for extracting PD signals from extensive noise and interference. A sparsity time-frequency (TF) map is proposed for improved multi-PD source separation. Various artificial neural networks (ANNs) and support vector machine (SVM)-based PD pattern recognition algorithms are developed for classifying different types of PD source that may exist in a transformer tank. A number of PD signal representation techniques based on phase-resolved PD (PRPD), phase-resolved pulse sequence (PRPS), and spectral kurtosis are also presented. Various feature extraction and dimensionality reduction methods – including statistical operators, principle component analysis (PCA), and discrete wavelet transform (DWT) – are also discussed.

Frequency response analysis (FRA) plays an important role in transformer winding mechanical deformation/displacement diagnosis. FRA measurement can be useful to detect damage on the core and localized damage on the winding. A number of international standards for FRA measurements and their interpretation schemes are discussed. Case studies are presented using several schemes available to analyze the FRA response. Finally, non-mechanical factors which are influential on the FRA measurement are also discussed.

Moisture is detrimental because it reduces the remaining life of a transformer's paper insulation and decreases the maximum operating temperature that the transformer can continue to operate at without any major risk. The importance of moisture assessment is demonstrated using an online sensor (water activity probe)-based technique to measure the water content of transformer insulation. Processing data from water activity probes is performed using adsorption isotherms of cellulose and compared with Karl Fischer titration and dielectric response methods. An advantage of water activity probes over the other techniques is that the insulation does not need to be in moisture equilibrium for the measurement to be accurate. Several case studies in this book have demonstrated the importance of this technique for life assessment of transformer insulation. Comparison of sensor-based moisture measurements data with dielectric response methods should be continued to provide better confidence on dielectric response methods.

Biodegradable (ester) oils have been promoted recently for use in transformers, primarily for environmental reasons, but also to reduce the fire risk of mineral oil-based transformers. Similar to degradation of mineral oil, thermal, hydrolytic, and oxidation processes are mainly responsible for ageing of ester oil. However, oxidation has a more pronounced ageing effect on ester oil. Standards for analysis and interpretation of dissolved fault gases in ester-based insulating oils are yet to be developed. The existing FDS-based moisture determination tool is built considering the effects of moisture, temperature, and ageing on the dielectric response behavior of mineral oil-impregnated cellulose paper insulation, and hence these tools may not be fully suitable for estimating the moisture content of an ester oil–paper insulation system. More field experience and refinement of current condition monitoring techniques is required for ester oil-filled transformers.

An intelligent framework for transformer condition monitoring using online sensors and offline diagnosis is presented in this book. The objective of this framework is to collect online sensor-based measurements, use signal processing tools to extract measured signals from noise, then extract meaningful features and finally use pattern recognition techniques to identify fault types in transformers. A SmartBox platform – including hardware and software for a condition monitoring system – is proposed, and a concept of health index and its interpretation is also presented.

The transformer being one of the most costly pieces of equipment in the electricity grid, research and development is always in progress by utilities, research centers, and academic institutions. Although a number of important topics of condition monitoring are covered, still a number of topics could not be covered in this book. Condition monitoring of bushings and on-load tap changers (OLTCs) is equally important for maintaining the overall health of a transformer. It is necessary to develop appropriate diagnosis tools for both bushings and OLTCs to improve transformer operation. The application of fiber optics has recently been proposed for the direct measurement of moisture levels within the transformer. Future research may provide better evaluation of optical sensors for direct moisture measurement and develop a new thermal-driven moisture dynamic model for transformers. As transformers age, their clamping pressure drops. Understanding the relationship between the progressive loss of winding clamping pressure with ageing and the winding paper mechanical strength is an important research topic. In the current economic environment, managers and accountants are keen to understand the lifecycle cost of transformers to determine when to replace or repair a transformer. In a future edition of this book, we hope to address some of these topics.

Index

Transformer Ageing: Monitoring and Estimation Techniques, First Edition.
Edited by Tapan Kumar Saha and Prithwiraj Purkait.

e

f

r

s

www.ingramcontent.com/pod-product-compliance
Lightning Source LLC
LaVergne TN
LVHW062108200926
848484LV00007B/1616